INTERNATIONAL MATHEMATICAL SERIES

Series Editor: **Tamara Rozhkovskaya**
Novosibirsk, Russia

This series was founded in 2002 and is a joint publication of Springer and "Tamara Rozhkovskaya Publisher." Each volume presents contributions from the Volume Editors and Authors exclusively invited by the Series Editor Tamara Rozhkovskaya who also prepares the Camera Ready Manuscript. This volume is distributed by "Tamara Rozhkovskaya Publisher" (tamara@mathbooks.ru) in Russia and by Springer over all the world.

For other titles published in this series, go to
www.springer.com/series/6117

AROUND THE RESEARCH OF VLADIMIR MAZ'YA I

Function Spaces

Editor: **Ari Laptev**

Imperial College London, UK
Royal Institute of Technology, Sweden

SPRINGER
TAMARA ROZHKOVSKAYA PUBLISHER

Editor
Ari Laptev
Department of Mathematics
Imperial College London
Huxley Building, 180 Queen's Gate
London SW7 2AZ
United Kingdom
a.laptev@imperial.ac.uk

ISBN 978-1-4614-2547-2 e-ISBN 978-1-4419-1341-8

DOI 10.1007/978-1-4419-1341-8
Springer New York Dordrecht Heidelberg London

Printed on acid-free paper

Springer is part of Springer Science+Business Media (www.springer.com)

Vladimir Maz'ya was born on December 31, 1937, in Leningrad (present day, St. Petersburg) in the former USSR. His first mathematical article was published in Doklady Akad. Nauk SSSR when he was a fourth-year student of the Leningrad State University. From 1961 till 1986 V. Maz'ya held a senior research fellow position at the Research Institute of Mathematics and Mechanics of LSU, and then, during 4 years, he headed the Laboratory of Mathematical Models in Mechanics at the Institute of Engineering Studies of the Academy of Sciences of the USSR. Since 1990, V. Maz'ya lives in Sweden. At present, Vladimir Maz'ya is a Professor Emeritus at Linköping University and Professor at Liverpool University. He was elected a Member of Royal Swedish Academy of Sciences in 2002.

Sergey L. Sobolev (left) and Vladimir G. Maz'ya (right). Novosibirsk, 1978

The list of publications of V. Maz'ya contains 20 books and more than 450 research articles covering diverse areas in Analysis and containing numerous fundamental results and fruitful techniques. Research activities of Vladimir Maz'ya have strongly influenced the development of many branches in Analysis and Partial Differential Equations, which are clearly highlighted by the contributions to this collection of 3 volumes, where the world-recognized specialists present recent advantages in the following areas:

I. *Function Spaces.* Various aspects of the theory of Sobolev spaces, isoperimetric and capacitary inequalities, Hardy, Sobolev, and Poincaré inequalities in different contexts, sharp constants, extension operators, traces, weighted Sobolev spaces, Orlicz–Sobolev spaces, Besov spaces, etc.

Laurent Schwartz (left) and Vladimir Maz'ya (right). Paris, 1992

II. *Partial Differential Equations.* Asymptotic analysis, multiscale asymptotic expansions, homogenization, boundary value problems in domains with singularities, boundary integral equations, mathematical theory of water waves, Wiener regularity of boundary points, etc.

III. *Analysis and Applications.* Various problems including the oblique derivative problem, spectral properties of the Schrödinger operator, ill-posed problems, etc.

I. Function Spaces

Ari Laptev Ed.

Hardy Inequalities for Nonconvex Domains ... 1
Farit Avkhadiev and Ari Laptev

Distributions with Slow Tails and Ergodicity of Markov Semigroups in Infinite Dimensions ... 13
Sergey Bobkov and Boguslaw Zegarlinski

On Some Aspects of the Theory of Orlicz–Sobolev Spaces ... 81
Andrea Cianchi

Mellin Analysis of Weighted Sobolev Spaces with Nonhomogeneous Norms on Cones ... 105
Martin Costabel, Monique Dauge, and Serge Nicaise

Optimal Hardy–Sobolev–Maz'ya Inequalities with Multiple Interior Singularities ... 137
Stathis Filippas, Achilles Tertikas, and Jesper Tidblom

Sharp Fractional Hardy Inequalities in Half-Spaces ... 161
Rupert L. Frank and Robert Seiringer

Collapsing Riemannian Metrics to Sub-Riemannian and the Geometry of Hypersurfaces in Carnot Groups ... 169
Nicola Garofalo and Christina Selby

Sobolev Homeomorphisms and Composition Operators ... 207
Vladimir Gol'dshtein and Aleksandr Ukhlov

Extended L^p Dirichlet Spaces ... 221
Niels Jacob and René L. Schilling

Characterizations for the Hardy Inequality ... 239
Juha Kinnunen and Riikka Korte

Geometric Properties of Planar BV-Extension Domains ... 255
Pekka Koskela, Michele Miranda Jr., and Nageswari Shanmugalingam

On a New Characterization of Besov Spaces with Negative Exponents ... 273
Moshe Marcus and Laurent Véron

Isoperimetric Hardy Type and Poincaré Inequalities on Metric Spaces ... 285
Joaquim Martín and Mario Milman

Gauge Functions and Sobolev Inequalities on Fluctuating Domains ... 299
Eric Mbakop and Umberto Mosco

A Converse to the Maz'ya Inequality for Capacities under Curvature Lower Bound ... 321
Emanuel Milman

Pseudo-Poincaré Inequalities and Applications to Sobolev Inequalities ... 349
Laurent Saloff-Coste

The p-Faber-Krahn Inequality Noted ... 373
Jie Xiao

Index ... 391

References to Maz'ya's Publications Made in Volume I ... 393

II. Partial Differential Equations

Ari Laptev Ed.

Large Solutions to Semilinear Elliptic Equations with Hardy Potential and Exponential Nonlinearity ... 1
Catherine Bandle, Vitaly Moroz, and Wolfgang Reichel

Stability Estimates for Resolvents, Eigenvalues, and Eigenfunctions of Elliptic Operators on Variable Domains ... 23
Gerassimos Barbatis, Victor I. Burenkov, and Pier Domenico Lamberti

Operator Pencil in a Domain with Concentrated Masses. A Scalar Analog of Linear Hydrodynamics ... 61
Gregory Chechkin

Selfsimilar Perturbation near a Corner: Matching Versus Multiscale Expansions for a Model Problem ... 95
Monique Dauge, Sébastien Tordeux, and Grégory Vial

Stationary Navier–Stokes Equation on Lipschitz Domains in Riemannian Manifolds with Nonvanishing Boundary Conditions ... 135
Martin Dindoš

On the Regularity of Nonlinear Subelliptic Equations ... 145
András Domokos and Juan J. Manfredi

Rigorous and Heuristic Treatment of Sensitive Singular Perturbations Arising in Elliptic Shells ... 159
Yuri V. Egorov, Nicolas Meunier, and Evariste Sanchez-Palencia

On the Existence of Positive Solutions of Semilinear Elliptic Inequalities on Riemannian Manifolds ... 203
Alexander Grigor'yan and Vladimir A. Kondratiev

Recurrence Relations for Orthogonal Polynomials and Algebraicity of Solutions of the Dirichlet Problem ... 219
Dmitry Khavinson and Nikos Stylianopoulos

On First Neumann Eigenvalue Bounds for Conformal Metrics ... 229
Gerasim Kokarev and Nikolai Nadirashvili

Necessary Condition for the Regularity of a Boundary Point for Porous Medium Equations with Coefficients of Kato Class ... 239
Vitali Liskevich and Igor I. Skrypnik

The Problem of Steady Flow over a Two-Dimensional Bottom Obstacle ... 253
Oleg Motygin and Nikolay Kuznetsov

Well Posedness and Asymptotic Expansion of Solution of Stokes Equation Set in a Thin Cylindrical Elastic Tube ... 275
Grigory P. Panasenko and Ruxandra Stavre

On Solvability of Integral Equations for Harmonic Single Layer Potential on the Boundary of a Domain with Cusp ... 303
Sergei V. Poborchi

Hölder Estimates for Green's Matrix of the Stokes System in Convex Polyhedra ... 315
Jürgen Roßmann

Boundary Integral Methods for Periodic Scattering Problems 337
Gunther Schmidt

Boundary Coerciveness and the Neumann Problem for 4th Order Linear Partial Differential Operators 365
Gregory C. Verchota

Index 379

References to Maz'ya's Publications Made in Volume II 381

III. Analysis and Applications

Ari Laptev Ed.

Optimal Control of a Biharmonic Obstacle Problem 1
David R. Adams, Volodymyr Hrynkiv, and Suzanne Lenhart

Minimal Thinness and the Beurling Minimum Principle 25
Hiroaki Aikawa

Progress in the Problem of the L^p-Contractivity of Semigroups for Partial Differential Operators 47
Alberto Cialdea

Uniqueness and Nonuniqueness in Inverse Hyperbolic Problems and the Black Hole Phenomenon 77
Gregory Eskin

Global Green's Function Estimates 105
Michael W. Frazier and Igor E. Verbitsky

On Spectral Minimal Partitions: the Case of the Sphere 153
Bernard Helffer, Thomas Hoffmann-Ostenhof, and Susanna Terracini

Weighted Sobolev Space Estimates for a Class of Singular Integral Operators 179
Dorina Mitrea, Marius Mitrea, and Sylvie Monniaux

On general Cwikel–Lieb–Rozenblum and Lieb–Thirring Inequalities 201
Stanislav Molchanov and Boris Vainberg

Estimates for the Counting Function of the Laplace Operator on Domains with Rough Boundaries 247
Yuri Netrusov and Yuri Safarov

$W^{2,p}$-Theory of the Poincaré Problem 259
Dian K. Palagachev

Weighted Inequalities for Integral and Supremum Operators 279
Luboš Pick

Finite Rank Toeplitz Operators in the Bergman Space 331
Grigori Rozenblum

Resolvent Estimates for Non-Selfadjoint Operators via Semigroups 359
Johannes Sjöstrand

Index 385

References to Maz'ya's Publications Made in Volume III 387

Contributors

Editor

Ari Laptev

President
The European Mathematical Society

Professor
Head of Department
Department of Mathematics
Imperial College London
Huxley Building, 180 Queen's Gate
London SW7 2AZ, UK
a.laptev@imperial.ac.uk

Professor
Department of Mathematics
Royal Institute of Technology
100 44 Stockholm, Sweden
laptev@math.kth.se

Ari Laptev is a world-recognized specialist in Spectral Theory of Differential Operators. He discovered a number of sharp spectral and functional inequalities. In particular, jointly with his former student T. Weidl, A. Laptev proved sharp Lieb–Thirring inequalities for the negative spectrum of multidimensional Schrödinger operators, a problem that was open for more than twenty five years.

A. Laptev was brought up in Leningrad (Russia). In 1971, he graduated from the Leningrad State University and was appointed as a researcher and then as an Assistant Professor at the Mathematics and Mechanics Department of LSU. In 1982, he was dismissed from his position at LSU due to his marriage to a British subject. Only after his emigration from the USSR in 1987 he was able to continue his career as a mathematician. Then A. Laptev was employed in Sweden, first as a lecturer at Linköping University and then from 1992 at the Royal Institute of Technology (KTH). In 1999, he became a professor at KTH and also Vice Chairman of its Department of Mathematics. From January 2007 he is employed by Imperial College London where from September 2008 he is the Head of Department of Mathematics.

A. Laptev was the Chairman of the Steering Committee of the five years long ESF Programme SPECT, the President of the Swedish Mathematical Society from 2001 to 2003, and the President of the Organizing Committee of the Fourth European Congress of Mathematics in Stockholm in 2004. He is now the President of the European Mathematical Society for the period January 2007–December 2010.

Authors

David R. Adams Vol. III
University of Kentucky
Lexington, KY 40506-0027
USA
dave@ms.uky.edu

Hiroaki Aikawa Vol. III
Hokkaido University
Sapporo 060-0810
JAPAN
aik@math.sci.hokudai.ac.jp

Farit Avkhadiev Vol. I
Kazan State University
420008 Kazan
RUSSIA
favhadiev@ksu.ru

Catherine Bandle Vol. II
Mathematisches Institut
Universität Basel
Rheinsprung 21, CH-4051 Basel
SWITZERLAND
catherine.bandle@unibas.ch

Gerassimos Barbatis Vol. II
University of Athens
157 84 Athens
GREECE
gbarbatis@math.uoa.gr

Sergey Bobkov Vol. I
University of Minnesota
Minneapolis, MN 55455
USA
bobkov@math.umn.edu

Victor I. Burenkov Vol. II
Università degli Studi di Padova
63 Via Trieste, 35121 Padova
ITALY
burenkov@math.unipd.it

Grigori Chechkin Vol. II
Lomonosov Moscow State University
Vorob'evy Gory, Moscow
RUSSIA
chechkin@mech.math.msu.su

Alberto Cialdea Vol. III
Università della Basilicata
Viale dell'Ateneo Lucano 10,
85100, Potenza
ITALY
cialdea@email.it

Andrea Cianchi Vol. I
Università di Firenze
Piazza Ghiberti 27, 50122 Firenze
ITALY
cianchi@unifi.it

Martin Costabel Vol. I
Université de Rennes 1
Campus de Beaulieu
35042 Rennes
FRANCE
martin.costabel@univ-rennes1.fr

Monique Dauge Vols. I, II
Université de Rennes 1
Campus de Beaulieu
35042 Rennes
FRANCE
monique.dauge@univ-rennes1.fr

Martin Dindoš Vol. II
Maxwell Institute of
Mathematics Sciences
University of Edinburgh
JCMB King's buildings Mayfield Rd
Edinbugh EH9 3JZ
UK
m.dindos@ed.ac.uk

András Domokos Vol. II
California State
University Sacramento
Sacramento 95819
USA
domokos@csus.edu

Yuri V. Egorov Vol. II
Université Paul Sabatier
118 route de Narbonne
31062 Toulouse Cedex 9
FRANCE
egorov@mip.ups-tlse.fr

Gregory Eskin Vol. III
University of California
Los Angeles, CA 90095-1555
USA
eskin@math.ucla.edu

Nicola Garofalo Vol. I
Purdue University
West Lafayette, IN 47906
USA
garofalo@math.purdue.edu
and

Università di Padova
35131 Padova
ITALY
garofalo@dmsa.unipd.it

Vladimir Gol'dshtein Vol. I
Ben Gurion University of the Negev
P.O.B. 653, Beer Sheva 84105
ISRAEL
vladimir@bgu.ac.il

Alexander Grigor'yan Vol. II
Bielefeld University
Bielefeld 33501
GERMANY
grigor@math.uni-bielefeld.de

Stathis Filippas Vol. I
University of Crete
71409 Heraklion
Institute of Applied and
Computational Mathematics
71110 Heraklion
GREECE
filippas@tem.uoc.gr

Rupert L. Frank Vol. I
Princeton University
Washington Road
Princeton, NJ 08544
USA
rlfrank@math.princeton.edu

Michael W. Frazier Vol. III
University of Tennessee
Knoxville, Tennessee 37922
USA
frazier@math.utk.edu

Bernard Helffer Vol. III
Université Paris-Sud
91 405 Orsay Cedex
FRANCE
Bernard.Helffer@math.u-psud.fr

Thomas Hoffmann-Ostenhof Vol. III
Institut für Theoretische Chemie
Universität Wien
Währinger Strasse 17, and
International Erwin
Schrödinger Institute
for Mathematical Physics
Boltzmanngasse 9
A-1090 Wien
AUSTRIA
thoffman@esi.ac.at

Volodymyr Hrynkiv Vol. III
University of Houston-Downtown
Houston, TX 77002-1014
USA
HrynkivV@uhd.edu

Niels Jacob Vol. I
Swansea University
Singleton Park
Swansea SA2 8PP
UK
n.jacob@swansea.ac.uk

Dmitry Khavinson Vol. II
University of South Florida
4202 E. Fowler Avenue, PHY114
Tampa, FL 33620-5700
USA
dkhavins@cas.usf.edu

Juha Kinnunen Vol. I
Institute of Mathematics
Helsinki University of Technology
P.O. Box 1100, FI-02015
FINLAND
juha.kinnunen@tkk.fi

Gerasim Kokarev Vol. II
University of Edinburgh
King's Buildings, Mayfield Road
Edinburgh EH9 3JZ
UK
G.Kokarev@ed.ac.uk

Vladimir A. Kondratiev Vol. II
Moscow State University
119992 Moscow
RUSSIA
vla-kondratiev@yandex.ru

Riikka Korte Vol. I
University of Helsinki
P.O. Box 68
Gustaf Hällströmin katu 2 b
FI-00014
FINLAND
riikka.korte@helsinki.fi

Pekka Koskela Vol. I
University of Jyväskylä
P.O. Box 35 (MaD), FIN-40014
FINLAND
pkoskela@maths.jyu.fi

Nikolay Kuznetsov Vol. II
Institute for Problems
in Mechanical Engineering
Russian Academy of Sciences
V.O., Bol'shoy pr. 61
199178 St. Petersburg
RUSSIA
nikolay.g.kuznetsov@gmail.com

Pier Domenico Lamberti Vol. II
Universitá degli Studi di Padova
63 Via Trieste, 35121 Padova
ITALY
lamberti@math.unipd.it

Ari Laptev Vol. I
Imperial College London
Huxley Building, 180 Queen's Gate
London SW7 2AZ
UK
a.laptev@imperial.ac.uk
and
Royal Institute of Technology
100 44 Stockholm
SWEDEN
laptev@math.kth.se

Suzanne Lenhart Vol. III
University of Tennessee
Knoxville, TN 37996-1300
USA
lenhart@math.utk.edu

Vitali Liskevich Vol. II
Swansea University
Singleton Park
Swansea SA2 8PP
UK
v.a.liskevich@swansea.ac.uk

Juan J. Manfredi Vol. II
University of Pittsburgh
Pittsburgh, PA 15260
USA
manfredi@pitt.edu

Moshe Marcus Vol. I
Israel Institute of Technology-Technion
33000 Haifa
ISRAEL
marcusm@math.technion.ac.il

Joaquim Martín Vol. I
Universitat Autònoma de Barcelona
Bellaterra, 08193 Barcelona
SPAIN
jmartin@mat.uab.cat

Eric Mbakop Vol. I
Worcester Polytechnic Institute
100 Institute Road
Worcester, MA 01609
USA
steve055@WPI.EDU

Nicolas Meunier Vol. II
Université Paris Descartes (Paris V)
45 Rue des Saints Pères
75006 Paris
FRANCE
nicolas.meunier@math-info.univ-paris5.fr

Emanuel Milman Vol. I
Institute for Advanced Study
Einstein Drive, Simonyi Hall
Princeton, NJ 08540
USA
emilman@math.ias.edu

Mario Milman Vol. I
Florida Atlantic University
Boca Raton, Fl. 33431
USA
extrapol@bellsouth.net

Michele Miranda Jr. Vol. I
University of Ferrara
via Machiavelli 35
44100, Ferrara
ITALY
michele.miranda@unife.it

Dorina Mitrea Vol. III
University of Missouri at Columbia
Columbia, MO 65211
USA
mitread@missouri.edu

Marius Mitrea Vol. III
University of Missouri
at Columbia
Columbia, MO 65211
USA
mitream@missouri.edu

Stanislav Molchanov Vol. III
University of North Carolina
at Charlotte
Charlotte, NC 28223
USA
smolchan@uncc.edu

Sylvie Monniaux Vol. III
Université Aix-Marseille 3
F-13397 Marseille Cédex 20
FRANCE
sylvie.monniaux@univ.u-3mrs.fr

Vitaly Moroz Vol. II
Swansea University
Singleton Park
Swansea SA2 8PP
UK
v.moroz@swansea.ac.uk

Umberto Mosco Vol. I
Worcester Polytechnic Institute
100 Institute Road
Worcester, MA 01609
USA
mosco@WPI.EDU

Oleg Motygin Vol. II
Institute for Problems
in Mechanical Engineering
Russian Academy of Sciences
V.O., Bol'shoy pr. 61
199178 St. Petersburg
RUSSIA
o.v.motygin@gmail.com

Nikolai Nadirashvili Vol. II
Centre de Mathématiques
et Informatique
Université de Provence
39 rue F. Joliot-Curie
13453 Marseille Cedex 13
FRANCE
nicolas@cmi.univ-mrs.fr

Yuri Netrusov Vol. III
University of Bristol
University Walk
Bristol BS8 1TW
UK
y.netrusov@bristol.ac.uk

Serge Nicaise Vol. I
Université Lille Nord de France
UVHC, 59313
Valenciennes Cedex 9
FRANCE
snicaise@univ-valenciennes.fr

Dian K. Palagachev Vol. III
Technical University of Bari
Via E. Orabona 4
70125 Bari
ITALY
palaga@poliba.it

Grigory P. Panasenko Vol. II
University Jean Monnet
23, rue Dr Paul Michelon
42023 Saint-Etienne
FRANCE
Grigory.Panasenko@univ-st-etienne.fr

Luboš Pick Vol. III
Charles University
Sokolovská 83, 186 75 Praha 8
CZECH REPUBLIC
pick@karlin.mff.cuni.cz

Sergei V. Poborchi Vol. II
St. Petersburg State University
28, Universitetskii pr., Petrodvorets
St. Petersburg 198504
RUSSIA
poborchi@mail.ru

Wolfgang Reichel Vol. II
Universität Karlsruhe (TH)
D-76128 Karlsruhe
GERMANY
wolfgang.reichel@math.uni-karlsruhe.de

Jürgen Roßmann Vol. II
University of Rostock
Institute of Mathematics
D-18051 Rostock
GERMANY
juergen.rossmann@uni-rostock.de

Grigori Rozenblum Vol. III
Chalmers University of Technology
University of Gothenburg
S-412 96, Gothenburg
SWEDEN
grigori@math.chalmers.se

Yuri Safarov Vol. III
King's College London
Strand, London WC2R 2LS
UK
yuri.safarov@kcl.ac.uk

Laurent Saloff-Coste Vol. I
Cornell University
Mallot Hall, Ithaca, NY 14853
USA
lsc@math.cornell.edu

Evariste Sanchez-Palencia Vol. II
Université Pierre et Marie Curie
4 place Jussieu
75252 Paris
FRANCE
sanchez@lmm.jussieu.fr

René L. Schilling Vol. I
Technische Universität Dresden
Institut für Stochastik
D-01062 Dresden
GERMANY
rene.schilling@tu-dresden.de

Gunther Schmidt Vol. II
Weierstrass Institute of
Applied Analysis and Stochastics
Mohrenstr. 39, 10117 Berlin
GERMANY
schmidt@wias-berlin.de

Robert Seiringer Vol. I
Princeton University
P. O. Box 708
Princeton, NJ 08544
USA
rseiring@princeton.edu

Christina Selby Vol. I
The Johns Hopkins University
11100 Johns Hopkins Road
Laurel, MD 20723.
USA
Christina.Selby@jhuapl.edu

Nageswari Shanmugalingam Vol. I
University of Cincinnati
Cincinnati, OH 45221-0025
USA
nages@math.uc.edu

Johannes Sjöstrand Vol. III
Université de Bourgogne
9, Av. A. Savary, BP 47870
FR-21078 Dijon Cédex
and UMR 5584, CNRS
FRANCE
johannes@u-bourgogne.fr

Igor I. Skrypnik Vol. II
Institute of Applied
Mathematics and Mechanics
Donetsk
UKRAINE
iskrypnik@iamm.donbass.com

Ruxandra Stavre Vol. II
Institute of Mathematics
"Simion Stoilow" Romanian Academy
P.O. Box 1-764
014700 Bucharest
ROMANIA
Ruxandra.Stavre@imar.ro

Nikos Stylianopoulos Vol. II
University of Cyprus
P.O. Box 20537
1678 Nicosia
CYPRUS
nikos@ucy.ac.cy

Susanna Terracini Vol. III
Università di Milano Bicocca
Via Cozzi, 53 20125 Milano
ITALY
susanna.terracini@unimib.it

Achilles Tertikas Vol. I
University of Crete and
Institute of Applied and
Computational Mathematics
71110 Heraklion
GREECE
tertikas@math.uoc.gr

Jesper Tidblom Vol. I
The Erwin Schrödinger Institute
Boltzmanngasse 9
A-1090 Vienna
AUSTRIA
Jesper.Tidblom@esi.ac.at

Sébastien Tordeux Vol. II
Institut de Mathématiques de Toulouse
31077 Toulouse
FRANCE
sebastien.tordeux@insa-toulouse.fr

Alexander Ukhlov Vol. I
Ben Gurion University of the Negev
P.O.B. 653, Beer Sheva 84105
ISRAEL
ukhlov@math.bgu.ac.il

Boris Vainberg Vol. III
University of North Carolina
at Charlotte
Charlotte, NC 28223
USA
brvainbe@uncc.edu

Igor E. Verbitsky Vol. III
University of Missouri
Columbia, Missouri 65211
USA
verbitskyi@missouri.edu

Gregory C. Verchota Vol. II
Syracuse University
215 Carnegie
Syracuse NY 13244
USA
gverchot@syr.edu

Laurent Véron Vol. I
Université de Tours
Parc de Grandmont
37200 Tours
FRANCE
veronl@univ-tours.fr

Grégory Vial Vol. II
IRMAR, ENS Cachan Bretagne
CNRS, UEB
35170 Bruz
FRANCE
gvial@bretagne.ens-cachan.fr

Jie Xiao Vol. I
Memorial University
of Newfoundland
St. John's, NL A1C 5S7
CANADA
jxiao@mun.ca

Boguslaw Zegarlinski Vol. I
Imperial College London
Huxley Building
180 Queen's Gate
London SW7 2AZ
UK
b.zegarlinski@imperial.ac.uk

Function Spaces

Contributions of Vladimir Maz'ya

One of the crucial contributions of V. Maz'ya to the theory of function spaces concerns the general understanding of relations between properties of embedding maps of function spaces and geometrical conditions on the domains, where the functions are defined. In 1960, V. Maz'ya discovered the equivalence of embedding and compactness theorems for Sobolev spaces and properties of isoperimetric and isocapacitory functions introduced by him. In particular, he found the best constant in the Gagliardo–Nirenberg inequality (this result was obtained independently by H. Federer and W. Fleming in 1960). V. Maz'ya studied sharp Sobolev type inequalities and discovered what is called now the *Hardy–Maz'ya–Sobolev inequalities.* The approach of Maz'ya did not use specific properties of the Euclidean space, which was remarked by him already in the 1960's. Recently Maz'ya's results were extended to Dirichlet forms and Markov operators on measure spaces.
— V. Maz'ya introduced the notions of *p-capacity* (1960) and *polyharmonic capacity* (1963). In 1970, together with V. Khavin, he initiated the so-called *nonlinear potential theory* based on L^p-norms.
— Using capacity approach, V. Maz'ya found necessary and sufficient conditions for fundamental spectral properties of elliptic operators. In 1962, he obtained two-sided estimates for the first Laplace eigenvalue stronger than the celebrated Cheeger inequality of 1970.
— Various contributions of V. Maz'ya to the development of the theory of function spaces concern, in particular, some aspects of Orlicz–Sobolev spaces (the characterization of Sobolev type inequalities involving Orlicz norms), spaces of functions with bounded variation, weighted Sobolev spaces with applications to boundary value problems in domains with conical points, Sobolev spaces in singularly perturbed domains, etc.
— Recently V. Maz'ya has shown how to generalize his capacitary inequalities to functions defined on topological spaces, which yields, in particular, sharp forms of the classical Sobolev and Moser inequalities.

Main Topics

In this volume, the following topics are discussed:

- Hardy inequalities in nonconvex domains. Hardy–Sobolev inequalities for symmetric functions in a ball. /Avkhadiev–Laptev/
- Weak Poincaré type inequalities, isoperimetric and capacitary inequalities of Sobolev type and their applications, Mazya's capacitary analogue of the coarea inequality in setting of metric probability spaces. /Bobkov–Zegarlinski/

- Orlicz–Sobolev spaces, optimal Sobolev embeddings and related inequalities in Orlicz spaces, classical and approximate differentiability properties of Orlicz–Sobolev functions, trace inequalities on the boundary /Cianchi/
- Weighted Sobolev spaces with nonhomogeneous norms on cones and optimal characterization of the structure of such spaces by Mellin transformation in noncritical and in critical cases /Costabel–Dauge–Nicaise/
- Hardy–Sobolev–Maz'ya inequalities. Necessary and sufficient conditions. /Filippas–Tertikas–Tidblom/
- Hardy inequality for fractional Sobolev spaces on half-spaces. Sharp constant. /Frank–Seiringer/
- Geometry of hypersurfaces in Carnot groups. The limit behavior of a family of left-invariant Riemannian metrics on a Carnot group with step two. /Garofalo–Selby/
- Sobolev homeomorphisms and the invertibility of bounded composition operators of Sobolev spaces. /Gol'dshtein–Ukhlov/
- L^p versions of extended Dirichlet spaces in the context of generalized Bessel potential spaces, and related variational capacities. /Jacob–Schilling/
- Multidimensional Hardy inequalities. Necessary and sufficient conditions. /Kinnunen–Korte/
- Geometric properties of planar domains admitting extension for functions with bounded variation. Necessary condition for a bounded, simply connected domain to be a $W^{1,1}$-extension domain. /Koskela–Miranda Jr.–Shanmugalingam/
- A new description of Besov spaces $B^{-s,q}(\Sigma)$ with negative exponents by means of an integrability condition on the Poisson potential of its elements. /Marcus–Véron/
- Isoperimetric Hardy type and Poincaré inequalities on metric spaces. Manifolds where Hardy type operators characterize Poincaré inequalities. /Martín–M. Milman/
- Sobolev and Poincaré inequalities on domains with fluctuating geometry, in particular, fractal domains displaying random self-similarity. /Mbakop–Mosco/
- Classical inequalities, including Maz'ya's isocapacitary inequalities, with their functional and isoperimetric counterparts in a measure-metric space setting. A converse to the Maz'ya inequality for capacities. /E. Milman/
- Pseudo-Poincaré inequalities and applications to Sobolev inequalities in different contexts, including setting on Riemannian manifolds. /Saloff-Coste/
- The p-Faber-Krahn inequality. Improvement and characterization by means of Maz'ya's capacity method, the Euclidean volume, the Sobolev type inequality and Moser–Trudinger's inequality. /Xiao/

Contents

Hardy Inequalities for Nonconvex Domains 1
Farit Avkhadiev and Ari Laptev
1 Introduction 1
2 Main Results 2
3 Proof of the Main Results 5
4 Remarks 9
References 11

Distributions with Slow Tails and Ergodicity of Markov Semigroups in Infinite Dimensions 13
Sergey Bobkov and Boguslaw Zegarlinski
1 Weak Forms of Poincaré Type Inequalities 13
2 L^p-Embeddings under Weak Poincaré 17
3 Growth of Moments and Large Deviations 20
4 Relations for L^p-Like Pseudonorms 22
5 Isoperimetric and Capacitary Conditions 24
6 Convex Measures 33
7 Examples. Perturbation 36
8 Weak Poincaré with Oscillation Terms 39
9 Convergence of Markov Semigroups 45
10 Markov Semigroups and Weak Poincaré 51
11 L^2 Decay to Equilibrium in Infinite Dimensions 58
11.1 Basic inequalities and decay to equilibrium in the product case 58
11.2 Semigroup for an infinite system with interaction. .. 60
11.3 L^2 decay 62
12 Weak Poincaré Inequalities for Gibbs Measures 66
References 78

On Some Aspects of the Theory of Orlicz–Sobolev Spaces 81
Andrea Cianchi
1 Introduction 81
2 Background 82
3 Orlicz–Sobolev Embeddings 85
3.1 Embeddings into Orlicz spaces 86
3.2 Embeddings into rearrangement invariant spaces 90
4 Modulus of Continuity 92
5 Differentiability Properties 95
6 Trace Inequalities 98
References 102

Mellin Analysis of Weighted Sobolev Spaces with Nonhomogeneous Norms on Cones 105
Martin Costabel, Monique Dauge, and Serge Nicaise
1 Introduction 105
2 Notation: Weighted Sobolev Spaces on Cones 107
2.1 Weighted spaces with homogeneous norms 108
2.2 Weighted spaces with nonhomogeneous norms 109
3 Characterizations by Mellin Transformation Techniques 111
3.1 Mellin characterization of spaces with homogeneous norms 111
3.2 Mellin characterization of seminorms 114
3.3 Spaces defined by Mellin norms 119
3.4 Spaces defined by weighted seminorms 122
3.5 Mellin characterization of spaces with nonhomogeneous norms 125
4 Structure of Spaces with Nonhomogeneous Norms in the Critical Case 129
4.1 Weighted Sobolev spaces with analytic regularity 129
4.2 Mellin regularizing operator in one dimension 130
4.3 Generalized Taylor expansions 132
5 Conclusion 135
References 136

Optimal Hardy–Sobolev–Maz'ya Inequalities with Multiple Interior Singularities 137
Stathis Filippas, Achilles Tertikas, and Jesper Tidblom
1 Introduction 137
2 Improved Hardy Inequalities with Multiple Singularities 142
3 Hardy–Sobolev–Maz'ya Inequalities 150
References 158

Sharp Fractional Hardy Inequalities in Half-Spaces 161
Rupert L. Frank and Robert Seiringer
1 Introduction and Main Results 161
2 Proofs 163
2.1 General Hardy inequalities 163
2.2 Proof of Theorem 1.1 165
References 167

Collapsing Riemannian Metrics to Sub-Riemannian and the Geometry of Hypersurfaces in Carnot Groups 169
Nicola Garofalo and Christina Selby
1 Introduction 169
2 Hypersurfaces in Carnot Groups of Step 2 172
3 The Limit as $\epsilon \to 0$ of the Rescaled ϵ-Volume Forms on M .. 176
4 Orthonormal Basis 179
5 Geometric Guantities with respect to the Collapsing Metrics 184
6 First and Second Variation Formulas for H-Perimeter 197
References 205

Sobolev Homeomorphisms and Composition Operators 207
Vladimir Gol'dshtein and Alexander Ukhlov
1 Introduction 207
2 Composition Operators in Sobolev Spaces 209
3 Proof of the Main Result 216
References 219

Extended L^p Dirichlet Spaces 221
Niels Jacob and René L. Schilling
1 Introduction 221
2 The Case for an L^p Theory for Dirichlet Forms 222
3 Bessel Potential Spaces and (r, p)-Capacities 224
4 Extended L^p-Dirichlet Spaces 227
5 The Limiting Case $\lambda \to 0$ and Transience 235
References 237

Characterizations for the Hardy Inequality 239
Juha Kinnunen and Riikka Korte
1 Introduction 239
2 Maz'ya Type Characterization 241
3 The Capacity Density Condition 243
4 Characterizations in the Borderline Case 247
5 Eigenvalue Problem 249
References 252

Geometric Properties of Planar BV-Extension Domains 255
Pekka Koskela, Michele Miranda Jr., and Nageswari Shanmugalingam
1 Introduction ... 255
2 Preliminaries ... 257
3 Proofs of the Results ... 266
4 Examples ... 269
References ... 271

On a New Characterization of Besov Spaces with Negative Exponents ... 273
Moshe Marcus and Laurent Véron
1 Introduction ... 273
2 The Left-Hand Side Inequality (1.4) ... 274
3 The Right-Hand Side Inequality (1.4) ... 278
3.1 The case $0 < s < 2$ ... 278
3.2 The general case ... 280
4 A Regularity Result for the Green Operator ... 283
References ... 284

Isoperimetric Hardy Type and Poincaré Inequalities on Metric Spaces ... 285
Joaquim Martín and Mario Milman
1 Introduction ... 285
2 Background ... 287
3 Hardy Isoperimetric Type ... 289
4 Model Riemannian Manifolds ... 291
5 E. Milman's Equivalence Theorems ... 293
6 Some Spaces That Are not of Isoperimetric Hardy Type ... 294
References ... 297

Gauge Functions and Sobolev Inequalities on Fluctuating Domains ... 299
Eric Mbakop and Umberto Mosco
1 Introduction ... 299
2 Gauge Functions ... 301
3 Gauged Poincaré Inequalities ... 307
4 Gauged Capacitary Inequalities ... 314
5 Fluctuating Domains ... 317
References ... 319

A Converse to the Maz'ya Inequality for Capacities under Curvature Lower Bound ... 321
Emanuel Milman
1 Introduction ... 321
2 Definitions and Preliminaries ... 324
2.1 Isoperimetric inequalities ... 324

2.2 Functional inequalities 325
2.3 Known connections 327
3 Capacities 328
3.1 1-Capacity and isoperimetric profiles 328
3.2 q-Capacitary and weak Orlicz–Sobolev inequalities . 329
3.3 q-Capacitary and strong Orlicz–Sobolev inequalities 331
3.4 Passing between q-capacitary inequalities 333
3.5 Combining everything 335
4 The Converse Statement 336
4.1 Case $q \geqslant 2$ 339
4.2 Case $1 < q \leqslant 2$ 342
References 347

Pseudo-Poincaré Inequalities and Applications to Sobolev Inequalities 349
Laurent Saloff-Coste
1 Introduction 349
2 Sobolev Inequality and Volume Growth 351
3 The Pseudo-Poincaré Approach to Sobolev Inequalities..... 352
4 Pseudo-Poincaré Inequalities 354
5 Pseudo-Poincaré Inequalities and the Liouville Measure 357
6 Homogeneous Spaces 361
7 Ricci Curvature Bounded Below 364
8 Domains with the Interior Cone Property 367
References 371

The p-Faber-Krahn Inequality Noted 373
Jie Xiao
1 The p-Faber-Krahn Inequality Introduced 373
2 The p-Faber-Krahn Inequality Improved 376
3 The p-Faber-Krahn Inequality Characterized 382
References 389

Index 391

References to Maz'ya's Publications Made in Volume I 393

Hardy Inequalities for Nonconvex Domains

Farit Avkhadiev and Ari Laptev

Abstract We obtain a series of Hardy type inequalities for domains involving both distance to the boundary and distance to the origin. In particular, we obtain the Hardy–Sobolev inequality for the class of symmetric functions in a ball and prove that for $d \geqslant 3$ the Hardy inequality involving the distance to the boundary holds with the constant $1/4$ in a large family of domains not necessarily convex. We also present an example showing that for any positive fixed constant there is an ellipsoid layer such that the Hardy inequality with the distance to the boundary fails.

1 Introduction

The classical Hardy inequality states that if $d \geqslant 3$ then for any function u such that $\nabla u \in L^2(\mathbb{R}^d)$

$$\Big(\frac{d-2}{2}\Big)^2 \int_{\mathbb{R}^d} \frac{|u(x)|^2}{|x|^2}\, dx \leqslant \int_{\mathbb{R}^d} |\nabla u(x)|^2\, dx. \tag{1.1}$$

The constant $((d-2)/2)^2$ in (1.1) is sharp, but is not achieved. The Hardy inequality for convex domains in $\mathbb{R}^d$ is usually formulated in terms of the distance to the boundary. Let $\Omega \subset \mathbb{R}^d$ be a convex domain, and let $\delta(x) = \operatorname{dist}(x, \partial\Omega)$ be the distance from $x \in \Omega$ to $\partial\Omega$. The following Hardy inequality

Farit Avkhadiev
Department of Mechanics and Mathematics, Kazan State University, 420008 Kazan, Russia
e-mail: favhadiev@ksu.ru

Ari Laptev
Department of Mathematics, Imperial College London, London SW7 2AZ, UK and Royal Institute of Technology, 100 44 Stockholm, Sweden
e-mail: a.laptev@imperial.ac.uk, laptev@math.kth.se

A. Laptev (ed.), *Around the Research of Vladimir Maz'ya I: Function Spaces*, International Mathematical Series 11, DOI 10.1007/978-1-4419-1341-8_1,

is well known (cf. [5, 6]):

$$\int_{\Omega} |\nabla u|^2 \, dx \geqslant \frac{1}{4} \int_{\Omega} \frac{|u(x)|^2}{\delta^2(x)} \, dx, \quad u \in H_0^1(\Omega). \tag{1.2}$$

In this case, the equality is not achieved either, which gives room for improvements of the last inequality. As was shown by Bresis and Marcus [4],

$$\int_{\Omega} |\nabla u|^2 \, dx \geqslant \frac{1}{4} \int_{\Omega} \frac{|u(x)|^2}{\delta^2(x)} \, dx + \lambda(\Omega) \int_{\Omega} |u(x)|^2 \, dx, \quad u \in H_0^1(\Omega),$$

where $\lambda(\Omega) \geqslant \dfrac{1}{4\mathrm{diam}^2(\Omega)}$. M. Hoffmann-Ostenhof, Th. Hoffmann-Ostenhof and Laptev [11] estimated $\lambda(\Omega)$ via the Lebesgue measure of Ω:

$$\lambda(\Omega) \geqslant \frac{3}{4} \left(\frac{v_d}{|\Omega|} \right)^{2/d},$$

where v_d denotes the volume of unit ball.

Avkhadiev [1, 2] and Filippas, Maz'ya, and Tertikas [7] proved that

$$\lambda(\Omega) \geqslant 3D_{\mathrm{int}}^{-2}(\Omega),$$

where $D_{\mathrm{int}}(\Omega) = 2 \sup \{\delta(x) : x \in \Omega\}$ denotes the interior diameter of Ω.

Recently, Avkhadiev and Wirths [3] proved that

$$[\lambda(\Omega) \geqslant 4\lambda_0 D_{\mathrm{int}}^{-2}(\Omega),$$

where $\lambda_0 = 0.940\ldots$ is the first positive root of the equation $J_0(\lambda) + 2\lambda J_0'(\lambda) = 0$ for the Bessel function; moreover this constant is sharp in all dimensions.

Filippas, Moschini, and Tertikas [8] established some results on sharp Hardy inequalities in the case of nonconvex domains. Note that Theorem 3.2 in [8] is generalized by our result presented in this paper: Theorem 2.4 below implies that the sharp Hardy inequality holds in $\Omega \setminus B_\rho$, where Ω is a convex domain and $B_\rho = \{x : |x| \leqslant \rho\} \subset \Omega$ satisfying the condition (2.4).

In Section 2.3, for a given positive constant we present two ellipsoids E_1, E_2, $E_2 \subset E_1$ such that the Hardy inequality in $\Omega = E_1 \setminus E_2$ is not valid with this constant.

2 Main Results

The first result deals with a Hardy inequality in nonconvex domains.

Theorem 2.1. *Let $\Omega \subset \mathbb{R}^d$, $d \geqslant 1$, be a bounded domain, and let $B_\rho = \{x : |x| \leqslant \rho\} \subset \Omega$. For $x \in \Omega \setminus B_\rho$ denote $\delta_\rho(x) = \mathrm{dist}\,(x, \partial B_\rho) = |x| - \rho$ and $\delta_\Omega = \mathrm{dist}\,(x, \partial\Omega)$. Then for any $u \in H_0^1(\Omega \setminus B_\rho)$*

$$\int_{\Omega\setminus B_\rho} |\nabla u(x)|^2\,dx \geqslant \frac{1}{4}\int_{\Omega\setminus B_\rho}\Big(\frac{(d-1)(d-3)}{|x|^2}+\frac{1}{\delta_\rho^2(x)}+\frac{1}{\delta_\Omega^2(x)}$$

$$-2\,\frac{\Delta\delta_\Omega(x)}{\delta_\Omega(x)}-2\,\frac{x\cdot\nabla\delta_\Omega(x)}{|x|\delta_\rho(x)\delta_\Omega(x)}+2(d-1)\,\frac{x\cdot\nabla\delta_\Omega(x)}{|x|^2\delta_\Omega(x)}\Big)\,|u(x)|^2\,dx.$$

From Theorem 2.1 we obtain several consequence of independent interest. The following assertion follows from Theorem 2.1 because

$$-\frac{\Delta\delta_R(x)}{\delta_R(x)}=\frac{d-1}{|x|\delta_R(x)},\qquad x\cdot\nabla\delta_R(x)=-|x|.$$

Corollary 2.1. *Let $\Omega = B_R = \{x : |x| < R\}$, $R > \rho$, and let $\delta_R = \delta_\Omega = \operatorname{dist}(x, \partial B_R) = R - |x|$. Then for any $u \in H_0^1(\Omega \setminus B_\rho)$*

$$\int_{B_R\setminus B_\rho} |\nabla u(x)|^2\,dx$$

$$\geqslant \frac{1}{4}\int_{B_R\setminus B_\rho}\Big(\frac{(d-1)(d-3)}{|x|^2}+\frac{1}{\delta_\rho^2(x)}+\frac{1}{\delta_R^2(x)}+\frac{2}{\delta_\rho(x)\delta_R(x)}\Big)\,|u(x)|^2\,dx.$$

Remark 2.1. In particular, as $R \to \infty$, Corollary 2.1 provides us with the Hardy inequality for the exterior of the ball B_ρ:

$$\int_{B_R\setminus B_\rho} |\nabla u(x)|^2\,dx \geqslant \frac{1}{4}\int_{B_R\setminus B_\rho}\Big(\frac{(d-1)(d-3)}{|x|^2}+\frac{1}{\delta_\rho^2(x)}\Big)\,|u(x)|^2\,dx.$$

Letting $\rho \to 0$ and using $(d-1)(d-3)+1 = (d-2)^2$, we obtain the following assertion.

Corollary 2.2.

$$\int_{B_R} |\nabla u(x)|^2\,dx \geqslant \frac{1}{4}\int_{B_R}\Big(\frac{(d-2)^2}{|x|^2}+\frac{1}{\delta_R^2(x)}+\frac{2}{|x|\delta_R(x)}\Big)\,|u(x)|^2\,dx,$$

where $u \in H_0^1(B_R)$ if $d \geqslant 2$ and $u \in H_0^1(B_R \setminus \{0\})$ if $d = 1$.

Remark 2.2. The last inequality without the term $2/(|x|\delta_R(x))$ follows from Theorem 3.2 in [8]. Note that both constants at the first two terms on the right-hand side of this inequality are sharp.

Now, assuming $-\Delta\delta_\Omega(x) + (d-2)\,x\cdot\nabla\delta_\Omega(x)/|x|^2 \geqslant 0$ and setting $\rho = 0$, we recover Theorem 3.2 in [8].

Corollary 2.3. *Let $\Omega \subset \mathbb{R}^d$ be a bounded domain. Assume that $-\Delta\delta_\Omega(x) + (d-2)\,x\cdot\nabla\delta_\Omega(x)/|x|^2 \geqslant 0$. Then for any $u \in H_0^1(\Omega \setminus \{0\})$*

$$\int_{\Omega\setminus B_\rho} |\nabla u(x)|^2\, dx \geqslant \frac{1}{4} \int_{\Omega\setminus B_\rho} \Big(\frac{(d-2)^2}{|x|^2} + \frac{1}{\delta_\Omega^2(x)}\Big) |u(x)|^2\, dx.$$

As was noted by Frank and Seiringer [9, Lemma 4.3], for any nonnegative symmetric decreasing function u in $\mathbb{R}^d$

$$\|u\|_{2^*,2}^2 = v_d^{-2/d} \int_{\mathbb{R}^d} \frac{u^2(x)}{|x|^2}\, dx, \tag{2.1}$$

where $\|u\|_{2^*,2}$ is the Lorentz norm and $2^* = 2d/(d-2)$. Taking into account this fact, we obtain the following assertion.

Theorem 2.2. *Suppose that* $B_R = \{x \in \mathbb{R}^d : |x| < R\}$, $d \geqslant 3$, *and* $\delta_R(x) = R - |x|$ *is the distance from* x *to* ∂B_R. *Then for any nonnegative symmetric decreasing function* $u \in H_0^1(B_R)$ *the following Hardy–Sobolev inequalities hold:*

$$\int_{B_R} |\nabla u(x)|^2\, dx \geqslant \frac{1}{4} \int_{B_R} \frac{u^2(x)}{\delta_R^2(x)}\, dx + v_d^{2/d} \Big(\frac{(d-2)}{2}\Big)^2 \|u\|_{2^*,2}^2 \tag{2.2}$$

and

$$\int_{B_R} |\nabla u(x)|^2\, dx \geqslant \frac{1}{4} \Big[\int_{B_R} \frac{u^2(x)}{\delta_R^2(x)}\, dx + d(d-2)\, v_d^{2/d}\, \|u\|_{2^*}^2 \Big],$$

where the constants are independent of the radius of the ball.

In the case $d \geqslant 3$, there is a large family of domains with the constant $1/4$ in the Hardy inequality involving the distance to the boundary. To show this fact, we first consider spherical ring domains. More precisely, we generalize the basic inequality for convex domains by considering $B_R \setminus B_\rho$, where $B_R := \{x \in \mathbb{R}^d \mid |x| < R\}$, $B_\rho := \{x \in \mathbb{R}^d \mid |x| \leqslant \rho\}$, $0 \leqslant \rho < R \leqslant +\infty$.

Theorem 2.3. *Let* $B_R \setminus B_\rho \subset \mathbb{R}^d$, $d \geqslant 2$. *Then for any* $u \in H_0^1(B_R \setminus B_\rho)$

$$\int_{B_R\setminus B_\rho} |\nabla u|^2\, dx \;\geqslant\; \frac{1}{4} \int_{B_R\setminus B_\rho} \Big(\frac{1}{\delta^2} + \frac{(d-1)(d-3)}{|x|^2}\Big) |u|^2\, dx + \sigma(u), \tag{2.3}$$

where $\delta = \operatorname{dist}(x, \partial(B_R\setminus B_\rho))$ *and* $\sigma(f)$ *is the integral over the middle surface*

$$\Sigma_m = \{x \in B_R \setminus B_\rho \mid |x| = m := (\rho + R)/2\}$$

defined by

$$\sigma(f) = \frac{2m^{d-1}}{R-\rho} \int_{\Sigma_m} |u|^2\, d\omega \geqslant 0, \quad (\omega = x/|x|).$$

Consider a convex open domain Ω in $\mathbb{R}^d$, $d \geqslant 3$, such that $B_\rho \subset \Omega$ and

$$(d-2)\ \operatorname{dist}(x, \partial(\Omega \setminus B_\rho) \leqslant \rho. \tag{2.4}$$

For instance, if $d = 3$ and $\Omega = B_R$, then the condition (2.4) is equivalent to the inequality $3\rho \geqslant R$.

Theorem 2.4. *Let $d \geqslant 3$, and let Ω be a convex open set in $\mathbb{R}^d$ satisfying the condition* (2.4). *Then*

$$\int_{\Omega\setminus B_\rho} |\nabla u|^2\, dx \geqslant \frac{1}{4}\int_{\Omega\setminus B_\rho} \frac{|u|^2}{\delta^2}\, dx \quad \forall u \in H_0^1(\Omega \setminus B_\rho), \tag{2.5}$$

where $\delta = \operatorname{dist}(x, \partial(\Omega \setminus B_\rho))$.

3 Proof of the Main Results

Proof of Theorems 2.1 and 2.2. We begin by proving a simple consequence of the Cauchy–Schwarz inequality and integration by parts. Different versions of this assertion can be found, for example, in [11, 12].

Lemma 3.1. *Let $\Omega \subset \mathbb{R}^d$, and let $\mathcal{F}(x)$ be a vector field of $x \in \Omega$. If $\mathcal{F}(x)$ and $\operatorname{div}\mathcal{F}(x)$ are finite for $x \in \operatorname{supp} u$, then*

$$\frac{1}{4}\int_\Omega \Big(2\operatorname{div}\mathcal{F}(x) - |\mathcal{F}(x)|^2\Big)|u(x)|^2\, dx \leqslant \int_\Omega |\nabla u|^2\, dx, \quad u \in C_0^\infty(\Omega). \tag{3.1}$$

Proof. Indeed,

$$\begin{aligned}\Big(\int_\Omega \operatorname{div}\mathcal{F}(x)|u(x)|^2\, dx\Big)^2 &= \Big(\int_\Omega \mathcal{F}(x)\cdot\nabla(|u(x)|^2)\, dx\Big)^2\\ &\leqslant 4\int_\Omega |\nabla u|^2\, dx \int_\Omega |\mathcal{F}(x)|^2|u(x)|^2\, dx.\end{aligned}$$

It remains to note that

$$\begin{aligned}&\frac{1}{4}\int_\Omega \Big(2\operatorname{div}\mathcal{F}(x) - |\mathcal{F}(x)|^2\Big)|u(x)|^2\, dx\\ &\qquad\leqslant \frac{1}{4}\,\frac{\Big(\int_\Omega \operatorname{div}\mathcal{F}(x)|u(x)|^2\, dx\Big)^2}{\int_\Omega |\mathcal{F}(x)|^2|u(x)|^2\, dx} \leqslant \int_\Omega |\nabla u|^2\, dx.\end{aligned}$$

□

We need the following simple geometrical lemma for bounded convex domains.

Lemma 3.2. *Let $\Omega \subset \mathbb{R}^d$ be a convex bounded domain with C^1 boundary. Then the value of the interior radius $R_{\text{int}} = \sup\{\delta_\Omega(x) : x \in \Omega\}$ is achieved at some point $x_0 \in \Omega$ and for any $x \in \Omega$*

$$\nabla \delta_\Omega(x) \cdot (x - x_0) \leqslant 0. \tag{3.2}$$

Proof. The function $\delta_\Omega : \overline{\Omega} \to \mathbb{R}_+$ defined on the compact set $\overline{\Omega}$ is continuous. Thus, R_{int} is attained at some point $x_0 \in \Omega$.

Introduce Ω_r, $0 < r < R_{\mathrm{int}}$, such that

$$\Omega_r = \{x \in \Omega : \delta_\Omega(x) > r\}.$$

The domains Ω_r are convex, and $x_0 \in \Omega_r$ for $0 < r \leqslant R_{\mathrm{int}}$. Hence for any $x \in \partial\Omega_r$ the vector $-\nabla\delta_\Omega(x)$ is the outward unit normal to $\partial\Omega_r$ at the point x. Then $-\nabla\delta_\Omega(x) \cdot (x - x_0) \geqslant 0$, which completes the proof. □

Remark 3.1. Lemma 3.2 remains valid for an arbitrary bounded convex domain. In this case, at points where $\nabla\delta_\Omega(x)$ is not uniquely defined, one should consider in (3.2) a normal to any of the supporting hyperplanes to Ω_r at $x \in \partial\Omega_r$.

The following result concerning the Laplacian of the distance function for a convex domain is well known.

Lemma 3.3. *Let $\Omega \subset \mathbb{R}^d$, $d \geqslant 2$, be a convex domain, and let δ_Ω be the distance function to its boundary. Then $-\Delta\,\delta_\Omega(x)$ is a positive measure.*

Proof. Let $y_0 \in \Omega$. Assume that $\nabla\delta_\Omega(y_0)$ exists. Consider orthonormal coordinates $\{y_1, y_2, \dots, y_{d-1}\}$ in the hyperplane defined by $y_0 \in \Omega$ and $-\nabla\delta_\Omega(y)$. In these coordinates, $\partial\Omega$ can be represented by a concave function of class Lip_1. Note that $\partial_{y_d}\delta_\Omega(y_0) = 0$, where y_d is the coordinate along $\nabla\delta_\Omega(y)$. Since Δ is invariant under rotations and parallel shifts, we have $\Delta\delta_\Omega(y_0) \leqslant 0$. □

Let

$$\mathcal{F}(x) = \frac{(d-1)x}{|x|^2} - \frac{\nabla\delta_\rho(x)}{\delta_\rho(x)} - \frac{\nabla\delta_\Omega(x)}{\delta_\Omega(x)}.$$

Since $|\nabla\delta_\rho(x)| = |\nabla\delta_\Omega(x)| = 1$, we obtain

$$\operatorname{div}\mathcal{F}(x) = \frac{(d-1)(d-2)}{|x|^2} + \frac{1}{\delta_\rho^2(x)} + \frac{1}{\delta_\Omega^2(x)} - \frac{(d-1)}{|x|\delta_\rho(x)} - \frac{\Delta\delta_\Omega(x)}{\delta_\Omega(x)}. \tag{3.3}$$

Since $\nabla\delta_\rho(x) = x/|x|$, we have

$$\begin{aligned} |\mathcal{F}(x)|^2 = {} & \frac{(d-1)^2}{|x|^2} + \frac{1}{\delta_\rho^2(x)} + \frac{1}{\delta_\Omega^2(x)} \\ & - 2\frac{(d-1)}{|x|\delta_\rho(x)} - 2\frac{(d-1)x\cdot\nabla\delta_\Omega(x)}{|x|^2\delta_\Omega(x)} + 2\frac{x\cdot\nabla\delta_\Omega(x)}{|x|\delta_\rho(x)\delta_\Omega(x)}. \end{aligned} \tag{3.4}$$

Substituting (3.3) and (3.4) into the left-hand side of (3.1), we obtain the assertion of Theorem 2.1.

Combining (2.1) with Corollary 2.2, we obtain the first statement of Theorem 2.2.

In order to prove the second inequality in Theorem 2.2, it suffices to use (2.2) and note that (cf. [9])

$$\|u\|_{2^*}^2 \leqslant \frac{d-2}{d}\|u\|_{2^*,2}^2.$$

Theorems 2.1 and 2.2 are proved. □

Proof of Theorem 2.3. Without loss of generality, we can assume that $0 < \rho < R < \infty$. Consider the function

$$\psi := r^{-\alpha}\sqrt{\delta}, \quad r = |x| \in (\rho, m) \cup (m, R),$$

where α is a real parameter. It is clear that δ is equal to $r - \rho$ in (ρ, m) and to $R - r$ in (m, R) respectively. A direct computation shows that

$$\lim_{r \to m^-} \frac{d \log \psi}{dr} - \lim_{r \to m^+} \frac{d \log \psi}{dr} = \frac{2}{R - \rho} \tag{3.5}$$

and

$$-\frac{\Delta\psi}{\psi} = \frac{1}{4\delta^2} + \frac{n-1-2\alpha}{2\delta r}\operatorname{sgn}(r-m) + \frac{\alpha(n-2-\alpha)}{r^2}, \tag{3.6}$$

where $r = |x| \in (\rho, m) \cup (m, R)$.

Let $f \in C_0^\infty(B_R \setminus B_\rho)$. It is obvious that

$$\begin{aligned} 0 &\leqslant \int_{B_R \setminus B_\rho} \Big(\nabla f - \frac{f}{\psi}\nabla\psi\Big)^2 dx \\ &= \int_{B_R \setminus B_\rho} \Big(|\nabla f|^2 - (\nabla f^2, \nabla \log \psi) + \frac{f^2}{\psi^2}|\nabla\psi|^2\Big) dx. \end{aligned}$$

To transform the middle term, we apply the Green formula to the domains $\Omega_- = \{x \in B_R \setminus B_\rho \mid |x| < m\}$ and $\Omega_+ = \{x \in B_R \setminus B_\rho \mid |x| > m\}$ separately. Summing up the results and using (3.5), we find

$$\int_{B_R \setminus B_\rho} ((\nabla f^2, \nabla \log \psi) + f^2 \Delta\psi))\, dx = \frac{2m^{d-1}}{R-\rho}\int_{\Sigma_m} |f|^2\, d\omega \geqslant 0.$$

Using these formulas and the identity

$$\Delta \log \psi + (\nabla \log \psi)^2 = (\Delta\psi)/\psi,$$

we obtain

$$\int_{B_R \setminus B_\rho} |\nabla f|^2\, dx \geqslant \int_{B_R \setminus B_\rho} \Big(-\frac{\Delta\psi}{\psi}\Big)|f|^2\, dx + \frac{2m^{d-1}}{R-\rho}\int_{\Sigma_m} |f|^2\, d\omega. \tag{3.7}$$

Using (3.6) with $\alpha = (d-1)/2$ and (3.7), we obtain (2.3), which completes the proof of Theorem 2.3. □

Proof of Theorem 2.4. To prove (2.5) for a given function $f \in C_0^\infty(\Omega \setminus B_\rho)$, it suffices to prove (2.5) for any convex d-dimensional polytope containing B_ρ and the support of f. This is a simple consequence of a result by Hadwiger [10] on approximations of convex domains by polytopes (cf. also [1, 2] for applications of this idea to Hardy type inequalities).

Without loss of generality, we can assume that Ω is a convex d-dimensional polytope. Introduce the subdomans

$$\Omega_- = \{x \in \Omega \setminus B_\rho \mid |x| - \rho < \operatorname{dist}(x, \partial\Omega)\}$$

and

$$\Omega_+ = \{x \in \Omega \setminus B_\rho \mid |x| - \rho > \operatorname{dist}(x, \partial\Omega)\}.$$

We also introduce the so-called "middle" surface Σ_m by the formula

$$\Sigma_m = \{x \in \Omega \setminus B_\rho \mid |x| - \rho = \operatorname{dist}(x, \partial\Omega)\}.$$

Let $f \in C_0^\infty(\Omega \setminus B_\rho)$. Using $\psi = r^{-(d-1)/2}\sqrt{|x| - \rho}$ in the same way as in the proof of Theorem 2.3, we find

$$\int_{\Omega_-} |\nabla f|^2\, dx \geqslant \frac{1}{4}\int_{\Omega_-}\left(\frac{1}{\delta^2} + \frac{(d-1)(d-3)}{|x|^2}\right)|f|^2\, dx + \int_{\Sigma_m} \varphi(x)|f|^2\, dS, \tag{3.8}$$

where

$$\varphi(x) = \frac{\partial \log \psi}{\partial \nu} = \left(\frac{1}{2\delta} - \frac{d-1}{2|x|}\right)\frac{\partial |x|}{\partial \nu} \tag{3.9}$$

and ν is the exterior normal to the boundary of Ω_-. Taking into account that $\delta = |x| - \rho$ and $\partial|x|/\partial\nu \geqslant 0$ on Σ_m and using (2.4), we have $\varphi(x) \geqslant 0$ on Σ_m. Hence

$$\int_{\Omega_-} |\nabla f|^2\, dx \geqslant \frac{1}{4}\int_{\Omega_-}\left(\frac{1}{\delta^2} + \frac{(d-1)(d-3)}{|x|^2}\right)|f|^2\, dx. \tag{3.10}$$

In order to prove Theorem 2.4, it remains to show that

$$\int_{\Omega_+} |\nabla f|^2\, dx \geqslant \frac{1}{4}\int_{\Omega_+}\frac{|f|^2}{\delta^2}\, dx. \tag{3.11}$$

Let $S_j \subset \partial\Omega$ be a face of the polytope. Then

$$\Omega_+ = \cup_j \Omega_j,$$

where $\Omega_j = \{(x', t) \in \Omega_+ : x' \in S_j, 0 < t < l_j(x')\}$ and $t = l_j(x')$ is the equation for equidistant points. By the standard one-dimensional Hardy inequality, we obtain

$$\int_0^{l_j(x')} \Big|\frac{\partial f(x',t)}{\partial t}\Big|^2 dt \geqslant \frac{1}{4}\int_0^{l_j(x')} \frac{|f(x',t)|^2}{t^2}\, dt.$$

Therefore,

$$\int_{\Omega_j} |\nabla f|^2\, dx \geqslant \int_{S_j}\int_0^{l_j(x')} \Big|\frac{\partial f(x',t)}{\partial t}\Big|^2 dtdx'$$
$$\geqslant \frac{1}{4}\int_{S_j}\int_0^{l_j(x')} \frac{|f(x',t)|^2}{t^2}\, dtdx' = \frac{1}{4}\int_{\Omega_j} \frac{|f|^2}{\delta^2}\, dx.$$

Summing up these inequalities, we obtain (3.11). □

4 Remarks

1. If the condition (2.4) is not satisfied, then (3.8) and (3.9) imply the inequality

$$\int_{\Omega\setminus B_\rho} |\nabla u|^2\, dx + \frac{d-1}{2}\int_{\Sigma_m} \frac{|u|^2}{|x|}\, dS \geqslant \frac{1}{4}\int_{\Omega\setminus B_\rho} \frac{|u|^2}{\delta^2}\, dx \quad \forall u \in H_0^1(\Omega\setminus B_\rho)$$

for any convex open set Ω containing the closed ball B_ρ provided that $d \geqslant 3$.

2. The following statement is a consequence of Theorem 2.1 and is complementary to Theorem 2.4.

Corollary 4.1. *Let Ω be a bounded convex domain. Assume that the origin is chosen in such a way that* $\max\{\delta_\Omega(x) : x \in \Omega\}$ *is achieved at* 0. *Suppose that* $B_\rho \subset \Omega \subset \{x : (d-2)|x| < (d-1)\rho\}$, $d \geqslant 2$. *Then for any* $u \in H_0^1(\Omega\setminus B_\rho)$

$$\int_{\Omega\setminus B_\rho} |\nabla u(x)|^2\, dx \geqslant \frac{1}{4}\int_{\Omega\setminus B_\rho} \Big(\frac{(d-1)(d-3)}{|x|^2} + \frac{1}{\delta_\rho^2(x)} + \frac{1}{\delta_\Omega^2(x)}\Big) |u(x)|^2\, dx.$$

Proof. By Lemma 3.3, for convex domains we have $\Delta\delta_\Omega(x) \leqslant 0$. By Lemma 3.2, $x\cdot\nabla\delta_\Omega \leqslant 0$. If $(d-2)|x| < (d-1)\rho$, then

$$\frac{2}{\delta_\rho(x)\delta_\Omega(x)} \geqslant \frac{2(d-1)}{|x|\delta_\Omega(x)},$$

which implies the required assertion. □

3. In particular, in the case $d = 2$, for any convex domain satisfying assumptions of Corollary 4.1

$$\int_{\Omega\setminus B_\rho} |\nabla u(x)|^2\,dx \geqslant \frac{1}{4}\int_{\Omega\setminus B_\rho}\Big(\frac{1}{(|x|-\rho)^2} - \frac{1}{|x|^2} + \frac{1}{\delta_\Omega^2(x)}\Big)|u(x)|^2\,dx.$$

Hence we obtain (1.2) as $\rho \to 0$.

4. Note that for $d = 2$ and $\Omega = \Omega_2 := B_R \setminus B_\rho \subset \mathbb{R}^2$ Theorem 2.4 is not true. As is shown in [1], the best constant $\lambda(\Omega_2)$ in the Hardy inequality

$$\int_{\Omega_2} |\nabla u|^2\,dx \geqslant \lambda(\Omega_2)\int_{\Omega_2}\frac{|u(x)|^2}{\delta^2(x)}\,dx, \quad u \in H_0^1(\Omega_2), \tag{4.1}$$

satisfies the inequalities

$$\frac{2}{\pi}\log\frac{R}{\rho} \leqslant \frac{1}{\lambda(\Omega_2)} \leqslant \log\frac{R}{\rho} + k_0, \tag{4.2}$$

where $k_0 = (\Gamma(1/4))^4/(2\pi^2) = 8.75\ldots$. Clearly, $\lambda(\Omega_2) \to 0$ as $R/\rho \to \infty$.

5. Theorem 2.4 fails if B_ρ is replaced with an arbitrary convex compact subset of Ω. We show that for any $d \geqslant 3$ and $\varepsilon > 0$ there exist d-dimensional ellipsoids E_1, E_2, $\overline{E_2} \subset E_1$ and a function $f_e \in C_0^1(\Omega_e)$ such that

$$\int_{\Omega_e} |\nabla f_e(x)|^2\,dx \leqslant \varepsilon\int_{\Omega_e}\frac{|f_e(x)|^2}{\delta^2(x)}\,dx, \tag{4.3}$$

where $\Omega_e = E_1 \setminus \overline{E_2}$ and $\delta(x)$ is the distance from $x \in \Omega_e$ to $\partial\Omega_e$.

Indeed, using (4.1) and (4.2) for a given $\varepsilon > 0$ we choose $R/\rho > \exp(\pi\varepsilon)$ such that $\lambda(\Omega_2) < \varepsilon/2$. Then for $\Omega_2 := B_R\setminus B_\rho \subset \mathbb{R}^2$ there exists $f_2 \in C_0^1(\Omega_2)$ such that

$$\int_{\Omega_2} |\nabla f_2(x')|^2\,dx' < \frac{\varepsilon}{2}\int_{\Omega_2}\frac{|f_2(x')|^2}{\delta^2(x')}\,dx', \quad x' = (x_1, x_2). \tag{4.4}$$

Let $s > 0$. Define $g_s \in C_0^1(\mathbb{R})$ by the formula

$$g_s(t) = \begin{cases} 1, & |t| \leqslant s, \\ 0, & |t| > 1+s, \\ (1-(|t|-s)^2)^2, & s < |t| \leqslant 1+s. \end{cases} \tag{4.5}$$

Define $\Omega_d \subset \mathbb{R}^d$ as the product $\Omega_2 \times \mathbb{R}^{d-2}$, and set

$$f_e(x) := f_2(x_1, x_2)\,\Pi_{j=3}^d\, g_s(x_j).$$

It is obvious that

$$\int_{\Omega_d} |\nabla f_e(x)|^2\,dx = (2s)^{d-2}\int_{\Omega_2} |\nabla f_2(x')|^2\,dx' + A.$$

Since dist $(\mathrm{x}, \partial\Omega_{\mathrm{d}}) = \mathrm{dist}\,((\mathrm{x}_1, \mathrm{x}_2), \partial\Omega_2) =: \delta(\mathrm{x}')$, we have

$$\int_{\Omega_d} \frac{|f_e(x)|^2}{\mathrm{dist}^2(\mathrm{x}, \partial\Omega_{\mathrm{d}})}\, dx = (2s)^{d-2} \int_{\Omega_2} \frac{|f_2(x')|^2}{\delta^2(x')}\, dx' + B,$$

where A and B are constants independent on s. By these equations and (4.4), for $d \geqslant 3$ and sufficiently large s we have

$$\int_{\Omega_d} |\nabla f_e(x)|^2\, dx < \varepsilon \int_{\Omega_d} \frac{|f_e(x)|^2}{\mathrm{dist}^2(\mathrm{x}, \partial\Omega_{\mathrm{d}})}\, dx.$$

Taking the ellipsoids E_j, $j = 1, 2$, defined by

$$\frac{x_1^2 + x_2^2}{a_j^2} + \frac{x_3^2 + .. + x_d^2}{b^2} < 1, \quad a_1 = R, \quad a_2 = \frac{1}{2}(\rho + \min_{x' \in \mathrm{supp}\, f_2} |x'|)$$

with sufficiently large b, we obtain (4.3).

Acknowledgments. The authors acknowledge a partial support by the ESF Programme SPECT. F.G. Avkhadiev was supported by the Russian Foundation for Basic Research (grant 08-01-00381) and by the Göran Gustafssons Stiftelse in Sweden.

References

1. Avkhadiev, F.G.: Hardy type inequalities in higher dimensions with explicit estimate of constants. Lobachevskii J. Math. **21**, 3–31 (2006)
2. Avkhadiev, F.G.: Hardy-type inequalities on planar and spatial open sets. Proc. Steklov Inst. Math. **255**, no. 1, 2–12 (2006)
3. Avkhadiev, F.G., Wirths, K.-J.: Unified Poincaré and Hardy inequalities with sharp constants for convex domains. Z. Angew. Math. Mech. **87**, 632–642 (2007)
4. Brezis, H., Marcus, M.: Hardy's inequalities revisited. Ann. Scuola Norm. Sup. Pisa Cl. Sci. (4) **25**, no. 1-2, 217–237 (1997/98)
5. Davies, E.B.: Spectral Theory and Differential Operators. Cambridge Univ. Press, Cambridge (1995)
6. Davies, E.B.: A review of Hardy inequalities. In: The Maz'ya Anniversary Collection 2. Oper. Theory Adv. Appl. **110**, 55–67 (1999)
7. Filippas, S., Maz'ya, V., Tertikas, A.: Sharp Hardy–Sobolev inequalities. C. R., Math., Acad. Sci. Paris **339**, no. 7, 483–486 (2004)
8. Filippas, S., Moschini, L., Tertikas, A.: Sharp two-sided heat kernel estimates for critical Schrödinger operators on bounded domains. Commun. Math. Phys. **273**, 237–281 (2007)
9. Frank, R.L., Seiringer, R.: Non-linear ground state representations and sharp Hardy inequalities. arXiv:0803.0503v1 [math.AP] 4 Mar. 2008

10. Hadwiger, H.: Vorlesungen über Inhalt, Oberfläsche und Isoperimetrie. Springer (1957)
11. Hoffmann-Ostenhof, M., Hoffmann-Ostenhof, Th., Laptev, A.: A geometrical version of Hardy's inequalities. J. Funct. Anal. **189**, no 2, 539–548 (2002)
12. Hoffmann-Ostenhof, M., Hoffmann-Ostenhof, Th., Laptev, A., Tidblom, J.: Many particle Hardy inequalities. [To appear]

Distributions with Slow Tails and Ergodicity of Markov Semigroups in Infinite Dimensions

Sergey Bobkov and Boguslaw Zegarlinski

Abstract We discuss some geometric and analytic properties of probability distributions that are related to the concept of weak Poincaré type inequalities. We deal with isoperimetric and capacitary inequalities of Sobolev type and applications to finite-dimensional convex measures with weights and infinite-dimensional Gibbs measures. As one of the basic tools, V. G. Mazya's capacitary analogue of the co-area inequality is adapted to the setting of metric probability spaces.

1 Weak Forms of Poincaré Type Inequalities

In this paper, we discuss some geometric and analytic properties of probability distributions, such as embeddings, concentration, and convergence of the associated semigroups, that are related to the concept of weak Poincaré type inequalities. Such inequalities may have different forms and appear in different contexts and settings. We mainly restrict ourselves to the setting of an arbitrary metric probability space, say, (M, d, μ) (keeping in mind the Euclidean space $\mathbf{R}^n$ as a basic space and source for various examples). We will be focusing on the following definition.

Definition. We say that (M, d, μ) satisfies a *weak Poincaré type inequality* with rate function $C(p)$, $1 \leqslant p < 2$, if for any bounded, locally Lipschitz function f on M with μ-mean zero

Sergey Bobkov
School of Mathematics, University of Minnesota, Minneapolis, MN 55455, USA
e-mail: `bobkov@math.umn.edu`

Boguslaw Zegarlinski
Department of Mathematics, Imperial College London, Huxley Building, 180 Queen's Gate, London SW7 2AZ, UK
e-mail: `b.zegarlinski@imperial.ac.uk`

A. Laptev (ed.), *Around the Research of Vladimir Maz'ya I: Function Spaces*,
International Mathematical Series 11, DOI 10.1007/978-1-4419-1341-8_2,

$$\|f\|_p \leqslant C(p)\,\|\nabla f\|_2 \quad \forall\, p \in [1,2). \tag{1.1}$$

More precisely, (1.1) involves a parameter family of Poincaré type inequalities that are controlled by a certain parameter function. Here, we use the standard notation $\|f\|_p = \Big(\int |f|^p\,d\mu\Big)^{1/p}$ for the L^p-norm, as well as $\|\nabla f\|_2 = \Big(\int |\nabla f|^2\,d\mu\Big)^{1/2}$. Note that it is a rather convenient way to understand the modulus of the gradient in general as the function

$$|\nabla f(x)| = \limsup_{y \to x} \frac{|f(x) - f(y)|}{d(x,y)}, \qquad x \in M$$

(with the convention that $|\nabla f(x)| = 0$ if x is an isolated point in M). By saying that f is "locally Lipschitz" we mean that f has a finite Lipschitz seminorm on every ball in M, so that $|\nabla f(x)|$ is everywhere finite. Once (1.1) holds for all bounded locally Lipschitz f, it continues to hold for all unbounded locally Lipschitz functions with μ-mean zero, as long as the right-hand side of (1.1) is finite. (The latter implies the finiteness of $\|f\|_p$ for all $p < 2$.)

As a more general scheme, one could start from a probability space (M,μ), equipped with some (local or discrete) Dirichlet form $\mathcal{E}(f,f)$, and to consider the inequalities

$$\|f\|_p \leqslant C(p)\,\sqrt{\mathcal{E}(f,f)}, \quad 1 \leqslant p < 2, \tag{1.2}$$

within the domain $\mathcal{D}$ of the Dirichlet form. Within the metric probability space framework, we thus put $\mathcal{E}(f,f) = \|\nabla f\|_2^2$. But one may also study (1.2) in the setting of a finite graph or, more generally, of Markov kernels, or the setting of Gibbs measures.

The main idea behind (1.1)–(1.2) is to involve in analysis more probability distributions and to quantify their possible analytic and other properties by means of the rate function. Indeed, if $C(p)$ may be chosen to be a constant, we arrive at the usual form of the Poincaré type inequality

$$\lambda_1\,\mathrm{Var}_\mu(f) \leqslant \mathcal{E}(f,f), \tag{1.3}$$

where

$$\mathrm{Var}_\mu(f) = \int f^2\,d\mu - \Big(\int f\,d\mu\Big)^2$$

stands for the variance of f under μ. This inequality itself poses a rather strict constraint on the measure μ. For example, under (1.3) in the setting of a metric probability space (M,d,μ), any Lipschitz function f on M must have a finite exponential moment. This property, discovered by Herbst [22] and later by Gromov and Milman [20] and by Borovkov and Utev [13], may be stated as a deviation inequality

$$\mu\{|f| > t\} \leqslant C\,e^{-c\sqrt{\lambda_1}\,t}, \quad t > 0, \tag{1.4}$$

with some positive absolute constants C and c, where for normalization reasons it is supposed that $\|f\|_{\mathrm{Lip}} \leqslant 1$ and $\int f\,d\mu = 0$. (The best constant in the exponent is known to be $c = 2$ and it is attained for the one-sided exponential distribution on the real line $M = \mathbf{R}$, cf. [7].) With a proper understanding of the Lipschitz property, discrete and more general analogues of (1.4) also hold under (1.3) (cf., for example, [2, 1, 24, 25]).

Another classical line of applications of the usual Poincaré type inequality deals with the Markov semigroup P_t of linear operators associated to μ on $\mathbf{R}^n$ (or other Riemannian manifold). This semigroup has a generator L, which may be introduced via the equality

$$\mathcal{E}(f,g) = -\int f\,Lg\,d\mu, \quad f,g \in \mathcal{D},$$

so that $P_t = e^{tL}$ in the operator sense. Under (1.3) and mild technical assumptions, every P_t represents a contraction on $L^2(\mu)$, i.e., P_t may be extended from $\mathcal{D}$ as a linear continuous operator acting on the whole space $L^2(\mu)$ with the operator norm $\|P_t\| \leqslant 1$. Moreover, for any $f \in L^2(\mu)$,

$$\mathrm{Var}_\mu(P_t f) \leqslant \mathrm{Var}_\mu(f)\,e^{-\lambda_1 t}, \quad t > 0, \tag{1.5}$$

which expresses the $L^2(\mu)$ exponential ergodicity property of the Markov semigroup.

The exponential bounds such as (1.4)–(1.5) do not hold any longer without the hypothesis on the presence of the usual Poincaré type inequality. However, one may hope to get weaker conclusions under weaker assumptions, such as the weak Poincaré type inequality (1.1). In the latter case, the rate of growth of $C(p)$ as $p \to 2$ turns out to be responsible for the strength of deviations of Lipschitz functions and for the rate of convergence of $P_t f$ to a constant function, as well.

As a result, in the general situation more freedom in choosing suitable rate function $C(p)$ will allow us to involve more interesting probability spaces, especially those without finite exponential moments. In this connection it should be noted that another kind of inequalities, which serve this aim, is described by the weak forms of Poincaré type inequalities, that involve an oscillation term $\mathrm{Osc}\,(f) = \operatorname{ess\,sup} f - \operatorname{ess\,inf} f$ with respect to μ. Namely, one considers

$$\mathrm{Var}_\mu(f) \leqslant \beta(s)\mathcal{E}(f,f) + s\,\mathrm{Osc}\,(f)^2. \tag{1.6}$$

These inequalities are supposed to hold for all $s > 0$ with some function β, so that the case of the constant function $\beta(s) = 1/\lambda_1$ also returns us to the usual Poincaré inequalities.

The inequalities with a free parameter have a long history in analysis, including, for example, [32, 17, 16, 18, 27, 21, 5, 6] and many others. The weak Poincaré type inequalities (1.6) have recently been studied by Röckner and Wang [35, 37], as an approach to the problem on the slow rates in the convergence of the associated Markov semigroups in $\mathbf{R}^n$. This work was motivated by Liggett [27], who considered similar multiplicative forms of (1.6). In the setting of Riemannian manifolds, Barthe, Cattiaux, and Roberto [4] studied the weak Poincaré type inequalities from the point of view of concentration and connected them with the family of capacitary inequalities, a classical object in the theory of Sobolev spaces. Such inequalities go back to the pioneering works of V. G. Maz'ya in the 60s and 70s; let us only mention [28], his book [29], and a nice exposition given by A. Grigoryan in [19]. See also [15], where entropic versions of (1.6) are treated. On the other hand, although weak Poincaré type inequalities (1.1) should certainly be of independent alternative interest, they seemed to attract much less attention. And for several reasons one may wonder how to fill this gap.

We explore, how the weak forms of Poincaré type inequalities (1.1)/(1.3) and (1.5) are related to each other (Section 8). Note that for probability measures on the real line, all the forms may be reduced to Hardy type inequalities with weights and this way they may be characterized explicitly in terms of the density of a measure (cf. [31, 29]). One obvious advantage of (1.1) over (1.6) is that one may freely apply (1.1) to unbounded functions, while (1.6) is more delicate in this respect. In fact, the relation (1.1), taken as a potential "nice" hypothesis, gives rise to a larger family of Poincaré type inequalities between the norms of f and $|\nabla f|$ in Lebesgue spaces. This property, which we briefly discuss in Section 2, is usually interpreted as kind of embedding theorems. It is illustrated in Section 3 in the problem of large deviations of Lipschitz functionals. Sections 4 and 5 are technical, with aim to create tools to estimate the rate function for classes of measures on the Euclidean space under certain convexity conditions (cf. Sections 6 and 7). In Section 10, we discuss consequences of our weak Poincaré inequalities for the L^p decay to equilibrium of Markov semigroups in $\mathbf{R}^n$. But before, in Section 9, we introduce the notation and recall classical arguments, that are used in the presence of the usual Poincaré inequalities.

Later we extend the corresponding idea to infinite dimensional situation in Section 11, where, in particular, we prove a stretched exponential decay for a product case. As we demonstrate there, it is the infinite dimensional case in which our more general than (1.6) inequalities play an important role in estimates of the decay rates. Finally, in the last section, we prove a weak Poincaré inequality for Gibbs measures with slowly decaying tails in the region of strong mixing. Using this result, we obtain an estimate for the decay to equilibrium in L^2 for all Lipschitz cylinder functions with the same stretched exponential rate.

2 L^p-Embeddings under Weak Poincaré

Let us start with an abstract metric probability space (M, d, μ) satisfying the weak Poincaré type inequality

$$\|f - \mathbf{E}f\|_p \leqslant C(p)\, \|\nabla f\|_2, \quad 1 \leqslant p < 2, \tag{2.1}$$

with a (finite) rate function $C(p)$. For definiteness, we may put $C(p) = C(1)$ for $0 < p < 1$, although in some places we will consider (2.1) for all $p \in (0, 2)$ with rate function defined in $0 < p < 1$ in a different way. Here and in the sequel, we use the standard notation $\mathbf{E}f = \mathbf{E}_\mu f = \int f\, d\mu$ for the expectation of f under the measure μ.

Let $W^q(\mu)$ denote the space of all locally Lipschitz functions g on M, equipped with the norm

$$\|f\|_{W^q} = \|\nabla f\|_q + \|f\|_1.$$

Clearly, the norm is getting stronger with the growing parameter q. From (2.1) it follows that $\|f\|_p \leqslant (1 + C(p))\, \|f\|_{W^2}$, which means that all $L^p(\mu)$, $1 \leqslant p < 2$, are embedded in $W^2(\mu)$. Therefore, one may wonder whether this property may be sharpened by replacing $W^2(\mu)$ with other spaces $W^q(\mu)$. The answer is affirmative and is given by the following assertion.

Theorem 2.1. *Given $1 \leqslant p < q \leqslant +\infty$, $q \geqslant 2$, for any locally Lipschitz f on M*

$$\|f - \mathbf{E}f\|_p \leqslant C(p, q)\, \|\nabla f\|_q, \tag{2.2}$$

with constants $C(p, q) = \frac{12\, C(r)}{r}\, p$, where $\frac{1}{r} = \frac{1}{2} + \frac{1}{p} - \frac{1}{q}$.

Thus, $L^p(\mu)$ may be embedded in $W^q(\mu)$, whenever $1 \leqslant p < q \leqslant +\infty$ and $q \geqslant 2$.

In particular, $\frac{1}{r} = 1 - \frac{1}{q}$ for $p = 2$, so r represents the dual exponent $q^* = \frac{q}{q-1}$ and $C(2, q) = \frac{24\, C(q^*)}{q^*} \leqslant 24\, C(q^*)$. Hence we obtain a dual variant of (2.1).

Corollary 2.2. *Under* (2.1), *for any bounded, locally Lipschitz function f on M*

$$\|f - \mathbf{E}f\|_2 \leqslant 24\, C(q^*)\, \|\nabla f\|_q, \quad q > 2. \tag{2.3}$$

Now, let us turn to the proofs, which actually contain standard arguments. In the sequel, we will use the following elementary:

Lemma 2.3. *For any measurable function f on a probability space (M,μ) with a median m and for any $p \geqslant 1$*

$$\|f-m\|_p \leqslant 3 \inf_{c\in\mathbf{R}} \|f-c\|_p.$$

Proof. One may assume that the norm $\|f\|_p$ is finite and non-zero. Note that, in general, the median is not determined uniquely. Nevertheless, by the monotonicity of this multi-valued functional, for any median $m = m(f)$ of f there is a median $m(|f|)$ of $|f|$ such that $|m(f)| \leqslant m(|f|)$. On the other hand, by the Chebyshev inequality,

$$\mu\{|f|>t\} \leqslant \frac{\|f\|_p^p}{t^p} < \frac{1}{2},$$

as long as $t > 2^{1/p}\,\|f\|_p$, so $m(|f|) \leqslant 2^{1/p}\,\|f\|_p$ for any median of $|f|$. The two bounds yield

$$\|f-m(f)\|_p \leqslant \|f\|_p + |m(f)| \leqslant (1+2^{1/p})\,\|f\|_p \leqslant 3\,\|f\|_p.$$

Applying this to $f-c$ and noting that $m(f)-c$ is one of the medians of $f-c$, we arrive at the desired conclusion. □

Lemma 2.3 allows us to freely interchange medians and expectations in the weak Poincaré type inequality. This can be stated as follows.

Lemma 2.4. *Under the hypothesis* (2.1), *for any locally Lipschitz function f on M with median $m(f)$*

$$\|f-m(f)\|_p \leqslant C'(p)\,\|\nabla f\|_2, \quad 1 \leqslant p < 2, \tag{2.4}$$

where $C'(p) = 3C(p)$. In turn, (2.4) *implies* (2.1) *with $C(p) = 2C'(p)$.*

Indeed, by Lemma 2.3, $\|f-m(f)\|_p \leqslant 3\,\|f-\mathbf{E}f\|_p$, and thus (2.1) implies (2.4). On the other hand, assuming that $m(f)=0$ and starting from (2.4), we get

$$\|f-\mathbf{E}f\|_p \leqslant \|f\|_p + |\mathbf{E}f| \leqslant 2\,\|f\|_p = 2\,\|f-m(f)\|_p \leqslant 2C'(p)\,\|\nabla f\|_2.$$

Lemma 2.5. *Assume that the metric probability space (M,d,μ) satisfies*

$$\|f\|_p \leqslant A(p)\,\|\nabla f\|_2, \quad 0<p<2, \tag{2.5}$$

in the class of all locally Lipschitz functions f on M with median $m(f)=0$. Then in the same class,

$$\|f\|_p \leqslant 2A(p,q)\,\|\nabla f\|_q, \quad 0<p<q\leqslant +\infty,\ q\geqslant 2, \tag{2.6}$$

with constants $A(p,q) = \frac{A(r)}{r}\,p$, *where* $\frac{1}{r} = \frac{1}{2} + \frac{1}{p} - \frac{1}{q}$.

Clearly, $A(p,2) = A(p)$, so that (2.6) generalizes (2.5) within a factor of 2. If (2.5) is only given for the range $1 \leqslant p < 2$, one may just put $A(p) = A(1)$ for $0 < p < 1$.

Note that, due to the assumption $p < q$, we always have $0 < r < 2$. The assumption $q \geqslant 2$ guarantees that $r \leqslant p$.

Proof of Lemma 2.5. We may assume that $2 < q < +\infty$ and $\|f\|_q < +\infty$. First, let $f \geqslant 0$ and $m(f) = 0$. Hence $\mu\{f=0\} \geqslant \frac{1}{2}$. By the hypothesis (2.5), for any $r \in (0,2)$

$$\mathbf{E} f^r \leqslant A(r)^r\,(\mathbf{E}\,|\nabla f|^2)^{r/2}.$$

Apply this inequality to the function $f^{p/r}$, which is nonnegative and has median zero, as well as f. Then, using the Hölder inequality with exponents $\alpha, \beta > 1$ such that $\frac{1}{\alpha} + \frac{1}{\beta} = 1$, we get

$$\begin{aligned}\mathbf{E} f^p = \mathbf{E}\,(f^{p/r})^r &\leqslant A(r)^r \left(\frac{p}{r}\right)^r \left(\mathbf{E}\, f^{2(\frac{p}{r}-1)}\,|\nabla f|^2\right)^{r/2} \\ &\leqslant A(r)^r \left(\frac{p}{r}\right)^r \left(\mathbf{E}\, f^{2\alpha\,(\frac{p}{r}-1)}\right)^{r/2\alpha} \left(\mathbf{E}\,|\nabla f|^{2\beta}\right)^{r/2\beta}. \end{aligned} \tag{2.7}$$

Now let $\frac{1}{r} = \frac{1}{2} + \frac{1}{p} - \frac{1}{q}$ and choose α so that $2\alpha\,(\frac{p}{r} - 1) = p$, i.e., $\frac{1}{2\alpha} = \frac{1}{r} - \frac{1}{p}$. Since $q > 2$, we have $r < p$, so $\alpha > 0$. Moreover, $\alpha > 1 \Leftrightarrow \frac{1}{r} < \frac{1}{2} + \frac{1}{p}$ which is fulfilled. Also, put $\frac{1}{2\beta} = \frac{1}{q}$, so that $\beta = \frac{q}{2} > 1$. Then (2.7) turns into

$$\mathbf{E} f^p \leqslant A(r)^r \left(\frac{p}{r}\right)^r (\mathbf{E}\, f^p)^{r/2\alpha}\,(\mathbf{E}\,|\nabla f|^q)^{r/2\beta},$$

which is equivalent to

$$\|f\|_p \leqslant A(p,q)\,\|\nabla f\|_q. \tag{2.8}$$

In the general case, we split $f = f^+ - f^-$ with $f^+ = \max\{f,0\}$ and $f^- = \max\{-f,0\}$. Without loss of generality, let $|\nabla f(x)| = 0$, when $f(x) = 0$ (otherwise, we may work with functions of the form $T(f)$ with smooth T, approximating the identity function and satisfying $T'(0) = 0$). Then both f^+ and f^- are nonnegative, have median at zero, and $|\nabla f^+| = |\nabla f|\,1_{\{f>0\}}$, $|\nabla f^-| = |\nabla f|\,1_{\{f<0\}}$. Hence, by the previous step (2.8) applied to these functions,

$$\int_{\{f>0\}} |f|^p\,d\mu \leqslant A(p,q)^p \left(\int_{\{f>0\}} |\nabla f|^q\,d\mu\right)^{p/q},$$

$$\int_{\{f<0\}} |f|^p \, d\mu \leqslant A(p,q)^p \left(\int_{\{f<0\}} |\nabla f|^q \, d\mu \right)^{p/q}.$$

Finally, adding these inequalities and using an elementary bound $a^s + b^s \leqslant 2(a+b)^s$ $(a, b \geqslant 0,\ 0 \leqslant s \leqslant 1)$, we arrive at the desired estimate (2.6). □

Proof of Theorem 2.1. By Lemmas 2.4 and 2.5, for any locally Lipschitz f on M with median $m(f)$, whenever $1 \leqslant p < q \leqslant +\infty$ and $q \geqslant 2$, we have

$$\|f - m(f)\|_p \leqslant \frac{6p\, C(r)}{r} \, \|\nabla f\|_q.$$

Another application of Lemma 2.4 doubles the constant on the right-hand side. □

3 Growth of Moments and Large Deviations

As another immediate consequence of Theorem 2.1, we consider the case $q = +\infty$. Then $\frac{1}{r} = \frac{p+2}{2p}$, and we obtain the following assertion.

Corollary 3.1. *Under the weak Poincaré type inequality* (2.1), *any Lipschitz function f on M has finite L^p-norms and, if $\|f\|_{\mathrm{Lip}} \leqslant 1$ and $\mathbf{E}f = 0$,*

$$\|f\|_p \leqslant 6(p+2)\, C\left(\frac{2p}{p+2}\right), \quad p \geqslant 1. \tag{3.1}$$

In the case of the usual Poincaré type inequality,

$$\lambda_1 \operatorname{Var}_\mu(f) \leqslant \int |\nabla f|^2 \, d\mu,$$

we have $C(p) = \frac{1}{\sqrt{\lambda_1}}$, and the inequality (3.1) gives

$$\|f\|_p \leqslant \frac{6(p+2)}{\sqrt{\lambda_1}}, \quad p \geqslant 1.$$

Up to a universal constant $c > 0$, the latter may also be stated as a large deviation bound $\mu\{\sqrt{\lambda_1}\, |f| \geqslant t\} \leqslant 2\, e^{-ct}$ $(t \geqslant 0)$ or, equivalently, as

$$\|f\|_{\psi_1} \leqslant \frac{1}{c\sqrt{\lambda_1}} \tag{3.2}$$

in terms of the Orlicz norm generated by the Young function $\psi_1(t) = e^{|t|} - 1$.

Thus, Theorem 2.1 may be viewed as a generalization of the Gromov–Milman theorem [20] on the concentration in the presence of the Poincaré type inequality. Let us give one more specific example by imposing the condition

$$C(p) \leqslant \frac{a}{(2-p)^{\gamma}}, \quad 1 \leqslant p < 2, \tag{3.3}$$

with some parameters $a, \gamma \geqslant 0$. In particular, if $\gamma = 0$, we return to the usual Poincaré type inequality.

Corollary 3.2. *Let (M, d, μ) satisfy the weak Poincaré type inequality (2.1) with a rate function admitting the polynomial growth (3.3). For any function f on M with $\|f\|_{\mathrm{Lip}} \leqslant 1$ and $\mathbf{E} f = 0$*

$$\mu\{|f| \geqslant t\} \leqslant 2^{\gamma+1} \exp\bigg\{ - c_1(\gamma+1) \Big(\frac{c_2 t}{a}\Big)^{1/(\gamma+1)} \bigg\}, \quad t > 0, \tag{3.4}$$

where c_1 and c_2 are positive numerical constants.

Proof. According to Corollary 3.1, for any $p \geqslant 1$

$$\|f\|_p \leqslant 6\,(p+2)\,\frac{a}{(2-\frac{2p}{p+2})^{\gamma}} = 6a \cdot 4^{-\gamma}(p+2)^{\gamma+1} \leqslant 18\,a \cdot (3/4)^{\gamma} p^{\gamma+1},$$

where we used $p + 2 \leqslant 3p$ at the last step. Hence, in the range $p \geqslant 1$, we have got the bound $\mathbf{E}\,|f|^p \leqslant (Cp)^{p(\gamma+1)}$ with a constant given by $C^{\gamma+1} = 18 \cdot (3/4)^{\gamma}\, a$. This bound may a little be weakened as

$$\mathbf{E}\,|f|^p \leqslant 2^{\gamma+1}(Cp)^{p(\gamma+1)} \tag{3.5}$$

to serve also the values $0 < p < 1$. Indeed, then we may use $\|f\|_p \leqslant \|f\|_1 \leqslant C^{\gamma+1}$, so $\mathbf{E}\,|f|^p \leqslant C^{p(\gamma+1)}$. Hence (3.5) would follow from $1 \leqslant 2p^p$, which is true since, on the positive half-axis, the function $2p^p$ is minimized at $p = \frac{1}{e}$ and has the minimum value $2e^{-1/e} > 1$.

Thus, (3.5) holds in the range $p > 0$. Now, by the Chebyshev inequality, for any $t > 0$

$$\mu\{|f| \geqslant t\} \leqslant \frac{\mathbf{E}\,|f|^p}{t^p} \leqslant 2^{\gamma+1}\,\frac{(Cp)^{p(\gamma+1)}}{t^p} = 2^{\gamma+1}\,(Dq)^q,$$

where $q = p\,(\gamma+1)$ and $D = \frac{C}{(\gamma+1)\,t^{1/(\gamma+1)}}$. The quantity $(Dq)^q$ is minimized, when $q = 1/(De)$, and the minimum is

$$e^{-1/(De)} = \exp\bigg\{ - \frac{(\gamma+1)\,t^{1/(\gamma+1)}}{Ce} \bigg\} = \exp\bigg\{ - \frac{4\,(\gamma+1)\,t^{1/(\gamma+1)}}{3e\,(24\,a)^{1/(\gamma+1)}} \bigg\}.$$

Thus, we arrive at (3.4) with $c_1 = 4/(3e)$ and $c_2 = 1/24$. □

In analogue with the usual Poincaré type inequality and similarly (3.2), the deviation inequality (3.4) of Corollary 3.2 may be restated equivalently in terms of the Orlicz norm generated by the Young function

$$\psi_{1/(\gamma+1)}(t) = \exp\{|t|^{1/(\gamma+1)}\} - 1.$$

Indeed, arguing in one direction, we consider $\xi = |f|^{1/(\gamma+1)}$ as a random variable on $(\mathbf{R}^n, \mu)$ and write (3.4) as

$$\mu\{\xi > t\} \leqslant Ae^{-Bt}, \quad t > 0,$$

with parameters $A = 2^{\gamma+1}$, $B = c_1(\gamma+1)\,(\frac{c_2}{a})^{1/(\gamma+1)}$. Then for any $r \in (0, B)$

$$\mathbf{E}\, e^{r\xi} - 1 = r\int_0^{+\infty} e^{rt}\mu\{\xi > t\}\, dt \leqslant Ar\int_0^{+\infty} e^{-(B-r)t}\, dt = \frac{Ar}{B-r} = 1$$

if $r = r_0 = \frac{B}{A+1}$. Hence $\mathbf{E}\exp\{r_0|f|^{1/(\gamma+1)}\} \leqslant 2$, which means that

$$\|f\|_{\psi_{1/(\gamma+1)}} \leqslant \frac{1}{r_0^{\gamma+1}} = \frac{(A+1)^{\gamma+1}}{B^{\gamma+1}} = \frac{a}{c_2}\,\frac{(2^{\gamma+1}+1)^{\gamma+1}}{(c_1(\gamma+1))^{\gamma+1}}.$$

Thus, under (3.3), up to some constant c_γ depending on γ only, we get

$$\|f\|_{\psi_{1/(\gamma+1)}} \leqslant c_\gamma a.$$

4 Relations for L^p-Like Pseudonorms

To give some examples of metric probability spaces satisfying weak Poincaré type inequalities, we need certain relations for L^p-like pseudonorms, which we discuss in this section. For a measurable function f on the probability space (M, μ) and $q, r > 0$ we introduce the following standard notation. Put

$$\|f\|_q = \Big(\int |f|^q\, d\mu\Big)^{1/q}$$

and

$$\|f\|_{r,1} = \int_0^{+\infty} \mu\{|f| > t\}^{1/r}\, dt, \quad \|f\|_{r,\infty} = \sup_{t>0}\Big[t\,\mu\{|f| > t\}^{1/r}\Big].$$

As for how these quantities are related, there is the following elementary (and apparently well-known) statement: If $0 < q < r$, then

$$\|f\|_{r,1} \geqslant \|f\|_{r,\infty} \geqslant \Big(\frac{r-q}{r}\Big)^{1/q}\|f\|_q.$$

In particular,

$$\|f\|_{r,1} \geqslant \left(\frac{r-q}{r}\right)^{1/q} \|f\|_q. \tag{4.1}$$

However, the constant on the right-hand side is not optimal and may be improved, when q and r approach 1.

Lemma 4.1. *If $0 < q < r \leqslant 1$, then*

$$\|f\|_{r,1} \geqslant \left(\frac{r-q}{r}\right)^{1/q-1} \|f\|_q. \tag{4.2}$$

To see the difference between (4.1) and (4.2), we note that $\|f\|_{r,1} = \|f\|_1$ for the value $r = 1$ and, letting $q \to 1^-$, we obtain equality in (4.2), but not in (4.1).

Proof. Introduce the distribution function $F(t) = \mu\{|f| \leqslant t\}$ and put $u(t) = 1 - F(t)$. Since $u \leqslant 1$, for any $t > 0$

$$\|f\|_q^q = \int_0^{+\infty} s^q\, dF(s) = \int_0^t u(s)\, ds^q + \int_t^{+\infty} u(s)\, ds^q \leqslant t^q + q \int_t^{+\infty} s^{q-1}\, u(s)\, ds.$$

Let $0 < r < 1$. By the Hölder inequality with exponents $p = \frac{1}{r}$ and $p^* = \frac{p}{p-1}$, we have

$$\begin{aligned}\int_t^{+\infty} s^{q-1}\, u(s)\, ds &\leqslant \|s^{q-1}\|_{L^{p^*}(t,+\infty)}\, \|u(s)\|_{L^p(t,+\infty)} \\ &= \left(\int_t^{+\infty} s^{p^*(q-1)}\, ds\right)^{1/p^*} \left(\int_t^{+\infty} u(s)^p\, ds\right)^{1/p}.\end{aligned}$$

The last integral may be bounded from above just by $\|f\|_{r,1} = \int_0^{+\infty} u(s)^p\, ds$. Note that $p^*(q-1) < -1$; moreover,

$$\begin{aligned} p^*(q-1) + 1 &= -\frac{1-pq}{p-1} = -\frac{r-q}{1-r}, \\ \frac{p^*(q-1)+1}{p^*} &= -\frac{1-pq}{p-1}\,\frac{p-1}{p} = -(r-q). \end{aligned}$$

Hence the pre-last integral is convergent and

$$\left(\int_t^{+\infty} s^{p^*(q-1)}\, ds\right)^{1/p^*} = \frac{t^{\frac{p^*(q-1)+1}{p^*}}}{(-p^*(q-1)-1)^{1/p^*}} = \left(\frac{1-r}{r-q}\right)^{1-r} t^{-(r-q)}$$

since $\frac{1}{p^*} = \frac{p-1}{p} = 1 - r$. Thus,

$$\|f\|_q^q \leqslant t^q + q\left(\frac{1-r}{r-q}\right)^{1-r} t^{-(r-q)} \|f\|_{r,1}^r.$$

It remains to optimize over all $t > 0$ on the right-hand side. Changing the variable $t^q = s$, we write

$$\|f\|_q^q \leqslant \varphi(s) \equiv s + \frac{C}{\alpha} s^{-\alpha},$$

where $\alpha = \frac{r}{q} - 1$ and

$$C = q\left(\frac{1-r}{r-q}\right)^{1-r} \left(\frac{r}{q} - 1\right) \|f\|_{r,1}^r = (1-r)^{1-r}(r-q)^r \|f\|_{r,1}^r.$$

Since $\alpha > 0$, the function φ is minimized at $s_0 = C^{1/(\alpha+1)}$ and, at this point,

$$\varphi(s_0) = C^{1/(\alpha+1)} + \frac{C}{\alpha} C^{-\alpha/(\alpha+1)} = \left(1 + \frac{1}{\alpha}\right) C^{1/(\alpha+1)}.$$

Note that $\alpha + 1 = \frac{r}{q}$ and $\frac{\alpha+1}{\alpha} = \frac{r}{r-q}$, so

$$C^{1/(\alpha+1)} = \left[(1-r)^{1-r}(r-q)^r \|f\|_{r,1}^r\right]^{q/r} = (1-r)^{q(\frac{1}{r}-1)} (r-q)^q \|f\|_{r,1}^q$$

and

$$\varphi(s_0) = \frac{r}{r-q} (1-r)^{q(\frac{1}{r}-1)} (r-q)^q \|f\|_{r,1}^q.$$

Therefore,

$$\|f\|_q \leqslant \varphi(s_0)^{1/q} = \left(\frac{r}{r-q}\right)^{\frac{1}{q}-1} r(1-r)^{\frac{1}{r}-1} \|f\|_{r,1}.$$

It remains to note that $r(1-r)^{\frac{1}{r}-1} \leqslant 1$, whenever $0 < r < 1$. □

5 Isoperimetric and Capacitary Conditions

Here, we focus on general necessary and sufficient conditions for weak Poincaré type inequalities to hold on a metric probability space (M, d, μ). Sufficient conditions are usually expressed in terms of the isoperimetric function of the measure μ, so it is natural to explore the role of isoperimetric inequalities. By an *isoperimetric inequality* one means any relation

$$\mu^+(A) \geqslant I(\mu(A)), \quad A \subset M, \tag{5.1}$$

connecting the outer Minkowski content or μ-perimeter

$$\mu^+(A) = \liminf_{\varepsilon \to 0+} \frac{\mu(A^\varepsilon) - \mu(A)}{\varepsilon}$$
$$= \liminf_{\varepsilon \to 0+} \frac{\mu\{x \in M \setminus A : \exists a \in A,\ d(x,a) < \varepsilon\}}{\varepsilon}$$

with μ-size in the class of all Borel sets A in M with measure $0 < \mu(A) < 1$. (Here A^ε denotes an open ε-neighborhood of A.)

The function I, appearing in (5.1), may be an arbitrary nonnegative function, defined on the unit interval (0,1). If this function is optimal, it is often referred to as the isoperimetric function or the isoperimetric profile of the measure μ.

To any nonnegative function I on $(0, \frac{1}{2}]$ we associate a nondecreasing function $C_I(r)$ given by

$$\frac{1}{C_I(r)} = \inf_{0<t\leqslant\frac{1}{2}} \left[I(t)\, t^{-1/r} \right], \quad 0 < r < 1. \tag{5.2}$$

One of our aims is to derive the following assertion.

Theorem 5.1. *In the presence of the isoperimetric inequality* (5.1), *the space* (M, d, μ) *satisfies the weak Poincaré type inequality*

$$\|f - \mathbf{E}f\|_p \leqslant C(p)\, \|\nabla f\|_2, \quad 0 < p < 2,$$

with rate function

$$C(p) = 8 \inf_{\frac{2p}{2+p}<r<1} \left[C_I(r) \left(\frac{r}{r - \frac{2p}{2+p}} \right)^{\frac{2-p}{2p}} \right]. \tag{5.3}$$

We first consider one important particular case.

Lemma 5.2. *Given* $c > 0$ *and* $0 < r \leqslant 1$, *we assume that* (M, d, μ) *satisfies*

$$\mu^+(A) \geqslant c\,\mu(A)^{1/r} \tag{5.4}$$

for all Borel sets $A \subset M$ *with* $0 < \mu(A) \leqslant \frac{1}{2}$. *Then for any locally Lipschitz function* $f \geqslant 0$ *on* M *with median zero and for all* $q \in (0, r)$

$$\|f\|_q \leqslant \frac{1}{c} \left(\frac{r}{r-q} \right)^{1/q-1} \int |\nabla f|\, d\mu. \tag{5.5}$$

For the proof, we recall the well-known co-area formula which remains to hold in the form of an inequality for arbitrary metric probability spaces (cf. [10]). Namely, for any function f on M having a finite Lipschitz seminorm

$$\int |\nabla f| \, d\mu \geqslant \int_{-\infty}^{+\infty} \mu^+\{f > t\} \, dt.$$

Note that the function $t \to \mu^+\{f > t\}$ is always Borel measurable for continuous f, so the second integral makes sense. Hence, by Lemma 4.1, if $\|f\|_{\mathrm{Lip}} < +\infty$,

$$\int |\nabla f| \, d\mu \geqslant c \int_0^{+\infty} \mu\{f > t\}^{1/r} \, dt = c\|f\|_{r,1} \geqslant c \left(\frac{r-q}{r}\right)^{1/q-1} \|f\|_q.$$

A simple truncation argument extends this inequality to all locally Lipschitz $f \geqslant 0$.

Proof of Theorem 5.1. By the definition (5.2), whenever $0 < r < 1$, the space (M, d, μ) satisfies the isoperimetric inequality (5.4) with $c = 1/C_I(r)$, so the functional inequality (5.5) holds.

Let $f \geqslant 0$ be locally Lipschitz on M with median zero. Given $0 < q < r < 1$, apply (5.5) to $f^{p/q}$ with $p > 0$ to be specified later on. Then

$$\begin{aligned}
\int f^p \, d\mu = \int (f^{p/q})^q \, d\mu &\leqslant \frac{1}{c^q} \left(\frac{r}{r-q}\right)^{1-q} \left(\int |\nabla f^{p/q}| \, d\mu\right)^q \\
&= \frac{1}{c^q} \left(\frac{r}{r-q}\right)^{1-q} \left(\frac{p}{q}\right)^q \left(\int f^{\frac{p}{q}-1} |\nabla f| \, d\mu\right)^q \\
&\leqslant \frac{1}{c^q} \left(\frac{r}{r-q}\right)^{1-q} \left(\frac{p}{q}\right)^q \left(\int f^{2(\frac{p}{q}-1)} \, d\mu\right)^{q/2} \left(\int |\nabla f|^2 \, d\mu\right)^{q/2},
\end{aligned}$$

where we used the Cauchy inequality at the last step. Choose p so that $2\,(\frac{p}{q}-1) = p$, i.e., $p = 2q/(2-q)$ or $q = 2p/(2+p)$. Then the obtained bound becomes

$$\left(\int f^p \, d\mu\right)^{1-q/2} \leqslant \frac{1}{c^q} \left(\frac{r}{r-q}\right)^{1-q} \left(\frac{2}{2-q}\right)^q \left(\int |\nabla f|^2 \, d\mu\right)^{q/2},$$

and, using $\frac{1-q/2}{q} = \frac{1}{p}$ and $\frac{2}{2-q} < 2$, we get

$$\left(\int f^p \, d\mu\right)^{1/p} \leqslant \frac{2}{c} \left(\frac{r}{r-q}\right)^{1/q-1} \left(\int |\nabla f|^2 \, d\mu\right)^{1/2}.$$

By doubling the expression on the right-hand side like in the proof of Lemma 2.5, we may remove the condition $f \geqslant 0$ and thus get in the general locally Lipschitz case

$$\|f - m(f)\|_p \leqslant \frac{4}{c} \left(\frac{r}{r-q}\right)^{1/q-1} \|\nabla f\|_2$$

with $q = \frac{2p}{2+p}$, where $m(f)$ is a median of f under μ. Note that $q < 1 \Leftrightarrow p < 2$. Finally, by Lemma 2.4,

$$\|f - \mathbf{E}_\mu f\|_p \leqslant \frac{8}{c} \left(\frac{r}{r-q}\right)^{1/q-1} \|\nabla f\|_2.$$

It remains to take the infimum over all $r \in (q, 1)$, and we arrive at the desired Poincaré type inequality with rate function (5.4). □

REMARK 5.1. In order to get a simple upper bound for the rate function

$$C(p) = 8 \inf_{q<r<1} \left[C_I(r) \left(\frac{r}{r-q}\right)^{\frac{1}{q}-1}\right], \quad \text{where} \quad q = \frac{2p}{2+p},$$

in many interesting cases, one may just take

$$r = \frac{1+q}{2} = \frac{3p+2}{2(p+2)},$$

for example. In this case,

$$\left(\frac{r}{r-q}\right)^{\frac{1}{q}-1} = \left(\frac{1+q}{1-q}\right)^{\frac{1}{q}-1} = (1+s)^{2/s} < e^2$$

for $s = 2q/(1-q)$. Hence we obtain the following assertion.

Corollary 5.3. *In the presence of the isoperimetric inequality* (5.1) *with the associated function $C_I(r)$, the space (M, d, μ) satisfies the weak Poincaré type inequality*

$$\|f - \mathbf{E} f\|_p \leqslant C(p) \, \|\nabla f\|_2, \quad 0 \leqslant p < 2,$$

with rate function $C(p) = 8e^2 \, C_I(\frac{3p+2}{2(p+2)})$.

In particular, if μ satisfies a Cheeger type isoperimetric inequality $\mu^+(A) \geqslant c\,\mu(A)$ $(0 < \mu(A) \leqslant \frac{1}{2})$, then $C_I(r)$ is bounded by $1/c$, and Corollary 5.3 yields the usual Poincaré type inequality

$$\|f - \mathbf{E} f\|_2 \leqslant \frac{C}{c} \|\nabla f\|_2$$

with a universal constant C. Thus, Theorem 5.1 includes the Maz'ya–Cheeger theorem (up to a multiplicative factor).

Consider a more general class of isoperimetric inequalities.

Corollary 5.4. *Assume that the metric probability space (M, d, μ) satisfies, for some $\alpha \geqslant 0$ and $c > 0$, an isoperimetric inequality*

$$\mu^+(A) \geqslant c \frac{t}{\log^{1/\alpha}(\frac{4}{t})}, \quad t = \mu(A),\ 0 < t \leqslant \frac{1}{2}.$$

Then for some universal constant C it satisfies the weak Poincaré type inequality with rate function

$$C(p) = \frac{C}{c}\left(\frac{3}{2-p}\right)^{1/\alpha}, \quad 1 \leqslant p < 2.$$

Proof. First we show that, given $p > 1$, for all $t \in (0,1)$

$$\frac{t}{\log^{1/\alpha}(\frac{4}{t})} \geqslant \frac{[\alpha e(p-1)]^{1/\alpha}}{4^{p-1}}\, t^p. \tag{5.6}$$

Indeed, for any $C > 0$, replacing $t = 4s$, we can write

$$C \frac{t}{\log^{1/\alpha}(\frac{4}{t})} \geqslant t^p \Longleftrightarrow s^{\alpha(p-1)} \log \frac{1}{s^{\alpha(p-1)}} \leqslant \alpha(p-1)\Big(\frac{C}{4^{p-1}}\Big)^\alpha.$$

But $\sup_{u>0}\,[u \log \frac{1}{u}] = \frac{1}{e}$, so we are reduced to $\frac{1}{e} \leqslant \alpha(p-1)\,(\frac{C}{4^{p-1}})^\alpha$, where the best constant is $C = \frac{4^{p-1}}{[\alpha e(p-1)]^{1/\alpha}}$.

Now, using the definition (5.2) with $r = 1/p$ and applying (5.6), we conclude that (M, d, μ) satisfies an isoperimetric inequality with the associated function

$$C_I(r) = \frac{C}{c} \frac{4^{\frac{1}{r}-1}}{(\frac{1}{r}-1)^{1/\alpha}}, \quad \text{where } C = \frac{1}{(\alpha e)^{1/\alpha}}.$$

Take $r = \frac{1+q}{2} = \frac{3p+2}{2(p+2)}$ with $1 \leqslant p < 2$ as in Corollary 5.3 ($q = \frac{2p}{2+p}$). Since $r \geqslant \frac{5}{6}$, we have $4^{\frac{1}{r}-1} \leqslant 4^{1/5}$. Also $\frac{1}{r} - 1 = \frac{1+q}{1-q} = \frac{2-p}{2+3p} \geqslant \frac{2-p}{8}$ and $\alpha^{1/\alpha} \geqslant e^{-e}$. Therefore,

$$C_I(r) \leqslant \frac{4^{1/5}\, e^e}{c}\left(\frac{8/e}{2-p}\right)^{1/\alpha}.$$

It remains to apply Corollary 5.3. □

Although the isoperimetric inequalities may serve as convenient sufficient conditions for the week Poincaré type inequalities, in general they are not necessary. To speak about both necessary and sufficient conditions expressed in terms of geometric characteristics of a measure μ, one has to involve the concept of the capacity of sets, which is close to, but different than the concept of the μ-perimeter.

Given a metric space (M, d) with a Borel (positive) measure μ and a pair of sets $A \subset \Omega \subset M$ such that A is closed and Ω is open in M, the relative μ-capacity of A with respect to Ω is defined as

$$\mathrm{cap}_\mu(A, \Omega) = \inf \int |\nabla f|^2 \, d\mu,$$

where the infimum is taken over all locally Lipschitz functions f on M, such that $f \geqslant 1$ on A and $f = 0$ outside Ω. The capacity of the set A is $\mathrm{cap}_\mu(A) = \inf_\Omega \mathrm{cap}_\mu(A, \Omega)$. This definition is usually applied, when M is the Euclidean space $\mathbf{R}^n$ equipped with the Lebesgue measure μ (or for Riemannian manifolds, cf. [29, 19]). To make the definition workable in the setting of a metric probability space (M, d, μ), so that to efficiently relate it to the energy functional $\int \nabla f|^2 \, d\mu$, the relative capacity should be restricted to the cases such as $\mu(\Omega) \leqslant 1/2$.

Thus, let (M, d, μ) be a metric probability space and A a closed set in M of measure $\mu(A) \leqslant 1/2$. Following [4], we define the μ-capacity of A by

$$\mathrm{cap}_\mu(A) \;=\; \inf_{\mu(\Omega) \leqslant 1/2} \mathrm{cap}_\mu(A, \Omega) = \inf \left\{ \int |\nabla f|^2 \, d\mu : 1_A \leqslant f \leqslant 1_\Omega \right\}, \quad (5.7)$$

where the first infimum runs over all open sets $\Omega \subset M$ containing A and with measure $\mu(\Omega) \leqslant 1/2$, and the second one is taken over all such Ω's and all locally Lipschitz functions $f : M \to [0, 1]$ such that $f = 1$ on A and $f = 0$ outside Ω.

Note that, by the regularity of measure, we have $\mu(A^\varepsilon) \downarrow \mu(A)$ as $\varepsilon \downarrow 0$. Hence, if $\mu(A) < 1/2$, open sets Ω such that $A \subset \Omega$, $\mu(\Omega) \leqslant 1/2$ do exist, so the second infimum is also well defined and the definition makes sense. If $\mu(A) = 1/2$ and Ω does not exist, let us agree that the capacity is undefined (actually, this case does not appear when dealing with functional inequalities).

With this definition the measure capacity inequalities on (M, d, μ) take the form

$$\mathrm{cap}_\mu(A) \geqslant J(\mu(A)), \quad (5.8)$$

where J is a nonnegative function defined on $(0, \frac{1}{2}]$ and A is any closed subset of M with $\mu(A) \leqslant 1/2$, for which the capacity is defined.

To see, how (5.8) is related to the weak Poincaré type inequality

$$\|f - \mathbf{E}f\|_p \leqslant C(p) \, \|\nabla f\|_2, \quad 1 \leqslant p < 2, \quad (5.9)$$

we take a pair of sets $A \subset \Omega \subset M$ and a function f as in the definition (5.7). Then f has median zero under μ and, by Lemma 2.3,

$$\|f - \mathbf{E}f\|_p \geqslant \frac{1}{3} \, \|f\|_p \geqslant \frac{1}{3} \, (\mu(A))^{1/p}.$$

Therefore, by (5.9),

$$\int |\nabla f|^2 \, d\mu \geqslant \frac{1}{9C(p)^2} \, (\mu(A))^{1/p}.$$

Taking the infimum over all admissible f and the supremum over all p, we get the following elementary assertion.

Theorem 5.5. *Under the Poincaré type inequality* (5.9), *the measure capacity inequality* (5.8) *holds with*

$$J(t) = \frac{1}{9} \sup_{1 \leqslant p < 2} \left[\frac{t^{1/p}}{C(p)}\right]^2, \quad 0 < t \leqslant \frac{1}{2}.$$

In particular, the usual Poincaré type inequality, when $C(p) = 1/\sqrt{\lambda_1}$ is constant, implies that $\mathrm{cap}_\mu(A) \geqslant c\lambda_1 \mu(A)$ with a numerical constant $c > 0$ (cf. [4]).

To move in the opposite direction from (5.8) to (5.9), we need a capacitary analogue of the co-area formula or co-area inequality, which was used in the proof of Lemma 5.2. It has indeed been known since the works by Maz'ya [28, 29], and below we just adapt his result and the argument of [30] to the setting of a metric probability space.

Lemma 5.6. *For any locally Lipschitz function $f \geqslant 0$ on M with μ-dedian zero*

$$\int_{\{f>0\}} |\nabla f|^2 \, d\mu \geqslant \frac{1}{5} \int_0^{+\infty} \mathrm{cap}_\mu\{f \geqslant t\} \, dt^2. \tag{5.10}$$

Note that the capacity functional $A \to \mathrm{cap}_\mu(A)$ is nondecreasing, so the second integrand in (5.10) represents a nonincreasing function in $t > 0$. For a proof of (5.10), we consider (locally Lipschitz) functions of the form

$$g = \frac{1}{c_1 - c_0} \max\{\min\{f, c_1\} - c_0, 0\}, \quad \text{where } c_1 > c_0 > 0.$$

We have $g = 1$ on the closed set $A = \{f \geqslant c_1\}$ and $g = 0$ outside the open set $\Omega = \{f > c_0\}$. Since $\mu(\Omega) \leqslant 1/2$, by the definition of the capacity,

$$\int |\nabla g|^2 \, d\mu \geqslant \mathrm{cap}_\mu(A, \Omega) \geqslant \mathrm{cap}_\mu(A).$$

On the other hand, since the function $(c_1 - c_0)\, g$ represents a Lipschitz transform of f, we have $(c_1 - c_0)\, |\nabla g(x)| \leqslant |\nabla f(x)|$ for all $x \in M$. In addition, g is constant on the open sets $\{f < c_0\}$ and $\{f > c_1\}$, so $|\nabla g| = 0$ on these sets. Therefore,

$$\int |\nabla g|^2 \, d\mu = \int_{\{c_0 \leqslant f \leqslant c_1\}} |\nabla g|^2 \, d\mu \leqslant \frac{1}{(c_1 - c_0)^2} \int_{\{c_0 \leqslant f \leqslant c_1\}} |\nabla f|^2 \, d\mu.$$

The two estimates yield

$$(c_1 - c_0)^2 \operatorname{cap}_\mu\{f \geqslant c_1\} \leqslant \int_{\{c_0 \leqslant f \leqslant c_1\}} |\nabla f|^2 \, d\mu$$

or, given $a \in (0,1)$, for any $t > 0$

$$\int_{\{at \leqslant f \leqslant t\}} |\nabla f|^2 \, d\mu \geqslant t^2 (1-a)^2 \operatorname{cap}_\mu\{f \geqslant t\}.$$

Now, we divide both sides by t and integrate over $(0, +\infty)$. This leads to

$$\int_{\{f>0\}} |\nabla f|^2 \, d\mu \geqslant \frac{(1-a)^2}{\log(1/a)} \int_0^{+\infty} t \operatorname{cap}_\mu\{f \geqslant t\} \, dt.$$

The coefficient on the right-hand side is greater than $2/5$ for almost an optimal choice $a = 0.3$.

Now, we are prepared to derive from the capacitary inequality (5.8) a certain weak Poincaré type inequality. This may be done with arguments similar to the ones used in the proof of Lemma 5.2. To get an estimate of the rate function, consistent with what we have got in Theorem 5.5, let us assume that (M, d, μ) satisfies

$$\operatorname{cap}_\mu(A) \geqslant \sup_{0 \leqslant p < 2} \left[\frac{\mu(A)^{1/p}}{C(p)} \right]^2, \quad 0 < \mu(A) \leqslant \frac{1}{2}, \tag{5.11}$$

with a given positive function $C(p)$ defined in $0 < p < 2$. Equivalently, we could start with the measure capacity inequality (5.8) with a "capacitary" function $J(t)$ and then (5.11) holds with

$$C_J(p) = \sup_{0 < t \leqslant 1/2} \frac{t^{1/p}}{\sqrt{J(t)}}, \quad 0 < p < 2. \tag{5.12}$$

Let $r = p/2$ and $q < r < 1$. Given a locally Lipschitz function $f \geqslant 0$ on M, we may combine Lemma 5.6 with Lemma 4.1, to get from (5.11) that

$$\begin{aligned} \int_{\{f>0\}} |\nabla f|^2 \, d\mu &\geqslant c \int_0^{+\infty} \mu\{f \geqslant t\}^{1/r} \, dt^2 = c \int_0^{+\infty} \mu\{f^2 \geqslant t\}^{1/r} \, dt \\ &= c \, \|f^2\|_{r,1} \geqslant c \left(\frac{r-q}{r} \right)^{1/q - 1} \|f^2\|_q, \end{aligned}$$

where $c = \frac{1}{5C(p)^2} = \frac{1}{5C(2r)^2}$. Equivalently,

$$\|f\|_{2q}^2 \leqslant 5\,C(2r)^2\Big(\frac{r}{r-q}\Big)^{\frac{1-q}{q}} \int_{\{f>0\}} |\nabla f|^2\,d\mu. \tag{5.13}$$

If f is not necessarily nonnegative, but has median zero, one may apply (5.13) to the functions f^+ and f^-, and, summing the corresponding inequalities, we will be led again to (5.13) for f. Moreover, by doubling the constant on the right, the assumption $m(f) = 0$ may be replaced with $\mathbf{E}f = 0$. Thus, in general,

$$\|f - \mathbf{E}f\|_{2q}^2 \leqslant 10\,C(2r)^2\Big(\frac{r}{r-q}\Big)^{\frac{1-q}{q}} \int |\nabla f|^2\,d\mu, \quad 0 < q < r < 1.$$

Finally, replacing $2q$ with the variable p, we arrive at the following assertion.

Theorem 5.7. *Under the hypothesis* (5.11), *the weak Poincaré type inequality*

$$\|f - \mathbf{E}f\|_p \leqslant C'(p)\,\|\nabla f\|_2, \quad 0 < p < 2,$$

holds with rate function

$$C'(p) = \sqrt{10}\,\inf_{\frac{p}{2}<r<1}\bigg[C(2r)\Big(\frac{r}{r-p/2}\Big)^{\frac{1-p/2}{p}}\bigg]. \tag{5.14}$$

Alternatively, if we start with the measure capacity inequality (5.8) with a function $J(t)$, one may associate to it the function C_J defined in (5.12), and then the rate function of the theorem will take the form

$$C'(p) = \sqrt{10}\,\inf_{\frac{p}{2}<r<1}\,\sup_{0<t\leqslant 1/2}\bigg[\frac{t^{1/(2r)}}{\sqrt{J(t)}}\Big(\frac{r}{r-p/2}\Big)^{\frac{1-p/2}{p}}\bigg].$$

In particular, like in Corollary 5.3, choosing in (5.14) the value $r = (1+q)/2$ with $q = p/2$ and using the bounds $(\frac{r}{r-q})^{\frac{1-q}{2q}} < e$ and $\sqrt{10}\,e < 9$ (just to simplify the numerical constant), one may take

$$C'(p) = 9\,C(1+p/2) = 9\sup_{0<t\leqslant 1/2}\bigg[\frac{t^{\frac{1}{1+p/2}}}{\sqrt{J(t)}}\bigg], \quad 0 < p < 2. \tag{5.15}$$

Thus, starting with the weak Poincaré type inequality (5.9) with rate function $C(p)$, we obtain a geometric (capacity) inequality of the form (5.11), which in turn leads to (5.9), however, with a somewhat worse rate function $C'(p)$. Nevertheless, in some interesting cases, these two rate functions are in essence equivalent as $p \to 2$. For example, as in Corollary 5.4, if $C(p) = C \cdot (2-p)^{-1/\alpha}$, then $C'(p) = 9 \cdot 2^{1/\alpha} C \cdot (2-p)^{-1/\alpha}$, which is of the same order. It is in this sense one may say that weak Poincaré type inequalities have an equivalent capacitary description.

6 Convex Measures

Here, we illustrate Theorem 5.1 and especially its Corollary 5.4 on the example of probability distributions on the Euclidean space $M = \mathbf{R}^n$ possessing certain convexity properties. The obtained results will be applied to the so-called convex measures introduced and studied in the works of Borell [11, 12].

A Borel probability measure μ is called *$\varkappa$-concave*, where $-\infty \leqslant \varkappa \leqslant 1$, if for all $t \in (0,1)$ it satisfies a Brunn–Minkowski type inequality

$$\mu(tA + (1-t)B) \geqslant \left[t\mu(A)^{\varkappa} + (1-t)\mu(B)^{\varkappa}\right]^{1/\varkappa} \tag{6.1}$$

in the class of all nonempty Borel sets $A, B \subset \mathbf{R}^n$.

When $\varkappa = 0$, the right-hand side of (6.1) is understood as $\mu(A)^t \mu(B)^{1-t}$ and then we arrive at the notion of a log-concave measure, previously considered by Prékopa [33, 34] and Leindler [26] (cf. also [14]). When $\varkappa = -\infty$, the right-hand side is understood as $\min\{\mu(A), \mu(B)\}$. The inequality (6.1) is getting stronger, as the parameter $\varkappa$ is increasing, so the case $\varkappa = -\infty$ describes the largest class, whose members are called *convex* or *hyperbolic probability measures.*

Borell gave a complete characterization of such measures. If μ is absolutely continuous with respect to the Lebesgue measure and is supported on some open convex set $K \subset \mathbf{R}^n$, the necessary and sufficient condition for μ to satisfy (6.1) is that it has a positive density p on K such that for all $t \in (0,1)$ and $x, y \in K$

$$p(tx + (1-t)y) \geqslant \left[tp(x)^{\varkappa_n} + (1-t)p(y)^{\varkappa_n}\right]^{1/\varkappa_n}, \tag{6.2}$$

where $\varkappa_n = \frac{\varkappa}{1-n\varkappa}$ (necessarily $\varkappa \leqslant \frac{1}{n}$). Thus, the $\varkappa$-concavity with $\varkappa < 0$ means that the density is representable in the form $p = V^{-\beta}$ for some positive convex function V on $\mathbf{R}^n$, possibly taking an infinite value, where $\beta \geqslant n$ and $\varkappa = -\frac{1}{\beta - n}$.

Below we consider $\varkappa$-concave probability measures with $\varkappa < 0$. As was shown in [23] for the convex body case ($\varkappa = \frac{1}{n}$) and then in [8] for the general log-concave case ($\varkappa = 0$), any log-concave probability measure shares the usual Poincaré type inequality. This property fails when $\varkappa < 0$ even under strong integrability hypotheses. Nevertheless, with such additional hypotheses one may reach weak Poincaré type inequalities! More precisely, we will involve the condition that the distribution function $F(r) = \mu\{|x| \leqslant r\}$ of the

Euclidean norm has the tails $1 - F(r)$ decreasing to zero, as $r \to +\infty$, at worst as e^{-ct^α}. As long as the parameter of the convexity $\varkappa$ is negative, there is no reason to distinguish between the case corresponding to the exponential tails with $\alpha \geqslant 1$ (which is typical for log-concave distributions) and the case of (relatively) heavy or slow tails, when $\alpha < 1$.

We need some preparations. Denote by B_ρ an open Euclidean ball of radius $\rho > 0$ with center at the origin.

Lemma 6.1. *Any $\varkappa$-concave probability measure, $-\infty < \varkappa \leqslant 1$, satisfies the isoperimetric inequality*

$$2\rho\, \mu^+(A) \geqslant \frac{1 - [t^{1-\varkappa} + (1-t)^{1-\varkappa}]\, \mu(B_\rho)}{-\varkappa}, \tag{6.3}$$

where $t = \mu(A)$, $0 < t < 1$, with arbitrary $\rho > 0$.

In the log-concave case, the inequality (6.3) should read as

$$2\rho\, \mu^+(A) \geqslant t \log \frac{1}{t} + (1-t) \log \frac{1}{1-t} + \log \mu(B_\rho). \tag{6.4}$$

By the Prékopa–Leindler functional form of the Brunn–Minkowski inequality, (6.4) was derived in [8]. The arbitrary $\varkappa$-concave case was considered by Barthe [3], who applied an extension of the Prékopa–Leindler theorem in the form of Borell and Brascamp–Lieb. The inequality (6.3) was used in [3] to study the isoperimetric dimension of $\varkappa$-concave measures with $\varkappa > 0$. A direct proof of (6.3), not appealing to any functional form was given in [9].

To make the exposition self-contained, let us briefly remind the argument, which is based on the following representation for the μ-perimeter, explicitly relating it to measure convexity properties. Namely, let a probability measure μ on $\mathbf{R}^n$ be absolutely continuous and have a continuous density $p(x)$ on an open supporting convex set, say K. It is easy to check that for any sufficiently "nice" set A, for example, a finite union of closed balls in K or the complement in $\mathbf{R}^n$ to the finite union of such balls

$$\mu^+(A) = \lim_{\varepsilon \to 0+} \frac{\mu((1-\varepsilon)A + \varepsilon B_\rho) + \mu((1-\varepsilon)\overline{A} + \varepsilon B_\rho) - 1}{2r\varepsilon}, \tag{6.5}$$

where $\overline{A} = \mathbf{R}^n \setminus A$. In the case of a $\varkappa$-concave μ, it remains to apply the original convexity property (6.1) to the right-hand side of (6.5) to get

$$\mu^+(A) \geqslant \lim_{\varepsilon \to 0+} \frac{((1-\varepsilon)\mu(A)^\varkappa + \varepsilon\mu(B_\rho)^\varkappa)^{1/\varkappa} + ((1-\varepsilon)\mu(\overline{A})^\varkappa + \varepsilon\mu(B_\rho)^\varkappa)^{1/\varkappa} - 1}{2r\varepsilon},$$

which is exactly (6.3). Note that, by the Borell characterization, we do not lose generality by assuming that μ is full-dimensional (i.e., absolutely continuous).

From Lemma 6.1 we can now derive the following assertion.

Lemma 6.2. *Let μ be a $\varkappa$-concave probability measure on $\mathbf{R}^n$, $-\infty < \varkappa < 0$, and let A be a Borel subset of $\mathbf{R}^n$ of measure $t = \mu(A) \leqslant \frac{1}{2}$. If $\rho > 0$ satisfies*

$$\mu\{|x| > \rho\} \leqslant \frac{t}{2}, \tag{6.6}$$

then

$$\mu^+(A) \geqslant \frac{c(\varkappa)}{\rho}\, t, \quad \textit{where}\ c(\varkappa) = \frac{1 - (2/3)^{-\varkappa}}{-2\varkappa}. \tag{6.7}$$

Proof. By Lemma 6.1, since $\mu(B_\rho) \geqslant 1 - \frac{t}{2}$,

$$\begin{aligned} -2\rho\varkappa\,\mu^+(A) &\geqslant 1 - [t^{1-\varkappa} + (1-t)^{1-\varkappa}]\left(1 - \frac{t}{2}\right)^{\varkappa} \\ &= 1 - \left[t\left(\frac{t}{1-t/2}\right)^{-\varkappa} + (1-t)\left(\frac{1-t}{1-t/2}\right)^{-\varkappa}\right]. \end{aligned}$$

Clearly, on the interval $0 \leqslant t \leqslant 1/2$, the ratio $\frac{t}{1-t/2}$ is increasing and so bounded by $2/3$. Also $\frac{1-t}{1-t/2} \leqslant 1$, so

$$-2\rho\varkappa\,\mu^+(A) \geqslant 1 - \left[t\left(\frac{2}{3}\right)^{-\varkappa} + (1-t)\right] = t\left[1 - \left(\frac{2}{3}\right)^{-\varkappa}\right],$$

which is the claim (6.7). □

Note that $c(\varkappa)$ continuously depends on $\varkappa$ and $\lim_{\varkappa\to 0} c(\varkappa) = c(0) = \frac{1}{2}\log\frac{3}{2}$, while $c(\varkappa) \sim \frac{1}{-2\varkappa}$ as $\varkappa \to -\infty$. In particular, $c(\varkappa) \geqslant \frac{c}{1-\varkappa}$ for $\varkappa \leqslant 0$. As a result, we obtain the following assertion.

Theorem 6.3. *Let μ be a $\varkappa$-concave probability measure on $\mathbf{R}^n$, $-\infty < \varkappa < 0$, such that*

$$\int \Phi(|x|)\, d\mu(x) \leqslant D \tag{6.8}$$

for some increasing continuous function $\Phi : [0,+\infty) \to [0,+\infty)$. For any Borel set A in $\mathbf{R}^n$ of measure $t = \mu(A) \leqslant \frac{1}{2}$

$$\mu^+(A) \geqslant \frac{c}{1-\varkappa}\,\frac{t}{\Phi^{-1}(\frac{2D}{t})}, \tag{6.9}$$

where c is a positive universal constant and Φ^{-1} is the inverse function.

Indeed, by the Chebyshev inequality and the hypothesis (6.8),

$$\mu\{|x| > \rho\} \leqslant \frac{D}{\Phi(\rho)} \leqslant \frac{t}{2},$$

where the last bound is obviously fulfilled for $\rho \geqslant \Phi^{-1}(\frac{2D}{t})$. By Lemma 6.2, we get

$$\mu^+(A) \geqslant c(\varkappa)\,\frac{t}{\Phi^{-1}(\frac{2D}{t})},$$

and the theorem follows.

As a basic example, we consider the function $\Phi(x) = \exp\{(x/\lambda)^\alpha\}$ with parameters $\alpha, \lambda > 0$, which has the inverse $\Phi^{-1}(y) = \lambda \log^{1/\alpha} y$, $y \geqslant 1$. Then the hypothesis (6.8) with $D = 2$ is equivalent to saying that the Orlicz norm generated by the Young function $\psi_\alpha(x) = e^{|x|^\alpha} - 1$, $x \in \mathbf{R}$, is bounded by λ for the Euclidean norm, i.e., $\|\,|x|\,\|_{\psi_\alpha} \leqslant \lambda$ in the Orlicz space $L^{\psi_\alpha}(\mathbf{R}^n, \mu)$.

Corollary 6.4. *Let μ be a $\varkappa$-concave probability measure on $\mathbf{R}^n$, $-\infty < \varkappa < 0$, such that, for some $\alpha > 0$ and $\lambda > 0$,*

$$\int \exp\Big\{\Big(\frac{|x|}{\lambda}\Big)^\alpha\Big\}\,d\mu(x) \leqslant 2. \tag{6.10}$$

Then for any Borel set A in $\mathbf{R}^n$ of measure $t = \mu(A) \leqslant \frac{1}{2}$ with some universal constant $c > 0$

$$\mu^+(A) \geqslant \frac{c}{1-\varkappa}\,\frac{t}{\lambda \log^{1/\alpha}(4/t)}.$$

Now, we may recall Corollary 5.4.

Corollary 6.5. *Any $\varkappa$-concave probability measure μ on $\mathbf{R}^n$, $-\infty < \varkappa < 0$, such that*

$$\int \exp\Big\{\Big(\frac{|x|}{\lambda}\Big)^\alpha\Big\}\,d\mu(x) \leqslant 2, \quad \alpha, \lambda > 0,$$

satisfies the weak Poincaré type inequality with rate function

$$C(p) = C\lambda\,(1-\varkappa)\,\Big(\frac{3}{2-p}\Big)^{1/\alpha},$$

where C is a universal constant.

7 Examples. Perturbation

Given a spherically invariant, absolutely continuous probability measure μ on $\mathbf{R}^n$, we write its density in the form

$$p(x) = \frac{1}{Z}\,e^{-V(|x|)}, \quad x \in \mathbf{R}^n,$$

where $V = V(t)$ is defined and finite for $t > 0$ and Z is a normalizing factor. If V is convex and nondecreasing, then μ is log-concave (and conversely). If not, one may hope that μ will be $\varkappa$-concave for some $\varkappa < 0$. Namely, by the Borell characterization (6.2) with $\varkappa < 0$, the $\varkappa$-concavity of μ is equivalent to the convexity of the function $p^{\varkappa_n}$, where $\varkappa_n = \frac{\varkappa}{1-n\varkappa}$. In other words, μ is $\varkappa$-concave if and only if

1) the function $V(t)$ is nondecreasing in $t > 0$;
2) the function $e^{-\varkappa_n V(t)}$ is convex on $(0, +\infty)$.

If V is twice continuously differentiable, the second property is equivalent to

2') $V''(t) - \varkappa_n V'(t)^2 \geqslant 0$ for all $t > 0$

As a more specific example, we consider densities of the form

$$p(x) = \frac{1}{Z}\, e^{-(a+b|x|)^\alpha}, \quad x \in \mathbf{R}^n, \tag{7.1}$$

with parameters $a, b > 0$ and $\alpha > 0$, which corresponds to $V(t) = (a + bt)^\alpha$.

It is clear that property 1) is fulfilled. If $\alpha \geqslant 1$, V is convex and the measure μ is log-concave. So, assume that $0 < \alpha < 1$, in which case V is not convex. It is easy to verify, the inequality of property 2') holds for all $t > 0$ if and only it holds for $t = 0$, and then it reads as

$$(\alpha - 1) - \alpha\varkappa_n a^\alpha \geqslant 0.$$

Hence an optimal choice is $\varkappa_n = -\frac{1-\alpha}{\alpha a^\alpha}$ or, equivalently,

$$\varkappa = -\frac{1-\alpha}{\alpha a^\alpha - n(1-\alpha)} \quad \text{provided that} \quad \alpha a^\alpha - n(1-\alpha) > 0. \tag{7.2}$$

Conclusion 1. *The probability measure μ with density* (7.1) *is convex if and only if $\alpha a^\alpha - n(1-\alpha) \geqslant 0$, in which case it is $\varkappa$-concave with the convexity parameter $\varkappa$ given by* (7.2).

In other words, μ is convex only if the parameter a is sufficiently large. By Corollary 6.5, if $\varkappa > -\infty$, i.e., if $\alpha a^\alpha - n(1-\alpha) > 0$, the measure μ satisfies the weak Poincaré type inequality

$$\|f - \mathbf{E}f\|_p \leqslant C(p)\, \|\nabla f\|_2, \quad 1 \leqslant p < 2, \tag{7.3}$$

with rate function

$$C(p) = C\left(\frac{3}{2-p}\right)^{1/\alpha}, \tag{7.4}$$

where C depends on the parameters a, b, α and the dimension n.

However, it is unlikely that the requirement (7.2), $a > a_0 > 0$, is crucial for (7.3) to hold with some rate function. To see this, a perturbation argument may be used to prove the following elementary:

Theorem 7.1. *Assume that a metric probability space (M, d, μ) satisfies the weak Poincaré type inequality (7.3). Let ν be a probability measure on M, which is absolutely continuous with respect to μ and has density $w = \frac{d\nu}{d\mu}$ such that*

$$c_1 \leqslant w(x) \leqslant c_2, \quad x \in M, \tag{7.5}$$

for some $c_1, c_2 > 0$. Then (M, d, ν) also satisfies (7.3) with rate function $C'(p) = \frac{2c_2}{\sqrt{c_1}}\, C(p)$.

Proof. Indeed, assume that f is bounded and locally Lipschitz on M with

$$\mathbf{E}f = \int f \, d\mu = 0.$$

Then, by (7.3) and (7.5), for any $p \in [1, 2)$

$$\begin{aligned} \|f\|_{L^p(\nu)}^p = \int |f|^p \, d\nu \ &\leqslant \ c_2 \int |f|^p \, d\mu \\ &\leqslant c_2\, C(p)^p \Big(\int |\nabla f|^2 \, d\mu \Big)^{p/2} \ \leqslant \ \frac{c_2}{c_1^{p/2}}\, C(p)^p \Big(\int |\nabla f|^2 \, d\nu \Big)^{p/2}, \end{aligned}$$

so

$$\|f\|_{L^p(\nu)} \leqslant \frac{c_2^{1/p}}{c_1^{1/2}}\, C(p)\, \|\nabla f\|_{L^2(\nu)}.$$

Since $c_2 \geqslant 1$, we find

$$\inf_{c \in \mathbf{R}} \|f - c\|_{L^p(\nu)} \leqslant \frac{c_2}{\sqrt{c_1}}\, C(p)\, \|\nabla f\|_{L^2(\nu)}.$$

But, in general, $\|f - \mathbf{E}f\|_p \leqslant 2\, \|f - c\|_p$ for any $c \in \mathbf{R}$. □

Let us return to the measure $\mu = \mu_a$ with density (7.2). Write $V_a(x) = (a + b|x|)^\alpha$ and write the normalizing constant as a function of a, $Z = Z(a)$, although it depends also on the remaining parameters $b > 0$ and $\alpha \in (0, 1)$. For all $a_1, a_2 \geqslant 0$ we have

$$|V_{a_1}(x) - V_{a_2}(x)| \leqslant |a_1 - a_2|^\alpha.$$

Therefore, the density $w(x) = \dfrac{d\mu_{a_1}(x)}{d\mu_{a_2}(x)}$ satisfies $c \leqslant w(x) \leqslant 1/c$ with

$$c = \frac{\min\{Z(a_1), Z(a_2)\}}{\max\{Z(a_1), Z(a_2)\}}\, e^{-|a_1 - a_2|^\alpha},$$

so that the condition (7.4) is fulfilled. Hence, by Theorem 7.1, the weak Poincaré type inequality (7.3) holds for all measures μ_a simultaneously with rate function of the form (7.4), as long as it holds for at least one measure μ_a. But, as we have already observed, the latter is true under (7.2) by the convexity property of such measures. Thus, Conclusion 1 may be complemented with the following one.

CONCLUSION 2. *Probability measures μ having densities* (7.1) *with arbitrary parameters $a, b \geqslant 0$, $\alpha \in (0,1)$ satisfy the weak Poincaré type inequality* (7.3) *with rate function $C(p) = C \cdot (\frac{3}{2-p})^{1/\alpha}$, where C depends on a, b, α, and n.*

8 Weak Poincaré with Oscillation Terms

Let us return to the setting of an abstract metric probability space (M, d, μ). It is now a good time to look at the relationship between the weak Poincaré type inequalities

$$\|f - \mathbf{E}f\|_p \leqslant C(p)\, \|\nabla f\|_2, \quad 1 \leqslant p < 2, \tag{8.1}$$

which is our main object of research, and Poincaré type inequalities

$$\mathrm{Var}_\mu(f) \ \leqslant\ \beta(s)\, \|\nabla f\|_2^2 + s\, \mathrm{Osc}\,(f)^2, \quad s > 0, \tag{8.2}$$

that involve an oscillation term $\mathrm{Osc}\,(f) = \mathrm{ess\,sup}\, f - \mathrm{ess\,inf}\, f$ and some nonnegative function $\beta(s)$. (Note that we always have $\mathrm{Var}_\mu(f) \leqslant \frac{1}{4}\, \mathrm{Osc}\,(f)^2$, so for $s \geqslant 1/4$ (8.2) is automatically fulfilled.)

In both cases, f represents an arbitrary locally Lipschitz function with a possible reasonable constraint that the right-hand sides should be finite. Hence, from the point of view of direct applications, (8.2) makes sense only for bounded f, while (8.1) may also be used for many unbounded functions. Nevertheless, both forms are in a certain sense equivalent, i.e., there is some relationship between $C(p)$ and $\beta(s)$. To study this type of connections, we first note the following elementary inequality of Nash type.

Theorem 8.1. *Under the weak Poincaré type inequality* (8.1), *for all bounded locally Lipschitz f on M and any $p \in [1,2)$*

$$\mathrm{Var}_\mu(f) \leqslant C(p)^p\ \mathrm{Osc}\,(f)^{2-p}\, \|\nabla f\|_2^p. \tag{8.3}$$

Indeed, since (8.3) is translation invariant, we may assume $\mathbf{E}f = 0$. Then it is obvious that $\operatorname{ess\,inf} f \leqslant 0 \leqslant \operatorname{ess\,sup} f$, so μ-almost everywhere $\operatorname{Osc}(f) \geqslant \|f\|_\infty \geqslant |f|$. By (8.1),

$$\mathbf{E}|f|^2 = \mathbf{E}|f|^p\,|f|^{2-p} \leqslant \mathbf{E}|f|^p\,\operatorname{Osc}(f)^{2-p} \leqslant C(p)^p(\mathbf{E}|\nabla f|^2)^{p/2}\operatorname{Osc}(f)^{2-p},$$

where all expectations are with respect to μ.

From Theorem 8.1 we derive an additive form of (8.3).

Theorem 8.2. *Under the weak Poincaré type inequality* (8.1), (8.2) *holds with*

$$\beta(s) = \inf_{1 \leqslant p < 2}\left[C(p)^2\, s^{1-\frac{2}{p}}\right]. \tag{8.4}$$

Proof. Using the Young inequality $xy \leqslant \frac{x^\alpha}{\alpha} + \frac{y^\beta}{\beta}$, where $x, y \geqslant 0$, $\alpha, \beta > 1$, $\frac{1}{\alpha} + \frac{1}{\beta} = 1$, for any $\varepsilon > 0$ we can estimate the right-hand side of (8.3) by

$$C(p)^p\left[\frac{[\frac{1}{\varepsilon}\,(\mathbf{E}\,|\nabla f|^2)^{p/2}]^\alpha}{\alpha} + \frac{[\varepsilon\,\operatorname{Osc}(f)^{2-p}]^\beta}{\beta}\right].$$

Choose $\alpha = \frac{2}{p}$ and $\beta = \frac{2}{2-p}$, to get

$$\operatorname{Var}_\mu(f) \leqslant \frac{C(p)^p}{\alpha\varepsilon^\alpha}\,\mathbf{E}\,|\nabla f|^2 + \frac{C(p)^p\,\varepsilon^\beta}{\beta}\,\operatorname{Osc}(f)^2. \tag{8.5}$$

Put

$$s = \frac{C(p)^p\,\varepsilon^\beta}{\beta}, \quad \text{so that} \quad \varepsilon = \left[\frac{\beta s}{C(p)^p}\right]^{1/\beta}.$$

Then the coefficient in front of $\mathbf{E}\,|\nabla f|^2$ in (8.5) becomes

$$\frac{C(p)^p}{\alpha\varepsilon^\alpha} = \frac{C(p)^p}{\alpha}\left[\frac{\beta s}{C(p)^p}\right]^{-\alpha/\beta} = C(p)^{p(1+\frac{\alpha}{\beta})}\,\frac{1}{\alpha\beta^{\alpha/\beta}}\,s^{-\alpha/\beta}.$$

The first exponent on the right is

$$p(1 + \frac{\alpha}{\beta}) = p(1 + \frac{2}{p}\,\frac{2-p}{2}) = 2.$$

For the second term we have

$$\frac{1}{\alpha\beta^{\alpha/\beta}} = \frac{p}{2}\left(\frac{2-p}{2}\right)^{(2-p)/p} \leqslant 1.$$

Also $\frac{\alpha}{\beta} = \frac{2}{p} - 1$, and (8.5) yields $\operatorname{Var}_\mu(f) \leqslant C(p)^2\, s^{1-2/p}\,\mathbf{E}\,|\nabla f|^2 + s\,\operatorname{Osc}(f)^2$. $\square$

Corollary 8.3. *If for some $a, b \geqslant 0$ and $\alpha > 0$, the rate function in the weak Poincaré type inequality* (8.1) *admits the bound*

$$C(p) \leqslant a \left(\frac{b}{2-p}\right)^{1/\alpha}, \quad 1 \leqslant p < 2, \tag{8.6}$$

then, with some numerical constants $\beta_0, \beta_1 > 0$, (8.2) *holds with*

$$\beta(s) = \beta \log^{2/\alpha} \frac{1}{s}, \quad s > 0, \tag{8.7}$$

where $\beta = \beta_0 a^2 (\beta_1 b)^{2/\alpha}$.

Proof. We may and do assume that $s < \frac{1}{4}$. Write $p = 2 - \varepsilon$, so that $0 < \varepsilon \leqslant 1$ and $\frac{2}{p} - 1 = \frac{\varepsilon}{2-\varepsilon}$. By Theorem 8.2, the hypothesis (8.6), and the inequality $\frac{\varepsilon}{2-\varepsilon} \leqslant \varepsilon$, for the optimal value of $\beta(s)$ we have

$$\beta(s) \leqslant C(p)^2 s^{1-\frac{2}{p}} \leqslant a^2\, b^{2/\alpha}\, \frac{1}{\varepsilon^{2/\alpha}}\, \frac{1}{s^{\frac{\varepsilon}{2-\varepsilon}}} \leqslant a^2\, b^{2/\alpha}\, \frac{1}{\varepsilon^{2/\alpha} s^{\varepsilon}}$$

for all $\varepsilon \in (0, 1]$. To optimize over all such ε, we consider the function $\varphi(\varepsilon) = \varepsilon^{2/\alpha} s^{\varepsilon}$. Then $\varphi(0) = 0$, $\varphi(1) = s$, and $\varphi'(\varepsilon) = \varepsilon^{2/\alpha} s^{\varepsilon} \left(\frac{2}{\alpha\varepsilon} - \log \frac{1}{s}\right)$. Hence the (unique) point of maximum of φ on $[0, +\infty)$ is $\varepsilon_0 = \frac{2}{\alpha \log \frac{1}{s}}$ and, at this point,

$$\varphi(\varepsilon_0) = \left(\frac{2}{\alpha \log \frac{1}{s}}\right)^{2/\alpha} e^{-2/\alpha} = \left(\frac{2}{\alpha e}\right)^{2/\alpha} \frac{1}{\log^{2/\alpha} \frac{1}{s}}.$$

Hence, if $\varepsilon_0 \leqslant 1$, i.e., $s \leqslant e^{-2/\alpha}$, then

$$\beta(s) \leqslant a^2\, b^{2/\alpha} \frac{1}{\varphi(\varepsilon_0)} = a^2\, b^{2/\alpha} \left(\frac{\alpha e}{2}\right)^{2/\alpha} \log^{2/\alpha} \frac{1}{s} \leqslant e^{1/e} a^2 (be)^{2/\alpha} \log^{2/\alpha} \frac{1}{s}.$$

Note that, since $s < 1/4$, the requirement $s \leqslant e^{-2/\alpha}$ is automatically fulfilled, as long as $\alpha \geqslant 1/\log 2$. In that case, (8.7) is thus proved with constants $\beta_0 = e^{1/e}$ and $\beta_1 = e$.

Now, let $\alpha < 1/\log 2$ and $s \geqslant e^{-2/\alpha}$. Then φ is increasing and is maximized on [0,1] at $\varepsilon = 1$, which gives $\beta(s) \leqslant a^2\, b^{2/\alpha}\, \frac{1}{s}$. So, we need the bound

$$\frac{1}{s} \leqslant A \log^{2/\alpha} \frac{1}{s}$$

in the interval $e^{-2/\alpha} \leqslant s \leqslant 1/4$. Since the function $t \log \frac{1}{t}$ is decreasing in $t \geqslant 1/e$, the optimal value of A is attained at $s = 1/4$, so $A = 4/\log^{2/\alpha} 4$. Therefore, (8.7) is valid with $\beta_0 = 4\, e^{1/e}$ and $\beta_1 = e/\log 4$. Corollary 8.3 is proved. □

In particular, we have the following assertion.

Corollary 8.4. *Any $\varkappa$-concave probability measure μ on $\mathbf{R}^n$, $-\infty < \varkappa < 0$, such that $\int \exp\{(\frac{|x|}{\lambda})^\alpha\}\, d\mu(x) \leqslant 2$ $(\alpha, \lambda > 0)$, satisfies* (8.2) *with*

$$\beta(s) = \beta \log^{2/\alpha} \frac{1}{s}, \quad s > 0,$$

where $\beta = \beta_0 \lambda^2 (1-\varkappa)^2\, \beta_1^{2/\alpha}$, $\beta_0, \beta_1 > 0$ are numerical constants.

On the basis of (8.1) one may also consider a more general type of "oscillations," for example, Poincaré type inequalities of the form

$$\mathrm{Var}_\mu(f) \leqslant \beta_q(s)\, \|\nabla f\|_2^2 + s\, \|f - \mathbf{E} f\|_q^2, \quad s > 0, \tag{8.8}$$

with a fixed finite parameter $q > 2$. As we will see, this form is natural in the study of the slow rates of convergence of the associated semigroups $P_t f$, when f is unbounded, but is still in $L^q(\mu)$. Note that (8.8) is automatically fulfilled for $s \geqslant 1$ (since β_q is nonnegative), so one may restrict oneself to the values $s < 1$. We prove the following assertion.

Theorem 8.5. *Under the weak Poincaré type inequality* (8.1) *with rate function $C(p)$,* (8.8) *holds with*

$$\beta_q(s) = \inf_{1 \leqslant p < 2} \left[C(p)^2\, s^{-\frac{q}{q-2}\frac{2-p}{p}} \right]. \tag{8.9}$$

Proof. The argument is very similar to the one used in the proof of Theorem 8.2. Given $p \in [1,2)$ and $q > 2$, by the Hölder inequality, we have

$$\mathbf{E}\, |f|^2 \leqslant \|f\|_p^r\, \|f\|_q^{2-r},$$

where $r = \frac{p(q-2)}{q-p}$. Therefore, if $\mathbf{E} f = 0$ (which we assume), by the hypothesis (8.1),

$$\mathbf{E}\, |f|^2 \leqslant C(p)\, \|f\|_q^{2-r}\, \|\nabla f\|_2^r. \tag{8.10}$$

Using the Young inequality with exponents $\alpha, \beta > 1$, $\frac{1}{\alpha} + \frac{1}{\beta} = 1$, for any $\varepsilon > 0$ we can estimate the right-hand side of (8.10) by

$$C(p)^r \left[\frac{[\frac{1}{\varepsilon}\, \|\nabla f\|_2^r]^\alpha}{\alpha} + \frac{[\varepsilon\, \|f\|_q^{2-r}]^\beta}{\beta} \right].$$

Choose $\alpha = \frac{2}{r}$ and $\beta = \frac{2}{2-r}$ to get

$$\mathbf{E}\,|f|^2 \leqslant \frac{C(p)^r}{\alpha\varepsilon^\alpha}\,\mathbf{E}\,|\nabla f|^2 + \frac{C(p)^r\,\varepsilon^\beta}{\beta}\,\|f\|_q^2. \tag{8.11}$$

Put

$$s = \frac{C(p)^r\,\varepsilon^\beta}{\beta}, \quad \text{so that} \quad \varepsilon = \left[\frac{\beta s}{C(p)^r}\right]^{1/\beta}.$$

Then the coefficient in front of $\mathbf{E}\,|\nabla f|^2$ in (8.11) becomes

$$\frac{C(p)^r}{\alpha\varepsilon^\alpha} = \frac{C(p)^r}{\alpha}\left[\frac{\beta s}{C(p)^r}\right]^{-\alpha/\beta} = C(p)^{r(1+\frac{\alpha}{\beta})}\,\frac{1}{\alpha\beta^{\alpha/\beta}}\,s^{-\alpha/\beta}.$$

The first exponent on the right is $r(1+\frac{\alpha}{\beta}) = r(1+\frac{2}{r}\,\frac{2-r}{2}) = 2$. For the second term we have

$$\frac{1}{\alpha\beta^{\alpha/\beta}} = \frac{r}{2}\left(\frac{2-r}{2}\right)^{(2-r)/r} \leqslant 1.$$

Also $\frac{\alpha}{\beta} = \frac{2}{r} - 1 = \frac{q}{q-2}\,\frac{2-p}{p}$, and we arrive at

$$\mathbf{E}\,|f|^2 \leqslant C(p)^2\,s^{-\frac{q}{q-2}\,\frac{2-p}{p}}\,\mathbf{E}\,|\nabla f|^2 + s\,\|f\|_q^2,$$

which is the claim. □

Now, we can strengthen Corollaries 8.3 and 8.4.

Corollary 8.6. *If the rate function in the weak Poincaré type inequality* (8.1) *admits the bound* (8.6), *then* (8.8) *holds with*

$$\beta_q(s) = \beta\log^{\frac{2}{\alpha}}\frac{2}{s}, \quad s > 0, \tag{8.12}$$

where $\beta = 2a^2\,(4b\,\frac{q}{q-2})^{2/\alpha}$.

Proof. As in the proof of Corollary 8.3, we assume that $s < 1$ and write $p = 2-\varepsilon$, so that $0 < \varepsilon \leqslant 1$ and $\frac{2}{p} - 1 = \frac{\varepsilon}{2-\varepsilon}$. Put $Q = \frac{q}{q-2}$. By Theorem 8.5 and the inequality $\frac{\varepsilon}{2-\varepsilon} \leqslant \varepsilon$, for the optimal value of $\beta_q(s)$ we have

$$\beta_q(s) \leqslant C(p)^2\,s^{Q(1-\frac{2}{p})} \leqslant a^2\,b^{2/\alpha}\frac{1}{\varepsilon^{2/\alpha}}\,\frac{1}{s^{Q\frac{\varepsilon}{2-\varepsilon}}} \leqslant a^2\,b^{2/\alpha}\,\frac{1}{\varepsilon^{2/\alpha}s^{Q\varepsilon}}$$

for all $\varepsilon \in (0,1]$. To optimize over all such ε, we consider the function $\varphi(\varepsilon) = \varepsilon^{2/\alpha}s^{Q\varepsilon}$. We have $\varphi(0) = 0$ and $\varphi(1) = s^Q$. As we know, the (unique) point of maximum of φ on $[0,+\infty)$ is $\varepsilon_0 = \frac{2}{Q\alpha\log\frac{1}{s}}$ and, at this point,

$$\varphi(\varepsilon_0) = \left(\frac{2}{Q\alpha\log\frac{1}{s}}\right)^{2/\alpha} e^{-2/\alpha} = \left(\frac{2}{Q\alpha e}\right)^{2/\alpha}\frac{1}{\log^{2/\alpha}\frac{1}{s}}.$$

Hence, if $\varepsilon_0 \leqslant 1$, i.e., $s \leqslant e^{-2/Q\alpha}$, then

$$\beta_q(s) \leqslant \frac{a^2\, b^{2/\alpha}}{\varphi(\varepsilon_0)} = a^2\, b^{2/\alpha} \left(\frac{Q\alpha e}{2}\right)^{2/\alpha} \log^{2/\alpha}\frac{1}{s} \leqslant 2a^2\,(Qbe)^{2/\alpha} \log^{2/\alpha}\frac{1}{s},$$

where we used $(\frac{\alpha}{2})^{2/\alpha} \leqslant e^{1/e} < 2$. Thus, for this range of s, (8.12) is proved.

Now, we assume that $s \geqslant e^{-2/Q\alpha}$. Then φ is increasing and is maximized on [0,1] at $\varepsilon = 1$, which gives $\beta_q(s) \leqslant a^2\, b^{2/\alpha} s^{-Q}$. So, we need a bound of the form

$$s^{-Q} \leqslant A \log^{2/\alpha}(2/s)$$

or, equivalently,

$$A^{-1/Q} \leqslant s \log^{2/Q\alpha}(2/s)$$

in the interval $e^{-2/Q\alpha} \leqslant s \leqslant 1$. The function $s\log^c(2/s)$ with parameter $c > 0$ is increasing in $0 < s \leqslant 2e^{-c}$ and decreasing in $s \geqslant 2e^{-c}$, so we only need to consider the endpoints of that interval. For the point $s = 1$ we get

$$A = 1/\log^{2/\alpha} 2,$$

while for $s = e^{-2/Q\alpha}$ we get

$$A = \frac{e^{2/\alpha}}{\log^{2/\alpha}(2\, e^{2/Q\alpha})} \leqslant \left(\frac{e}{\log 2}\right)^{2/\alpha} < 4^{2/\alpha}.$$

The corollary is proved. □

REMARK 8.1. As a result, one may also generalize Corollary 8.4. Namely, any $\varkappa$-concave probability measure μ on $\mathbf{R}^n$ with $\varkappa < 0$ and

$$\int \exp\Big\{\Big(\frac{|x|}{\lambda}\Big)^\alpha\Big\}\, d\mu(x) \leqslant 2, \quad \alpha, \lambda > 0,$$

satisfies the weak Poincaré type inequality (8.8) with

$$\beta_q(s) = \beta \log^{2/\alpha}\frac{2}{s}, \quad q > 2,$$

where β depends on λ, α, $\varkappa$, and q.

REMARK 8.2. It is also possible to derive a weak Poincaré type inequality (8.1) from (8.2) or (8.8) with some rate functions $C(p)$ explicitly in terms of $\beta(s)$ or $\beta_q(s)$. This may be done by virtue of the measure capacity inequalities

$$\mathrm{cap}_\mu(A) \geqslant J(\mu(A)),$$

which we discussed in Section 5. As was shown in [4], the latter is fulfilled with $J(t) = t/(4\beta(t/4))$ in the presence of (8.2). Hence, applying Theorem

5.7 in a somewhat weaker form (5.15), we conclude that (8.1) holds with

$$C(p)^2 = 81 \sup_{0<t\leqslant 1/2} \left[\frac{t^{\frac{4}{2+p}}}{J(t)}\right] = 81 \sup_{0<t\leqslant 1/2} \left[4^{\frac{4}{2+p}}\, t^{\frac{2-p}{2+p}}\, \beta(t/4)\right].$$

Hence we arrive at the following assertion.

Theorem 8.7. *In the presence of* (8.2), *the weak Poincaré type inequality* (8.1) *holds with rate function given by*

$$C(p)^2 = C^2 \sup_{0<s\leqslant 1/8} \left[s^{\frac{2-p}{2+p}}\, \beta(s)\right],$$

where C is a universal constant.

9 Convergence of Markov Semigroups

Let μ be an absolutely continuous Borel probability measure on $\mathbf{R}^n$. We assume that the measure is regular enough in the following sense: There exists a family of operators $(P_t)_{t\geqslant 0}$, acting on some space $\mathcal{D}$ of bounded smooth functions f on $\mathbf{R}^n$ with bounded partial derivatives, dense in all $L^p(\mu)$, $p \geqslant 1$, such that

1) $P_t f \in \mathcal{D}$ for all $f \in \mathcal{D}$,
2) P_0 is the identity operator, i.e., $P_0 f = f$ for all $f \in \mathcal{D}$,
3) P_t forms a semigroup, i.e., $P_t(P_s f) = P_{t+s} f$ for all $t, s \geqslant 0$,
4) for any $f \in \mathcal{D}$, in the space $L^\infty(\mu)$, we have $\|P_t f - f\|_\infty \to 0$ as $t \to 0^+$,
5) for any $f \in \mathcal{D}$, in the space $L^1(\mu)$, the limit $Lf = \lim_{t\to 0^+} \frac{P_t f - f}{t}$ exists,
6) for all $f, g \in \mathcal{D}$

$$\int \langle \nabla f, \nabla g\rangle\, d\mu = -\int f\, Lg\, d\mu. \tag{9.1}$$

Equality in 5) expresses the property that L represents the generator of the semigroup P_t. This is usually denoted by $P_t = e^{tL}$, where the exponential function is understood in the operator sense. Owing to 1) and 3), it may be generalized as the property that for any $f \in \mathcal{D}$ and $t \geqslant 0$, in the space $L^1(\mu)$,

$$L(P_t f) = \lim_{\varepsilon\to 0^+} \frac{P_{t+\varepsilon} f - P_t f}{\varepsilon}. \tag{9.2}$$

In other words, the L^1-valued map $t \to P_t f$ is differentiable from the right and has the right derivative $L(P_t f)$. The equalities (9.1) and (9.2) may be used to prove, in particular, the following assertion.

Lemma 9.1. *Given a twice continuously differentiable function u on the real line, for any $f \in \mathcal{D}$ the function $t \to \int u(P_t f)\, d\mu$ is differentiable from the right and has the right derivative*

$$\frac{d}{dt} \int u(P_t f)\, d\mu = - \int u''(P_t f)\, |\nabla P_t f|^2\, d\mu. \tag{9.3}$$

To illustrate classical applications, we assume that a measure μ satisfies a Poincaré type inequality

$$\lambda_1 \operatorname{Var}_\mu(f) \leqslant \int |\nabla f|^2\, d\mu \tag{9.4}$$

for some $\lambda_1 > 0$ in the class of all smooth f on $\mathbf{R}^n$.

For $u(x) = x$ the equality (9.3) implies that the function $\varphi(t) = \int P_t f\, d\mu$, where $f \in \mathcal{D}$, has the right derivative zero at every point $t \geqslant 0$. Since this function is also continuous, it must be equal to a constant, i.e., $\int f\, d\mu$. Taking $u(x) = x^2$ and assuming that $\int f\, d\mu = 0$, from (9.3) and (9.4) we have

$$\frac{d}{dt} \int |P_t f|^2\, d\mu = -2 \int |\nabla P_t f|^2\, d\mu \leqslant -2\lambda_1 \int |P_t f|^2\, d\mu.$$

Thus, the function $\varphi(t) = \int |P_t f|^2\, d\mu$ is continuous and has the right derivative satisfying $\varphi'(t) \leqslant -2\lambda_1 \varphi(t)$. It is a simple calculus exercise to derive from this differential inequality the bound on the rate of convergence, $\varphi(t) \leqslant \varphi(0) e^{-2\lambda_1 t}$. Therefore,

$$\int |P_t f|^2\, d\mu \leqslant e^{-2\lambda_1 t} \int |f|^2\, d\mu, \quad t \geqslant 0. \tag{9.5}$$

In particular, we obtain a contraction property $\|P_t f\|_2 \leqslant \|f\|_2$ for all $f \in \mathcal{D}$, which allows us to extend P_t to all $L^2(\mu)$ as a linear contraction. Moreover, by continuity, (9.5) extends to all $f \in L^2(\mu)$ with μ-mean zero, and we also have

$$\int P_t f\, d\mu = \int f\, d\mu.$$

Our next natural step is to generalize (9.5) to L^p-spaces.

Theorem 9.2. *For all $f \in L^p(\mu)$, $p > 1$, and $t \geqslant 0$*

$$\iint |P_t f(x) - P_t f(y)|^p\, d\mu(x) d\mu(y) \leqslant e^{-\frac{4(p-1)}{p} \lambda_1 t} \iint |f(x) - f(y)|^p\, d\mu(x) d\mu(y). \tag{9.6}$$

Proof. As in the previous example of the quadratic function, for any twice continuously differentiable, convex function u on the real line and any $f \in \mathcal{D}$, by (9.3) and (9.4), we have

$$\frac{d}{dt} \int u(P_t f)\, d\mu = -\int u''(P_t f)\, |\nabla P_t f|^2\, d\mu$$
$$= -\int |\nabla v(P_t f)|^2\, d\mu \leqslant -\lambda_1 \operatorname{Var}_\mu[v(P_t f)], \tag{9.7}$$

where the derivative is understood as the derivative from the right and v is a differentiable function satisfying $v'^2 = u''$. In particular, we may take $u(z) = |z|^p$ with $p \geqslant 2$, so that $u''(z) = p(p-1)|z|^{p-2}$ and

$$v(z) = 2\sqrt{\frac{p-1}{p}}\, \operatorname{sign}(z)\, |z|^{p/2},$$

to get

$$\frac{d}{dt} \int |P_t f|^p\, d\mu \leqslant -4\lambda_1\, \frac{p-1}{p} \int |P_t f|^p\, d\mu \tag{9.8}$$

provided that

$$\int \operatorname{sign}(P_t f)\, |P_t f|^{p/2}\, d\mu = 0.$$

The last equality holds, for example, when $P_t f$ has a distribution under μ, symmetric about zero. Moreover, a slight modification of $u(z) = |z|^p$ near zero allows us to replace the constraint $p \geqslant 2$ in (9.7) and (9.8) by the weaker condition $p > 1$ (cf. details at the end of the proof).

Now, on $M = \mathbf{R}^n \times \mathbf{R}^n$, we consider the product measure $\mu \otimes \mu$. By the subadditivity property of the variance functional, it also satisfies the Poincaré type inequality (9.4) with the same constant λ_1. In addition, with this measure one may associate the semigroup $\overline{P}_t$, $t \geqslant 0$, acting on a certain space $\overline{\mathcal{D}}$ of bounded smooth functions on $\mathbf{R}^n \times \mathbf{R}^n$ with bounded partial derivatives containing functions of the form

$$\overline{f}(x, y) = f(x) - f(y), \quad x, y \in \mathbf{R}^n, \quad f \in \mathcal{D}.$$

It easy to see that for such functions

$$(\overline{P}_t \overline{f})(x, y) = P_t f(x) - P_t f(y), \quad (\overline{L}\overline{f})(x, y) = Lf(x) - Lf(y),$$

where $\overline{L}$ is the generator of $\overline{P}_t$. Apply (9.8) to the product space $(\mathbf{R}^n \times \mathbf{R}^n, \mu \otimes \mu)$. Since $\overline{P}_t \overline{f}$ has a symmetric distribution under $\mu \otimes \mu$ about zero, the function

$$\varphi(t) = \iint |P_t f(x) - P_t f(y)|^p\, d\mu(x) d\mu(y)$$

is continuous and has the right derivative at every point $t \geqslant 0$ satisfying the differential inequality

$$\varphi'(t) \leqslant -C\,\varphi(t) \tag{9.9}$$

with $C = 4\lambda_1\,\frac{p-1}{p}$. Then $\varphi(t) \leqslant \varphi(0)e^{-Ct}$, which is the claim.

Thus, when $p \geqslant 2$, every P_t represents a continuous linear operator on $\mathcal{D}$ with respect to the L^p-norm, so it may be extended to the whole $L^p(\mu)$; moreover, the inequality (9.6) remains valid for all functions f in $L^p(\mu)$.

Now, let us see what modifications may be made in the case $1 < p < 2$. Given a fixed natural number N, define a convex, twice continuously differentiable, even function u_N through its second derivative

$$u_N''(z) = p(p-1)\,\min\{|z|^{p-2}, N\}, \quad z \in \mathbf{R},$$

and by requiring that $u_N(0) = u_N'(0) = 0$. Also, define an odd function v_N through its first derivative

$$v_N'(z) = \mathrm{sign}(z)\,\sqrt{u_N''(z)} = \mathrm{sign}(z)\,\sqrt{p(p-1)}\,\min\left\{|z|^{\frac{p}{2}-1}, \sqrt{N}\right\}, \quad z \neq 0,$$

or, equivalently,

$$v_N(z) = \sqrt{p(p-1)}\int_0^z \min\{|y|^{\frac{p}{2}-1}, \sqrt{N}\}\,dy.$$

We note that v_N is differentiable everywhere, except for $z = 0$, at which point the left and right derivatives exist, but do not coincide. On the other hand, $|v_N'(z)|$ is continuous everywhere, including the origin point $z = 0$, so that, in the class of all smooth g on $\mathbf{R}^n$, we always have a chain rule

$$u_N''(g(x))|\nabla g(x)|^2 = |\nabla v_N(g(x))|^2, \quad x \in \mathbf{R}^n,$$

even if $g(x) = 0$. Thus, the first part of (9.7) remains valid for u_N, i.e.,

$$\frac{d}{dt}\int u_N(P_t f)\,d\mu = -\int |\nabla v_N(P_t f)|^2\,d\mu.$$

We also recall that the Poincaré type inequality (9.4) extends to all locally Lipschitz functions f on $\mathbf{R}^n$. In particular, by the chain rule,

$$\lambda_1\,\mathrm{Var}_\mu(T(g)) \leqslant \int |T'(g)|^2\,|\nabla g|^2\,d\mu$$

if g is smooth on $\mathbf{R}^n$ and T on $\mathbf{R}$. For a fixed g this inequality may be written in dimension one as

$$\lambda_1\,\mathrm{Var}_\nu(T) \leqslant \int |T'|^2\,d\pi$$

with respect to the distribution ν of g under μ and the distribution π of g under the finite measure $|\nabla g|^2\, d\mu$. At this step, it is only required that T be locally Lipschitz on the line, and this is indeed true for $T = v_N$. Therefore, the second part of (9.7) also holds for v_N, and we get

$$\frac{d}{dt}\int u_N(P_t f)\, d\mu \leqslant -\lambda_1 \int v_N(P_t f)^2\, d\mu \tag{9.10}$$

provided that

$$\int v_N(P_t f)\, d\mu = 0.$$

Now, to estimate further the right-hand side of (9.10), we use the integral description of v_N to see that for $z > 0$

$$\begin{aligned} 0 \leqslant\ v(z) - v_N(z) &= \sqrt{p(p-1)}\int_0^z \Big[y^{\frac{p}{2}-1} - \min\{y^{\frac{p}{2}-1}, \sqrt{N}\}\Big]\, dy \\ &\leqslant \sqrt{p(p-1)}\int_0^{+\infty} y^{\frac{p}{2}-1}\, 1_{\{y^{\frac{p}{2}-1} > \sqrt{N}\}}\, dy \\ &= 2\sqrt{\frac{p-1}{p}}\, N^{-\frac{p}{2(2-p)}}. \end{aligned}$$

Hence

$$v(z)^2 - v_N(z)^2 \leqslant 2\, v(z)(v(z) - v_N(z)) \leqslant \frac{8(p-1)}{p}\, z^{\frac{p}{2}}\, N^{-\frac{p}{2(2-p)}} \leqslant \frac{4}{\sqrt{N}}\, z^{\frac{p}{2}},$$

so that

$$v_N(z)^2 \geqslant v(z)^2 - \frac{4}{\sqrt{N}}\, z^{\frac{p}{2}} = \frac{4(p-1)}{p}\, u(z) - \frac{4}{\sqrt{N}}\, z^{\frac{p}{2}},$$

and thus, for all $z \in \mathbf{R}$,

$$v_N(z)^2 \geqslant \frac{4(p-1)}{p}\, u_N(z) - \frac{4}{\sqrt{N}}\, |z|^{\frac{p}{2}}.$$

Therefore, (9.10) may be continued as

$$\frac{d}{dt}\int u_N(P_t f)\, d\mu \leqslant -4\lambda_1\, \frac{p-1}{p}\int u_N(P_t f)\, d\mu + \frac{4}{\sqrt{N}}\int |P_t f|^{\frac{p}{2}}\, d\mu,$$

where we assumed that $\int v_N(P_t f)\, d\mu = 0$. Since f is bounded, all $P_t f$ are uniformly bounded (cf. Corollary 9.3 concerning large values of p), so the above estimate yields

$$\frac{d}{dt}\int u_N(P_t f)\,d\mu \leqslant -4\lambda_1\,\frac{p-1}{p}\int u_N(P_t f)\,d\mu + \frac{A}{\sqrt{N}} \tag{9.11}$$

with some constant A independent of t. Applying (9.11) in the product space to functions of the form $f(x) - f(y)$, as in the case $p \geqslant 2$, we find that the function

$$\varphi_N(t) = \iint u_N(P_t f(x) - P_t f(y))\,d\mu(x)d\mu(y)$$

is continuous and has the right derivative satisfying at every point $t \geqslant 0$ the following modified form of (9.9):

$$\varphi_N'(t) \leqslant -C\varphi_N(t) + \varepsilon_N,$$

where $\varepsilon_N = \frac{A}{\sqrt{N}}$ and $C = 4\lambda_1\,\frac{p-1}{p}$, as above. In terms of $\psi_N(t) = \varphi_N(t)\,e^{Ct}$ this differential inequality takes a simpler form $\psi_N'(t) \leqslant \varepsilon_N\,e^{Ct}$, which is easily solved as

$$\psi_N(t) \leqslant \psi_N(0) + \frac{\varepsilon_N}{C}\,(e^{Ct} - 1).$$

Equivalently,

$$\varphi_N(t) \leqslant \varphi_N(0)\,e^{-Ct} + \frac{\varepsilon_N}{C}\,(1 - e^{-Ct}),$$

so

$$\iint u_N(P_t f(x) - P_t f(y))\,d\mu(x)d\mu(y) \leqslant e^{-\frac{4(p-1)}{p}\lambda_1 t}\iint u_N(f(x) - f(y))\,d\mu(x)d\mu(y) + \frac{\varepsilon_N}{C}.$$

It remains to let $N \to \infty$ and use the property that $u_N \to u$ uniformly on bounded intervals of the line. Thus, (9.6) holds for all functions f in $\mathcal{D}$ and therefore for all f from the whole space $L^p(\mu)$. Theorem 9.2 is proved. □

REMARK 9.1. Let us describe several immediate applications of Theorem 9.2.

1. Thus, every P_t represents a linear contraction in $L^p(\mu)$. Note that if $\int f\,d\mu = 0$, by the Jensen inequality, the left-hand side of (9.6) majorizes $\|P_t f\|_p^p$ and the integral on the right-hand side is majorized by $2^p\|f\|_p^p$. Hence we get a hypercontractive inequality

$$\|P_t f\|_p \leqslant 2e^{-\frac{4(p-1)}{p^2}\lambda_1 t}\,\|f\|_p.$$

2. Similarly, one may consider Orlicz norms different from L^p-norms. For example, using the Taylor expansion for $\psi_2(z) \equiv e^{z^2} - 1$, from (9.6) we get that for any $\alpha > 0$

$$\iint \psi_2(\alpha|P_t f(x) - P_t f(y)|)\, d\mu(x) d\mu(y) \leqslant e^{-2\lambda_1 t} \iint \psi_2(\alpha|f(x) - f(y)|)\, d\mu(x) d\mu(y).$$

Hence the operator P_t continuously acts on $L^{\psi_2}(\mu)$.

3. Letting $p \to +\infty$ in (9.6), we conclude that for any bounded measurable function f on $\mathbf{R}^n$ and for any $t \geqslant 0$

$$\operatorname{Osc}(P_t f) \leqslant \operatorname{Osc}(f). \tag{9.12}$$

In particular, P_t represents a contraction in $L^\infty(\mu)$, while for finite $p > 1$ these operators are hypercontractive.

4) Since the inequality (9.12) does not involve λ_1, it remains valid in the case $\lambda_1 = 0$. Such properties may be seen with the help of Lemma 9.1. Namely, from (9.3) it follows that, if u is additionally convex, then the function $t \to \int u(P_t f)\, d\mu$ is nonincreasing, so that

$$\int u(P_t f)\, d\mu \leqslant \int u(f)\, d\mu. \tag{9.13}$$

For example, the case $u(z) = |z|^p$, $p > 1$, yields

$$\|P_t f\|_p \leqslant \|f\|_p. \tag{9.14}$$

By the continuity of P_t on L^p, this inequality extends from $\mathcal{D}$ to the whole space $L^p(\mu)$. Note that, in Lemma 9.1, it is assumed that u is twice continuously differentiable and this is fulfilled as long as $p \geqslant 2$. However, the range $1 < p \leqslant 2$ may be treated with the help of a smooth approximation, such as in the proof of Theorem 9.2. Moreover, (9.14) remains valid for $p = 1$. We also note that, applying (9.14) in product spaces with $p = +\infty$, we arrive at (9.12).

10 Markov Semigroups and Weak Poincaré

As the next natural step, one may wonder what a weak Poincaré type inequality

$$\|f - \mathbf{E}f\|_p \leqslant C(p)\, \|\nabla f\|_2, \quad 1 \leqslant p < 2, \tag{10.1}$$

is telling us about possible contractivity property of the semigroup $(P_t)_{t \geqslant 0}$ associated to the Borel probability measure μ on $\mathbf{R}^n$. As in the previous section, we assume that properties 1)–6) are fulfilled, so that one may develop analysis, such as the basic identity (9.3) of Lemma 9.1.

Since (10.1) is weaker than the usual Poincaré type inequality (9.4), it is natural to expect to get a weak version of Theorem 9.2 on the rate of convergence of $P_t f$ to the constant function. In the classical case $p = 2$, lower rate of convergence have been studied by many authors. In particular, for this aim, developing the ideas of Ligget [27], Röckner and Wang [35] proposed to use a weak Poincaré type inequality with the generalized "oscillation term"

$$\mathrm{Var}_\mu(f) \leqslant \beta(s)\,\|\nabla f\|_2^2 + s\,\Phi(f)^2, \quad s > 0, \tag{10.2}$$

where Φ is a nonnegative functional on $\mathcal{D}$ satisfying

$$\Phi(P_t f) \leqslant \Phi(f) \quad \text{for all } \ t \geqslant 0. \tag{10.3}$$

Indeed, by Lemma 9.1, applied to $u(z) = z^2$, we have

$$\frac{d}{dt}\int |P_t f|^2\,d\mu = -2\int |\nabla P_t f|^2\,d\mu.$$

Hence, by (10.2) and (10.3), if $\int f\,d\mu = 0$, the function $\varphi(t) = \int |P_t f|^2\,d\mu$ has the right derivative satisfying

$$\varphi'(t) \leqslant -\frac{2}{\beta(s)}\,\varphi(t) + \frac{2s}{\beta(s)}\,\Phi(f)^2.$$

This differential inequality is solved as

$$\varphi(t) \leqslant \varphi(0)\,e^{-2t/\beta(s)} + s\,(1 - e^{-2t/\beta(s)})\,\Phi(f)^2,$$

so

$$\int |P_t f|^2\,d\mu \leqslant \inf_{s>0}\left[e^{-2t/\beta(s)}\int |f|^2\,d\mu + s\,\Phi(f)^2\right]. \tag{10.4}$$

Thus, we get a more general statement on the rate of convergence than the classical inequality (9.5), when $\beta(s) = 1/\lambda_1$, which is obtained from (10.4) by letting $s \to 0$. In applications, the right-hand side of (10.4) can be simplified as

$$\int |P_t f|^2\,d\mu \leqslant \xi(t)\left[\frac{1}{2}\int |f|^2\,d\mu + \Phi(f)^2\right], \tag{10.5}$$

where $\xi(t) = \inf\{s > 0 : \beta(s)\log\frac{2}{s} \leqslant 2t\}$.

As the most interesting examples, one may apply this scheme to the functionals $\Phi(f) = \mathrm{Osc}\,(f)$, or more generally $\Phi(f) = \|f - \mathbf{E}f\|_q$ or just $\Phi(f) = \|f\|_q$. Then, by the continuity of P_t, the resulting inequalities (10.4) and (10.5) extend from $\mathcal{D}$ to L^q-spaces.

In the presence of (10.1), we look for a corresponding expression for the bound on the rate of convergence explicitly in terms of the function $C(p)$. For this aim, we may appeal to Theorem 8.5, which relates (10.1) to (10.2) in the case $\Phi(f) = \|f - \mathbf{E}f\|_q$, $q > 2$. Indeed, by (8.9), the inequality (10.2) holds with

$$\beta(s) = \inf_{1 \leqslant p < 2} \left[C(p)^2 \, s^{\frac{q}{q-2}(1-2/p)} \right],$$

so the right-hand side of (10.3) is bounded from above by

$$\inf_{1 \leqslant p < 2} \inf_{s > 0} \left[\exp\left\{ -\frac{2t}{C(p)^2} \, s^{\frac{q}{q-2}(2/p-1)} \right\} \int |f|^2 \, d\mu + s \, \|f - \mathbf{E}f\|_q^2 \right].$$

In particular, we have the following assertion.

Theorem 10.1. *Assume that for some $a, b \geqslant 0$ and $\alpha > 0$ the rate function in the weak Poincaré type inequality* (10.1) *admits a polynomial bound*

$$C(p) \leqslant a \left(\frac{b}{2-p} \right)^{1/\alpha}, \quad 1 \leqslant p < 2. \tag{10.6}$$

Then for any $f \in L^q(\mu)$, $q > 2$, such that $\int f \, d\mu = 0$ and for all $t \geqslant 0$

$$\int |P_t f|^2 \, d\mu \ \leqslant \ 3 \exp\{-c t^{\frac{\alpha}{\alpha+2}}\} \|f\|_q^2, \tag{10.7}$$

where the constant $c > 0$ depends on the parameters a, b, α, and q only.

Indeed, by Corollary 8.6, the hypothesis (10.6) implies $\beta(s) \leqslant \beta \log^{2/\alpha}(2/s)$, where $\beta = \beta_0 a^2 \, (\beta_1 b \, \frac{q}{q-2})^{2/\alpha}$ with some positive absolute constants β_0 and β_1. Hence, in order to estimate $\xi(t)$ from above, it remains to solve

$$\beta \log^{1+\frac{2}{\alpha}}(2/s) \leqslant 2t,$$

and we arrive at

$$\xi(t) \leqslant 2 \exp\Big\{ -\Big(\frac{2t}{\beta}\Big)^{\frac{\alpha}{\alpha+2}} \Big\}.$$

Finally, apply (10.5).

Now, recalling Corollary 8.3 and Remark 8.1, we obtain the hypercontractivity property (10.7) for a large family of convex probability measures.

Corollary 10.2. *If a probability measure μ is $\varkappa$-concave for some $\varkappa < 0$ and*

$$\int \exp\left\{ \left(\frac{|x|}{\lambda} \right)^\alpha \right\} d\mu(x) \leqslant 2, \quad \alpha, \lambda > 0,$$

then it satisfies (10.7) *for any $f \in L^q(\mu)$, $q > 2$, such that $\int f \, d\mu = 0$.*

Using a perturbation argument, one may obtain other interesting examples. In particular, they include all probability measures μ on $\mathbf{R}^n$ with densities of the form (7.1), i.e.,

$$\frac{d\mu(x)}{dx} = \frac{1}{Z}\, e^{-(a+b|x|)^\alpha}, \quad x \in \mathbf{R}^n,$$

with parameters $a \geqslant 0$, $b > 0$, and $\alpha > 0$.

At the next step, we generalize the previous results to L^p-spaces, so that to control the rate of convergence of $P_t f$ for norms different than L^2-norms.

We start with the weak Poincaré type inequality (10.2) for the functional $\Phi(f) = \|f - \mathbf{E}f\|_r$, i.e., with the family of inequalities

$$\mathrm{Var}_\mu(f) \ \leqslant\ \beta_r(s)\, \|\nabla f\|_2^2 + s\, \|f - \mathbf{E}f\|_r^2, \quad s > 0, \tag{10.8}$$

where f is an arbitrary locally Lipschitz function on $\mathbf{R}^n$, $r > 2$, and β_r is a function of the parameter s.

Theorem 10.3. *Under* (10.8), *given* $q > p > 1$ *such that* $\frac{pr}{2} = q$, *for all* $f \in L^q(\mu)$ *and* $t, s \geqslant 0$

$$\begin{aligned}
&\iint |P_t f(x) - P_t f(y)|^p \, d\mu(x) d\mu(y) \\
&\qquad \leqslant \exp\Big\{ -\frac{4(p-1)}{p}\, \frac{t}{\beta_r(s)} \Big\} \iint |f(x) - f(y)|^p \, d\mu(x) d\mu(y) \\
&\qquad + s \cdot \frac{p}{2(p-1)} \Big(\iint |f(x) - f(y)|^q \, d\mu(x) d\mu(y) \Big)^{p/q}.
\end{aligned} \tag{10.9}$$

Proof. The argument represents a slight modification of the proof of Theorem 9.2. By (9.3), given a twice continuously differentiable, convex function u on the real line and a differentiable function v such that $v'^2 = u''$, we have for any $f \in \mathcal{D}$ such that $\int f\, d\mu = 0$ and for all $t, s > 0$

$$\begin{aligned}
\frac{d}{dt} \int u(P_t f)\, d\mu &= - \int u''(P_t f)\, |\nabla P_t f|^2 \, d\mu \\
&= - \int |\nabla v(P_t f)|^2 \, d\mu \\
&\leqslant -\frac{1}{\beta(s)}\, \mathrm{Var}_\mu[v(P_t f)] + \frac{s}{\beta(s)}\, \|v(P_t f) - \mathbf{E}_\mu v(P_t f)\|_r^2,
\end{aligned}$$

where the derivative is understood as the derivative from the right. In particular, we may take $u(z) = |z|^p$ with $p \geqslant 2$, so that $u''(z) = p(p-1)|z|^{p-2}$, and $v(z) = 2\sqrt{\frac{p-1}{p}}\, \mathrm{sign}(z)\, |z|^{p/2}$, to get

$$\frac{d}{dt}\int |P_t f|^p\,d\mu \leqslant -\frac{4(p-1)}{p}\,\frac{1}{\beta(s)}\int |P_t f|^p\,d\mu + \frac{s}{\beta(s)}\,\|\,|P_t f|^{p/2}\|_r^2 \quad (10.10)$$

provided that

$$\int \mathrm{sign}(P_t f)\,|P_t f|^{p/2}\,d\mu = 0.$$

Note that the latter holds when $P_t f$ has a distribution under μ, which is symmetric about zero. A slight modification of $u(z) = |z|^p$ near zero, described in the proof of Theorem 9.2, allows one to replace the constraint $p \geqslant 2$ by the weaker condition $p > 1$. Note that, by the contraction property (9.14),

$$\|\,|P_t f|^{p/2}\|_r^2 = \|P_t f\|_{pr/2}^p = \|P_t f\|_q^p \leqslant \|f\|_q^p,$$

so (10.10) yields

$$\frac{d}{dt}\int |P_t f|^p\,d\mu \leqslant -\frac{4(p-1)}{p}\,\frac{1}{\beta_r(s)}\int |P_t f|^p\,d\mu + \frac{s}{\beta_r(s)}\,\|f\|_q^p. \quad (10.11)$$

Now, to guarantee that $P_t f$ has a symmetric distribution, we consider the product measure $\mu\otimes\mu$ on $M = \mathbf{R}^n\times\mathbf{R}^n$. With this measure we associate the semigroup $\overline{P}_t$, $t \geqslant 0$, acting on a certain space $\overline{\mathcal{D}}$ of bounded smooth functions on $\mathbf{R}^n\times\mathbf{R}^n$ with bounded partial derivatives, containing all functions of the form

$$\overline{f}(x,y) = f(x) - f(y), \quad x, y \in \mathbf{R}^n, \quad f \in \mathcal{D}.$$

It easy to see that for such functions

$$(\overline{P}_t\overline{f})(x,y) = P_t f(x) - P_t f(y), \quad (\overline{L}\overline{f})(x,y) = Lf(x) - Lf(y),$$

where $\overline{L}$ is the generator of $\overline{P}_t$.

We are going to apply (10.11) to $\overline{f}$ on the product space $(\mathbf{R}^n\times\mathbf{R}^n, \mu\otimes\mu)$, so we need a hypothesis of the form (10.8) with respect to the product measure. Note that

$$\mathrm{Var}_{\mu\otimes\mu}(\overline{f}) = 2\ \mathrm{Var}_\mu(f), \quad \mathbf{E}_{\mu\otimes\mu}|\nabla\overline{f}(x,y)|^2 = 2\mathbf{E}_\mu|\nabla f|^2$$

and, by the Jensen inequality,

$$\mathbf{E}_\mu|f - \mathbf{E}_\mu f|^r \leqslant \mathbf{E}_{\mu\otimes\mu}|\overline{f}|^r.$$

Hence (10.8) implies

$$\mathrm{Var}_{\mu\otimes\mu}(\overline{f}) \leqslant \beta_r(s)\|\nabla\overline{f}\|_2^2 + 2s\|\overline{f}\|_r^2, \quad s > 0.$$

As a result, we obtain a slightly weakened form of (10.11), namely,

$$\frac{d}{dt}\int |\overline{P_t f}|^p \, d\mu\otimes\mu \leqslant -\frac{4(p-1)}{p}\,\frac{1}{\beta_r(s)}\int |\overline{P_t f}|^p \, d\mu\otimes\mu + \frac{2s}{\beta_r(s)}\,\|\overline{f}\|_q^p. \tag{10.12}$$

Let us note that, by virtue of the subadditivity property of the variance functional, (10.12) may be extended to the whole space $\overline{\mathcal{D}}$, however, with a worse constant in place of 2.

Thus, the function

$$\varphi(t) = \int |\overline{P_t f}|^p \, d\mu\otimes\mu = \iint |P_t f(x) - P_t f(y)|^p \, d\mu(x) d\mu(y)$$

is continuous and has the right derivative at every point $t \geqslant 0$ satisfying the differential inequality

$$\varphi'(t) \leqslant -A\,\varphi(t) + B$$

with

$$A = \frac{4(p-1)}{\beta_r(s)p}, \quad B = \frac{2s}{\beta_r(s)}\,\|\overline{f}\|_q^p.$$

Using the change $\varphi(t) = \psi(t)e^{-At}$, we obtain

$$\varphi(t) \leqslant \varphi(0)e^{-At} + \frac{B}{A}\,(1-e^{-At}) \leqslant \varphi(0)e^{-At} + \frac{B}{A},$$

i.e.,

$$\iint |P_t f(x) - P_t f(y)|^p \, d\mu(x)d\mu(y) \leqslant e^{-At}\iint |f(x) - f(y)|^p \, d\mu(x)d\mu(y) + \frac{B}{A}.$$

But

$$\frac{B}{A} = \frac{ps}{2(p-1)}\,\|\overline{f}\|_q^p,$$

so we arrive at the desired inequality (10.9). Finally, by continuity of P_t, this inequality extends from $\mathcal{D}$ to the whole space $L^q(\mu)$. □

At the expense of some constants, depending on p and q, the inequality (10.9) may be simplified. Namely, if $\int f d\mu = 0$, the left-hand side of (10.9) majorizes $\|P_t f\|_p^p$, while the integrals on the right-hand side are bounded by $2^p\|f\|^p$ and $2^q\|f\|^q$ respectively. Hence

$$\|P_t f\|_p^p \leqslant 2^p\, e^{-\frac{4(p-1)}{p}\,\frac{t}{\beta_r(s)}}\,\|f\|_p^p + s\cdot\frac{2^p\,p}{2(p-1)}\,\|f\|_q^p.$$

Corollary 10.4. *Given* $q > p > 1$, *under* (10.8) *with* $r = \frac{2q}{p}$, *for all* $f \in L^q(\mu)$ *with mean zero and for all* $t \geqslant 0$,

$$\|P_t f\|_p^p \leqslant 2^p \|f\|_q^p \inf_{s>0} \left[e^{-\frac{4(p-1)}{p} \frac{t}{\beta_r(s)}} + \frac{p}{2(p-1)} s \right]. \tag{10.13}$$

To further simplify this bound, define the function

$$\xi(t) = \inf\left\{ s > 0 : \beta_r(s) \log \frac{2}{s} \leqslant \frac{4p}{p-1} t \right\}$$

depending also on the parameters p and r. Then the expression in the square brackets in (10.13) is bounded by

$$\frac{s}{2} + \frac{p}{2(p-1)} s \leqslant \frac{p}{p-1} s.$$

Therefore,

$$\|P_t f\|_p^p \leqslant \frac{p}{p-1} 2^p \|f\|_q^p \xi(t). \tag{10.14}$$

Now, let us start with the weak Poincaré type inequality (10.1) with rate function $C(p)$ satisfying the bound (10.6), as in Theorem 10.1. Then, as we know from Corollary 8.6, the hypothesis (10.8) holds with

$$\beta_r(s) = \beta \log^{\frac{2}{\alpha}} \frac{2}{s},$$

where

$$\beta = 2a^2 \left(4b \frac{r}{r-2}\right)^{2/\alpha}.$$

Since $r = \frac{2q}{p}$, the coefficient is

$$\beta = 2a^2 \left(4b \frac{q}{q-p}\right)^{2/\alpha}.$$

We also find that

$$\xi(t) \leqslant 2 \exp\Big\{ - \Big(\frac{4p}{\beta(p-1)} t \Big)^{\alpha/(\alpha+2)} \Big\}.$$

As a result, we obtain the following generalization of Theorem 10.1.

Theorem 10.5. *Assume that the weak Poincaré type inequality* (10.1) *holds with rate function $C(p)$ satisfying the bound* (10.6) *with some parameters $a, b \geqslant 0$ and $\alpha > 0$. Given $q > p > 1$, for all $f \in L^q(\mu)$ with mean zero and for all $t \geqslant 0$*

$$\int |P_t f|^p \, d\mu \leqslant \frac{p}{p-1} 2^{p+1} \exp\{-ct^{\frac{\alpha}{\alpha+2}}\} \left(\int |f|^q \, d\mu \right)^{p/q}, \tag{10.15}$$

where the constant $c > 0$ depends on a, b, α, p, and q only.

More precisely, we may put

$$c = \left(\frac{4p}{p-1}\right)^{\frac{\alpha}{\alpha+2}} \frac{\left(\frac{q-p}{q}\right)^{\frac{2}{\alpha+2}}}{(2a^2)^{\frac{2}{\alpha}}\,(4b)^{\frac{2}{\alpha+2}}}.$$

11 L^2 Decay to Equilibrium in Infinite Dimensions

11.1 Basic inequalities and decay to equilibrium in the product case

In this and next sections, we further simplify the notation for the expectation setting $\mu f \equiv \mathbf{E}_\mu f = \int f\, d\mu$ for the expectation of f under a probability measure μ. This will prove to be useful when we have to deal with more involved mathematical expressions.

Consider a probability measure on the real line of the form

$$\nu_0(dx) \equiv \frac{1}{Z} e^{-V(x)} dx$$

with $V(x) \equiv \varsigma(1+x^2)^{\frac{\alpha}{2}}$, where $0 < \alpha \leqslant 1$ and $\varsigma \in (0,\infty)$, while Z denotes a normalization constant. Since $|x| \leqslant (1+x^2)^{\frac{1}{2}} \leqslant 1 + |x|$, by Theorem 7.1 and Corollary 8.6, we have the following assertion.

Lemma 11.1. *For any $p \in (2,\infty)$ there exists $\beta \in (0,\infty)$ such that for any $s \in (0,1)$*

$$\nu_0 \left|f - \nu_0 f\right|^2 \leqslant \overline{\beta}(s) \nu_0 \left|\nabla f\right|^2 + s\left(\nu_0 \left|f - \nu_0 f\right|^p\right)^{\frac{2}{p}} \tag{11.1}$$

with $\overline{\beta}(s) \equiv \beta\left(\log \frac{2}{s}\right)^{\frac{2}{\alpha}}$ for any function f, for which the right hand side is well defined.

By a simple inductive argument, one gets the following property for corresponding product measures.

Proposition 11.2 (product property). *Suppose that ν_i, $i \in \mathbb{N}$, satisfy*

$$\nu_i \left|f - \nu_i f\right|^2 \leqslant \overline{\beta}(s) \nu_i \left|\nabla_i f\right|^2 + s\left(\nu_i \left|f - \nu_i f\right|^p\right)^{\frac{2}{p}}. \tag{11.2}$$

Then the product measure $\mu_0 \equiv \otimes_{i\in\mathbb{N}} \nu_i$ also satisfies

$$\mu_0 |f - \mu_0 f|^2 \leqslant \overline{\beta}(s) \sum_{i \in \mathbb{N}} \mu_0 |\nabla_i f|^2 + s A_{p,\mu_0}(f) \tag{11.3}$$

with

$$A_{p,\mu_0}(f) \equiv \sum_{i \in \mathbb{N}} \mu_0 \left(\nu_i |f - \nu_i f|^p\right)^{\frac{2}{p}}. \tag{11.4}$$

Proof. Note that for $f_i \equiv \nu_i f_{i-1} \equiv \nu_{\leqslant i} f$, $i \in \mathbb{N}$, with $f_0 \equiv f$, we have

$$\mu_0 |f - \mu_0 f|^2 = \sum_{i \in \mathbb{N}} \mu_0 \nu_i |f_{i-1} - \nu_i f_{i-1}|^2 .$$

Hence, applying (11.2) to each term, we arrive at

$$\mu_0 |f - \mu_0 f|^2 \leqslant \sum_{i \in \mathbb{N}} \mu_0 \left(\overline{\beta}(s) \nu_i |\nabla_i f_{i-1}|^2 + s \left(\nu_i |f_{i-1} - \nu_i f_{i-1}|^p\right)^{\frac{2}{p}} \right).$$

Next, we note that (by using the Minkowski and Schwartz inequalities)

$$\left(\nu_i |f_{i-1} - \nu_i f_{i-1}|^p\right)^{\frac{2}{p}} \leqslant \nu_{\leqslant i-1} \left(\nu_i |f - \nu_i f|^p\right)^{\frac{2}{p}}$$

and

$$\nu_i |\nabla_i f_{i-1}|^2 \leqslant \nu_{\leqslant i} |\nabla_i f|^2 .$$

Thus, taking into the account the fact that $\mu_0 \nu_{\leqslant i} F = \mu_0 F$ (and similarly, with ν_i in place of $\nu_{\leqslant i}$), we arrive at

$$\mu_0 |f - \mu_0 f|^2 \leqslant \overline{\beta}(s) \sum_{i \in \mathbb{N}} \mu_0 |\nabla_i f|^2 + s \sum_{i \in \mathbb{N}} \mu_0 \left(\nu_i |f - \nu_i f|^p\right)^{\frac{2}{p}}. \tag{11.5}$$

This ends the proof of the proposition. □

The Dirichlet form defines the following Markov generator:

$$L^{(0)} \equiv \sum_{i \in \mathbb{N}} L_i^{(0)},$$

with $L_i^{(0)} \equiv \Delta_i - V'(x_i)\nabla_i$, where Δ_i and ∇_i denote the Laplace operator and derivative with respect to the ith variable respectively. It is well defined on a dense domain in $L^2(\mu_0)$. As in our situation V is smooth and V' is bounded, the corresponding semigroup $P_t^{(0)}$ in $L^2(\mu_0)$ extends nicely to a C_0-semigroup onto the space of continuous functions $\mathcal{C}(\Omega)$, where $\Omega \equiv \mathbb{R}^{\mathbb{N}}$. Using Proposition 11.2 and the fact that functional

$$A_{p,\mu_0}(f) \equiv \sum_{i \in \mathbb{N}} \mu_0 \left(\nu_i |f - \nu_i f|^p\right)^{\frac{2}{p}}$$

is monotone with respect to the semigroup (in the sense of (10.3)), one can see that Theorems 10.1 and 10.4 hold. In particular, we have

$$\mu_0 \left| P_t^{(0)} f - \mu_0 f \right|^2 \leqslant C e^{-ct^{\frac{\alpha}{\alpha+2}}} A_{p,\mu_0}(f)$$

with some constants $C, c \in (0, \infty)$ independent of f.

In the rest of the paper, we prove that an inequality of a similar shape remains true for infinite systems described by nontrivial Gibbs measures. Although the corresponding functional A_p may no longer be monotone, with extra work we show that the corresponding semigroups also satisfy stretched exponential decay estimate. We begin from presenting the necessary elements of the construction of the semigroups.

11.2 Semigroup for an infinite system with interaction.

Let $\Omega \equiv \mathbb{R}^{\mathcal{R}}$, with a countable connected graph $\mathcal{R}$ furnished with the natural metric (given by the number of edges in the shortest path connecting two points) and with at most stretched exponential volume growth.

Let $V \equiv \varsigma(1 + x^2)^{\frac{\alpha}{2}}$, with $0 < \alpha \leqslant 1$ and $\varsigma \in (0, \infty)$. Then

$$||V'||_\infty, \quad ||V''||_\infty < \infty.$$

We set $V_i(\omega) \equiv V(\omega_i)$. Let $U_i(\omega) \equiv V_i(\omega) + u_i(\omega)$, where u_i is a smooth function. Later on we set

$$a \equiv \sup_i \Big(2\gamma_{ii} + \sum_{j \neq i} \gamma_{ij} \Big), \tag{11.6}$$

$$\gamma_{ij} \equiv ||\nabla_i \nabla_j u_j||_\infty \tag{11.7}$$

and assume that $a \in (0, \infty)$. We note that, by the definition of local interaction V_i, we automatically have $||\nabla_i^2 V_i||_\infty < \infty$, so our assumption is only about u_j's. For simplicity of exposition, we assume that $\mathcal{R} = \mathbb{Z}^d$ and that the interaction is of finite range, i.e., for some $R \in (0, \infty)$ and all vertices i one has $\nabla_k u_i = 0$ when dist $(i, k) \geqslant R$.

Let P_t^Λ be a Markov semigroup associated to the generator

$$\mathcal{L}_\Lambda \equiv \sum_{i \in \mathcal{R}} L_i^{(0)} - \sum_{i \in \Lambda} \nabla_i u_i \cdot \nabla_i,$$

where

$$L_i^{(0)} \equiv \Delta_i - \nabla_i V_i(\omega) \nabla_i = \Delta_i - V'(\omega_i) \nabla_i$$

and the index i indicates that derivatives are taken with respect to ω_i, and $\Lambda \subset\subset \mathcal{R}$ (i.e., Λ is a bounded subset of $\mathcal{R}$). The following lemma will play later a crucial role in the control of decay to equilibrium. Naturally, it holds for P_t^Λ as well and is essential in defining the infinite volume semigroup as follows:

$$P_t f \equiv \lim_{\Lambda \to \mathcal{R}} P_t^\Lambda f$$

on the space of bounded continuous functions (cf., for example, [21]).

Lemma 11.3 (finite speed of propagation of information estimate). *There exist $A, B, C \in (0, \infty)$ such that for any smooth cylinder function f and any $i \in \mathcal{R}$*

$$||\nabla_i P_t f||^2 \leqslant C e^{At - Bd(i,\Lambda_f)} |||f|||^2, \tag{11.8}$$

where $\Lambda_f \subset \mathcal{R}$ is the smallest set $\mathcal{O} \subset \mathcal{R}$ such that f depends only on $\{\omega_i : i \in \mathcal{O}\}$ and

$$|||f|||^2 \equiv \sum_{i \in \mathcal{R}} ||\nabla_i f||_\infty^2.$$

The proof is based on the following arguments (note that, under our smoothness assumptions on the interaction, the pointwise operations are well justified):

$$\begin{aligned}
\frac{d}{d\tau} P_\tau |\nabla_i P_{t-\tau} f|^2 &= P_\tau \Big(\mathcal{L} |\nabla_i P_{t-\tau} f|^2 - 2\nabla_i P_{t-\tau} f \cdot \mathcal{L} \nabla_i P_{t-\tau} f \Big) \\
&\quad + 2P_\tau \left(\nabla_i P_{t-\tau} f \cdot [\mathcal{L}, \nabla_i] P_{t-\tau} f \right) \\
&\geqslant 2P_\tau \Big(\nabla_i P_{t-\tau} f \cdot [\mathcal{L}, \nabla_i] P_{t-\tau} f \Big) \\
&= P_\tau \Big(-2\nabla_i^2 U_i |\nabla_i P_{t-\tau} f|^2 - 2\sum_{j \neq i} \nabla_i \nabla_j u_j \nabla_i P_{t-\tau} f \nabla_j P_{t-\tau} f \Big) \\
&\geqslant -(2||\nabla_i^2 U_i||_\infty + \sum_{j \neq i} ||\nabla_i \nabla_j u_j||_\infty) \cdot P_\tau |\nabla_j P_{t-\tau} f|^2 \\
&\quad - \sum_{j \neq i} ||\nabla_i \nabla_j u_j||_\infty P_\tau |\nabla_j P_{t-\tau} f|^2.
\end{aligned}$$

Hence, with the notation introduced in (11.6) before the lemma, we have

$$||\nabla_i P_t f||^2 \leqslant e^{at} ||\nabla_i f||^2 + \sum_{j \neq i} \gamma_{ij} \int_0^t e^{a(t-\tau)} ||\nabla_j P_\tau f||^2 \, d\tau.$$

In particular, if $i \notin \Lambda_f$, we get

$$||\nabla_i P_t f||^2 \leqslant \sum_{j \neq i} \gamma_{ij} \int_0^t e^{a(t-\tau)} ||\nabla_j P_\tau f||^2 \, d\tau.$$

By standard arguments (cf. [21] and the references therein), this leads to the desired estimate of final speed of propagation of information (11.8).

11.3 L^2 decay

Our way to study the L^2 decay of the semigroup is as follows. Suppose that μ satisfies $\mu E_i f = \mu f$ for any $i \in \mathcal{R}$ with the following probability kernels:

$$E_i(f) \equiv E_i^\omega(f) \equiv \delta_\omega \left(\frac{\int fe^{-U_i} d\omega_i}{\int e^{-U_i} d\omega_i} \right) = \delta_\omega \left(\frac{\int fe^{-u_i} d\nu_i}{\int e^{-u_i} d\nu_i} \right), \tag{11.9}$$

where δ_ω denotes the Dirac mass concentrated at ω and, by definition, ν_i is an isomorphic copy of the probability measure ν_0. Then P_t is a symmetric semigroups in $L^2(\mu)$ with quadratic form of the generator given by

$$\mu|\nabla f|^2 \equiv \sum_{i\in\mathcal{R}} \mu|\nabla_i f|^2.$$

Let

$$A_p(f) \equiv A_{p,\mu}(f) \equiv \sum_{i\in\mathcal{R}} \mu\left(\nu_i\,|f - \nu_i f|^p\right)^{\frac{2}{p}}.$$

With this notation, we have the following assertion.

Lemma 11.4. *Assume that, with a positive function $\beta(s)$,*

$$\mu(f - \mu f)^2 \leqslant \beta(s)\ \mu|\nabla f|^2 + sA_p(f).$$

Then

$$\mu\left(P_t f - \mu f\right)^2 \leqslant \inf_s \left\{ e^{-\frac{t}{\beta(s)}} \mu\left(f - \mu f\right)^2 + s \sup_{0\leqslant\tau\leqslant t} A_p(P_\tau f) \right\}.$$

The above follows from the following simple arguments (cf., for example, [35] and the references therein) similar to those in Section 10. For $f_t \equiv P_t f$ we have

$$\frac{d}{dt}\mu\left(f_t - \mu f\right)^2 = -2\mu|\nabla f_t|^2 \leqslant -\frac{2}{\beta(s)}\mu\left(f_t - \mu f\right)^2 + \frac{2s}{\beta(s)} A_p(f_t).$$

Hence

$$\mu\left(f_t-\mu f\right)^2 \leqslant e^{-\frac{2t}{\beta(s)}}\mu\left(f-\mu f\right)^2+\int_0^t e^{-\frac{2(t-\tau)}{\beta(s)}}\frac{2s}{\beta(s)}A_p(f_\tau)d\tau$$

$$\leqslant e^{-\frac{2t}{\beta(s)}}\mu\left(f-\mu f\right)^2+s\sup_{0\leqslant\tau\leqslant t}A_p(f_\tau).$$

To go the route based on Lemma 11.4, we need an estimate for the functional A_p.

Proposition 11.5 (estimate of $A_p(f_\tau)$). *Suppose that* $\Lambda\equiv\Lambda_t\subset\subset\mathcal{R}$ *satisfies*

$$\operatorname{dist}\left(\Lambda^c,\Lambda_f\right)\geqslant\frac{1}{4}\operatorname{diam}\left(\Lambda\right)$$

with $\operatorname{diam}(\Lambda)=16\frac{A}{B}t$. *Then*

$$\sup_{0\leqslant\tau\leqslant t}A_p(f_\tau)\leqslant|\Lambda_t|\cdot G\left(\mu\left|f-\mu f\right|^p\right)^{\frac{2}{p}}+De^{-A\,t}\cdot|||f|||^2$$

with some constants $D,G\in(0,\infty)$ *independent of* f.

Proof. For any $\Lambda\subset\subset\mathcal{R}$ we have

$$A_p(f_\tau)\equiv\sum_{i\in\mathcal{R}}\mu\left(\nu_i\left|f_\tau-\nu_i f_\tau\right|^p\right)^{\frac{2}{p}}$$

$$=\sum_{i\in\Lambda}\mu\left(\nu_i\left|f_\tau-\nu_i f_\tau\right|^p\right)^{\frac{2}{p}}+\sum_{i\in\Lambda^c}\mu\left(\nu_i\left|f_\tau-\nu_i f_\tau\right|^p\right)^{\frac{2}{p}}.$$

Since for $p>2$ we have

$$\mu\left(\left(\nu_i\left|f_\tau-\nu_i f_\tau\right|^p\right)^{\frac{2}{p}}\right)\leqslant 4\mu\left(\left(\nu_i\left|f_\tau-\mu f_\tau\right|^p\right)^{\frac{2}{p}}\right)$$

$$\leqslant 4e^{\frac{4}{p}\sup_i||u_i||_\infty}\mu\left(\left(E_i\left|f_\tau-\mu f_\tau\right|^p\right)^{\frac{2}{p}}\right)$$

$$\leqslant 4e^{\frac{4}{p}\sup_i||u_i||_\infty}\left(\mu\left|f_\tau-\mu f_\tau\right|^p\right)^{\frac{2}{p}}$$

$$\leqslant 4e^{\frac{4}{p}\sup_i||u_i||_\infty}\left(\mu\left|f-\mu f\right|^p\right)^{\frac{2}{p}},$$

where we used the triangle and Hölder inequalities and gained a factor $e^{\frac{4}{p}\sup_i||u_i||_\infty}$ while passing from expectations with the measure ν_i to the expectation with the conditional expectation E_i. Thus,

$$A_p(f_\tau)\leqslant|\Lambda|\cdot G\left(\mu\left|f-\mu f\right|^p\right)^{\frac{2}{p}}+\sum_{i\in\Lambda^c}\mu\left(\nu_i\left|f_t-\nu_i f_t\right|^p\right)^{\frac{2}{p}}$$

with the constant $G\equiv 4e^{\frac{4}{p}\sup_i||u_i||_\infty}$. To estimate the sum over $i\in\Lambda^c$, we note that for $\omega,\widetilde{\omega}\in\Omega$ satisfying $\omega_j=\widetilde{\omega}_j$ for $j\neq i$

$$|f_\tau(\omega) - f_\tau(\widetilde{\omega})| = \left| \int_{\widetilde{\omega}_i}^{\omega_i} dx \; \nabla_i f_\tau(x \bullet \omega_{\mathcal{R}\setminus i}) \right| \leqslant |\omega_i - \widetilde{\omega}_i| \cdot \|\nabla_i f_\tau\|_\infty$$

with configuration

$$[x \bullet \omega_{\mathcal{R}\setminus i}]_j \equiv \delta_{ij} x + (1 - \delta_{ij}) \omega_j.$$

Thus, we get

$$|f_\tau(\omega) - f_\tau(\widetilde{\omega})| \leqslant |\omega_i - \widetilde{\omega}_i| \cdot C^{\frac{1}{2}} e^{\frac{A}{2}s - \frac{B}{2}d(i,\Lambda_f)} |||f|||$$

which implies

$$\mu \left(\nu_i \left|f_\tau - \nu_i f_\tau\right|^p\right)^{\frac{2}{p}} \leqslant 4Ce^{A\tau - Bd(i,\Lambda_f)} \left(\nu_0 |\omega_0|^p\right)^{\frac{2}{p}} \cdot |||f|||^2.$$

Since

$$\nu_0(dx) = \frac{1}{Z} e^{-\varsigma(1+x^2)^{\frac{\alpha}{2}}} \, dx,$$

one can obtain the following estimate (using the Stirling bound):

$$\left(\nu_0 |\omega_0|^p\right)^{\frac{2}{p}} \leqslant C' e^{\frac{4}{\alpha} \log p}$$

with some constant $C' \equiv C'(\alpha, \varsigma) \in (0, \infty)$ independent of $p \in (2, \infty)$. (This is an important place where we take advantage of oscillations in L^p; would we have the functional Osc as in (1.6), we would be in trouble.) Hence we obtain the following bound:

$$\sum_{i \in \Lambda^c} \mu \left(\nu_i \left|f_\tau - \nu_i f_\tau\right|^p\right)^{\frac{2}{p}} \leqslant De^{A\tau - \frac{B}{2}d(\Lambda^c,\Lambda_f)} \cdot |||f|||^2$$

with

$$D \equiv 4CC' e^{\frac{4}{\alpha} \log p} \sum_{i \in \mathcal{R} : \mathrm{dist}\,(i,\Lambda_f) \geqslant d(\Lambda^c,\Lambda_f)+1} e^{-\frac{B}{2}d(i,\Lambda_f)}$$

with the series being convergent due to our assumption about slower than exponential volume growth of $\mathcal{R}$. For $\tau \in [0, t]$, choosing $\Lambda \equiv \Lambda_t$ such that $\mathrm{dist}\,(\Lambda^c, \Lambda_f) \geqslant \frac{1}{4} \mathrm{diam}\,(\Lambda)$ with $\mathrm{diam}\,(\Lambda) = 16\frac{A}{B}t$, we get

$$\sum_{i \in \Lambda^c} \mu \left(\nu_i \left|f_\tau - \nu_i f_\tau\right|^p\right)^{\frac{2}{p}} \leqslant De^{-A\,t} \cdot |||f|||^2.$$

Combining all the above, we arrive at the following estimate:

$$\sup_{0 \leqslant \tau \leqslant t} A_p(f_\tau) \leqslant |\Lambda_t| \cdot G \left(\mu \left|f - \mu f\right|^p\right)^{\frac{2}{p}} + De^{-A\,t} \cdot |||f|||^2.$$

The proposition is proved. □

Given the above estimate for $A_p(f_t)$, we conclude with the following result.

Theorem 11.6. *Let $\Lambda \equiv \Lambda_t$ be an increasing family of bounded subsets of $\mathcal{R}$ such that* $\mathrm{dist}\,(\Lambda^c, \Lambda_f) \geqslant \frac{1}{4}\,\mathrm{diam}\,(\Lambda)$, *with* $\mathrm{diam}\,(\Lambda) = 16\frac{A}{B}t$ *and* $|\Lambda_t| \leqslant e^{\mathrm{diam}\,(\Lambda_t)^\theta}$, *with some $\theta \in (0,1)$ for all sufficiently large Λ_t. Assume that for a positive function $\beta(s) = \xi^{-1}\,(\log(1/s))^\eta$ defined with some $\xi, \eta \in (0,\infty)$*

$$\mu(f-\mu f)^2 \leqslant \beta(s)\;\mu|\nabla f|^2 + sA_p(f)$$

for each $s \in (0,1)$.

If $\theta \in (0, 1/\eta)$, then there exist constant $\zeta, J \in (0,\infty)$, and $\varepsilon \in (0,1)$ such that

$$\mu(f_t - \mu f_t)^2 \leqslant e^{-\zeta t^\varepsilon}\; J\left(\mu\,(f-\mu f)^2 + (\mu\,|f-\mu f|^p)^{\frac{2}{p}} + |||f|||^2\right).$$

REMARK 11.1. In the case of a regular lattice $\mathbb{Z}^d$ and finite range interactions, one would have $|\Lambda_t| \sim t^d$. Our weaker growth assumption allows one to include more general graphs, as well as interactions which are not of finite range.

We note that for our considerations it is relevant only what is the behavior of $\beta(s)$ for small s. This determines the long time behavior (while the short time estimates can be compensated by a choice of constant J). This allows us to disregard factor 2 (or any similar numerical factor) from within the log in $\beta(s)$ as compared to estimates used in the product case.

Proof of Theorem 11.6. By Lemma 11.4 and Proposition 11.5, we have

$$\begin{aligned}&\mu(f_t - \mu f_t)^2\\ &\leqslant \inf_s \left\{e^{-\frac{t}{\beta(s)}}\mu\,(f-\mu f)^2 + s\left(|\Lambda_t|\cdot G\,(\mu\,|f-\mu f|^p)^{\frac{2}{p}} + De^{-A\,t}\cdot|||f|||^2\right)\right\}.\end{aligned}$$

Hence, choosing $s = e^{-t^\sigma}$ with $\sigma \in (\theta, 1/\eta)$, we obtain

$$\begin{aligned}\mu(f_t - \mu f_t)^2 &\leqslant \exp\{-\xi t^{1-\sigma\eta}\}\mu\,(f-\mu f)^2\\ &\quad + e^{-t^\sigma}\left(e^{(16\frac{A}{B}t)^\theta}\cdot G\,(\mu\,|f-\mu f|^p)^{\frac{2}{p}} + De^{-A\,t}e^{\frac{4}{\alpha}\log p}\cdot|||f|||^2\right).\end{aligned}$$

Thus, if $\sigma \in (\theta, 1/\eta)$, then there exists a constant $J \in (0,\infty)$ such that with $\varepsilon \equiv \min(1-\sigma\eta,\ \sigma-\theta)$ and any $\zeta \in (0, \min(1,\xi))$ we have

$$\mu(f_t - \mu f_t)^2 \leqslant e^{-\zeta t^\varepsilon}\; J\left(\mu\,(f-\mu f)^2 + (\mu\,|f-\mu f|^p)^{\frac{2}{p}} + |||f|||^2\right).$$

The theorem is proved. □

12 Weak Poincaré Inequalities for Gibbs Measures

In this section, we prove a weak Poincaré inequality for Gibbs measures with slowly decaying tails in the region of strong mixing property. Using this result, we obtain an estimate for the decay to equilibrium in L_2 for all Lipschitz cylinder functions with the same stretched exponential rate.

For $\Lambda \subset\subset \mathcal{R}$ we define the following conditional expectations (generalizing the E_i introduced in (11.9))

$$dE_\Lambda \equiv \delta_\omega \left(\frac{\displaystyle\int f e^{-u_\Lambda} d\nu_\Lambda}{\displaystyle\int e^{-u_\Lambda} d\nu_\Lambda} \right)$$

with some smooth function u_Λ and $\nu_\Lambda \equiv \otimes_{i\in\Lambda}\nu_i$, so that for $\Lambda_0 \subset \Lambda$ we have

$$dE_{\Lambda|\Sigma_{\Lambda_0}} \equiv \rho_{\Lambda_0} d\nu_{\Lambda_0}$$

with $\|\log \rho_{\Lambda_0}\|_\infty \leqslant \phi|\Lambda_0|$ with some numerical constant $\phi \in (0,\infty)$. Recall that, by definition, a Gibbs measure satisfies

$$\mu E_\Lambda(f) = \mu f$$

for each finite Λ and any integrable function f (cf., for example, [21]).

We begin from the following lemma.

Lemma 12.1 (perturbation lemma). *Suppose that ν_i, $i \in \mathbb{N}$, satisfy*

$$\nu_i |f - \nu_i f|^2 \leqslant \overline{\beta}(s)\nu_i |\nabla_i f|^2 + s\left(\nu_i |f - \nu_i f|^p\right)^{\frac{2}{p}}$$

with a function $\overline{\beta} : (0, s_0) \to \mathbb{R}^+$, for some $s_0 > 0$. Then the conditional expectation $E_i \equiv \frac{1}{Z_i} e^{-u_i} d\nu_i$ satisfies

$$E_i |f - E_i f|^2 \leqslant \widetilde{\beta}(s) E_i |\nabla_i f|^2 + s\left(\nu_i |f - \nu_i f|^p\right)^{\frac{2}{p}}$$

with

$$\widetilde{\beta}(s) \equiv e^{\operatorname{osc}(u_i)}\overline{\beta}(s e^{-\operatorname{osc}(u_i)})$$

for $s \in (0, s_0 e^{\operatorname{osc} u_i})$, where $\operatorname{osc}(u_i) \equiv \sup u_i - \inf u_i$.

If f depends on ω_Γ, with $\Gamma \cap \Lambda \equiv \Lambda_0$, then

$$E_\Lambda |f - E_\Lambda f|^2 \leqslant \widetilde{\beta}_\Lambda(s) E_\Lambda |\nabla_\Lambda f|^2 + s e^{2\phi|\Lambda_0|} \sum_{i\in\Lambda_0} E_\Lambda \left(\nu_i |f - \nu_i f|^p\right)^{\frac{2}{p}}$$

with $\widetilde{\beta}_\Lambda(s) \equiv e^{2\phi|\Lambda_0|}\overline{\beta}(e^{-2\phi|\Lambda_0|}s)$ for $s \in (0, s_0 e^{2\phi|\Lambda_0|})$.

Proof. We have

$$E_i |f - E_i f|^2 \leqslant E_i |f - \nu_i f|^2 = \int |f - \nu_i f|^2 \frac{1}{Z_i} e^{-u_i} d\nu_i$$
$$\leqslant \frac{1}{Z_i} e^{-\inf u_i} \nu_i |f - \nu_i f|^2 .$$

Hence, by the assumed inequality for ν_i, we get

$$E_i |f - E_i f|^2 \leqslant \frac{1}{Z_i} e^{-\inf u_i} \overline{\beta}(s) \nu_i |\nabla_i f|^2 + s \frac{1}{Z_i} e^{-\inf u_i} (\nu_i |f - \nu_i f|^p)^{\frac{2}{p}}$$
$$\leqslant e^{\sup u_i - \inf u_i} \overline{\beta}(s) E_i |\nabla_i f|^2 + s e^{\sup u_i - \inf u_i} (\nu_i |f - \nu_i f|^p)^{\frac{2}{p}} .$$

Hence

$$E_i |f - E_i f|^2 \leqslant \widetilde{\beta}(s) E_i |\nabla_i f|^2 + s (\nu_i |f - \nu_i f|^p)^{\frac{2}{p}}$$

with

$$\widetilde{\beta}(s) \equiv e^{\operatorname{osc} u_i} \overline{\beta}(s e^{-\operatorname{osc} u_i})$$

for $s \in (0, s_0 e^{\operatorname{osc} u_i})$, where $\operatorname{osc} u_i \equiv \sup u_i - \inf u_i$. Similarly,

$$E_\Lambda |f - E_\Lambda f|^2 \leqslant E_\Lambda |f - \nu_\Lambda f|^2 = \int |f - \nu_\Lambda f|^2 \frac{1}{Z_\Lambda} e^{-u_\Lambda} d\nu_\Lambda$$
$$\leqslant \frac{1}{Z_\Lambda} e^{-\inf u_\Lambda} \nu_\Lambda |f - \nu_\Lambda f|^2$$

and therefore (using the product property of Weak Poincaré inequality as in Proposition 11.2),

$$E_\Lambda |f - E_\Lambda f|^2 \leqslant \frac{1}{Z_\Lambda} e^{-\inf u_\Lambda} \overline{\beta}(s) \nu_\Lambda |\nabla_\Lambda f|^2$$
$$+ s \frac{1}{Z_\Lambda} e^{-\inf u_\Lambda} \sum_{i \in \Lambda} \nu_\Lambda (\nu_i |f - \nu_i f|^p)^{\frac{2}{p}}$$
$$\leqslant e^{\operatorname{osc}(u_\Lambda)} \overline{\beta}(s) E_\Lambda |\nabla_\Lambda f|^2 + s e^{\operatorname{osc}(u_\Lambda)} \sum_{i \in \Lambda} E_\Lambda (\nu_i |f - \nu_i f|^p)^{\frac{2}{p}}$$

with $\operatorname{osc}(u_\Lambda) \equiv \sup u_\Lambda - \inf u_\Lambda$. In the case where f depends on ω_Γ, with $\Gamma \cap \Lambda \equiv \Lambda_0$, one can stream-line the above arguments as follows. Noting that $dE_{\Lambda|\Sigma_{\Lambda_0}} \equiv \rho_{\Lambda_0} d\nu_{\Lambda_0}$ with $|| \log \rho_{\Lambda_0} ||_\infty \leqslant \phi |\Lambda_0|$ with some numerical constant $\phi \in (0, \infty)$, by similar arguments as above, we obtain

$$E_\Lambda \big|f - E_\Lambda f\big|^2 = E_{\Lambda|\Sigma_{\Lambda_0}} \big|f - E_{\Lambda|\Sigma_{\Lambda_0}} f\big|^2 \leqslant E_{\Lambda|\Sigma_{\Lambda_0}} \big|f - \nu_{\Lambda_0} f\big|^2$$
$$\leqslant e^{2\phi|\Lambda_0|} \overline{\beta}(s) E_\Lambda \big|\nabla_\Lambda f\big|^2 + s e^{2\phi|\Lambda_0|} \sum_{i \in \Lambda_0} E_\Lambda \big(\nu_i |f - \nu_i f|^p\big)^{\frac{2}{p}} .$$

The lemma is proved. □

Later on we consider a given set $\Lambda \subset\subset \mathcal{R}$ (for example, a ball of radius $L \in \mathbb{N}$ in a suitable metric of the graph) and write $\Lambda + j$ to denote a similar set around a point $j \in \mathcal{R}$. (If the graph $\mathcal{R}$ admits a structure of a linear space, this will coincide with a translation of Λ by the vector j.)

Lemma 12.2 (product property bis). *Suppose that*

$$E_\Lambda |f - E_\Lambda f|^2 \leqslant \widetilde{\beta}_\Lambda(s) E_\Lambda |\nabla_\Lambda f|^2 + s \sum_{i \in \Lambda_0} E_\Lambda \left(\nu_i |f - \nu_i f|^p\right)^{\frac{2}{p}}.$$

Let $\Gamma \equiv \bigcup_{l \in \mathbb{N}} \Lambda + j_l$ *with* j_l *such that* $\operatorname{dist}(\Lambda + j_l, \Lambda + j_{l'}) \geqslant 2R$, *for* $l \neq l'$. *Assume that* E_Λ *satisfy the following local Markov property:*

$$\forall f \in \Sigma_\Lambda \quad \Longrightarrow \quad E_\Lambda(f) \in \Sigma_{\Lambda_R},$$

where $\Lambda_R \equiv \{j \in \mathcal{R} : \operatorname{dist}(j, \Lambda) \leqslant R\}$ *for a given* $R \geqslant 1$. *Then*

$$E_\Gamma |f - E_\Gamma f|^2 \leqslant \widetilde{\beta}_\Lambda(s) E_\Gamma |\nabla_\Gamma f|^2 + s \sum_{i \in \Gamma} E_\Gamma \left(\nu_i |f - \nu_i f|^p\right)^{\frac{2}{p}}.$$

If $f \in \Sigma_\Theta$ *(i.e.,* $\Lambda_f \subseteq \Theta$*) and* $\Lambda_f \cap \Gamma \subset \bigcup_l \Lambda_0 + j_l$, *then the above inequality holds with* $\widetilde{\beta}(s) \equiv \widetilde{\beta}_{\Lambda_0}(s)$.

REMARK 12.1. The local Markov property is true when the interaction is of finite range R.

Because of the local Markov property, in our setup E_Γ acts as a product measure. Therefore, the proof is similar to the proof of Proposition 11.2 (product property).

Later on we consider a family of $\Gamma_k \subset \mathcal{R}$, $k \in \mathbb{N}$. Let $\boldsymbol{\Pi}_n(f) \equiv E_{\Gamma_n} \dots E_{\Gamma_1}(f)$. We note that, as in [38, 39], setting

$$f_0 \equiv f \quad \text{and} \quad f_n \equiv E_{\Gamma_n} f_{n-1} = \boldsymbol{\Pi}_n(f),$$

we have

$$\mu\left(f - \mu f\right)^2 = \sum_{n \in \mathbb{N}} \mu \, E_{\Gamma_n} \left(f_{n-1} - E_{\Gamma_n} f_{n-1}\right)^2.$$

Hence, by Lemma 12.2, we get

$$\begin{aligned} E_{\Gamma_n} \left(f_{n-1} - E_{\Gamma_n} f_{n-1}\right)^2 &\leqslant \widetilde{\beta}_\Lambda(s) E_{\Gamma_n} |\nabla_{\Gamma_n} f_{n-1}|^2 \\ &\quad + s \sum_{i \in \Gamma_n} E_{\Gamma_n} \left(\nu_i |f_{n-1} - \nu_i f_{n-1}|^p\right)^{\frac{2}{p}}. \end{aligned}$$

Now, we prove the following bound for expectation of terms involving the pth norms.

Lemma 12.3.

$$\sum_{i\in\Gamma_{n+1}} \mu\big(\nu_i|E_{\Gamma_n}F-\nu_iE_{\Gamma_n}F|^p\big)^{\frac{2}{p}} \leqslant \sum_{i\in\Gamma_{n+1},\ j} \eta_{ij}\mu\big(\nu_j|F-\nu_jF|^p\big)^{\frac{2}{p}},$$

where $\eta_{ii} \equiv 2e^{6\|u_i\|}$ *and*

$$\eta_{ij} \equiv 2\sum_{k(i):j\in\Lambda_{k(i)}} \big[\mathrm{osc}\ {}_{\Lambda_{k(i)}\cap\Lambda_F}\big(E_{\Lambda_{k(i)}\setminus\Lambda_F}(D_i)\big)\big]^2\cdot e^{4\phi(|\Lambda_{k(i)}\cap\Lambda_F|+\frac{1}{2})}|\Lambda_{k(i)}\cap\Lambda_F|$$

with

$$D_i \equiv \frac{\rho_{\Lambda_{k(i)}}(\omega_{\Lambda_{k(i)}}\bullet\omega_i\bullet\omega_{\mathcal{R}\setminus\Lambda_{k(i)}\cup\{i\}})}{\rho_{\Lambda_{k(i)}}(\omega_{\Lambda_{k(i)}}\bullet\widetilde{\omega}_i\bullet\omega_{\mathcal{R}\setminus\Lambda_{k(i)}\cup\{i\}})}-1,$$

where $\Lambda_{k(i)}\subset\Gamma_n$, $i\in\Gamma_{n+1}$, *is such that* $i\in\partial_R\Lambda_{k(i)}\equiv\{j\in\Lambda^c_{k(i)}:\mathrm{dist}\,(j,\Lambda_{k(i)})\leqslant R\}$.

Proof. First we note that for $i\in\Gamma_n\cap\Gamma_{n+1}$ the quantity $E_{\Gamma_n}F-\nu_iE_{\Gamma_n}F$ vanishes. For $i\in\Gamma_{n+1}$ let $\Lambda_{k(i)}\subset\Gamma_n$ be such that

$$i\in\partial_R\Lambda_{k(i)}\equiv\{j\in\Lambda^c_{k(i)}:\mathrm{dist}\,(j,\Lambda_{k(i)})\leqslant R\}.$$

Let $\widetilde{\Gamma}_n^{(i)}\equiv\Gamma_n\setminus\widetilde{\Lambda}^{(i)}$ and $\widetilde{\Lambda}^{(i)}\equiv\cup\Lambda_{k(i)}$. Note that $E_{\Gamma_n}F=E_{\widetilde{\Gamma}_n^{(i)}}E_{\widetilde{\Lambda}^{(i)}}F$ and $\nu_iE_{\widetilde{\Gamma}_n^{(i)}}=E_{\widetilde{\Gamma}_n^{(i)}}\nu_i$. Hence, using the Minkowski inequality for the $L_p(\nu_i)$ norm and the Schwartz inequality for $E_{\widetilde{\Gamma}_n^{(i)}}$, we get

$$\big(\nu_i|E_{\widetilde{\Gamma}_n^{(i)}}E_{\widetilde{\Lambda}^{(i)}}F-\nu_iE_{\widetilde{\Gamma}_n^{(i)}}E_{\widetilde{\Lambda}^{(i)}}F|^p\big)^{\frac{2}{p}}\leqslant E_{\widetilde{\Gamma}_n^{(i)}}\big(\nu_i|E_{\widetilde{\Lambda}^{(i)}}F-\nu_iE_{\widetilde{\Lambda}^{(i)}}F|^p\big)^{\frac{2}{p}}.$$

On the other hand, we have

$$\begin{aligned}\big(\nu_i\,|E_{\widetilde{\Lambda}^{(i)}}F-\nu_iE_{\widetilde{\Lambda}^{(i)}}F|^p\big)^{\frac{2}{p}}&\leqslant 2\,\big(\nu_i\,|E_{\widetilde{\Lambda}^{(i)}}(F-\nu_iF)|^p\big)^{\frac{2}{p}}\\&\quad+2\,\big(\nu_i\,|[E_{\widetilde{\Lambda}^{(i)}},\nu_i]F|^p\big)^{\frac{2}{p}},\end{aligned}\tag{12.1}$$

where

$$[E_{\widetilde{\Lambda}^{(i)}},\nu_i]F\equiv E_{\widetilde{\Lambda}^{(i)}}\nu_iF-\nu_iE_{\widetilde{\Lambda}^{(i)}}F.$$

The first term on the right-hand side of (12.1) can be bounded as follows:

$$\begin{aligned}2\big(\nu_i|E_{\widetilde{\Lambda}^{(i)}}(F-\nu_iF)|^p\big)^{\frac{2}{p}}&\leqslant 2e^{4\|u_i\|}\big(\nu_i|E'_{\widetilde{\Lambda}^{(i)}}|F-\nu_iF|^p\big)^{\frac{2}{p}}\\&\leqslant 2e^{6\|u_i\|}E_{\widetilde{\Lambda}^{(i)}}\,(\nu_i|F-\nu_iF|^p)^{\frac{2}{p}},\end{aligned}\tag{12.2}$$

where $E'_{\widetilde{\Lambda}^{(i)}}$ denotes an expectation with interaction u_i removed so it commutes with ν_i expectation, and we can apply the Minkowski inequality (for the $L_p(\nu_i)$ norm) and the Schwartz inequality for $E'_{\widetilde{\Lambda}^{(i)}}$ at the end inserting back the interaction u_i.

The second term on the right-hand side of (12.1) is estimated as follows. First we note that

$$2\big(\nu_i\big|[E_{\widetilde{\Lambda}^{(i)}},\nu_i]F\big|^p\big)^{\frac{2}{p}} \leqslant 2\sum \big(\nu_i\big|[E_{\Lambda_{k(i)}},\nu_i]F\big|^p\big)^{\frac{2}{p}}. \tag{12.3}$$

Next, we observe that

$$[E_{\Lambda_{k(i)}},\nu_i]F = \int \nu_i(d\widetilde{\omega}_i)\big\{E_{\Lambda_{k(i)}}\big(D_i(F - E_{\Lambda_{k(i)}}F)\big)\big\}, \tag{12.4}$$

where

$$D_i \equiv \frac{\rho_{\Lambda_{k(i)}}(\omega_{\Lambda_{k(i)}} \bullet \omega_i \bullet \omega_{\mathcal{R}\setminus\Lambda_{k(i)}\cup\{i\}})}{\rho_{\Lambda_{k(i)}}(\omega_{\Lambda_{k(i)}} \bullet \widetilde{\omega}_i \bullet \omega_{\mathcal{R}\setminus\Lambda_{k(i)}\cup\{i\}})} - 1.$$

If F depends on variables $\Lambda_{k(i)} \cap \Lambda_F$, then

$$\begin{aligned}
&\big|E_{\Lambda_{k(i)}}\big(D_i(F - E_{\Lambda_{k(i)}}F)\big)\big| \\
&\quad = \big|E_{\Lambda_{k(i)}}\big(E_{\Lambda_{k(i)}\setminus\Lambda_F}(D_i)(F - E_{\Lambda_{k(i)}}F)\big)\big| \\
&\quad \leqslant \operatorname{osc}\big(E_{\Lambda_{k(i)}\setminus\Lambda_F}(D_i)\big)\cdot E_{\Lambda_{k(i)}}\big|F - E_{\Lambda_{k(i)}}F\big|
\end{aligned} \tag{12.5}$$

with oscillation over variables indexed by points in $\Lambda_{k(i)} \setminus \Lambda_F$. Thus,

$$\begin{aligned}
&\big(\nu_i\big|[E_{\Lambda_{k(i)}},\nu_i]F\big|^p\big)^{\frac{2}{p}} \\
&\qquad \leqslant \big[\operatorname{osc}\big(E_{\Lambda_{k(i)}\setminus\Lambda_F}(D_i)\big)\big]^2\cdot \nu_i E_{\Lambda_{k(i)}}\big|F - E_{\Lambda_{k(i)}}F\big|^2.
\end{aligned} \tag{12.6}$$

Using (12.3)–(12.6), we arrive at

$$\begin{aligned}
&2\big(\nu_i\big|[E_{\widetilde{\Lambda}^{(i)}},\nu_i]F\big|^p\big)^{\frac{2}{p}} \\
&\qquad \leqslant 2\sum \big[\operatorname{osc}\big(E_{\Lambda_{k(i)}\setminus\Lambda_F}(D_i)\big)\big]^2\cdot \nu_i E_{\Lambda_{k(i)}}\big|F - E_{\Lambda_{k(i)}}F\big|^2.
\end{aligned} \tag{12.7}$$

Next, we note that

$$E_{\Lambda_{k(i)}}\big|F - E_{\Lambda_{k(i)}}F\big|^2 \leqslant e^{2\phi|\Lambda_{k(i)}\cap\Lambda_F|}\nu_{\Lambda_{k(i)}\cap\Lambda_F}\big|F - \nu_{\Lambda_{k(i)}\cap\Lambda_F}F\big|^2.$$

On the other hand, choosing a lexicographic order $\{j_l \in \Lambda\}_{l=1\ldots|\Lambda|}$, we have

$$\begin{aligned}
\nu_\Lambda|F - \nu_\Lambda F|^2 &= \nu_\Lambda\Big|\sum_{l=1\ldots|\Lambda|-1} \nu_{\Lambda_l}F - \nu_{\Lambda_{l+1}}F\Big|^2 \\
&\leqslant |\Lambda|\sum_{j\in\Lambda}\nu_\Lambda\nu_j|F - \nu_j F|^2 \leqslant |\Lambda|\sum_{j\in\Lambda}\nu_\Lambda\left(\nu_j|F - \nu_j F|^p\right)^{\frac{2}{p}}
\end{aligned} \tag{12.8}$$

with the convention that $\nu_{\Lambda_0} \equiv \mathbf{I}$ is the identity operator and $\Lambda_{l+1} = \Lambda_l \cup \{j_{l+1}\}$. Using this together with the previous inequality, we get the following assertion.

Lemma 12.4.

$$E_{\Lambda_{k(i)}}\left|F - E_{\Lambda_{k(i)}}F\right|^2 \leqslant e^{4\phi|\Lambda_{k(i)}\cap\Lambda_F|}|\Lambda_{k(i)}\cap\Lambda_F| \sum_{j\in\Lambda_{k(i)}} E_{\Lambda_{k(i)}}\left(\nu_j|F-\nu_jF|^p\right)^{\frac{2}{p}}.$$

Combining with (12.7), we arrive at

$$\begin{aligned}2\left(\nu_i\left|[E_{\widetilde{\Lambda}^{(i)}},\nu_i]F\right|^p\right)^{\frac{2}{p}} \leqslant 2\sum_{k(i)}\left[\operatorname{osc}_{\Lambda_{k(i)}\cap\Lambda_F}\left(E_{\Lambda_{k(i)}\setminus\Lambda_F}(D_i)\right)\right]^2 \\ \cdot e^{4\phi[|\Lambda_{k(i)}\cap\Lambda_F|+\frac{1}{2}}|\Lambda_{k(i)}\cap\Lambda_F| \\ \cdot\sum_{j\in\Lambda} E_iE_{\Lambda_{k(i)}}\left(\nu_j|F-\nu_jF|^p\right)^{\frac{2}{p}}.\end{aligned} \tag{12.9}$$

This ends the estimates for the second term on the right-hand side of (12.1).

Using (12.2) and (12.9), we find

$$\begin{aligned}\left(\nu_i\left|E_{\widetilde{\Gamma}_n^{(i)}}E_{\widetilde{\Lambda}^{(i)}}F - \nu_iE_{\widetilde{\Gamma}_n^{(i)}}E_{\widetilde{\Lambda}^{(i)}}F\right|^p\right)^{\frac{2}{p}} \leqslant 2e^{6\|u_i\|}E_{\widetilde{\Lambda}^{(i)}}\left(\nu_i\left|F-\nu_iF\right|^p\right)^{\frac{2}{p}} \\ +2\sum_{k(i)}\left[\operatorname{osc}_{\Lambda_{k(i)}\cap\Lambda_F}\left(E_{\Lambda_{k(i)}\setminus\Lambda_F}(D_i)\right)\right]^2\cdot e^{4\phi[|\Lambda_{k(i)}\cap\Lambda_F|+\frac{1}{2}]}|\Lambda_{k(i)}\cap\Lambda_F| \\ \cdot\sum_{j\in\Lambda_{k(i)}} E_{\widetilde{\Gamma}_n^{(i)}}E_iE_{\Lambda_{k(i)}}\left(\nu_j|F-\nu_jF|^p\right)^{\frac{2}{p}}.\end{aligned}$$

From this we conclude that

$$\begin{aligned}\sum_{i\in\Gamma_{n+1}}\mu\left(\nu_i\left|E_{\Gamma_n}F-\nu_iE_{\Gamma_n}F\right|^p\right)^{\frac{2}{p}} \leqslant \sum_{i\in\Gamma_{n+1}} 2e^{6\|u_i\|}\mu\left(\nu_i\left|F-\nu_iF\right|^p\right)^{\frac{2}{p}} \\ +2\sum_{i\in\Gamma_{n+1}}\sum_{k(i)}\left[\operatorname{osc}_{\Lambda_{k(i)}\cap\Lambda_F}\left(E_{\Lambda_{k(i)}\setminus\Lambda_F}(D_i)\right)\right]^2 \\ \cdot e^{4\phi[|\Lambda_{k(i)}\cap\Lambda_F|+\frac{1}{2}}|\Lambda_{k(i)}\cap\Lambda_F|\cdot\sum_{j\in\Lambda_{k(i)}}\mu\left(\nu_j|F-\nu_jF|^p\right)^{\frac{2}{p}}.\end{aligned}$$

Lemma 12.3 is proved. □

Applying iteratively Lemma 12.3, we arrive at the following result.

Proposition 12.5. *Suppose that Γ_n, $n\in\mathbb{N}$, is a periodic sequence of period N such that $\bigcup_{l=1,\dots,N}\Gamma_l=\mathcal{R}$. Then there exists a constant $C\in(0,\infty)$ such that for any $p\in(2,\infty)$ and any $1\leqslant n\leqslant N-1$*

$$\sum_{i\in\Gamma_{n+1}} \mu\left(\nu_i|\boldsymbol{\Pi}_n f - \nu_i \boldsymbol{\Pi}_n f|^p\right)^{\frac{2}{p}}$$
$$\leqslant \sum_{i\in\Gamma_{n+1}} \sum_{j_n\in\Gamma_n\setminus\Gamma_{n+1},\ldots,j_1\in\Gamma_1\setminus\Gamma_2} \eta_{ij_n}\eta_{j_nj_{n-1}}\ldots\eta_{j_2j_1}\mu\left(\nu_j|f-\nu_j f|^p\right)^{\frac{2}{p}}$$
$$\leqslant C\sum_{i\in\mathcal{R}} \mu\left(\nu_j|f-\nu_j f|^p\right)^{\frac{2}{p}}.$$

Moreover, for $n = N$

$$\sum_{i\in\Gamma_{n+1}} \mu\left(\nu_i|\boldsymbol{\Pi}_N F - \nu_i \boldsymbol{\Pi}_N F|^p\right)^{\frac{2}{p}} \leqslant \lambda \sum_{j\in\mathcal{R}} \mu\left(\nu_j|F-\nu_j F|^p\right)^{\frac{2}{p}}$$

with

$$\lambda \equiv \sup_{j\in\mathcal{R}} \sum_{i\in\Gamma_{n+1}} \sum_{j_n\in\Gamma_n\setminus\Gamma_{n+1},\ldots,j_1\in\Gamma_1\setminus\Gamma_2} \eta_{ij_n}\eta_{j_nj_{n-1}}\ldots\eta_{j_2j}. \tag{12.10}$$

Therefore, for any $n\in\mathbb{N}$

$$\sum_{i\in\mathcal{R}} \mu\left(\nu_i|\boldsymbol{\Pi}_n f - \nu_i \boldsymbol{\Pi}_n f|^p\right)^{\frac{2}{p}} \leqslant C\lambda^{[\frac{n}{N}]} \sum_{i\in\mathcal{R}} \mu\left(\nu_j|f-\nu_j f|^p\right)^{\frac{2}{p}},$$

where $[n/N]$ *is the integer part of* n/N, *with some constant* $C\in(0,\infty)$.

REMARK 12.2. Because of our assumption that conditional expectations satisfy local Markov property, λ is defined by a finite sum and therefore is finite.

Lemma 12.6. *There exists a constant* $\gamma_0 \in (0,\infty)$ *such that for any* $p\in(2,\infty)$ *and* $s\in(0,s_0)$

$$\mu\left|\nabla_i E_{\Gamma_n} F\right|^2 \leqslant \gamma_0 \mu\left|\nabla_i F\right|^2 + \widetilde{\beta}(s)\sum_j \boldsymbol{\eta}_{ij}\mu|\nabla_j F|^2 + s\sum_j \boldsymbol{\eta}_{ij}\mu\left(\nu_j|F-\nu_j F|^p\right)^{\frac{2}{p}}$$

with

$$\boldsymbol{\eta}_{ij} \equiv \sum_{k(i):j\in\Lambda_{k(i)}} \gamma_0 \left\| \operatorname{osc}_{\Lambda_{k(i)}\cap\Lambda_F}\left(E_{\Lambda_{k(i)}\setminus\Lambda_F}(\nabla_i U_{\Lambda_{k(i)}})\right)\right\|_\infty^2 \cdot e^{4\phi|\Lambda_{k(i)}\cap\Lambda_F|}|\Lambda_{k(i)}\cap\Lambda_F|$$

defined for $j\in\widetilde{\Lambda}^{(i)}$ *and zero otherwise.*

Proof. For $i\in\Gamma_{n+1}$ let $\Lambda_{k(i)}\subset\Gamma_n$ be such that

$$i\in\partial_R\Lambda_{k(i)} \equiv \{j\in\Lambda^c_{k(i)} : \operatorname{dist}(j,\Lambda_{k(i)})\leqslant R\}.$$

Let $\widetilde{\Gamma}_n^{(i)} \equiv \Gamma_n \setminus \widetilde{\Lambda}^{(i)}$, where $\widetilde{\Lambda}^{(i)} \equiv \cup \Lambda_{k(i)}$. Note that

$$E_{\Gamma_n} F = E_{\widetilde{\Gamma}_n^{(i)}} E_{\widetilde{\Lambda}^{(i)}} F, \quad \nabla_i E_{\widetilde{\Gamma}_n^{(i)}} = E_{\widetilde{\Gamma}_n^{(i)}} \nabla_i.$$

Hence

$$\nabla_i E_{\Gamma_n} F = \nabla_i E_{\widetilde{\Gamma}_n^{(i)}} E_{\widetilde{\Lambda}^{(i)}} F = E_{\widetilde{\Gamma}_n^{(i)}} \nabla_i E_{\widetilde{\Lambda}^{(i)}} F.$$

On the other hand, we have

$$\nabla_i E_{\widetilde{\Lambda}^{(i)}} F = E_{\widetilde{\Lambda}^{(i)}} \nabla_i F + [\nabla_i, E_{\widetilde{\Lambda}^{(i)}}] F,$$

where $[\nabla_i, E_{\widetilde{\Lambda}^{(i)}}] F \equiv \nabla_i E_{\widetilde{\Lambda}^{(i)}} F - E_{\widetilde{\Lambda}^{(i)}} \nabla_i F$. We note that

$$[\nabla_i, E_{\widetilde{\Lambda}^{(i)}}] F = \sum_{k(i)} E_{\widetilde{\Lambda}^{(i)}} \big(E_{\Lambda_{k(i)}} \big(F; \nabla_i U_{\Lambda_{k(i)}} \big) \big),$$

where

$$E_{\Lambda_{k(i)}} \big(F; \nabla_i U_{\Lambda_{k(i)}} \big) \equiv E_{\Lambda_{k(i)}} \big(F \cdot \nabla_i U_{\Lambda_{k(i)}} \big) - E_{\Lambda_{k(i)}} (F) E_{\Lambda_{k(i)}} \big(\nabla_i U_{\Lambda_{k(i)}} \big).$$

If F depends on variables in $\Lambda_{k(i)} \cap \Lambda_F$, we have

$$\begin{aligned} \big| E_{\Lambda_{k(i)}} \big(F; \nabla_i U_{\Lambda_{k(i)}} \big) \big| &= \big| E_{\Lambda_{k(i)}} \big((F - E_{\Lambda_{k(i)}} F) \cdot E_{\Lambda_{k(i)} \setminus \Lambda_F} (\nabla_i U_{\Lambda_{k(i)}}) \big) \big| \\ &\leqslant \operatorname{osc}_{\Lambda_{k(i)} \cap \Lambda_F} \big(E_{\Lambda_{k(i)} \setminus \Lambda_F} (\nabla_i U_{\Lambda_{k(i)}}) \big) \cdot E_{\Lambda_{k(i)}} \big| F - E_{\Lambda_{k(i)}} F \big|. \end{aligned}$$

Thus, in this case,

$$\begin{aligned} \big| \nabla_i E_{\Gamma_n} F \big| &\leqslant E_{\Gamma_n} \big| \nabla_i F \big| + E_{\widetilde{\Gamma}_n^{(i)}} \big| [\nabla_i, E_{\widetilde{\Lambda}^{(i)}}] F \big| \\ &\leqslant E_{\Gamma_n} \big| \nabla_i F \big| + \sum_{k(i)} E_{\Gamma_n} \big| \big(E_{\Lambda_{k(i)}} \big(F; \nabla_i U_{\Lambda_{k(i)}} \big) \big) \big| \\ &\leqslant E_{\Gamma_n} \big| \nabla_i F \big| + \sum_{k(i)} E_{\Gamma_n} \operatorname{osc}_{\Lambda_{k(i)} \cap \Lambda_F} \big(E_{\Lambda_{k(i)} \setminus \Lambda_F} (\nabla_i U_{\Lambda_{k(i)}}) \big) \\ &\quad \cdot E_{\Lambda_{k(i)}} \big| F - E_{\Lambda_{k(i)}} F \big|. \end{aligned}$$

Therefore, there exists a constant $\gamma_0 \in (0, \infty)$ depending only on the number of $k(i)$'s such that

$$\begin{aligned} \mu \big| \nabla_i E_{\Gamma_n} F \big|^2 \leqslant \gamma_0 \mu \big| \nabla_i F \big|^2 + \sum_{k(i)} \gamma_0 \big\| \operatorname{osc}_{\Lambda_{k(i)} \cap \Lambda_F} \big(E_{\Lambda_{k(i)} \setminus \Lambda_F} (\nabla_i U_{\Lambda_{k(i)}}) \big) \big\|_\infty^2 \\ \cdot \mu \big(E_{\Lambda_{k(i)}} \big| F - E_{\Lambda_{k(i)}} F \big|^2 \big). \end{aligned}$$

Using Lemma 12.1, we obtain

$$\mu\left|\nabla_i E_{\Gamma_n} F\right|^2 \leqslant \gamma_0 \mu\left|\nabla_i F\right|^2 + \widetilde{\beta}(s) \sum_j \boldsymbol{\eta}_{ij} \mu |\nabla_j F|^2 + s \sum_j \boldsymbol{\eta}_{ij} \mu\big(\nu_j |F - \nu_j F|^p\big)^{\frac{2}{p}},$$

where $\boldsymbol{\eta}_{ij}$, for $i \in \mathbb{Z}^d \setminus \widetilde{\Lambda}^{(i)}$, $\operatorname{dist}(i, \widetilde{\Lambda}^{(i)}) \leqslant R$, $j \in \widetilde{\Lambda}^{(i)}$, are defined by

$$\boldsymbol{\eta}_{ij} \equiv \sum_{k(i): j \in \Lambda_{k(i)}} \gamma_0 \left\| \operatorname{osc}_{\Lambda_{k(i)} \cap \Lambda_F} \left(E_{\Lambda_{k(i)} \setminus \Lambda_F} (\nabla_i U_{\Lambda_{k(i)}})\right) \right\|_\infty^2 \cdot e^{4\phi |\Lambda_{k(i)} \cap \Lambda_F|} |\Lambda_{k(i)} \cap \Lambda_F|.$$

The lemma is proved. □

Proposition 12.7. *Suppose that Γ_n, $n \in \mathbb{N}$, is a periodic sequence of period N such that $\bigcup_{l=1\ldots N} \Gamma_l = \mathcal{R}$ and so for any $i \in \mathcal{R}$ there exists $1 \leqslant l(i) \leqslant N$ for which $\nabla_i E_{\Gamma_{l(i)}} f = 0$. Then for any $p \in (2, \infty)$ and any $1 \leqslant n \leqslant N - 1$*

$$\mu \left|\nabla_i E_{\Gamma_n} \ldots E_{\Gamma_1} F\right|^2 \leqslant X(s) \mu \left|\nabla_i F\right|^2 + s Z(s) \sum_j \mu \left(\nu_j |F - \nu_j F|^p\right)^{\frac{2}{p}}$$

with $X(s) \equiv X(1 + \widetilde{\beta}(s)^{N-1})$ and $Z(s) \equiv Z(1 + \widetilde{\beta}(s)^{N-1})$ with some constants $X, Z > 0$. Moreover, for $n \geqslant N$

$$\mu \left|\nabla_i \boldsymbol{\Pi}_n f\right|^2 \leqslant s Z(s) \lambda^{\frac{n}{N}} \sum_{j \in \mathcal{R}} \mu \left(\nu_j |f - \nu_j f|^p\right)^{\frac{2}{p}}.$$

Proof. For $n \leqslant N$, by Lemma 12.6, we have

$$\begin{aligned}
\mu \left|\nabla_{\Gamma_{n+1}} \boldsymbol{\Pi}_n F\right|^2 &= \sum_{i \in \Gamma_{n+1}} \mu \left|\nabla_i \boldsymbol{\Pi}_n F\right|^2 \\
&\leqslant \sum_{i, j_n} \mathbf{1}_{\{i \in \Gamma_{n+1} \setminus \Gamma_n\}} \boldsymbol{A}^{(n+1,n)}(s) \mathbf{1}_{\{j_n \in \Gamma_n \setminus \Gamma_{n-1}\}} \mu \left|\nabla_{j_n} \boldsymbol{\Pi}_{n-1} F\right|^2 \\
&+ s \sum_{i, j_n} \mathbf{1}_{\{i \in \Gamma_{n+1} \setminus \Gamma_n\}} \boldsymbol{\eta}^{(n+1,n)} \mathbf{1}_{\{j_n \in \Gamma_n \setminus \Gamma_{n-1}\}} \mu \left(\nu_{j_n} |\boldsymbol{\Pi}_{n-1} F - \nu_{j_n} \boldsymbol{\Pi}_{n-1} F|^p\right)^{\frac{2}{p}},
\end{aligned}$$

where

$$\boldsymbol{A}^{(n+1,n)}(s)_{ij} \equiv \left(a \boldsymbol{I} + \boldsymbol{\eta}^{(n+1,n)}\right)_{ij} \equiv \left(\gamma_0 \delta_{ij} + \widetilde{\beta}(s) \eta_{ij}^{(n+1,n)}\right),$$

where $\mathbf{1}_{\{j_n \in \Gamma_n \setminus \Gamma_{n-1}\}}$ denotes the characteristic function of the set $\Gamma_n \setminus \Gamma_{n-1}$ and $\eta_{ij}^{(n+1,n)}$ is provided by Lemma 12.6 (with $i \in \Gamma_{n+1} \setminus \Gamma_n$ and $j \in \Gamma_n$). By induction, we arrive at the bound

$$\mu \left|\nabla_{\Gamma_{n+1}} \boldsymbol{\Pi}_n F\right|^2 \leqslant \sum_{i,j} \boldsymbol{\Theta}_{ij}^{(n,1)} \mu \left|\nabla_j F\right|^2 + s \sum_{k=1\ldots n} \sum_{i,j} \boldsymbol{\Upsilon}_{ij}^{(k)} \mu \left(\nu_j \left|\boldsymbol{\Pi}_{n-k} F - \nu_j \boldsymbol{\Pi}_{n-k} F\right|^p\right)^{\frac{2}{p}},$$

where

$$\begin{aligned}\boldsymbol{\Theta}&^{(n,m)}\\ &\equiv \mathbf{1}_{\{i\in\Gamma_{n+1}\setminus\Gamma_n\}} \boldsymbol{A}^{(n+1,n)}(s) \mathbf{1}_{\{j_n\in\Gamma_n\setminus\Gamma_{n-1}\}} \boldsymbol{A}^{(n,n-1)}(s) \mathbf{1}_{\{j_{n-1}\in\Gamma_{n-1}\setminus\Gamma_{n-2}\}}\\ &\ldots \mathbf{1}_{\{j_{m+1}\in\Gamma_{m+1}\setminus\Gamma_m\}} \boldsymbol{A}^{(m+1,m)}(s) \mathbf{1}_{\{j_m\in\Gamma_m\setminus\Gamma_{m-1}\}}\end{aligned}$$

with the convention that $\Gamma_0 \equiv \varnothing$, and

$$\boldsymbol{\Upsilon}^{(k)} \equiv \boldsymbol{\Theta}^{(n,n-k)} \cdot \boldsymbol{\eta}^{(n+1-k,n-k)} \mathbf{1}_{\{j\in\Gamma_{n-k}\setminus\Gamma_{n-k-1}\}}$$

with the convention that $\boldsymbol{\Theta}^{(n,n)} \equiv \mathbf{1}_{\{\Gamma_{n+1}\setminus\Gamma_n\}}$.

Using Proposition 12.5, we can simplify the above estimate as follows:

$$\mu \left|\nabla_{\Gamma_{n+1}} \boldsymbol{\Pi}_n F\right|^2 \leqslant \sum_{i,j} \boldsymbol{\Theta}_{ij}^{(n,1)} \mu \left|\nabla_j F\right|^2 + s \sum_{i,j} \boldsymbol{\Upsilon}_{ij}^{(n)} \mu \left(\nu_j \left|F - \nu_j F\right|^p\right)^{\frac{2}{p}},$$

where

$$\boldsymbol{\Upsilon}^{(n)} \equiv \sum_{k=1\ldots n} \boldsymbol{\Upsilon}^{(k)} \mathbf{1}_{\{j\in\Gamma_{n-k}\setminus\Gamma_{n-k-1}\}} \boldsymbol{\eta}^{(n-k,n-k-1)} \ldots \mathbf{1}_{\{j\in\Gamma_2\setminus\Gamma_1\}} \boldsymbol{\eta}^{(2,1)}.$$

We note that there is a constant $X \in (0,\infty)$ such that for $n \leqslant N-1$

$$\sup_j \sum_i \boldsymbol{\Theta}_{ij}^{(n,1)} \leqslant X(1+\widetilde{\beta}(s)^{N-1}).$$

If we assume that for each i there is an $l \leqslant N$ such that $\nabla_i E_{\Gamma_l} = \mathbf{0}$, then we get

$$\mu \left|\nabla_{\Gamma_{N+1}} \boldsymbol{\Pi}_N F\right|^2 \leqslant s \sum_{i,j} \boldsymbol{\Upsilon}_{ij}^{(N)} \mu \left(\nu_j \left|F - \nu_j F\right|^p\right)^{\frac{2}{p}}$$

Since

$$\sup_j \sum_i \boldsymbol{\Upsilon}_{ij}^{(n)} \leqslant C'(1+\widetilde{\beta}(s)^{N-1})$$

with some constant $C' \in (0,\infty)$ independent of $n \leqslant N$, we get

$$\mu \left|\nabla_{\Gamma_{N+1}} \boldsymbol{\Pi}_N F\right|^2 \leqslant s\, C'(1+\widetilde{\beta}(s)^{N-1}) \sum_j \mu \left(\nu_j \left|F - \nu_j F\right|^p\right)^{\frac{2}{p}}.$$

As a consequence for $n \geqslant N$, setting $F \equiv \boldsymbol{\Pi}_{n-N} f$ and using Proposition 12.5, we conclude that

$$\mu \left|\nabla_{\Gamma_{n+1}} \boldsymbol{\Pi}_n f\right|^2 \leqslant s\, Z(1+\widetilde{\beta}(s)^{N-1})\lambda^{[\frac{n}{N}]} \sum_{j\in\mathcal{R}} \mu\left(\nu_j \left|f-\nu_j f\right|^p\right)^{\frac{2}{p}},$$

where $Z \equiv CC'$. □

Theorem 12.8. *Suppose that Γ_n, $n \in \mathbb{N}$, is a periodic sequence of period N such that $\bigcup_{l=1\ldots N} \Gamma_l = \mathcal{R}$ and so for any $i \in \mathcal{R}$ there exists $1 \leqslant l(i) \leqslant N$ for which $\nabla_i E_{\Gamma_{l(i)}} f = 0$. Suppose that the parameter λ introduced in Proposition 12.5 satisfies $\lambda \in (0,1)$. Then for any $p \in (2,\infty)$*

$$\mu\left|f-\mu f\right|^2 \leqslant \beta(s)\mu\left|\nabla f\right|^2 + s\sum_{i\in\mathcal{R}} \mu\left(\nu_i|f-\nu_i f|^p\right)^{\frac{2}{p}}$$

with

$$\beta(s) \equiv \widetilde{\beta}(\vartheta^{-1}(s))X(\vartheta^{-1}(s))N \equiv X(\widetilde{\beta}(\vartheta^{-1}(s)) + \widetilde{\beta}(\vartheta^{-1}(s))^N)N$$

for $s \in (0, \vartheta(s_0))$, where $\vartheta(s) \equiv sN(C+\widetilde{\beta}(s)Z(s))(1-\lambda)^{-1}$ with $Z(s) \equiv Z(1+\widetilde{\beta}(s)^{N-1})$, with some constants $X, Z, C > 0$.

Proof. By Lemma 12.2, we have

$$\begin{aligned}\mu E_{\Gamma_{n+1}} \left|\boldsymbol{\Pi}_n f - E_{\Gamma_{n+1}}\boldsymbol{\Pi}_n f\right|^2 \leqslant{}& \widetilde{\beta}(s)\mu\left|\nabla_{\Gamma_{n+1}}\boldsymbol{\Pi}_n f\right|^2 \\ &+ s\sum_{i\in\Gamma_{n+1}} \mu\left(\nu_i|\boldsymbol{\Pi}_n f - \nu_i\boldsymbol{\Pi}_n f|^p\right)^{\frac{2}{p}}.\end{aligned}$$

Hence, by Propositions 12.5, 12.7 and Lemma 12.6, for $n \leqslant N-1$ we have

$$\begin{aligned}\mu E_{\Gamma_{n+1}} \left|\boldsymbol{\Pi}_n f - E_{\Gamma_{n+1}}\boldsymbol{\Pi}_n f\right|^2 \leqslant{}& \widetilde{\beta}(s)X(s)\mu\left|\nabla f\right|^2 \\ &+ s(C+\widetilde{\beta}(s)Z(s))\sum_{i\in\mathcal{R}} \mu\left(\nu_i|f-\nu_i f|^p\right)^{\frac{2}{p}},\end{aligned}$$

while for $n \geqslant N$ we have

$$\mu E_{\Gamma_{n+1}} \left|\boldsymbol{\Pi}_n f - E_{\Gamma_{n+1}}\boldsymbol{\Pi}_n f\right|^2 \leqslant s(C+\widetilde{\beta}(s)Z(s))\lambda^{[\frac{n}{N}]}\sum_{i\in\mathcal{R}} \mu\left(\nu_i|f-\nu_i f|^p\right)^{\frac{2}{p}}.$$

Thus, provided that $\lambda \in (0,1)$, we arrive at

$$\begin{aligned}&\mu\left|f-\mu f\right|^2 = \sum_{n\in\mathbb{Z}^+} \mu E_{\Gamma_{n+1}} \left|\boldsymbol{\Pi}_n f - E_{\Gamma_{n+1}}\boldsymbol{\Pi}_n f\right|^2 \\ &\leqslant \widetilde{\beta}(s)X(s)N\mu\left|\nabla f\right|^2 + sN(C+\widetilde{\beta}(s)Z(s))(1-\lambda)^{-1}\sum_{i\in\mathcal{R}} \mu\left(\nu_i|f-\nu_i f|^p\right)^{\frac{2}{p}}.\end{aligned}$$

The theorem is proved. □

Examples. Suppose that $\mathcal{R} = \mathbb{Z}^d$, $d \in \mathbb{N}$. Then the corresponding covering Γ_n, $n = 1 \dots 2^d$, was introduced as a collection of suitable translates a sufficiently large cube Λ_0 for $d = 1$ in [40] and for general d in [36]. In the case where the local specification E_Λ, $\Lambda \subset\subset \mathbb{Z}^d$ satisfies the *strong mixing condition* (for cubes)

$$|E_\Lambda(f;g)| \leqslant \text{Const } |||f||| \cdot |||g|||e^{-M \text{ dist}(\Lambda_f,\Lambda_g)}$$

with some constant $M \in (0,\infty)$ independent of size of the cube, one shows (cf., for example, [40, 36, 21]) that, starting with a sufficiently large cube Λ_0, one can achieve $\lambda \in (0,1)$. In our case, the strong mixing condition holds at least for finite range sufficiently small interactions u_Λ.

In our setup, by Corollary 8.6 and Lemma 12.1, $\overline{\beta}(s) \equiv C_0(\log(1/s))^\delta$ with some positive C_0 and $\delta \in (0,\infty)$ for all sufficiently small $s > 0$. Hence

$$\beta(s) = C(\log(1/s))^{N\delta}$$

with some positive constant C for all sufficiently small $s > 0$. Thus, the above considerations (cf. Theorem 12.8) apply and we have the following result.

Theorem 12.9. *Let μ be a Gibbs measure on $\mathbb{R}^{\mathbb{Z}^d}$ corresponding to the reference product measure $\mu_0 \equiv \nu_0^{\otimes \mathbb{Z}^d}$, where the probability measure $d\nu_0 \equiv \frac{1}{Z}\exp\{-V\}dx$ on real line is defined with $V \equiv \varsigma(1+x^2)^{\frac{\alpha}{2}}$, with $0 < \alpha \leqslant 1$, $\varsigma \in (0,\infty)$, and a local finite range smooth interaction u_Λ, $\Lambda \subset\subset \mathbb{Z}^d$, which is sufficiently small or more generally such that the Strong Mixing Condition holds.*

Then μ satisfies the weak Poincaré inequality

$$\mu|f - \mu f|^2 \leqslant \beta(s)\mu|\nabla f|^2 + sA_p(f)^2$$

with $\beta(s) \equiv C(\log(1/s))^{N\delta}$ with some positive constant C and $N = 2^d$ for all $s \in (0,\overline{s})$ for some $\overline{s} > 0$. Hence there exists $\varepsilon \in (0,1)$ and constants $c, H \in (0,\infty)$ such that the semigroup $P_t \equiv e^{t\nabla^*\nabla}$ (with the generator corresponding to the Dirichlet form $\mu|\nabla f|^2$) satisfies

$$\mu(P_t f - \mu f)^2 \leqslant e^{-ct^\varepsilon} H\left(\mu(f-\mu f)^2 + (\mu|f-\mu f|^p)^{\frac{2}{p}} + |||f|||^2\right)$$

for each cylinder function for which the right hand side is well defined (with a constant $H \in (0,\infty)$ dependent on Λ_f).

Acknowledgments. Part of this work was done while one of us (B.Z.) was holding the *Chaire de Excellence Pierre de Fermat*, at LSP, UPS Toulouse, sponsored by the ADERMIP. Research of S.B. was supported in part by the NSF (grant DMS-0706866). Both authors would like to thank all members of LSP, and in particular D. Bakry and M. Ledoux, for discussions and hospitality in Toulouse.

References

1. Aida, S., Stroock, D.: Moment estimates derived from Poincaré and logarithmic Sobolev inequalities. Math. Research Letters, **1**, 75–86 (1994)
2. Alon, N., Milman, V.D.: λ_1, isoperimetric inequalities for graphs, and superconcentrators. J. Comb. Theory, Ser. B **38**, 75–86 (1985)
3. Barthe, F.: Log-concave and spherical models in isoperimetry. Geom. Funct. Anal. **12**, no. 1, 32–55 (2002)
4. Barthe, F., Cattiaux, P., Roberto, C.: Concentration for independent random variables with heavy tails. AMRX Appl. Math. Res. Express, no. 2, 39–60 (2005)
5. Bertini, L., Zegarlinski, B.: Coercive inequalities for Gibbs measures. J. Funct. Anal. **162**, no. 2, 257–286 (1999)
6. Bertini, L., Zegarlinski, B.: Coercive inequalities for Kawasaki dynamics. The product case. Markov Process. Related Fields **5**, no. 2, 125–162 (1999)
7. Bobkov, S.G.: Remarks on the Gromov–Milman inequality. Vestn. Syktyvkar Univ. Ser. 1 **3**, 15–22 (1999)
8. Bobkov, S.G.: Isoperimetric and analytic inequalities for log-concave probability distributions. Ann. Probab. **27**, no. 4, 1903–1921 (1999)
9. Bobkov, S.G.: Large deviations and isoperimetry over convex probability measures. Electr. J. Probab. **12**, 1072–1100 (2007)
10. Bobkov, S.G., Houdré, C.: Some connections between isoperimetric and Sobolev type inequalities. Memoirs Am. Math. Soc. **129**, no. 616 (1997)
11. Borell, C.: Convex measures on locally convex spaces. Ark. Math. **12**, 239–252 (1974)
12. Borell, C.: Convex set functions in *d*-space. Period. Math. Hungar. **6**, no. 2, 111–136 (1975)
13. Borovkov, A.A., Utev, S.A.: On an inequality and a characterization of the normal distribution connected with it. Probab. Theor. Appl. **28**, 209–218 (1983)
14. Brascamp, H.J., Lieb, E.H.: On extensions of the Brunn–Minkowski and Prekopa–Leindler theorems, including inequalities for log concave functions, and with an application to the diffusion equation. J. Funct. Anal. **22**, no. 4, 366–389 (1976)
15. Cattiaux, P., Gentil, I., Guillin, A.: Weak logarithmic Sobolev inequalities and entropic convergence. Probab. Theory Rel. Fields **139**, 563–603 (2007)
16. Davies, E.B.: Heat Kernels and Spectral Theory. Cambridge Univ. Press, Cambridge (1989)
17. Davies, E.B., Simon B.: Ultracontractivity and the heat kernel for Schrödinger operators and Dirichlet Laplacians. J. Funct. Anal. **59**, 335–395 (1984)
18. Deuschel, J.D.: Algebraic L^2 decay of attractive critical processes on the lattice. Ann. Probab. **22**, 335–395 (1991)
19. Grigor'yan, A.: Isoperimetric inequalities and capacities on Riemannian manifolds. In: The Maz'ya Anniversary Collection 1 (Rostock, 1998), pp. 139–153. Birkhäuser (1999)

20. Gromov, M.L., Milman, V.D.: A topological application of the isoperimetric inequality. Am. J. Math. **105**, 843–854 (1983)
21. Guionnet, A., Zegarlinski, B.: Lectures on logarithmic Sobolev inequalities. In: Séminaire de Probabilités, Vol. 36 Lect. Notes Math. **1801**, pp. 1–134. Springer, Berlin (2003)
22. Herbst, I.W.: On canonical quantum field theories. J. Math. Phys. **17**, 1210–1221 (1976)
23. Kannan, R., Lovász, L., Simonovits, M.: Isoperimetric problems for convex bodies and a localization lemma. Discrete Comp. Geom. **13**, 541–559 (1995)
24. Ledoux, M.: Concentration of measure and logarithmic Sobolev inequalities. In: Lect. Notes Math. **1709**, pp. 120–216. Springer, Berlin (1999)
25. Ledoux, M.: The Concentration of Measure Phenomenon. Am. Math. Soc., Providence, RI (2001)
26. Leindler, L.: On a certain converse of Hölder's inequality II. Acta Sci. Math. Szeged **33**, 217–223 (1972)
27. Liggett, T.M.: L_2 rates of convergence for attractive reversible nearest particle systems: the critical case. Ann. Probab. **19**, no. 3, 935–959 (1991)
28. Maz'ya, V.G.: On certain integral inequalities for functions of many variables (Russian). Probl. Mat. Anal. **3**, 33–68 (1972); English transl.: J. Math. Sci., New York **1**, 205–234 (1973)
29. Maz'ya, V.G.: Sobolev Spaces. Springer, Berlin etc. (1985)
30. Maz'ya, V.: Conductor inequalities and criteria for Sobolev type two-weight imbeddings. J. Comput. Appl. Math. **114**, no. 1, 94–114 (2006)
31. Muckenhoupt, B.: Hardy's inequality with weights. Studia Math. **44**, 31–38 (1972)
32. Nash, J.: Continuity of solutions of parabolic and elliptic equations. Am. J. Math. **80**, no. 4. 931–954 (1958)
33. Prékopa, A.: Logarithmic concave measures with applications to stochastic programming. Acta Sci. Math. Szeged **32**, 301–316 (1971)
34. Prékopa, A.: On logarithmic concave measures and functions. Acta Sci. Math. Szeged **34**, 335–343 (1973)
35. Röckner, M., Wang, F.-Y.: Weak Poincaré inequalities and L^2-convergence rates of Markov semigroups. J. Funct. Anal. **185**, 564–603 (2001)
36. Stroock, D.W., Zegarlinski, B.: The logarithmic Sobolev inequality for discrete spin systems on a lattice. Commun. Math. Phys. **149**, 175–193 (1992)
37. Wang, F.-Y.: Functional Inequalities, Markov Processes and Spectral Theory. Sci. Press, Beijing (2004)
38. Zegarlinski, B.: Gibbsian description and description by stochastic dynamics in statistical mechanics of lattice spin systems with finite range interactions. In: Stochastic Processes, Physics and Geometry. II, pp. 714–732. World Scientific, Singapore (1995)
39. Zegarlinski, B.: Isoperimetry for Gibbs measures. Ann. Probab. **29**, no. 2, 802–819 (2001)
40. Zegarlinski, B.: Log-Sobolev inequalities for infinite one-dimensional lattice systems. Commun. Math. Phys. **133**, no. 1, 147–162 (1990)

On Some Aspects of the Theory of Orlicz–Sobolev Spaces

Andrea Cianchi

A Vladimir Maz'ya, con stima e amicizia

Abstract We survey results, obtained by the author and his coauthors over the last fifteen years, on optimal Sobolev embeddings and related inequalities in Orlicz spaces. Some of the presented results are very recent and are not published yet. We recall basic properties concerning Orlicz and Orlicz–Sobolev spaces and then dicsuss embeddings of Sobolev type and embeddings into spaces of uniformly continuous functions, classical and approximate differentiability properties of Orlicz–Sobolev functions and also trace inequalities on the boundary.

1 Introduction

The Orlicz–Sobolev spaces extend the classical Sobolev spaces in that the role of the Lebesgue spaces in their definition is played by more general Orlicz spaces. Loosely speaking, this amounts to replacing the power function t^p in the definition of the Sobolev spaces by an arbitrary Young function. The study of Orlicz–Sobolev spaces is especially motivated by the analysis of variational problems and of partial differential equations whose nonlinearities are non-necessarily of polynomial type. Systematic investigations on these spaces were initiated some forty years ago by Donaldson [23] (1971), Donaldson and Trudinger [24] (1971), Maz'ya [33, 34] (1972/1973), Gosser [28] (1974), Adams[5] (1977) and have been the object of a number of subse-

Andrea Cianchi
Dipartimento di Matematica e Applicazioni per l'Architettura, Università di Firenze, Piazza Ghiberti 27, 50122 Firenze, Italy
e-mail: cianchi@unifi.it

A. Laptev (ed.), *Around the Research of Vladimir Maz'ya I: Function Spaces*,
International Mathematical Series 11, DOI 10.1007/978-1-4419-1341-8_3,

quent papers. Some of the main contributions on this topic can be found in the monographs [41, 42] by Rao and Ren.

Researches have pointed out that some properties of standard Sobolev spaces are preserved when powers are replaced by arbitrary Young functions, but others take a different form, and reproduce the usual results only under additional assumptions on the relevant Young functions. This witnesses a richer structure of the Orlicz–Sobolev spaces compared to that of the plain Sobolev spaces.

Interestingly, the theory of Orlicz–Sobolev furnishes a unified framework on several issues, which, by contrast, require separate formulations in the theory of classical Sobolev spaces. Under this respect, the former not only extends, but also provides further insight on the latter. Let us add that, unlike that of Lebesgue spaces, the class of Orlicz spaces turns out to be closed under certain operations, such as that of associating an optimal range in Sobolev and trace embeddings. In the light of these facts, one can be led to maintain the opinion that Orlicz spaces provide a more appropriate setting than Lebesgue spaces in dealing with Sobolev functions.

In the present paper, we survey several results, obtained by the author and his coauthors over the last fifteen years, on optimal Sobolev embeddings and related inequalities in Orlicz spaces. Let us emphasize that some of them are very recent and are contained in papers still in preprint form, or even in preparation. After recalling some basic notions and properties concerning Orlicz and more general rearrangement invariant spaces in Section 2, we give the definition of Orlicz–Sobolev space and present embeddings of Sobolev type in Section 3. Embeddings into spaces of uniformly continuous functions are the subject of Section 4, whereas Section 5 deals with classical and approximate differentiability properties of Orlicz–Sobolev functions. The last section is concerned with trace inequalities on the boundary.

2 Background

We recall here a few definitions and properties of function spaces to be used in what follows.

Let $(\mathcal{R}, \nu)$ be a positive, σ-finite and nonatomic measure space, and let f be a measurable function on $\mathcal{R}$. The *decreasing rearrangement* $f^* : [0, \nu(\mathcal{R})) \to [0, \infty]$ of f is defined as

$$f^*(s) = \sup\{t \geqslant 0 : \nu(\{x \in \mathcal{R} : |f(x)| > t\}) > s\} \qquad \text{for } s \in [0, \nu(\mathcal{R})).$$

In other words, f^* is the unique nonincreasing right-continuous function on $[0, \nu(\mathcal{R}))$ which is equimeasurable with f.

A Banach function space $X(\mathcal{R})$, in the sense of Luxemburg (cf., for example, [8]), is called *rearrangement invariant* (*r.i.*, for short) if

$$\|f\|_{X(\mathcal{R})} = \|g\|_{X(\mathcal{R})} \qquad \text{whenever} \qquad f^* = g^* . \tag{2.1}$$

If $\nu(\mathcal{R}) < \infty$, then

$$L^\infty(\mathcal{R}) \to X(\mathcal{R}) \to L^1(\mathcal{R}) \tag{2.2}$$

for every r.i. space $X(\mathcal{R})$. Here, the arrow " $\to$ " stands for continuous embedding. Thus, $L^\infty(\mathcal{R})$ and $L^1(\mathcal{R})$ are the smallest and the largest, respectively, r.i. space on $\mathcal{R}$.

The Orlicz spaces, the Lorentz (–Zygmund) spaces, and their combination, the Orlicz–Lorentz spaces, are special instances of r.i. spaces. Each one of these families of spaces provides a generalization of the Lebesgue spaces.

A function $A : [0, \infty) \to [0, \infty]$ is called a *Young function* if it is convex (non trivial), left-continuous and vanishes at 0; thus, any such function takes the form

$$A(t) = \int_0^t a(\tau) d\tau \qquad \text{for } t \geqslant 0 \tag{2.3}$$

for some nondecreasing, left-continuous function $a : [0, \infty) \to [0, \infty]$ which is neither identically equal to 0 nor to ∞. Since $\lim_{t\to\infty} A(t) = \infty$, in what follows any Young function A will be continued to $[0, \infty]$ on setting $A(\infty) = \infty$.

The Orlicz space $L^A(\mathcal{R})$ associated with a Young function A is the r.i. space equipped with the *Luxemburg norm* defined as

$$\|f\|_{L^A(\mathcal{R})} = \inf\left\{\lambda > 0 : \int_{\mathcal{R}} A\left(\frac{|f(x)|}{\lambda}\right) d\nu(x) \leqslant 1\right\}$$

for any measurable function f on $\mathcal{R}$. In particular, $L^A(\mathcal{R}) = L^p(\mathcal{R})$ if $A(t) = t^p$ for some $p \in [1, \infty)$, and $L^A(\mathcal{R}) = L^\infty(\mathcal{R})$ if $A(t) = \infty\chi_{(1,\infty)}(t)$.

The *Young conjugate* of A is the Young function $\widetilde{A}$ defined as

$$\widetilde{A}(t) = \sup\{\tau t - A(\tau) : \tau \geqslant 0\} \qquad \text{for} \qquad t \geqslant 0 .$$

A Young function A is said to dominate another Young function B near infinity if positive constants c and t_0 exist such that

$$B(t) \leqslant A(ct) \qquad \text{for} \quad t \geqslant t_0 . \tag{2.4}$$

Two functions A and B are called *equivalent near infinity* if they dominate each other near infinity. We write $B \preceq A$ to denote that A dominates B near infinity, and $A \approx B$ to denote that A and B are equivalent near infinity.

The function B is said to increase essentially more slowly than A near infinity if $B \preceq A$, but $A \not\preceq B$. This is equivalent to saying that

$$\lim_{r\to\infty} \frac{A^{-1}(r)}{B^{-1}(r)} = 0. \tag{2.5}$$

Here, A^{-1} and B^{-1} are the (generalized) right-continuous inverses of A and B respectively.

The notion of dominance comes into play in the description of inclusion relations between Orlicz spaces. Actually, when $\nu(\mathcal{R}) < \infty$,

$$L^A(\mathcal{R}) \to L^B(\mathcal{R}) \quad \text{if and only if} \quad B \preceq A. \tag{2.6}$$

The *Matuzewska-Orlicz upper index* $I_\infty(A)$ at infinity of a finite-valued Young function A is defined as

$$I_\infty(A) = \lim_{t\to\infty} \frac{\log\left(\limsup_{s\to\infty} \frac{A(ts)}{A(s)}\right)}{\log t}. \tag{2.7}$$

The *Lorentz space* $L^{p,q}(\mathcal{R})$, with $p, q \in [1, \infty]$, is the collection of all measurable functions f on $\mathcal{R}$ making the quantity

$$\|f\|_{L^{p,q}(\mathcal{R})} = \|s^{\frac{1}{p}-\frac{1}{q}} f^*(s)\|_{L^q(0,\nu(\mathcal{R}))} \tag{2.8}$$

finite. When $\nu(\mathcal{R}) < \infty$, the *Lorentz–Zygmund space* $L^{p,q;\gamma}(\mathcal{R})$, with $p, q \in [1, \infty]$ and $\gamma \in \mathbb{R}$ is defined as the set of measurable functions f for which the expression

$$\|f\|_{L^{p,q;\gamma}(\mathcal{R})} = \|s^{\frac{1}{p}-\frac{1}{q}}(1 + \log(\nu(\mathcal{R})/s))^\gamma f^*(s)\|_{L^q(0,\nu(\mathcal{R}))} \tag{2.9}$$

is finite. Generalized Lorentz–Zygmund spaces, involving iterated logarithmic weight functions, also naturally arise in applications (cf. Examples 3.15 and 6.8).

The class of Lorentz–Zygmund spaces not only extends those of the Lebesgue spaces ($p = q$, $\alpha = 0$) and of the Lorentz spaces ($\alpha = 0$), but also overlaps with that of the Orlicz spaces. Actually, the spaces $L^{p,q;\gamma}(\mathcal{R})$ reproduce (up to equivalent norms) the Orlicz spaces $L^p(\log L)^\alpha(\mathcal{R})$ ($1 < p = q$, $\gamma = \alpha/q$) associated with any Young function equivalent to $t^p(\log t)^\alpha$ near infinity, and the Orlicz spaces $\exp L^\beta(\mathcal{R})$ ($p = q = \infty$, $\gamma = -1/\beta$) associated with any Young function equivalent to $\exp(t^\beta)$ near infinity.

Note also that $L^{p,q;\gamma}(\mathcal{R})$ need not be an r.i. space for certain values of the parameters p, q and γ. A complete description of the admissible values of these parameters can be found in [7] (cf. also [37]). However, in the applications of the present paper, $L^{p,q;\gamma}(\mathcal{R})$ will always be an r.i. space, possibly up to equivalent norms.

Assume that $\nu(\mathcal{R}) < \infty$. Let $p \in (1, \infty]$, $q \in [1, \infty)$, and let D be a Young function. If $p < \infty$, assume that

$$\int^\infty \frac{D(t)}{t^{1+p}}\, dt < \infty.$$

We denote by $L(p,q,D)(\mathcal{R})$ the Orlicz–Lorentz space of all measurable functions f on $\mathcal{R}$ for which the quantity

$$\|f\|_{L(p,q,D)(\mathcal{R})} = \|s^{-\frac{1}{p}} f^*(\nu(\mathcal{R})s^{\frac{1}{q}})\|_{L^D(0,1)} \tag{2.10}$$

is finite. The expression $\|\cdot\|_{L(p,q,D)(\mathcal{R})}$ is a norm, and $L(p,q,D)(\mathcal{R})$, equipped with this norm, is an r.i. space (cf. the proof of [14, Proposition 2.1]). Note that the spaces $L(p,q,D)(\mathcal{R})$ include (up to equivalent norms) the Orlicz spaces and various instances of Lorentz and Lorentz–Zygmund spaces.

3 Orlicz–Sobolev Embeddings

Let Ω be an open subset of $\mathbb{R}^n$, $n \geqslant 2$. The mth order *Orlicz–Sobolev space* $W^{m,A}(\Omega)$ associated with a positive integer m and with a Young function A is defined as

$$\begin{aligned} W^{m,A}(\Omega) = \{u \in L^A(\Omega) : u \text{ is } m-\text{times weakly differentiable in } \Omega \\ \text{and } D^\alpha u \in L^A(\Omega) \text{ for every } \alpha \text{ such that } |\alpha| \leqslant m\}. \end{aligned}$$

Here, α is any multi-index having the form $\alpha = (\alpha_1, \dots, \alpha_n)$ for nonnegative integers $\alpha_1, \dots, \alpha_n$, $|\alpha| = \alpha_1 + \dots + \alpha_n$, and $D^\alpha u = \frac{\partial^{|\alpha|} u}{\partial x_1^{\alpha_1} \dots \partial x_n^{\alpha_n}}$. The space $W^{m,A}(\Omega)$ is a Banach space endowed with the norm

$$\|u\|_{W^{m,A}(\Omega)} = \Big\| \sum_{k=0}^{m} |\nabla^k u| \Big\|_{L^A(\Omega)}, \tag{3.1}$$

where $\nabla^k u$ stands for the vector of all the derivatives $D^\alpha u$ with $|\alpha| = k$, and $|\nabla^k u|$ denotes the Euclidean norm of $\nabla^k u$.

The space $W^{m,A}_{\mathrm{loc}}(\Omega)$ is defined accordingly as the collection of functions u in Ω such that $u \in W^{m,A}(\Omega')$ for every open set Ω' such that $\Omega' \subset\subset \Omega$.

It is clear that the choice $A(t) = t^p$, with $p \in [1,\infty)$, or $A(t) = \infty\chi_{(1,\infty)}(t)$, in $W^{m,A}(\Omega)$ and $W^{m,Ap}_{\mathrm{loc}}(\Omega)$ reproduces the classical Sobolev spaces $W^{m,p}(\Omega)$ and $W^{m,p}_{\mathrm{loc}}(\Omega)$, with $p \in [1,\infty)$, or $p = \infty$, respectively.

In the present section, we are concerned with optimal embeddings of $W^{m,A}(\Omega)$ into Orlicz spaces and into rearrangement invariant spaces. Here, we limit ourselves to considering the case when $|\Omega| < \infty$; results for domains with infinite volume are also available (cf. [11, 14]).

3.1 Embeddings into Orlicz spaces

We begin with a sharp embedding theorem for $W^{m,A}(\Omega)$ into Orlicz spaces, and with related Sobolev–Poincaré inequalities. Given a Young function A, we exhibit another Young function $A_{\frac{n}{m}}$, an mth order Sobolev conjugate of A, having the property that $L^{A_{\frac{n}{m}}}(\Omega)$ is the smallest Orlicz space into which $W^{m,A}(\Omega)$ is continuously embedded. The function $A_{\frac{n}{m}}$ is defined as follows.

Let $0 \leqslant m < n$, and let A be any Young function satisfying

$$\int_0 \left(\frac{t}{A(t)}\right)^{\frac{m}{n-m}} dt < \infty. \tag{3.2}$$

Let $H : [0, \infty) \to [0, \infty)$ be the function given by

$$H(r) = \left(\int_0^r \left(\frac{t}{A(t)}\right)^{\frac{m}{n-m}} dt\right)^{\frac{n-m}{n}} \qquad \text{for } r \geqslant 0. \tag{3.3}$$

Then $A_{\frac{n}{m}}$ is defined as

$$A_{\frac{n}{m}}(t) = A\big(H^{-1}(t)\big) \qquad \text{for } t \geqslant 0, \tag{3.4}$$

where H^{-1} denotes the (generalized) left-continuous inverse of H. The use of the index $\frac{n}{m}$ is due to the fact that $A_{\frac{n}{m}}$ depends on n and m only through their quotient.

Let us emphasize that assumption (3.2) is not restrictive. Indeed, by (2.6), replacing, if necessary, the function A by an equivalent Young function near infinity fulfilling (3.2) leaves the space $W^{m,A}(\Omega)$ unchanged (up to equivalent norms), since we are assuming that $|\Omega| < \infty$.

Our statements involve a standard notion of Lipschitz domain, namely a bounded set with a Lipschitz boundary (cf., for example, [35, Definition 1.1.9/1] for a precise definition).

Theorem 3.1. *Let Ω be a Lipschitz domain in $\mathbb{R}^n$. Let $1 \leqslant m < n$, let A be a Young function satisfying* (3.2), *and let $A_{\frac{n}{m}}$ be the Sobolev conjugate defined by* (3.4). *Then a constant $C = C(\Omega, m)$ exists such that*

$$\|u\|_{L^{A_{\frac{n}{m}}}(\Omega)} \leqslant C\|u\|_{W^{m,A}(\Omega)} \tag{3.5}$$

for every $u \in W^{m,A}(\Omega)$, or, equivalently,

$$\int_\Omega A_{\frac{n}{m}}\left(\frac{|u(y)|}{C\left(\int_\Omega A(\sum_{k=0}^m |\nabla^k u(x)|)\, dx\right)^{m/n}}\right) dy$$

$$\leqslant \int_\Omega A\Big(\sum_{k=0}^m |\nabla^k u(x)|\Big)\,dx \tag{3.6}$$

for every $u \in W^{m,A}(\Omega)$. The space $L^{A_{\frac{n}{m}}}(\Omega)$ is optimal among all Orlicz spaces, in the sense that if (3.5) *holds with $A_{\frac{n}{m}}$ replaced by another Young function B, then $L^{A_{\frac{n}{m}}}(\Omega) \to L^B(\Omega)$.*

Theorem 3.1 for $m = 1$ is established in [11] with A_n replaced by an equivalent Young function, and in [12] in the present form. The case $m > 1$ can be found in [15]. Earlier Orlicz–Sobolev embeddings involving a Sobolev conjugate of A different from $A_{\frac{n}{m}}$, which is not optimal in general, are contained in [24] and [4]. Sobolev inequalities of special form, involving Orlicz norms and general measures, have been characterized in terms of isocapacitary inequalities by [33].

Remark 3.2. In view of (2.6), the space $L^{A_{\frac{n}{m}}}(\Omega)$ is determined (up to equivalent norms) just by the asymptotic behavior of $A_{\frac{n}{m}}$ near infinity.

Remark 3.3. Note that Theorem 3.1 is only stated for $1 \leqslant m < n$ since if $m \geqslant n$, then trivially

$$W^{m,A}(\Omega) \to W^{m,1}(\Omega) \to L^\infty(\Omega)$$

(cf., for example, [4, Theorem 5.4]). On the other hand, Theorem 3.1 tells us that the embedding

$$W^{m,A}(\Omega) \to L^\infty(\Omega) \tag{3.7}$$

also holds if $1 \leqslant m < n$ provided that A grows so fast at infinity that

$$\int^\infty \left(\frac{t}{A(t)}\right)^{\frac{m}{n-m}} dt < \infty\,. \tag{3.8}$$

Indeed, if (3.8) is in force, then $A_{n/m}(t) = \infty$ for every $t > H_{n/m}(\infty)$. Hence, by Remark 3.2, $L^{A_{n/m}}(\Omega) = L^\infty(\Omega)$.

The embedding (3.7) for $m = 1$ was obtained, under a condition equivalent to (3.8), in [34], and independently rediscovered in [46]. The higher-order case can be found in [31].

Example 3.4. Assume that

$$A(t) \approx t^p (\log t)^\alpha,$$

where either $p > 1$ and $\alpha \in \mathbb{R}$, or $p = 1$ and $\alpha \geqslant 0$. From Theorem 3.1 and Remark 3.3 we deduce that

$$W^{m,A}(\Omega) \to \begin{cases} L^{\frac{pn}{n-mp}}(\log L)^{\frac{\alpha n}{n-mp}}(\Omega) & \text{if } 1 \leqslant p < \frac{n}{m}, \\ \exp L^{\frac{n}{n-m-\alpha m}}(\Omega) & \text{if } p = \frac{n}{m} \text{ and } \alpha < \frac{n-m}{m}, \\ \exp\exp L^{\frac{n}{n-m}}(\Omega) & \text{if } p = \frac{n}{m} \text{ and } \alpha = \frac{n-m}{m}, \\ L^{\infty}(\Omega) & \text{if either } p = \frac{n}{m} \\ & \quad \text{and } \alpha > \frac{n-m}{m}, \text{ or } p > \frac{n}{m}\,. \end{cases} \tag{3.9}$$

Moreover, all the range spaces are optimal in the class of Orlicz spaces. Here, $\exp\exp L^{\frac{n}{n-m}}(\Omega)$ denotes the Orlicz space on Ω associated with the Young function $\exp\big(\exp\big(t^{\frac{n}{n-m}}\big)\big) - e$.

When $p \neq \frac{n}{m}$, embedding (3.9) agrees with the standard Sobolev embedding if $\alpha = 0$, and overlaps with results from [24] if $\alpha \neq 0$. If $p = n$, $\alpha = 0$ and $m = 1$, (3.9) reproduces a result from [39, 47, 48]; the case where $m > 1$ is contained in [39, 45]. The embedding (3.9) for $p = n$, $\alpha \leqslant 0$ and $m = 1$ recovers [27], and, for $p = n$ and arbitrary α and m, overlaps with [25].

Example 3.5. Assume that

$$A(t) \approx t^p(\log(\log t))^{\alpha},$$

where p and α are as in Example 3.5. Then, we obtain from Theorem 3.1 and Remark 3.3 that

$$W^{m,A}(\Omega) \to L^{A_{\frac{n}{m}}}(\Omega),$$

where

$$A_{\frac{n}{m}}(t) \approx \begin{cases} t^{\frac{pn}{n-mp}}(\log(\log t))^{\frac{\alpha n}{n-mp}} & \text{if } 1 \leqslant p < \frac{n}{m}, \\ e^{t^{\frac{n}{n-m}}(\log t)^{\frac{\alpha m}{n-m}}} & \text{if } p = \frac{n}{m}. \end{cases}$$

If, instead, $p > \frac{n}{m}$, then

$$W^{m,A}(\Omega) \to L^{\infty}(\Omega).$$

Moreover, the range spaces are sharp in the framework of Orlicz spaces on Ω.

Notice that, unlike the case of Example 3.4, the space $L^{A_{\frac{n}{m}}}(\Omega)$ is never embedded into $L^{\infty}(\Omega)$ in the limiting case where $p = \frac{n}{m}$, whatever α is.

The condition (3.8) also characterizes the Young functions A and the integers n and m for which the Orlicz–Sobolev space $W^{m,A}(\Omega)$ is a Banach algebra ([16]). Recall that $W^{m,A}(\Omega)$ is an algebra if $uv \in W^{m,A}(\Omega)$ whenever $u, v \in W^{m,A}(\Omega)$, and

$$\|uv\|_{W^{m,A}(\Omega)} \leqslant C\|u\|_{W^{m,A}(\Omega)}\|v\|_{W^{m,A}(\Omega)} \tag{3.10}$$

for some constant C and for every $u, v \in W^{m,A}(\Omega)$.

Theorem 3.6. *Let Ω be a Lipschitz domain in $\mathbb{R}^n$. Let m be a nonnegative integer, and let A be a Young function. Then $W^{m,A}(\Omega)$ is a Banach algebra if and only if either $m \geqslant n$, or $1 \leqslant m < n$ and* (3.8) *is in force.*

We now present Sobolev–Poincaré type inequalities where the $L^{A_{\frac{n}{m}}}(\Omega)$ norm of a function in $W^{m,A}(\Omega)$ is estimated in terms of the $L^A(\Omega)$ norm of its derivatives of highest order m. Obviously, this is only possible under some normalization condition on u. We first consider the case where u is assumed to belong to the subspace $W_0^{m,A}(\Omega)$ of $W^{m,A}(\Omega)$ of those functions which vanish on $\partial\Omega$, together will all their derivatives of order less than m, in a suitable sense. Precisely, we define

$$\begin{aligned} W_0^{m,A}(\Omega) =\{u \in W^{m,A}(\Omega) :\ &\text{the continuation of } u \text{ by } 0 \text{ outside } \Omega \\ &\text{is an } m\text{-times weakly differentiable function} \\ &\text{in the whole of } \mathbb{R}^n\}. \end{aligned}$$

Note that, unlike the ordinary Sobolev spaces, taking the closure in $W^{m,A}(\Omega)$ of smooth compactly supported functions in Ω yields, in general, a subspace smaller than $W_0^{m,A}(\Omega)$, even in the case where Ω is a smooth domain. This is due to the fact that smooth functions are not dense in $W^{m,A}(\Omega)$, unless A satisfies the so-called Δ_2-condition (cf., for example, [4, Chapter 8]).

Theorem 3.7. *Let $1 \leqslant m < n$, and let A be a Young function satisfying* (3.2). *Then there exists a constant $C = C(n,m)$ such that for any open bounded subset Ω of $\mathbb{R}^n$*

$$\|u\|_{L^{A_{\frac{n}{m}}}(\Omega)} \leqslant C\|\nabla^m u\|_{L^A(\Omega)} \tag{3.11}$$

for every $u \in W_0^{m,A}(\Omega)$ or, equivalently,

$$\int_\Omega A_{\frac{n}{m}}\left(\frac{|u(y)|}{C\big(\int_\Omega A(|\nabla^m u(x)|)\,dx\big)^{m/n}}\right) dy \leqslant \int_\Omega A(|\nabla^m u(x)|)\,dx \tag{3.12}$$

for every $u \in W_0^{m,A}(\Omega)$.

The following result provides us with a Sobolev–Poincaré inequality in $W^{m,A}(\Omega)$. As in the case of standard Sobolev spaces, admissible functions can be normalized by subtracting a suitable polynomial of degree less than m. Given a nonnegative integer h, the class of polynomials of degree $\leqslant h$ is denoted by $\mathcal{P}_h$.

Theorem 3.8. *Let Ω, m and A be as in Theorem* 3.1. *Then there exists a constant $C = C(\Omega, m)$ such that for every $u \in W^{m,A}(\Omega)$ a polynomial $P_{m-1} \in \mathcal{P}_{m-1}$ exists satisfying*

$$\|u - P_{m-1}\|_{L^{A_{\frac{n}{m}}}(\Omega)} \leqslant C\|\nabla^m u\|_{L^A(\Omega)}, \tag{3.13}$$

or, equivalently,

$$\int_\Omega A_{\frac{n}{m}}\left(\frac{|u(y)-P_{m-1}(y)|}{C\left(\int_\Omega A(|\nabla^m u(x)|)\,dx\right)^{m/n}}\right)dy \leqslant \int_\Omega A(|\nabla^m u(x)|)\,dx\,. \tag{3.14}$$

We conclude this subsection with a compactness theorem of Rellich type.

Theorem 3.9. *Let Ω, m and A be as in Theorem* 3.1. *Let B be any Young function increasing essentially more slowly than $A_{\frac{n}{m}}$ near infinity. Then the following embedding is compact:*

$$W^{m,A}(\Omega) \to L^B(\Omega).$$

Remark 3.10. The embedding

$$W^{m,A}(\Omega) \to L^A(\Omega) \tag{3.15}$$

is compact for every Lipschitz domain Ω and for every Young function A. Indeed, it is easily verified from (3.4) that every finite-valued Young function A increases essentially more slowly than $A_{\frac{n}{m}}$ near infinity; thus, in this case, the compactness of embedding (3.15) follows from Theorem 3.9. If, instead, $A(t) = \infty$ for large t, then (3.15) is equivalent to the compact embedding $W^{m,\infty}(\Omega) \to L^\infty(\Omega)$.

Proofs of Theorems 3.7–3.9 can be found in [11, 12] for $m = 1$ and in [15] for $m > 1$. A version of Theorem 3.7, for $m = 1$, in anisotropic Orlicz–Sobolev spaces whose norm depends on the full gradient ∇u, and not only on its length $|\nabla u|$, is established in [13].

3.2 Embeddings into rearrangement invariant spaces

This subsection deals with an optimal Orlicz–Sobolev embedding into rearrangement invariant spaces. The solution to this problem involves an Orlicz–Lorentz space of the form $L(p,q,D)$, whose norm is defined as in (2.10).

Let m be any positive integer such that $m < n$, let A be any Young function satisfying (3.2), and let a be the function related to A as in (2.3). We call E the Young function given by

$$E(t) = \int_0^t e(\tau)d\tau \qquad \text{for } t \geqslant 0, \tag{3.16}$$

where e is the nondecreasing, left-continuous function in $[0,\infty)$ obeying

$$e^{-1}(s) = \left(\int_{a^{-1}(s)}^{\infty} \left(\int_0^\tau \left(\frac{1}{a(t)} \right)^{\frac{m}{n-m}} dt \right)^{-\frac{n}{m}} \frac{d\tau}{a(\tau)^{\frac{n}{n-m}}} \right)^{\frac{m}{m-n}} \quad \text{for } s \geqslant 0. \tag{3.17}$$

Here, a^{-1} and e^{-1} are the (generalized) left-continuous inverses of a and e respectively.

The optimal r.i. range space for embeddings of $W^{m,A}(\Omega)$ is either $L^\infty(\Omega)$ or $L(\frac{n}{m}, 1, E)(\Omega)$, namely the r.i. space equipped with norm

$$\|f\|_{L(\frac{n}{m},1,E)(\Omega)} = \|s^{-\frac{m}{n}} f^*(|\Omega|s)\|_{L^E(0,1)}. \tag{3.18}$$

Theorem 3.11. *Let Ω, m and A be as in Theorem* 3.1.

(i) *If*

$$\int^\infty \left(\frac{t}{A(t)} \right)^{\frac{m}{n-m}} dt = \infty, \tag{3.19}$$

then a constant $C = C(\Omega, m)$ exists such that

$$\|u\|_{L(\frac{n}{m},1,E)(\Omega)} \leqslant C\|u\|_{W^{m,A}(\Omega)} \tag{3.20}$$

for every $u \in W^{m,A}(\Omega)$, or, equivalently,

$$\int_0^1 E\Big(\frac{1}{C} s^{-\frac{m}{n}} u^*(|\Omega|s)\Big)\, ds \leqslant \int_\Omega A\Big(\sum_{k=0}^m |\nabla^k u(x)|\Big)\, dx \tag{3.21}$$

for every $u \in W^{m,A}(\Omega)$.

(ii) *If*

$$\int^\infty \left(\frac{t}{A(t)} \right)^{\frac{m}{n-m}} dt < \infty, \tag{3.22}$$

then a constant $C = C(\Omega, m, A)$ exists such that

$$\|u\|_{L^\infty(\Omega)} \leqslant C\|u\|_{W^{m,A}(\Omega)} \tag{3.23}$$

for every $u \in W^{m,A}(\Omega)$.

Both $L(\frac{n}{m}, 1, E)(\Omega)$ and $L^\infty(\Omega)$ are optimal among all r. i. spaces in (3.20) *and* (3.23) *respectively. Indeed, if A fulfills* (3.19), *and* (3.20) *holds with $L(\frac{n}{m}, 1, E)(\Omega)$ replaced by another r.i. space $X(\Omega)$, then $L(\frac{n}{m}, 1, E)(\Omega) \to X(\Omega)$; if A fulfills* (3.22), *and* (3.23) *holds with $L^\infty(\Omega)$ replaced by another r.i. space $X(\Omega)$, then (trivially) $L^\infty(\Omega) \to X(\Omega)$.*

The case where $m = 1$ of Theorem 3.11 is contained in [14], whereas the result for higher-order Orlicz–Sobolev spaces is proved in [15].

Remark 3.12. In analogy with Remark 3.2, note that $L(\frac{n}{m}, 1, E)(\Omega)$ depends (up to equivalent norms) only on the behavior of E near infinity.

Remark 3.13. One can show that the function A always dominates E [14, Proposition 5.1]. Moreover, A is equivalent to E near infinity if and only if the Matuzewska–Orlicz index $I_\infty(A) < n/m$ [14, Proposition 5.2]. Thus, when $I_\infty(A) < n/m$,

$$\|u\|_{L(\frac{n}{m},1,E)(\Omega)} = \|s^{-\frac{m}{n}} u^*(s)\|_{L^A(0,|\Omega|)} .$$

Remark 3.14. Sobolev–Poincaré inequalities involving $L(\frac{n}{m},1,E)(\Omega)$ hold in the same spirit as those considered in Subsection 3.1 ([15]).

Example 3.15. Let A be as in Example 3.4. Via Theorem 3.11 and Remark 3.13 one can show that

$$W^{m,A}(\Omega) \to \begin{cases} L^{\frac{pn}{n-mp},p;\frac{\alpha}{p}}(\Omega) & \text{if } 1 \leqslant p < \frac{n}{m}, \\ L^{\infty,\frac{n}{m};\frac{m\alpha}{n}-1}(\Omega) & \text{if } p = \frac{n}{m} \text{ and } \alpha < \frac{n-m}{m}, \\ L^{\infty,\frac{n}{m};-\frac{m}{n},-1}(\Omega) & \text{if } p = \frac{n}{m} \text{ and } \alpha = \frac{n-m}{m}, \end{cases} \tag{3.24}$$

up to equivalent norms, and that all the range spaces are optimal among r.i. spaces. Here, $L^{\infty,\frac{n}{m};-\frac{m}{n},-1}(\Omega)$ denotes a generalized Lorentz–Zygmund space on Ω endowed with the norm given by

$$\|f\|_{L^{\infty,\frac{n}{m};-\frac{m}{n},-1}(\Omega)} = \|s^{-\frac{m}{n}}(1+\log(|\Omega|/s))^{-\frac{m}{n}}$$

$$\times (1+\log(1+\log(|\Omega|/s)))^{-1} f^*(s)\|_{L^{\frac{n}{m}}(0,|\Omega|)} \tag{3.25}$$

for any measurable function f on Ω.

The embedding (3.24) overlaps with results scattered in various papers, including [7, 9, 22, 25, 26, 29, 30, 35, 36, 38]. In particular, the case where $p \neq n$ and $\alpha = 0$ is contained in [36, 38], and the case where $p = n$ and $\alpha = 0$ can be found in [9, 30, 35]; the sharpness of these results is proved in [26] and [22].

4 Modulus of Continuity

Results from the preceding section (cf. Remark 3.3, or Theorem 3.11 (ii)) tell us that if a Young function A fulfills (3.22), then any function in $W^{m,A}(\Omega)$ is essentially bounded. Here we are concerned with continuity properties of functions in $W^{m,A}(\Omega)$.

In what follows, we denote by $C^0(\Omega)$ and $C^1(\Omega)$ the space of continuous functions on Ω, and the space of continuously differentiable functions on Ω respectively, equipped with the usual norms. Moreover, given a modulus of continuity σ, namely an increasing function from $[0,\infty)$ into $[0,\infty)$ vanishing at 0, we denote by $C^\sigma(\Omega)$ the space of uniformly continuous functions on Ω whose modulus of continuity does not exceed σ. The space $C^\sigma(\Omega)$ is endowed with the norm defined for a function u by

$$\|u\|_{C^\sigma(\Omega)} = \|u\|_{C^0(\Omega)} + \sup_{x\neq y} \frac{|u(x)-u(y)|}{\sigma(|x-y|)}.$$

A first result asserts that, if the condition (3.22) is fulfilled, then functions in $W^{m,A}(\Omega)$ are, in fact, continuous (cf. [31] and, independently, [10] for $m=1$, and [15] for $m>1$).

Theorem 4.1. *Let Ω be an open set in $\mathbb{R}^n$. Let m be a nonnegative integer, and let A be a Young function. Then*

$$W^{m,A}(\Omega) \to C^0(\Omega') \tag{4.1}$$

for every open set $\Omega' \subset\subset \Omega$ if and only if either $m \geqslant n$, or $1 \leqslant m < n$ and (3.22) holds. Moreover, if either of these condition is in force and Ω is a Lipschitz domain, then (4.1) holds with $\Omega' = \Omega$.

The next theorem is concerned with embeddings of $W^{m,A}(\Omega)$ into spaces of uniformly continuous functions. Given $1 \leqslant m \leqslant n$ and a Young function A, define $\xi, \eta : (0,\infty) \to [0,\infty]$ as

$$\xi(t) = t^{\frac{n}{n-m}} \int_t^\infty \frac{\widetilde{A}(\tau)}{\tau^{1+\frac{n}{n-m}}}\, d\tau \quad \text{for } t>0, \tag{4.2}$$

and

$$\eta(t) = t^{\frac{n}{n-m+1}} \int_0^t \frac{\widetilde{A}(\tau)}{\tau^{1+\frac{n}{n-m+1}}}\, d\tau \quad \text{for } t>0. \tag{4.3}$$

Theorem 4.2. *Let Ω be an open subset of $\mathbb{R}^n$. Let $1 \leqslant m \leqslant n$, and let A be a Young function such that the integral on the right-hand side of (4.3) converges for $t>0$.*

If $1 \leqslant m \leqslant n-1$, assume that

$$\int^\infty \left(\frac{t}{A(t)}\right)^{\frac{m}{n-m}} dt < \infty. \tag{4.4}$$

If $m=n$, assume that

$$\liminf_{t\to\infty} \frac{t}{A(t)} = 0. \tag{4.5}$$

Then the function $\vartheta_A : [0,\infty) \to [0,\infty)$, given by

$$\vartheta_A(r) = \begin{cases} \frac{r^{1-n}}{\xi^{-1}(r^{-n})} & \text{if } m = 1, \\ r^{m-n}\left(\frac{1}{\xi^{-1}(r^{-n})} + \frac{1}{\eta^{-1}(r^{-n})}\right) & \text{if } 2 \leqslant m \leqslant n-1, \\ \frac{1}{\eta^{-1}(r^{-n})} & \text{if } m = n, \end{cases} \tag{4.6}$$

is a modulus of continuity, and

$$W^{m,A}(\Omega) \to C^{\vartheta_A}(\Omega') \tag{4.7}$$

for every open set $\Omega' \subset\subset \Omega$.

Moreover, the space $C^{\vartheta_A}(\Omega')$ *is optimal, in the sense that if* (4.7) *holds with* $C^{\vartheta_A}(\Omega')$ *replaced by some other space* $C^{\sigma}(\Omega')$, *then* $C^{\vartheta_A}(\Omega') \to C^{\sigma}(\Omega')$.

Conversely, if there exists a modulus of continuity σ *such that*

$$W^{m,A}(\Omega) \to C^{\sigma}(\Omega') \tag{4.8}$$

for every open set $\Omega' \subset\subset \Omega$, *then either of* (4.4) *or* (4.5) *holds.*

A proof of Theorem 4.2, as well as of the remaining results of this section, can be found in [19]. Related results for $m = 1$ are in [18].

Remark 4.3. The assumption that the integral on the right-hand side of (4.3) be convergent in Theorem 4.2 is not a restriction, by the same reason why the convergence of the integral in (3.3) is not a restriction in Theorem 3.1.

Remark 4.4. Assumption (4.5) is equivalent to

$$L^A(\Omega') \subsetneqq L^1(\Omega')$$

for every $\Omega' \subset\subset \Omega$.

Remark 4.5. The case where $m > n$ in Theorem 4.2 is uninteresting. Actually, in this case

$$W^{m,A}(\Omega) \to C^1(\Omega')$$

for every $\Omega' \subset\subset \Omega$, since, if Ω'' is any Lipschitz domain such that $\Omega' \subset\subset \Omega'' \subset\subset \Omega$, then

$$W^{m,A}(\Omega) \to W^{m,1}(\Omega'') \to W^{n+1,1}(\Omega'') \to C^1(\Omega'') \to C^1(\Omega').$$

A characterization of embeddings of $W^{m,A}(\Omega)$ into the space $\mathrm{Lip}(\Omega')$ of Lipschitz continuous functions on $\Omega' \subset\subset \Omega$, namely the space $C^{\sigma}(\Omega')$ with $\sigma(r) = r$, can be easily derived from Theorem 4.2.

Corollary 4.6. *Let* Ω *be an open subset of* $\mathbb{R}^n$. *Let* $1 \leqslant m \leqslant n$, *and let* A *be a Young function. Then*

$$W^{m,A}(\Omega) \to \mathrm{Lip}(\Omega') \tag{4.9}$$

for every open set $\Omega' \subset\subset \Omega$ *if and only if either*

$$m = 1 \quad \text{and} \quad A(t) = \infty \quad \text{for large } t,$$

or

$$2 \leqslant m \leqslant n \quad \text{and} \quad \int^{\infty} \left(\frac{t}{A(t)} \right)^{\frac{m-1}{n+1-m}} dt < \infty. \tag{4.10}$$

A global version of Theorem 4.2 holds provided that Ω is a Lipschitz domain.

Theorem 4.7. *Under the same assumptions as in Theorem* 4.2, *if, in addition,* Ω *is a Lipschitz domain, then*

$$W^{m,A}(\Omega) \to C^{\vartheta_A}(\Omega). \tag{4.11}$$

We conclude this section with a discussion of the compactness of the embeddings of Theorems 4.1 and 4.2.

Let σ_1 and σ_2 be moduli of continuity. Following [24], we say that σ_1 decays to 0 essentially faster than σ_2 if there exists a modulus of continuity σ such that $\sigma_1(s) \leqslant \sigma_2(s)\sigma(s)$ for $s \geqslant 0$.

Theorem 4.8. *Under the same assumptions as in Theorem* 4.2, *if, in addition,* Ω *is a Lipschitz domain, then the embedding*

$$W^{m,A}(\Omega) \to C^0(\Omega) \tag{4.12}$$

is compact. Moreover, if σ *is a modulus of continuity such that* ϑ_A *decays to* 0 *essentially faster than* σ, *then also the embedding*

$$W^{m,A}(\Omega) \to C^{\sigma}(\Omega) \tag{4.13}$$

is compact.

5 Differentiability Properties

In this section, differentiability properties of Orlicz–Sobolev functions are taken into account, both in the classical and in the approximate (integral) sense.

In the spirit of Rademacher's theorem (and its extensions) and of the approximate differentiability results for functions in the standard Sobolev spaces, it turns out that functions in the Orlicz–Sobolev space $W^{m,A}(\Omega)$ do posses a classical mth order differential at a.e. points in Ω, provided that A grows sufficiently fast at infinity. Otherwise, they are merely approximately differentiable at a.e. points in Ω in the Orlicz–Sobolev conjugate norm of $W^{m,A}(\Omega)$. The discriminant between these two situations is the same as that coming into play when the boundedness and continuity of Orlicz–Sobolev functions are in question, namely the convergence or divergence of the integral

$$\int^{\infty} \left(\frac{t}{A(t)}\right)^{\frac{m}{n-m}} dt. \tag{5.1}$$

Precisely, given $u \in W^{m,A}(\Omega)$, let $T_x^m(u)$ denote the Taylor polynomial of degree m of u, which is well-defined at a.e. $x \in \Omega$ as

$$T_x^m(u)(y) = \sum_{0 \leqslant |\alpha| \leqslant m} \frac{1}{\alpha!} D^\alpha u(x)(y-x)^\alpha \qquad \text{for } y \in \mathbb{R}^n.$$

Here and in what follows, $D^\alpha u$ denotes the precise representative of the weak derivative of u corresponding to the multi-index α, and $\alpha! = \alpha_1 \cdot \alpha_2 \cdots \alpha_n$.

The classical differentiability result reads as follows.

Theorem 5.1. *Let Ω be an open subset of $\mathbb{R}^n$, and let A be a Young function. Assume that either $m \geqslant n$, or $1 \leqslant m < n$ and*

$$\int^{\infty} \left(\frac{t}{A(t)}\right)^{\frac{m}{n-m}} dt < \infty. \tag{5.2}$$

If $u \in W^{m,A}(\Omega)$, then u has an mth order differential almost everywhere in Ω, namely for a.e. $x \in \Omega$

$$u(y) - T_x^m(u)(y) = o(|y-x|^m) \qquad \text{as } y \to x. \tag{5.3}$$

The next result provides us with an approximate differentiability result replacing (5.3) when condition (5.2) fails. In the statement we make us of the notation

$$\fint_{B_r(x)} f(y)\, dy = \frac{1}{|B_r(x)|} \int_{B_r(x)} f(y)\, dy$$

for a locally integrable function f in Ω, where $B_r(x)$ is the ball, centered at x and with radius r, and $|B_r(x)|$ is its Lebesgue measure. Moreover, we define the averaged norm

$$\|f\|_{\not{L}^{A_{\frac{n}{m}}}(B_r(x))} = \inf\left\{\lambda > 0 : \fint_{B_r(x)} A_{\frac{n}{m}}\left(\frac{|f(y)|}{\lambda}\right) dy \leqslant 1\right\}.$$

Theorem 5.2. *Let Ω be an open subset of $\mathbb{R}^n$. Let $1 \leqslant m < n$ and let A be a Young function. Assume that*

$$\int^{\infty} \left(\frac{t}{A(t)}\right)^{\frac{m}{n-m}} dt = \infty. \tag{5.4}$$

If $u \in W^{m,A}(\Omega)$, then for every $\sigma > 0$

$$\lim_{r\to 0^+} \fint_{B_r(x)} A_{\frac{n}{m}}\left(\frac{|u(y)-T_x^m(u)(y)|}{\sigma r^m}\right) dy = 0 \tag{5.5}$$

for a.e. $x \in \Omega$. *Hence*

$$\lim_{r\to 0^+} \left\| \frac{u(\cdot)-T_x^m(u)(\cdot)}{r^m} \right\|_{L^{A_{\frac{n}{m}}}(B_r(x))} = 0 \tag{5.6}$$

for a.e. $x \in \Omega$.

Theorem 5.2 can be slightly improved as follows.

Theorem 5.3. *Under the same assumptions of Theorem* 5.2, *for every* $\sigma > 0$

$$\lim_{r\to 0^+} \fint_{B_r(x)} A_{\frac{n}{m}}\left(\frac{|u(y)-T_x^m(u)(y)|}{\sigma |y-x|^m}\right) dy = 0 \tag{5.7}$$

for a. e. $x \in \Omega$. *Hence*

$$\lim_{r\to 0^+} \left\| \frac{u(\cdot)-T_x^m(u)(\cdot)}{|\cdot - x|^m} \right\|_{L^{A_{\frac{n}{m}}}(B_r(x))} = 0 \tag{5.8}$$

for a. e. $x \in \Omega$.

Theorems 5.1, 5.2, and 5.3 are the object of [6] in the case where $m = 1$, and of [20] for $m > 1$.

Example 5.4. In the special case where

$$A(t) = t^p,$$

with $p \geqslant 1$, via Theorems 5.1 and 5.2 one recovers the classical results to which we alluded above. Indeed, equation (5.2) holds in this case if and only if $p > \frac{n}{m}$. Thus, if either $m \geqslant n$ or $1 \leqslant m < n$ and $p > \frac{n}{m}$, then any function $u \in W^{p,m}(\Omega)$ has an mth order differential almost everywhere in Ω [44, Chapter 8]. If instead, $1 \leqslant p < \frac{n}{m}$, we have that for a.e. $x \in \Omega$

$$\lim_{r\to 0^+} \fint_{B_r(x)} \left(\frac{|u(y)-T_x^m(u)(y)|}{|x-y|^m}\right)^{\frac{np}{n-mp}} dy = 0 \tag{5.9}$$

(cf. [49, Chapter 3]).

This classical framework does not include the borderline case where $p = \frac{n}{m}$, which can be dealt with via Theorems 5.1, 5.2, and 5.3. We state the corresponding result in the more general situation when

$$A(t) \approx t^{\frac{n}{m}} (\log t)^{\alpha}$$

for some $\alpha \geqslant 0$. Let $u \in W^{m,A}(\Omega)$. If $\alpha > \frac{n}{m} - 1$, then Theorem 5.1 tells us that u has an mth order differential almost everywhere in Ω. If, instead,

$\alpha < \frac{n}{m} - 1$, then from Theorem 5.3 we infer that, for every $\sigma > 0$,

$$\lim_{r\to 0^+} \fint_{B_r(x)} \left(\exp\left(\frac{|u(y) - T_x^m(u)(y)|}{\sigma |y-x|^m} \right)^{\frac{n}{n-m-\alpha m}} - 1 \right) dy = 0 \tag{5.10}$$

for a.e. $x \in \Omega$. Finally, in the limiting case where $\alpha = \frac{n}{m} - 1$, again Theorem 5.3 entails that, for every $\sigma > 0$,

$$\lim_{r\to 0^+} \fint_{B_r(x)} \left(\exp\left(\exp\left(\frac{|u(y) - T_x^m(u)(y)|}{\sigma |y-x|^m} \right)^{\frac{n}{n-m}} \right) - e \right) dy = 0 \tag{5.11}$$

for a.e. $x \in \Omega$. Analogous results involving norms also follow from Theorem 5.3.

Conclusions (5.10) and (5.11), with $m = 1$, overlap with results from [2].

Let us also mention that fine properties of functions from Orlicz–Sobolev spaces involving capacities, such as quasi-continuity, are analyzed in [3, 21, 32, 43].

6 Trace Inequalities

If Ω is a Lipschitz domain in $\mathbb{R}^n$, then a linear bounded operator

$$\mathrm{Tr} : W^{1,1}(\Omega) \to L^1(\partial\Omega), \tag{6.1}$$

the trace operator, exists such that

$$\mathrm{Tr}\, u = u_{|\partial\Omega}$$

whenever u is a continuous function on $\overline{\Omega}$. In particular, the operator Tr is well-defined on the Orlicz–Sobolev space $W^{m,A}(\Omega)$ for every Young function A.

In this section, we are concerned with optimal trace embeddings for $W^{m,A}(\Omega)$ into Orlicz spaces and into r.i. spaces on $\partial\Omega$, with respect to the $(n-1)$-dimensional Hausdorff measure. The material that will be presented is taken from [17]; earlier trace embeddings with non optimal ranges were established in [5, 24, 40].

Embeddings into Orlicz spaces are contained in Theorem 6.1 below, where we associate with any Young function A another Young function A_T having the property that $L^{A_T}(\partial\Omega)$ is the smallest Orlicz space into which the operator Tr maps $W^{m,A}(\Omega)$ continuously.

Note that, as in the case of embeddings into function spaces defined on the whole of Ω, we may restrict our attention to the case where

$$1 \leqslant m < n.$$

Moreover, since we are dealing with Orlicz–Sobolev spaces on domains Ω with finite measure, we may assume, without loss of generality, that

$$\int_0 \left(\frac{t}{A(t)}\right)^{\frac{m}{n-m}} dt < \infty. \tag{6.2}$$

The Young function A_T coming into play in the trace embedding into Orlicz space is defined by

$$A_T(t) = \int_0^{H^{-1}(t)} \left(\frac{A(\tau)}{\tau}\right)^{\frac{n-1-m}{n-m}} H(\tau)^{\frac{1}{m-n}} \, d\tau \qquad \text{for } t \geqslant 0, \tag{6.3}$$

where $H : [0, \infty) \to [0, \infty)$ is given by (3.3).

Note that only the asymptotic behavior at infinity of the integral on the right-hand side of (6.3) is relevant in applications.

Theorem 6.1. *Let Ω be a Lipschitz domain in $\mathbb{R}^n$. Let $1 \leqslant m < n$ and let A be a Young function fulfilling* (6.2).

(i) *Assume that*

$$\int^{\infty} \left(\frac{t}{A(t)}\right)^{\frac{m}{n-m}} dt = \infty. \tag{6.4}$$

Then there exists a constant $C = C(\Omega, m)$ such that

$$\|\operatorname{Tr} u\|_{L^{A_T}(\partial\Omega)} \leqslant C\|u\|_{W^{m,A}(\Omega)} \tag{6.5}$$

for every $u \in W^{m,A}(\Omega)$. Moreover, $L^{A_T}(\partial\Omega)$ is the optimal Orlicz space in (6.5), *in the sense that if* (6.5) *holds with A_T replaced by another Young function B, then $L^{A_T}(\partial\Omega) \to L^B(\partial\Omega)$.*

(ii) *Assume that*

$$\int^{\infty} \left(\frac{t}{A(t)}\right)^{\frac{m}{n-m}} dt < \infty. \tag{6.6}$$

Then there exists a constant $C = C(\Omega, m, A)$ such that

$$\|\operatorname{Tr} u\|_{L^{\infty}(\partial\Omega)} \leqslant C\|u\|_{W^{m,A}(\Omega)} \tag{6.7}$$

for every $u \in W^{m,A}(\Omega)$. The space $L^{\infty}(\partial\Omega)$ is (trivially) the optimal Orlicz space in (6.7).

Remark 6.2. The inequality (6.5) implies (and is, in fact, equivalent to) the integral inequality

$$\int_{\partial\Omega} A_T\left(\frac{|\mathrm{Tr}\, u(y)|}{C\left(\int_\Omega A(\sum_{k=0}^m |\nabla^k u(x)|)\, dx\right)^{m/n}}\right) d\mathcal{H}^{n-1}(y)$$
$$\leqslant \left(\int_\Omega A\Big(\sum_{k=0}^m |\nabla^k u(x)|\Big)\, dx\right)^{\frac{1}{n'}} \tag{6.8}$$

for every $u \in W^{m,A}(\Omega)$. Here, $n' = \frac{n}{n-1}$.

Example 6.3. Let

$$A(t) \approx t^p (\log t)^\alpha,$$

where either $p > 1$ and $\alpha \in \mathbb{R}$, or $p = 1$ and $\alpha \geqslant 0$. An application of Theorem 6.1 tells us that

$$\mathrm{Tr} : W^{m,A}(\Omega) \to \begin{cases} L^{\frac{p(n-1)}{n-mp}} (\log L)^{\frac{\alpha(n-1)}{n-mp}} (\partial\Omega) & \text{if } 1 \leqslant p < \frac{n}{m}, \\ \exp L^{\frac{n}{n-m-\alpha m}} (\partial\Omega) & \text{if } p = \frac{n}{m} \text{ and } \alpha < \frac{n-m}{m}, \\ \exp\exp L^{\frac{n}{n-m}} (\partial\Omega) & \text{if } p = \frac{n}{m} \text{ and } \alpha = \frac{n-m}{m}, \\ L^\infty(\partial\Omega) & \text{if either } p = \frac{n}{m} \text{ and} \\ & \quad \alpha > \frac{n-m}{m}, \text{ or } p > \frac{n}{m}, \end{cases} \tag{6.9}$$

all the range spaces being optimal in the class of Orlicz spaces. The case where $p \neq n$ and $\alpha = 0$ in (6.9) reproduces the standard trace inequality in Sobolev spaces. When $p = n$ and $\alpha = 0$, embedding (6.9) is a special case of a result from [1] and [35].

Example 6.4. Assume that

$$A(t) \approx t^p (\log(\log t))^\alpha,$$

where p and α are as in Example 6.3. Then, we infer from Theorem 6.1 that

$$\mathrm{Tr} : W^{m,A}(\Omega) \to L^{A_T}(\partial\Omega),$$

where

$$A_T(t) \approx \begin{cases} t^{\frac{p(n-1)}{n-mp}} (\log(\log t))^{\frac{\alpha(n-1)}{n-mp}} & \text{if } 1 \leqslant p < \frac{n}{m}, \\ e^{t^{\frac{n}{n-m}} (\log t)^{\frac{\alpha m}{n-m}}} & \text{if } p = \frac{n}{m}, \end{cases}$$

whereas

$$\mathrm{Tr} : W^{m,A}(\Omega) \to L^\infty(\partial\Omega) \qquad \text{if } p > \frac{n}{m}.$$

Moreover, the range spaces are sharp in the framework of Orlicz spaces on $\partial\Omega$.

The optimal rearrangement invariant range space for trace embeddings of Orlicz–Sobolev spaces under assumption (6.4) is exhibited in the next theorem. Such a space turns out to belong to the family of Orlicz–Lorentz spaces defined as in (2.10). In fact, it agrees with the space $L(\frac{n}{m}, n', E)(\partial\Omega)$, where E is defined by (3.16)-(3.17). Hence $L(\frac{n}{m}, n', E)(\partial\Omega)$ is the Orlicz–Lorentz space equipped with the norm given by

$$\|f\|_{L(\frac{n}{m},n',E)(\partial\Omega)} = \|s^{-\frac{m}{n}} f^*(\mathcal{H}^{n-1}(\partial\Omega)s^{\frac{1}{n'}})\|_{L^E(0,1)} \tag{6.10}$$

for any measurable function f on $\partial\Omega$.

Note that, when (6.4) fails to hold, Theorem 6.1 (ii) already provides the optimal r.i. range in the trace embedding of $W^{m,A}(\Omega)$, since $L^\infty(\partial\Omega)$ is the smallest r.i. space on $\partial\Omega$ by (2.2).

Theorem 6.5. *Let Ω, m and A be as in Theorem* 6.1 (i). *Let E be the Young function defined by* (3.16)–(3.17), *and let $L(\frac{n}{m}, n', E)(\partial\Omega)$ be the r.i. space endowed with the norm defined as in* (6.10). *Then there exists a constant $C = C(\Omega, m)$ such that*

$$\|\mathrm{Tr}\, u\|_{L(\frac{n}{m},n',E)(\partial\Omega)} \leqslant C\|u\|_{W^{m,A}(\Omega)} \tag{6.11}$$

for every $u \in W^{m,A}(\Omega)$. Moreover, $L(\frac{n}{m}, n', E)(\partial\Omega)$ is the optimal r.i. space in (6.11), *in the sense that if* (6.11) *holds with $L(\frac{n}{m}, n', E)(\partial\Omega)$ replaced by another r.i. space $X(\partial\Omega)$, then $L(\frac{n}{m}, n', E)\ (\partial\Omega) \to X(\partial\Omega)$.*

In view of applications of Theorem 6.5, owing to (2.6) the function E comes into play only through its behavior at infinity.

Remark 6.6. The inequality (6.11) turns out to be equivalent to the integral inequality

$$\int_0^1 E(C^{-1}s^{-\frac{m}{n}}(\mathrm{Tr}\, u)^*(\mathcal{H}^{n-1}(\partial\Omega)s^{\frac{1}{n'}}))\, ds \leqslant \int_\Omega A\Big(\sum_{k=0}^n |\nabla^k u|\Big)\, dx \tag{6.12}$$

for $u \in W^{m,A}(\Omega)$.

One always has that $E \preceq A$ [14, Proposition 5.1]. Moreover, $E \approx A$ if and only if $I_\infty(A) < \frac{n}{m}$, and the latter inequality is in turn equivalent to the fact that $L(\frac{n}{m}, n', E)(\partial\Omega) = L(\frac{n}{m}, n', A)(\partial\Omega)$, up to equivalent norms [14, Proposition 5.2]. The norm in $L(\frac{n}{m}, n', A)(\partial\Omega)$ is defined as in (2.10), namely

$$\|f\|_{L(\frac{n}{m},n',A)(\partial\Omega)} = \|s^{-\frac{m}{n}} f^*(\mathcal{H}^{n-1}(\partial\Omega)s^{\frac{1}{n'}})\|_{L^A(0,1)}$$

for any measurable function f on $\partial\Omega$. Thus, we have the following corollary of Theorem 6.5.

Corollary 6.7. *Let Ω, m and A be as in Theorem* 6.1 (i). *There exists a constant $C = C(\Omega, m)$ such that*

$$\|\mathrm{Tr}\, u\|_{L(\frac{n}{m},n',A)(\partial\Omega)} \leqslant C\|u\|_{W^{m,A}(\Omega)} \tag{6.13}$$

for every $u \in W^{m,A}(\Omega)$ if and only if $I_\infty(A) < \frac{n}{m}$. Moreover, under this assumption, $L(\frac{n}{m}, n', A)(\partial\Omega)$ is the optimal r.i. space in (6.13).

Example 6.8. Let A be as in Example 6.3. From Theorem 6.5 and Corollary 6.7 one can deduce that

$$\mathrm{Tr} : W^{m,A}(\Omega) \to \begin{cases} L^{\frac{p(n-1)}{n-mp},p;\frac{\alpha}{p}}(\partial\Omega) & \text{if } 1 \leqslant p < \frac{n}{m}, \\ L^{\infty,\frac{n}{m};\frac{m\alpha}{n}-1}(\partial\Omega) & \text{if } p = \frac{n}{m} \text{ and } \alpha < \frac{n-m}{m}, \\ L^{\infty,\frac{n}{m};-\frac{m}{n},-1}(\partial\Omega) & \text{if } p = \frac{n}{m} \text{ and } \alpha = \frac{n-m}{m}, \end{cases} \tag{6.14}$$

up to equivalent norms, and that all the range spaces are optimal among r.i. spaces.

References

1. Adams, D.R.: Traces of potentials II. Indiana Univ. Math. J. **22**, 907–918 (1973)
2. Adams, D.R., Hurri-Syriänen, R.: Capacity estimates. Proc. Am. Math. Soc. **131**, 1159–1167 (2002)
3. Adams, D.R., Hurri-Syriänen, R.: Vanishing exponential integrability for functions whose gradients belong to $L^n(\log(e+L))^\alpha$. J. Funct. Anal. **197**, 162–178 (2003)
4. Adams, R.A.: Sobolev spaces. Academic Press, New York etc. (1975)
5. Adams, R.A.: On the Orlicz–Sobolev imbedding theorem. J. Funct. Anal. **24**, 241–257 (1977)
6. Alberico, A., Cianchi, A.: Differentiability properties of Orlicz–Sobolev functions. Ark. Math. **43**, 1–28 (2005)
7. Bennett, C., Rudnick, K.: On Lorentz–Zygmund spaces. Dissert. Math. **175**, 1–72 (1980)
8. Bennett, C., Sharpley, R.: Interpolation of Operators. Academic Press, Boston (1988)
9. Brezis, H., Wainger, S.: A note on limiting cases of Sobolev embeddings and convolution inequalities. Commun. Partial Differ. Equ. **5**, 773–789 (1980)
10. Cianchi, A.: Continuity properties of functions from Orlicz–Sobolev spaces and embedding theorems. Ann. Sc. Norm. Super. Pisa, Cl. Sci., Ser. IV **23**, 576–608 (1996)
11. Cianchi, A.: A sharp embedding theorem for Orlicz–Sobolev spaces. Indiana Univ. Math. J. **45**, 39–65 (1996)
12. Cianchi, A.: Boundedness of solutions to variational problems under general growth conditions. Commun. Partial Differ. Equ. **22**, 1629–1646 (1997)
13. Cianchi, A.: A fully anisotropic Sobolev inequality. Pacific J. Math. **196**, 283–295 (2000)
14. Cianchi, A.: Optimal Orlicz–Sobolev embeddings. Rev. Mat. Iberoam. **20**, 427–474 (2004)
15. Cianchi, A.: Higher-order Sobolev and Poincaré inequalities in Orlicz spaces. Forum Math. **18**, 745–767 (2006)
16. Cianchi, A.: Orlicz–Sobolev algebras. Potential Anal. **28**, 379–388 (2008)
17. Cianchi, A.: Orlicz–Sobolev boundary trace embeddings. Math. Z. [To appear]
18. Cianchi, A., Pick, L.: Sobolev embeddings into spaces of Campanato, Morrey, and Hölder type, J. Math. Anal. Appl. **282**, 128–150 (2003)

19. Cianchi, A., Randolfi, M.: On the Modulus of Continuity of Sobolev Functions. Preprint
20. Cianchi, A., Randolfi, M.: Higher-order differentiability properties of functions in Orlicz–Sobolev spaces [In preparation]
21. Cianchi, A., Stroffolini, B.: An extension of Hedberg's convolution inequality and applications. J. Math. Anal. Appl. **227**, 166–186 (1998)
22. Cwikel, M., Pustylnik, E.: Sobolev type embeddings in the limiting case. J. Fourier Anal. Appl. **4**, 433–446 (1998)
23. Donaldson, D.T.: Nonlinear elliptic boundary value problems in Orlicz–Sobolev spaces. J. Differ. Equ. **10**, 507–528 (1971)
24. Donaldson, D.T., Trudinger, N.S.: Orlicz–Sobolev spaces and embedding theorems. J. Funct. Anal. **8**, 52–75 (1971)
25. Edmunds, D.E., Gurka, P., Opic, B.: Double exponential integrability of convolution operators in generalized Lorentz–Zygmund spaces. Indiana Univ. Math. J. **44**, 19–43 (1995)
26. Edmunds, D.E., Kerman, R.A., Pick, L.: Optimal Sobolev imbeddings involving rearrangement invariant quasi-norms. J. Funct. Anal. **170**, 307–355 (2000)
27. Fusco, N., Lions, P.L., Sbordone, C.: Some remarks on Sobolev embeddings in borderline cases. Proc. Am. Math. Soc. **70**, 561–565 (1996)
28. Gossez, J.-P.: Nonlinear elliptic boundary value problems for equations with rapidly (or slowly) increasing coefficients. Trans. Am. Math. Soc. **190**, 163–205 (1974)
29. Greco, L., Moscariello, G.: An embedding theorem in Lorentz–Zygmund spaces. Potential Anal. **5**, 581–590 (1996)
30. Hansson, K.: Imbedding theorems of Sobolev type in potential theory. Math. Scand. **45**, 77–102 (1979)
31. Koronel, J.D.: Continuity and kth order differentiability in Orlicz–Sobolev spaces: W^kL_A. Israel J. Math. **24**, 119–138 (1976)
32. Malý, J., Swanson D., Ziemer, W.P.: Fine behaviour of functions with gradients in a Lorentz space. Studia Math. **190**, 33–71 (2009)
33. Maz'ya, V.G.: On certain integral inequalities for functions of many variables (Russian). Probl. Mat. Anal. **3**, 33–68 (1972); English transl.: J. Math. Sci. New York **1**, 205–234 (1973)
34. Maz'ya, V.G.: The continuity and boundedness of functions from Sobolev spaces (Russian). Probl. Mat. Anal. **4**, 46–77 (1973); English transl.: J. Math. Sci., New York **6**, 29–50 (1976)
35. Maz'ya, V.G.: Sobolev Spaces. Springer, Berlin etc. (1985)
36. O'Neil, R.: Convolution operators and $L(p,q)$ spaces. Duke Math. J. **30**, 129–142 (1963)
37. Opic, B., Pick, L.: On generalized Lorentz–Zygmund spaces. Math. Ineq. Appl. **2**, 391–467 (1999)
38. Peetre, J.: Espaces d' interpolation et théorème de Soboleff. Ann. Inst. Fourier **16**, 279–317 (1966)
39. Pokhozhaev, S.I.: On the imbedding Sobolev theorem for $pl = n$ (Russian). In: Dokl. Conference, Sec. Math. Moscow Power Inst., pp. 158–170 (1965)
40. Palmieri, G.: An approach to the theory of some trace spaces related to the Orlicz–Sobolev spaces. Boll. Un. Mat. Ital. B **16**, 100–119 (1979)
41. Rao, M.M., Ren, Z.D.: Theory of Orlicz Spaces. Marcel Dekker Inc., New York (1991)
42. Rao, M.M., Ren, Z.D.: Applications of Orlicz Spaces. Marcel Dekker Inc., New York (2002)

43. Rudd, N.: A direct approach to Orlicz–Sobolev capacity. Nonlinear Anal. **60**, 129–147 (2005)
44. Stein, E.M.: Singular Integrals and Differentiablity Properties of Functions. Princeton Univ. Press, Princeton, NJ (1970)
45. Strichartz, R.S.: A note on Trudinger' s extension of Sobolev' s inequality. Indiana Univ. Math. J. **21**, 841–842 (1972)
46. Talenti, G.: An embedding theorem. In: Partial Differential Equations and the Calculus of Variations II. pp. 919–924. Birkhäuser (1989)
47. Trudinger, N.S.: On imbeddings into Orlicz spaces and some applications. J. Math. Mech. **17**, 473–483 (1967)
48. Yudovich, V.I.: Some estimates connected with integral operators and with solutions of elliptic equations (Russian). Dokl. Akad. Nauk SSSR **138**, 805–808 (1961); English transl.: Sov. Math. Dokl. **2**, 746–749 (1961)
49. Ziemer, W.P.: Weakly Differentiable Functions. Spriger, New York (1989)

Mellin Analysis of Weighted Sobolev Spaces with Nonhomogeneous Norms on Cones

Martin Costabel, Monique Dauge, and Serge Nicaise

Abstract On domains with conical points, weighted Sobolev spaces with powers of the distance to the conical points as weights form a classical framework for describing the regularity of solutions of elliptic boundary value problems (cf. works of Kondrat'ev and Maz'ya–Plamenevskii). Two classes of weighted norms are usually considered: homogeneous norms, where the weight exponent varies with the order of derivatives, and nonhomogeneous norms, where the same weight is used for all orders of derivatives. For the analysis of the spaces with homogeneous norms, Mellin transformation is a classical tool. In this paper, we show how Mellin transformation can also be used to give an optimal characterization of the structure of weighted Sobolev spaces with nonhomogeneous norms on finite cones in the case of both noncritical and critical indices. This characterization can serve as a basis for the proof of regularity and Fredholm theorems in such weighted Sobolev spaces on domains with conical points, even in the case of critical indices.

1 Introduction

When analyzing elliptic regularity in a neighborhood of a conical point on the boundary of an otherwise smooth domain, one is faced with the following dilemma.

Martin Costabel
IRMAR, Université de Rennes 1, Campus de Beaulieu, 35042 Rennes, France
e-mail: martin.costabel@univ-rennes1.fr

Monique Dauge
IRMAR, Université de Rennes 1, Campus de Beaulieu, 35042 Rennes, France
e-mail: monique.dauge@univ-rennes1.fr

Serge Nicaise
LAMAV, FR CNRS 2956, Université Lille Nord de France, UVHC, 59313 Valenciennes Cedex 9, France
e-mail: snicaise@univ-valenciennes.fr

A. Laptev (ed.), *Around the Research of Vladimir Maz'ya I: Function Spaces*, International Mathematical Series 11, DOI 10.1007/978-1-4419-1341-8_4,

Near the singular point, the conical geometry suggests the use of estimates in weighted Sobolev spaces with homogeneous norms, and a well-known tool for analyzing them and for obtaining the estimates is the Mellin transformation. This analysis is carried out in the classical paper [3] by Kondrat'ev.

On the other hand, since this analysis corresponds to a blow-up of the corner, i.e., a diffeomorphism between the tangent cone and an infinite cylinder, the conical point moves to infinity, and therefore functions in this class of spaces always have trivial Taylor expansions at the corner. Depending on the weight index, they either have no controlled behavior at the corner at all or they tend to zero. If one wants to study inhomogeneous boundary value problems, then smooth right-hand sides and the corresponding solutions will require spaces that allow the description of nontrivial Taylor expansions at corner points.

Appropriate spaces have been analyzed using tools from real analysis by Maz'ya and Plamenevskii [6]. Such spaces can be defined by nonhomogeneous weighted norms, where the weight exponent is the same for all derivatives. The simplest examples are ordinary, nonweighted Sobolev norms. As presented in detail in the book [4] by Kozlov, Maz'ya, and Rossmann, the analysis of these spaces with nonhomogeneous norms shows several peculiarities:

1. For a given space dimension n and Sobolev order m, there is a finite set of exceptional, "*critical*" weight exponents β, characterized in our notation by the condition

$$-\beta - \tfrac{n}{2} = \eta \in \mathbb{N}\,; \quad 0 \leqslant \eta \leqslant m-1\,,$$

such that, in the *noncritical case*, the space with nonhomogeneous norm splits into the direct sum of a space with homogeneous norm and a space of polynomials, corresponding to the Taylor expansion at the corner. In the critical case, the splitting involves an infinite-dimensional space of generalized polynomials. The study of the critical cases is of practical importance, because for example in two-dimensional domains, the ordinary Sobolev spaces with integer order are all in the critical case $\eta = m-1$.

2. The relation of the spaces with nonhomogeneous norms with respect to Taylor expansions at the corner is somewhat complicated, depending on the weight and order. For $\eta < 0$, the space with nonhomogeneous norm coincides with the corresponding space with homogeneous norm and contains all polynomials, but has no controlled Taylor expansion. For $0 \leqslant \eta < m$, the nonhomogeneous norm still allows all polynomials and controls the Taylor expansion of order $[\eta]$ at the corner. If $\eta \geqslant m$, then the space with nonhomogeneous norm again coincides with the corresponding space with homogeneous norm and has vanishing Taylor expansion of order $m-1$. Thus, there are two (nondisjoint) classes of spaces involved, and the weighted Sobolev spaces with nonhomogeneous norms fall into one or the other of these classes, namely the class of spaces with *homogeneous norms* on one hand and a class of spaces with *weighted norms and nontrivial Taylor expansion* on the other hand.

3. Whereas the *definition* of the nonhomogeneous norms is simple, it turns out that for the *analysis* of the spaces one also needs descriptions by more complicated equivalent norms, where the weight exponent does depend, in a specific way, on the order of the derivatives. Such "step-weighted" Sobolev spaces were studied by Nazarov [7, 8].

In [4], the analysis of the weighted Sobolev spaces with nonhomogeneous norms is presented using real-variable tools, in particular techniques based on the Hardy inequality.

In this paper, we present an analysis of the spaces with nonhomogeneous norms based on Mellin transformation. We show how the three points described above can be achieved in an optimal way. In particular,

1) we characterize the spaces with nonhomogeneous norms via Mellin transformation in the noncritical and in the critical case,

2) we give a natural definition via Mellin transforms of the second class of spaces mentioned in point 2 above, namely the spaces with weighted norms and nontrivial Taylor expansions,

3) we show how the question of equivalent norms can be solved via Mellin transformation.

The analysis in this paper is a generalization of the Mellin characterization of standard Sobolev spaces that was introduced in [2] for the analysis of elliptic regularity on domains with corners. For the case of critical weight exponents, we give a Mellin description of the generalized Taylor expansion that was introduced and analyzed with real-variable techniques in [4]. Based on our Mellin characterization, one can obtain Fredholm theorems and elliptic regularity results, in particular analytic regularity results, on domains with conical points. This is developed in the forthcoming work [1].

2 Notation: Weighted Sobolev Spaces on Cones

A *regular cone* $K \subset \mathbb{R}^n, n \geqslant 2$ is an unbounded open set of the form

$$K = \Big\{ \mathsf{x} \in \mathbb{R}^n \setminus \{\mathbf{0}\} \ : \ \frac{\mathsf{x}}{|\mathsf{x}|} \in G \Big\}, \tag{2.1}$$

where G is a smooth domain of the unit sphere $\mathbb{S}^{n-1}$ called the *solid angle* of K. Note that if $n = 2$, this implies that K has a Lipschitz boundary (excluding domains with cracks), which is not necessarily the case if $n \geqslant 3$. Note further that our analysis below is also valid in the case of domains with cracks.

The *finite cone* S associated with K is simply

$$S = K \cap B(\mathbf{0}, 1). \tag{2.2}$$

In the one-dimensional case, we consider $K = \mathbb{R}_+$ and $S = (0,1)$, which corresponds to $G = \{1\}$.

For $k \in \mathbb{N}$, $\|\cdot\|_{k;O}$ denotes the standard Sobolev norm of $H^k(O)$.

2.1 Weighted spaces with homogeneous norms

The spaces on which relies a large part of our analysis are the "classical" weighted spaces of Kondrat'ev. The "originality" of our definition is a new convention for their notation.

Definition 2.1. • Let β be a real number, and let $m \geqslant 0$ be an integer.
• β is called the *weight exponent* and m the *Sobolev exponent.*
• The *weighted space with homogeneous norm* $\mathsf{K}^m_\beta(K)$ is defined by

$$\mathsf{K}^m_\beta(K) = \left\{u \in \mathsf{L}^2_{\mathrm{loc}}(K) \;:\; r^{\beta+|\alpha|}\partial^\alpha_{\mathbf{x}} u \in \mathsf{L}^2(K) \quad \forall\alpha,\ |\alpha| \leqslant m\right\} \tag{2.3}$$

and endowed with seminorm and norm respectively defined as

$$|u|^2_{\mathsf{K}^m_\beta(K)} = \sum_{|\alpha|=m} \|r^{\beta+|\alpha|}\partial^\alpha_{\mathbf{x}} u\|^2_{0;K}, \quad \|u\|^2_{\mathsf{K}^m_\beta(K)} = \sum_{k=0}^{m} |u|^2_{\mathsf{K}^k_\beta(K)}. \tag{2.4}$$

The weighted spaces introduced by Kondrat'ev in [3] are denoted by $\mathring{W}^m_\alpha(K)$. The correspondence with our notation is

$$\mathring{W}^m_\alpha(K) = \mathsf{K}^m_{\frac{\alpha}{2}-m}(K), \text{ i.e., } \mathsf{K}^m_\beta(K) = \mathring{W}^m_{2\beta+2m}(K).$$

These spaces are also of constant use in related works by Kozlov, Maz'ya, Nazarov, Plamenevskii, Rossmann (cf. the monographs [9, 4, 5] for example). They are denoted by $V^m_\beta(K)$ with the following correspondence with our spaces

$$V^m_\beta(K) = \mathsf{K}^m_{\beta-m}(K), \text{ i.e., } \mathsf{K}^m_\beta(K) = V^m_{\beta+m}(K).$$

We choose the convention in (2.3) because it simplifies some statements. An obvious, but fundamental property of the scale K_β is its monotonicity with respect to m

$$\mathsf{K}^{m+1}_\beta(K) \subset \mathsf{K}^m_\beta(K), \quad m \in \mathbb{N}.$$

This allows a simple definition of $\mathscr{C}^\infty$ and analytic functions with weight, see Definition 4.1. Also, in mapping properties of differential operators with constant coefficients, as well as in elliptic regularity theorems ("shift theorem"), the shift in the weight exponent β is independent of the regularity parameter m, in contrast to what happens with the Kondrat'ev or the V^m_β spaces.

The space $\mathsf{K}^m_\beta(S)$ with its seminorm $|\cdot|_{\mathsf{K}^m_\beta(S)}$ and norm $\|\cdot\|_{\mathsf{K}^m_\beta(S)}$ is defined similarly by replacing K by S.

2.2 Weighted spaces with nonhomogeneous norms

Definition 2.2. • Let β be a real number, and let $m \geqslant 0$ be an integer.
• The *weighted space with nonhomogeneous norm* $\mathsf{J}^m_\beta(S)$ is defined by

$$\mathsf{J}^m_\beta(S) = \{u \in \mathsf{L}^2_{\mathrm{loc}}(S) \ : \ r^{\beta+m}\partial^\alpha_{\mathbf{x}} u \in \mathsf{L}^2(S) \quad \forall\alpha,\ |\alpha| \leqslant m\} \tag{2.5}$$

with its norm

$$\|u\|^2_{\mathsf{J}^m_\beta(S)} = \sum_{|\alpha|\leqslant m} \|r^{\beta+m}\partial^\alpha_{\mathbf{x}} u\|^2_{0;S}\,.$$

The seminorm of $\mathsf{J}^m_\beta(S)$ coincides with the seminorm of $\mathsf{K}^m_\beta(S)$:

$$|u|^2_{\mathsf{J}^m_\beta(S)} = |u|^2_{\mathsf{K}^m_\beta(S)} = \sum_{|\alpha|=m} \|r^{\beta+|\alpha|}\partial^\alpha_{\mathbf{x}} u\|^2_{0;S}\,. \tag{2.6}$$

• The space $\mathsf{J}^m_\beta(K)$ with its norm and seminorm is defined in the same way.

Our space $\mathsf{J}^m_\beta(S)$ is the same as the space denoted by $W^m_{2,\beta+m}(S)$ in [4]. The following properties are obvious consequences of the definitions:

Lemma 2.3. (a) *For all $\beta < \beta'$ we have the embedding $\mathsf{J}^m_\beta(S) \subset \mathsf{J}^m_{\beta'}(S)$.*

(b) *We have the embeddings for all $\beta \in \mathbb{R}$ and $m \in \mathbb{N}$*

$$\mathsf{K}^m_\beta(S) \subset \mathsf{J}^m_\beta(S) \subset \mathsf{K}^m_{\beta+m}(S). \tag{2.7}$$

(c) *Let $\alpha \in \mathbb{N}^n$ be a multiindex of length $|\alpha| = k \leqslant m$. Then the partial differential operator $\partial^\alpha_{\mathbf{x}}$ is continuous from $\mathsf{J}^m_\beta(S)$ into $\mathsf{J}^{m-k}_{\beta+k}(S)$.*

In contrast to the scale K^m_β, we do not necessarily have the inclusion of $\mathsf{J}^m_\beta(S)$ in $\mathsf{J}^{m-1}_\beta(S)$. We will see (Corollary 3.19) that such an inclusion does hold when m is large enough, which allows the definition of $\mathsf{J}^\infty_\beta(S)$ and of the corresponding analytic class.

A remarkable and unusual property of the spaces $\mathsf{J}^m_\beta(S)$ is that we do not, in general, obtain an equivalent norm for $\mathsf{J}^m_\beta(S)$ if we retain in (2.5) only the seminorm ($|\alpha| = m$) and the L^2 norm ($|\alpha| = 0$). A counterexample for such an equivalence is obtained with the following choice

$$m \geqslant 2, \quad m < \eta = -\beta - \tfrac{n}{2} < m+1, \quad u = x_1\,. \tag{2.8}$$

Then $r^{\beta+m}\partial^\alpha_{\mathbf{x}} u$ is square integrable for $|\alpha| = 0$ and for $|\alpha| \geqslant 2$, but not for $\alpha = (1, 0, \ldots, 0)$ (cf. Subsection 3.4 for further details).

We need more precise comparisons between the K and J spaces than the embeddings (2.7). As we will show later on, the space $\mathsf{K}^m_\beta(S)$ may be closed with finite codimension in $\mathsf{J}^m_\beta(S)$ (noncritical case), or not closed with infinite codimension (critical case). In the following lemma, we compare the properties of inclusion of the space $\mathscr{C}^\infty(\overline{S})$ of smooth functions.

Lemma 2.4. *Let $\beta \in \mathbb{R}$, and let $m \in \mathbb{N}$. Let $\eta = -\beta - \frac{n}{2}$.*

(a) *The space $\mathscr{C}^\infty(\overline{S})$ is embedded in $\mathsf{K}^m_\beta(S)$ if and only if $\eta < 0$.*

(b) *The space $\mathscr{C}^\infty(\overline{S})$ is embedded in $\mathsf{J}^m_\beta(S)$ if and only if $\eta < m$.*

Proof. Using polar coordinates and the Cauchy–Schwarz inequality, we see that

$$\mathsf{L}^\infty(S) \subset \mathsf{L}^2_\beta(S) \quad \Longleftrightarrow \quad \beta > -\tfrac{n}{2}.$$

The sufficiency follows by using this for all derivatives of $u \in \mathscr{C}^\infty(\overline{S})$.

We find the necessity of the conditions on η by considering the constant function $u = 1$ in both cases. □

Concerning spaces of finite regularity, it follows from the definition that the standard Sobolev space H^m without weight coincides with J^m_{-m}. For the Sobolev spaces H^m we have the embeddings corresponding to (2.7), namely

$$\mathsf{K}^m_{-m}(S) \subset \mathsf{H}^m(S) \subset \mathsf{K}^m_0(S). \tag{2.9}$$

In addition, we know from the Sobolev embedding theorem that if k is a nonnegative integer such that $k < m - \frac{n}{2}$, we have the embeddings

$$\mathsf{H}^m(S) \subset \mathscr{C}^k(\overline{S}) \subset \mathsf{H}^k(S).$$

In particular, for elements of $\mathsf{H}^m(S)$ all derivatives of length $|\alpha| \leqslant k$ have a trace at the vertex $\mathbf{0}$. On the other hand, by the density of smooth functions which are zero at the vertex, the elements of $\mathsf{K}^m_\beta(S)$, as soon as they have traces, have *zero* traces at the vertex.

One can expect that the spaces J have vertex traces similar to the standard Sobolev spaces. The investigation of this question will be the key to the comparison between the J spaces and the K spaces.

Using the same simple argument as in the proof of Lemma 2.4, we find the conditions for the inclusion of polynomials in the weighted Sobolev spaces.

We denote by $\mathbb{P}^M(S)$ the space of *polynomial functions* of degree $\leqslant M$ on S and by $\mathsf{P}^M(S)$ the space of *homogeneous* polynomials of degree M.

Lemma 2.5. *Suppose that $\beta \in \mathbb{R}$, $m, k \in \mathbb{N}$, and $\eta = -\beta - \frac{n}{2}$.*

(a) $\mathbb{P}^k(S) \subset \mathsf{K}^m_\beta(S) \quad \Longleftrightarrow \quad \mathbb{P}^0(S) \subset \mathsf{K}^m_\beta(S) \quad \Longleftrightarrow \quad \eta < 0.$

(b) $\mathbb{P}^k(S) \subset \mathsf{J}^m_\beta(S) \quad \Longleftrightarrow \quad \mathbb{P}^0(S) \subset \mathsf{J}^m_\beta(S) \quad \Longleftrightarrow \quad \eta < m.$

This complete similarity between the K spaces and the J spaces is no longer present if we refine the probe by considering the space of homogeneous polynomials. Still using the same simple argument based on finiteness of norms, we now get

Lemma 2.6. *Suppose that $\beta \in \mathbb{R}$, $m, k \in \mathbb{N}$, and $\eta = -\beta - \frac{n}{2}$.*

(a) $\mathsf{P}^k(S) \subset \mathsf{K}^m_\beta(S) \quad \Longleftrightarrow \quad \eta < k.$

(b) *If $k \geqslant m$, then* $\mathsf{P}^k(S) \subset \mathsf{J}^m_\beta(S) \Longleftrightarrow \eta < k.$

(c) *If* $k \leqslant m-1$, *then* $\mathsf{P}^k(S) \subset \mathsf{J}^m_\beta(S) \iff \mathsf{P}^0(S) \subset \mathsf{J}^m_\beta(S) \iff \eta < m$.

As we will show in the following, the question of inclusion of polynomials completely characterizes the structure of the spaces $\mathsf{J}^m_\beta(S)$ and their corner behavior.

3 Characterizations by Mellin Transformation Techniques

The homogeneous weighted Sobolev norms can be expressed by Mellin transformation, which is the Fourier transformation associated with the group of dilations. We first recall this characterization from Kondrat'ev's classical work [3]. Then we generalize it to include nonhomogeneous weighted Sobolev norms, based on the observation that the nonhomogeneous norms are defined by sums of homogeneous seminorms.

3.1 Mellin characterization of spaces with homogeneous norms

In this section, we recall the basic results from [3].

For a function u in $\mathscr{C}_0^\infty((0,\infty))$ the Mellin transform $\mathscr{M}[u]$ is defined for any complex number λ by the integral

$$\mathscr{M}[u](\lambda) = \int_0^\infty r^{-\lambda} u(r)\, \frac{\mathrm{d}r}{r}. \tag{3.1}$$

The function $\lambda \mapsto \mathscr{M}[u](\lambda)$ is then *holomorphic* on the entire complex plane $\mathbb{C}$. Note that $\mathscr{M}[u](\lambda)$ coincides with the Fourier–Laplace transform at $i\lambda$ of the function $t \mapsto u(\mathrm{e}^t)$.

Now, any function u defined on our cone K can be naturally written in polar coordinates as

$$\mathbb{R}_+ \times G \ni (r,\vartheta) \longmapsto u(\mathbf{x}) = u(r\vartheta).$$

If u has compact support which does not contain the vertex $\mathbf{0}$, the Mellin transform of u at $\lambda \in \mathbb{C}$ is the function $\mathscr{M}[u](\lambda) : \vartheta \mapsto \mathscr{M}[u](\lambda,\vartheta)$ defined on G by

$$\mathscr{M}[u](\lambda,\vartheta) = \int_0^\infty r^{-\lambda} u(r\vartheta)\, \frac{\mathrm{d}r}{r}, \quad \vartheta \in G. \tag{3.2}$$

If we define the function $\widetilde{u}$ on the cylinder $\mathbb{R} \times G$ by $\widetilde{u}(t,\vartheta) = u(\mathrm{e}^t\vartheta)$, we see that the Mellin transform of u at λ is the partial Fourier–Laplace transform of $\widetilde{u}$ at $-i\lambda$.

Hence the Mellin transform of a function $u \in \mathscr{C}_0^\infty(K)$ is holomorphic with values in $\mathscr{C}_0^\infty(G)$. On the other hand, if u is simply in $\mathsf{L}^2(K)$, the function $\mathrm{e}^{\frac{n}{2}t}\widetilde{u}$ belongs to L^2 on the cylinder $\mathbb{R} \times G$ and $\lambda \mapsto \mathscr{M}[u](\lambda)$ therefore defines an L^2 function on the line $\operatorname{Re}\lambda = -\frac{n}{2}$, with values in $\mathsf{L}^2(G)$.

More generally, the Mellin transformation extends to functions u given in a weighted space $\mathsf{K}^0_\beta(K)$ with a fixed real number β: Since $r^\beta u$ belongs to $\mathsf{L}^2(K)$, the function $\widetilde{u}$ in turn satisfies that $\mathrm{e}^{(\beta+\frac{n}{2})t}\widetilde{u}$ belongs to $\mathsf{L}^2(\mathbb{R} \times G)$. Therefore, $\lambda \mapsto \mathscr{M}[u](\lambda)$ defines an L^2 function on the line $\operatorname{Re}\lambda = -\beta - \frac{n}{2}$. If u belongs to $\mathsf{K}^m_\beta(K)$, then there appear parameter-dependent H^m norms for its Mellin transform, which motivates the following definition.

Definition 3.1. Let G be the solid angle of a regular cone K, and let $m \in \mathbb{N}$.

- For $\lambda \in \mathbb{C}$, the parameter-dependent H^m norm on G is defined by

$$\|U\|^2_{m;\,G;\,\lambda} = \sum_{k=0}^{m} |\lambda|^{2m-2k} \|U\|^2_{k;\,G}\,. \tag{3.3}$$

- Let $\lambda \mapsto U(\lambda)$ be a function with values in $\mathsf{H}^m(G)$, defined for λ in a strip $b_0 < \operatorname{Re}\lambda < b_1$. Then for any $b \in (b_0, b_1)$ we set

$$\mathcal{N}^m_G(U, b) = \left\{ \int_{\operatorname{Re}\lambda = b} \|U(\lambda)\|^2_{m;\,G;\,\lambda} \,\mathrm{d}\operatorname{Im}\lambda \right\}^{\frac{1}{2}} \tag{3.4}$$

and

$$\mathcal{N}^m_G(U, [b_0, b_1]) = \sup_{b \in (b_0, b_1)} \mathcal{N}^m_G(U, b).$$

Later on, we will use the following observation: Let $\lambda \mapsto U(\lambda)$ be *meromorphic* for $b_0 < \operatorname{Re}\lambda < b_1$ with values in $\mathsf{H}^m(G)$. If $\mathcal{N}^m_G(U, [b_0, b_1])$ is finite, then U is actually *holomorphic*. In fact, if U has a pole in λ_0, then $\mathcal{N}^m_G(U, b)$ is bounded from below by $C\,|b - \operatorname{Re}\lambda_0|^{-1}$.

As a consequence of the isomorphism between $\mathsf{K}^m_\beta(K)$ and $\mathsf{H}^m_{\beta+\frac{n}{2}}(\mathbb{R} \times G)$, one gets the following theorem.

Theorem 3.2. *Let β be a real number, and let $m \in \mathbb{N}$. Let*

$$\eta := -\beta - \tfrac{n}{2}, \quad \mathfrak{R}[\eta] := \big\{\lambda \in \mathbb{C} : \operatorname{Re}\lambda = \eta\big\}.$$

The Mellin transformation (3.2) $u \mapsto \mathscr{M}[u]$ induces an isomorphism from $\mathsf{K}^m_\beta(K)$ onto the space of functions $U : \mathfrak{R}[\eta] \times G \ni (\lambda, \vartheta) \mapsto U(\lambda, \vartheta)$ with finite norm $\mathcal{N}^m_G(U, \eta)$. The inverse Mellin transform can be written as

$$u(\mathbf{x}) = \frac{1}{2i\pi} \int_{\operatorname{Re}\lambda = \eta} r^\lambda \mathscr{M}[u](\lambda)(\vartheta)\,\mathrm{d}\lambda, \quad \mathbf{x} = r\vartheta. \tag{3.5}$$

From this theorem, we see immediately that if u belongs to the intersection of two weighted spaces $\mathsf{K}^m_\beta(K)$ and $\mathsf{K}^m_{\beta'}(K)$ with $\beta < \beta'$, the Mellin transform of u is defined on two different lines in $\mathbb{C}$. Since u belongs also to all intermediate spaces $\mathsf{K}^m_{\beta''}(K)$ for $\beta \leqslant \beta'' \leqslant \beta'$, the Mellin transform is defined in a complex strip. In fact, the Mellin transform of u is holomorphic in this strip, and this characterizes the intersection of weighted spaces with different weights, as stated in the following theorem.

Theorem 3.3. *Let $\beta < \beta'$ be two real numbers, and let $m \in \mathbb{N}$. Let*

$$\eta := -\beta - \tfrac{n}{2}, \quad \eta' := -\beta' - \tfrac{n}{2}.$$

(a) *Let $u \in \mathsf{K}^m_\beta(K) \cap \mathsf{K}^m_{\beta'}(K)$. Then the Mellin transform $U := \mathscr{M}[u]$ of u is holomorphic in the open strip $\eta' < \operatorname{Re}\lambda < \eta$ with values in $\mathsf{H}^m(G)$ and satisfies the following boundedness condition:*

$$\mathcal{N}^m_G(U, [\eta', \eta]) \leqslant C\Big(\|u\|_{\mathsf{K}^m_\beta(K)} + \|u\|_{\mathsf{K}^m_{\beta'}(K)} \Big). \tag{3.6}$$

(b) *Let U be a holomorphic function in the open strip $\eta' < \operatorname{Re}\lambda < \eta$ with values in $\mathsf{H}^m(G)$, satisfying $\mathcal{N}^m_G(U, [\eta', \eta]) < \infty$. Then the mapping*

$$b \longmapsto \Big((\xi, \vartheta) \mapsto U(b + i\xi, \vartheta) \Big) \tag{3.7}$$

has limits as $b \to \eta$ and $b \to \eta'$, and the inverse Mellin transforms

$$u' = \frac{1}{2i\pi} \int_{\operatorname{Re}\lambda = \eta'} r^\lambda\, U(\lambda)\, \mathrm{d}\lambda, \quad u = \frac{1}{2i\pi} \int_{\operatorname{Re}\lambda = \eta} r^\lambda\, U(\lambda)\, \mathrm{d}\lambda, \tag{3.8}$$

coincide with each other and define an element of $\mathsf{K}^m_\beta(K) \cap \mathsf{K}^m_{\beta'}(K)$.

In the following theorem, we recall the close relation between asymptotic expansions and meromorphic Mellin transforms.

Theorem 3.4. *Let $\beta < \beta'$ be two real numbers, and let*

$$\eta = -\beta - \tfrac{n}{2} \quad \text{and} \quad \eta' = -\beta' - \tfrac{n}{2}\,.$$

Let λ_0 be a complex number such that $\eta' < \operatorname{Re}\lambda_0 < \eta$. Let q be a nonnegative integer, and let $\varphi_0, \ldots, \varphi_q$ be fixed elements of $\mathsf{L}^2(G)$.

(a) *Let $u' \in \mathsf{K}^0_{\beta'}(K)$ be such that the identity*

$$u(\mathbf{x}) = u'(\mathbf{x}) + r^{\lambda_0} \sum_{j=0}^{q} \frac{1}{j!} \log^j r\, \varphi_j(\vartheta) \tag{3.9}$$

defines a function u in $\mathsf{K}^0_\beta(K)$. Then the Mellin transform U of u', defined for $\operatorname{Re}\lambda = \eta'$, has a meromorphic extension to the strip $\eta' < \operatorname{Re}\lambda < \eta$ such that the function V defined as

$$V(\lambda) := U(\lambda) - \sum_{j=0}^{q} \frac{\varphi_j}{(\lambda - \lambda_0)^{j+1}} \tag{3.10}$$

is holomorphic in $\eta' < \operatorname{Re}\lambda < \eta$ *with values in* $\mathsf{L}^2(G)$ *and satisfies the boundedness condition*

$$\mathcal{N}^0_G(V, [\eta', \eta]) < \infty. \tag{3.11}$$

(b) *Conversely, let* U *be a meromorphic function with values in* $\mathsf{L}^2(G)$, *such that* V *defined by* (3.10) *is holomorphic in the strip* $\eta' < \operatorname{Re}\lambda < \eta$ *and satisfies the boundedness condition* (3.11). *Then, like in the holomorphic case, the mapping* (3.7) *has limits at* η *and* η', *and the inverse Mellin formulas* (3.8) *define* $u \in \mathsf{K}^0_\beta(K)$ *and* $u' \in \mathsf{K}^0_{\beta'}(K)$. *They satisfy the relation* (3.9), *which can be also written in the form of a residue formula*

$$u(\mathbf{x}) - u'(\mathbf{x}) = \frac{1}{2i\pi} \int_{\mathfrak{C}} r^\lambda\, U(\lambda)\, \mathrm{d}\lambda \tag{3.12}$$

for a contour $\mathfrak{C}$ *surrounding* λ_0 *and contained in the strip* $\eta' < \operatorname{Re}\lambda < \eta$.

3.2 Mellin characterization of seminorms

The principle of our Mellin analysis is to apply to a function u and some of its derivatives $\partial^\alpha_{\mathbf{x}} u$ the Mellin characterization of K-weighted spaces from Theorems 3.2, 3.3, and 3.4.

Definition 3.5. • For any $\alpha \in \mathbb{N}^n$, we denote by $\mathscr{D}^\alpha$ the differential operator in polar coordinates satisfying

$$r^{|\alpha|} \partial^\alpha_{\mathbf{x}} = \mathscr{D}^\alpha(\vartheta; r\partial_r, \partial_\vartheta). \tag{3.13}$$

• For any $m \in \mathbb{N}$ and $\lambda \in \mathbb{C}$, let the parameter dependent *seminorm* $|\cdot|_{m;\,G;\,\mathscr{D}(\lambda)}$ be defined on $\mathsf{H}^m(G)$ by

$$|V|^2_{m;\,G;\,\mathscr{D}(\lambda)} = \sum_{|\alpha|=m} \|\mathscr{D}^\alpha(\vartheta; \lambda, \partial_\vartheta)V\|^2_{0;\,G}. \tag{3.14}$$

Lemma 3.6. *Let* $\beta < \beta_0$ *be two real numbers, and let* $\eta = -\beta - \frac{n}{2}$ *and* $\eta_0 = -\beta_0 - \frac{n}{2}$. *Let* $m \in \mathbb{N}$. *Let* $u \in \mathsf{K}^0_{\beta_0}(K)$ *with support in* $B(\mathbf{0}, 1)$ *be such that its* $\mathsf{K}^m_\beta(K)$ *seminorm is finite.*

Then the Mellin transform of u *is holomorphic for* $\operatorname{Re}\lambda < \eta_0$ *and has a meromorphic extension* U *to the half-plane* $\operatorname{Re}\lambda < \eta$. *Its poles are contained in the set of integers*

$$\{0, \ldots, m-1\} \cap (\eta_0, \eta)$$

and U *satisfies the estimates, with two constants* $c, C > 0$ *independent of* u

$$c\,|u|_{\mathsf{K}^m_\beta(K)} \leqslant \sup_{b\in(\eta_0,\eta)} \Big(\int_{\operatorname{Re}\lambda=b} |U(\lambda)|^2_{m;\,G;\,\mathscr{D}(\lambda)}\, \mathrm{d}\operatorname{Im}\lambda\Big)^{\frac12} \leqslant C\,|u|_{\mathsf{K}^m_\beta(K)}\,. \tag{3.15}$$

Proof. As $u \in \mathsf{K}^0_{\beta_0}(K)$, by Theorem 3.2 its Mellin transform $\lambda \mapsto \mathscr{M}[u](\lambda)$ is defined for all λ on the line $\operatorname{Re}\lambda = \eta_0$. We set

$$v_m := r^m \partial_r^m u \quad \text{and} \quad w_\alpha := r^m \partial_{\mathbf{x}}^\alpha u, \quad |\alpha| = m.$$

By assumption, the functions w_α for $|\alpha| = m$ all belong to $\mathsf{K}^0_\beta(K)$. Using the identity $r^k\partial_r^k = \sum_{|\beta|=k} \frac{k!}{\beta!}\,\mathbf{x}^\beta\partial_{\mathbf{x}}^\beta$, we obtain

$$v_m = \sum_{|\alpha|=m} \frac{m!}{\alpha!}\,\vartheta^\alpha w_\alpha\,, \quad \text{with} \quad \vartheta^\alpha = \frac{\mathbf{x}^\alpha}{r^m}\,,$$

hence v_m belongs to $\mathsf{K}^0_\beta(K)$ too. Therefore, the Mellin transforms $\lambda \mapsto \mathscr{M}[v_m](\lambda)$ and $\lambda \mapsto \mathscr{M}[w_\alpha](\lambda)$ are defined for all λ on the line $\operatorname{Re}\lambda = \eta$, and we have the estimates

$$c\,|u|^2_{\mathsf{K}^m_\beta(K)} \leqslant \int_{\operatorname{Re}\lambda=\eta} \Big(\|\mathscr{M}[v_m](\lambda)\|^2_{0;\,G} + \sum_{|\alpha|=m} \|\mathscr{M}[w_\alpha](\lambda)\|^2_{0;\,G}\Big)\, \mathrm{d}\operatorname{Im}\lambda \tag{3.16a}$$

and

$$\int_{\operatorname{Re}\lambda=\eta} \Big(\|\mathscr{M}[v_m](\lambda)\|^2_{0;\,G} + \sum_{|\alpha|=m} \|\mathscr{M}[w_\alpha](\lambda)\|^2_{0;\,G}\Big)\, \mathrm{d}\operatorname{Im}\lambda \leqslant C\,|u|^2_{\mathsf{K}^m_\beta(K)}\,. \tag{3.16b}$$

Since u, and thus v_m and w_α, have compact support, their Mellin transforms extend holomorphically to the half-planes $\operatorname{Re}\lambda < \eta_0$ for u, and $\operatorname{Re}\lambda < \eta$ for v_m and w_α. Moreover, due to the condition of support, estimate (3.16b) holds with the same constant C if we replace the integral over the line $\operatorname{Re}\lambda = \eta$ with the integral over any line $\operatorname{Re}\lambda = b$, $\eta_0 < b < \eta$:

$$\sup_{b\in(\eta_0,\eta)} \int_{\operatorname{Re}\lambda=b} \Big(\|\mathscr{M}[v_m](\lambda)\|^2_{0;\,G} + \sum_{|\alpha|=m} \|\mathscr{M}[w_\alpha](\lambda)\|^2_{0;\,G}\Big)\, \mathrm{d}\operatorname{Im}\lambda \leqslant C|u|^2_{\mathsf{K}^m_\beta(K)}\,. \tag{3.17}$$

Using the identity

$$r^m\partial_r^m = r\partial_r(r\partial_r - 1)\cdots(r\partial_r - m + 1)\,, \tag{3.18}$$

we find for all λ, $\operatorname{Re}\lambda \leqslant \eta_0$, the following relation between Mellin transforms: $\mathscr{M}[v_m](\lambda) = \lambda(\lambda-1)\cdots(\lambda-m+1)\mathscr{M}[u](\lambda)$. Hence we define a meromorphic extension U of $\mathscr{M}[u]$ by setting

$$U(\lambda) = \frac{\mathscr{M}[v_m](\lambda)}{\lambda(\lambda-1)\cdots(\lambda-m+1)} \quad \text{for} \quad \operatorname{Re}\lambda \leqslant \eta. \tag{3.19}$$

Since $\mathscr{M}[w_\alpha](\lambda) = \mathscr{D}^\alpha(\vartheta;\lambda,\partial_\vartheta)\mathscr{M}[u](\lambda)$ for $\operatorname{Re}\lambda \leqslant \eta_0$, by meromorphic extension we find that

$$\mathscr{M}[w_\alpha](\lambda) = \mathscr{D}^\alpha(\vartheta;\lambda,\partial_\vartheta)U(\lambda), \quad \text{for} \quad \operatorname{Re}\lambda \leqslant \eta. \tag{3.20}$$

Putting (3.16a), (3.17), (3.20) together and using the seminorm $|\cdot|_{m;\,G;\,\mathscr{D}(\lambda)}$ we have proved the equivalence (3.15). □

In Theorem 3.12 below we will see that under the conditions of the Lemma, the poles of the Mellin transform of u are associated with polynomials, corresponding to the Taylor expansion of u at the origin.

If λ is not an integer in the interval $[0, m-1]$, the seminorm $|V|_{m;\,G;\,\mathscr{D}(\lambda)}$ defines a norm on $\mathsf{H}^m(G)$ equivalent to the parameter dependent norm $\|V\|_{m;\,G;\,\lambda}$ introduced in Definition 3.1. In order to describe this equivalence in a neighborhood of integers, we need to introduce a projection operator on polynomial traces on G:

Definition 3.7. Let $k \in \mathbb{N}$.

- By $\mathsf{P}^k(G)$ we denote the space of restrictions to G of homogeneous polynomial functions of degree k on K.
- Let $\big(\varphi^k_\gamma\big)_{|\gamma|=k}$ be the basis in $\mathsf{P}^k(G)$ dual in $\mathsf{L}^2(G)$ of the homogeneous monomials $\big(\vartheta^\alpha/\alpha!\big)_{|\alpha|=k}$ ($\vartheta = \frac{\mathsf{x}}{|\mathsf{x}|}$), i.e.,

$$\int_G \frac{\vartheta^\alpha}{\alpha!}\,\varphi^k_\gamma(\vartheta)\,\mathrm{d}\vartheta = \delta_{\alpha\gamma}\,, \quad |\alpha| = |\gamma| = k. \tag{3.21}$$

By $\mathfrak{P}^k$ we denote the projection operator $\mathsf{L}^2(G) \to \mathsf{P}^k(G)$ defined as

$$\mathfrak{P}^k U = \sum_{|\alpha|=k} \langle U, \varphi^k_\alpha\rangle_G \,\frac{\vartheta^\alpha}{\alpha!}\,. \tag{3.22}$$

Lemma 3.8. *Suppose that $m \in \mathbb{N}$ and η_0, η are real numbers such that $\eta_0 < 0 \leqslant m < \eta$. Let $\delta \in (0, \frac{1}{2})$. Then there exist two constants $C, c > 0$ such that for all $V \in \mathsf{H}^m(G)$ the following estimates hold:*

(a) *For λ satisfying $\operatorname{Re}\lambda \in [\eta_0, \eta]$ and $|\lambda - k| \geqslant \delta$ for all $k \in \{0, \dots, m-1\}$*

$$c|V|_{m;\,G;\,\mathscr{D}(\lambda)} \leqslant \|V\|_{m;\,G;\,\lambda} \leqslant C|V|_{m;\,G;\,\mathscr{D}(\lambda)}\,. \tag{3.23}$$

(b) *For λ satisfying $|\lambda - k| \leqslant \delta$ for a $k \in \{0, \dots, m-1\}$*

$$c|V|_{m;\,G;\,\mathscr{D}(\lambda)} \leqslant \|V - \mathfrak{P}^k V\|_{m;\,G} + |\lambda - k|\,\|\mathfrak{P}^k V\|_{m;\,G} \leqslant C|V|_{m;\,G;\,\mathscr{D}(\lambda)}\,. \tag{3.24}$$

Proof. Let A be an annulus of the form $\{\mathbf{x} \in K,\ \frac{1}{R} < |\mathbf{x}| < R\}$ for $R > 1$. It is not hard to see that one has the following equivalence of the norm $\|\cdot\|_{m;\,G;\,\lambda}$ and seminorm $|\cdot|_{m;\,G;\,\mathscr{D}(\lambda)}$ with the norm and seminorm of $\mathsf{H}^m(A)$ on its closed subspace $\overline{\mathsf{S}}^\lambda_m(A)$ of homogeneous functions of the form $r^\lambda V(\vartheta)$:

$$c\,\|r^\lambda V\|_{m;\,A} \leqslant \|V\|_{m;\,G;\,\lambda} \leqslant C\,\|r^\lambda V\|_{m;\,A}$$

$$c\,|r^\lambda V|_{m;\,A} \leqslant |V|_{m;\,G;\,\mathscr{D}(\lambda)} \leqslant C\,|r^\lambda V|_{m;\,A}\,.$$

Here, the equivalence constants can be chosen uniformly for λ in the whole strip $\eta_0 \leqslant \operatorname{Re}\lambda \leqslant \eta$.

- The well-known Bramble–Hilbert lemma implies that the seminorm $|\cdot|_{m;\,A}$ is equivalent to the norm $\|\cdot\|_{m;\,A}$ on $\overline{\mathsf{S}}^\lambda_m(A)$ if and only if $\overline{\mathsf{S}}^\lambda_m(A)$ does not contain any nonzero polynomial of degree $\leqslant m-1$. Thus, for all $\lambda \notin \{0,\ldots,m-1\}$ there exists C_λ such that

$$\|r^\lambda V\|_{m;\,A} \leqslant C_\lambda |r^\lambda V|_{m;\,A}\,,$$

and C_λ can be chosen uniformly on the set $\operatorname{Re}\lambda \in [\eta_0,\eta]$ with $|\lambda - k| \geqslant \delta$ for all $k \in \{0,\ldots,m-1\}$; whence estimates (3.23) in case (a) of the lemma.

- Let λ be such that $|\lambda - k| \leqslant \delta$ for a $k \in \{0,\ldots,m-1\}$. The left inequality in (3.24) is easy to prove with the help of the estimate

$$|\mathfrak{P}^k V|_{m;\,G;\,\mathscr{D}(\lambda)} \leqslant C|\lambda - k|\,\|\mathfrak{P}^k V\|_{m;\,G}\,, \tag{1}$$

which follows from $|\mathfrak{P}^k V|_{m;\,G;\,\mathscr{D}(k)} = 0$.

Concerning the right estimate of (3.24), the Bramble–Hilbert lemma argument implies the equivalence of the seminorm with the norm for functions V such that $\mathfrak{P}^k V = 0$. For all $V \in \mathsf{H}^m(G)$

$$\|V - \mathfrak{P}^k V\|_{m;\,G} \leqslant C|V - \mathfrak{P}^k V|_{m;\,G;\,\mathscr{D}(\lambda)}\,. \tag{2}$$

On the other hand, the operator $r^m\partial_r^m$ is a linear combination of the operators $r^m\partial_{\mathbf{x}}^\alpha$, $|\alpha| = m$, with coefficients bounded on G. Therefore,

$$\|r^m\partial_r^m(r^\lambda V)\|_{0;\,A} \leqslant C|r^\lambda V|_{m;\,A}\,.$$

From (3.18) we get

$$|\lambda(\lambda-1)\cdots(\lambda-m+1)|\,\|V\|_{0;\,G} \leqslant C\|r^m\partial_r^m(r^\lambda V)\|_{0;\,A}\,.$$

Hence

$$|\lambda - k|\,\|V\|_{0;\,G} \leqslant C\|r^m\partial_r^m(r^\lambda V)\|_{0;\,A} \leqslant C'|V|_{m;\,G;\,\mathscr{D}(\lambda)}\,.$$

By the continuity of $\mathfrak{P}^k$ in $\mathsf{L}^2(G)$, we deduce

$$|\lambda - k| \, \|\mathfrak{P}^k V\|_{0;\,G} \leqslant C|V|_{m;\,G;\,\mathscr{D}(\lambda)}.$$

The equivalence of norms in $L^2(G)$ and $\mathsf{H}^m(G)$ on the finite dimensional range of $\mathfrak{P}^k$ yields finally

$$\text{(3)} \qquad |\lambda - k| \, \|\mathfrak{P}^k V\|_{m;\,G} \leqslant C|V|_{m;\,G;\,\mathscr{D}(\lambda)}.$$

It remains to bound $\|V - \mathfrak{P}^k V\|_{m;\,G}$. Using (1), (2), and (3), we find

$$\begin{aligned}
\|V - \mathfrak{P}^k V\|_{m;\,G} &\leqslant C \, |V - \mathfrak{P}^k V|_{m;\,G;\,\mathscr{D}(\lambda)} \\
&\leqslant C \left(|V|_{m;\,G;\,\mathscr{D}(\lambda)} + |\mathfrak{P}^k V|_{m;\,G;\,\mathscr{D}(\lambda)} \right) \\
&\leqslant C \left(|V|_{m;\,G;\,\mathscr{D}(\lambda)} + |\lambda - k| \, \|\mathfrak{P}^k V\|_{m;\,G} \right) \leqslant C \, |V|_{m;\,G;\,\mathscr{D}(\lambda)},
\end{aligned}$$

which completes the proof of the lemma. □

Putting the norm equivalences (3.23) and (3.24) together, one is led to the following definition of norms of meromorphic $\mathsf{H}^m(G)$-valued functions.

Definition 3.9. Let $\lambda \mapsto U(\lambda)$ be a meromorphic function with values in $\mathsf{H}^m(G)$ for λ in a strip $b_0 < \operatorname{Re}\lambda < b_1$.

- For $b \in (b_0, b_1)$ and $k \in \mathbb{N}$, and with $\mathfrak{P}^k_G$ the projection operator (3.22) we set

$$\begin{aligned}
\mathcal{N}^m_G(U, b, k) = \Bigg\{ &\int_{\substack{|\operatorname{Im}\lambda| \leqslant 1 \\ \operatorname{Re}\lambda = b}} \|(\mathbb{I} - \mathfrak{P}^k_G) U(\lambda)\|^2_{m;\,G} \, \mathrm{d}\operatorname{Im}\lambda \\
&+ \int_{\substack{|\operatorname{Im}\lambda| \leqslant 1 \\ \operatorname{Re}\lambda = b}} |\lambda - k|^2 \|\mathfrak{P}^k_G U(\lambda)\|^2_{m;\,G} \, \mathrm{d}\operatorname{Im}\lambda + \int_{\substack{|\operatorname{Im}\lambda| \geqslant 1 \\ \operatorname{Re}\lambda = b}} \|U(\lambda)\|^2_{m;\,G;\,\lambda} \, \mathrm{d}\operatorname{Im}\lambda \Bigg\}^{\frac{1}{2}}.
\end{aligned} \tag{3.25a}$$

- For $\mathfrak{N} = \{k_1, \ldots, k_j\} \subset \mathbb{N} \cap [b_0, b_1]$, and using the norm (3.4), we introduce

$$\begin{aligned}
&\mathcal{N}^m_G(U, [b_0, b_1], \mathfrak{N}) \\
&\quad = \max\Big\{ \sup_{b \in B_0} \mathcal{N}^m_G(U, b), \ \sup_{b \in B_1} \mathcal{N}^m_G(U, b, k_1), \ldots, \sup_{b \in B_j} \mathcal{N}^m_G(U, b, k_j) \Big\},
\end{aligned} \tag{3.25b}$$

  with the sets $B_\ell = (k_\ell - \frac{1}{2}, k_\ell + \frac{1}{2}) \cap (b_0, b_1)$ for $\ell = 1, \ldots, j$ and

$$B_0 = (b_0, b_1) \setminus \cup_{\ell=1}^{j} B_\ell.$$

- If $\mathfrak{N} = \varnothing$, the definition (3.25b) becomes (cf. Definition 3.1)

$$\mathcal{N}_G^m(U,[b_0,b_1],\varnothing) = \sup_{b\in(b_0,b_1)} \mathcal{N}_G^m(U,b) = \mathcal{N}_G^m(U,[b_0,b_1]). \tag{3.25c}$$

Using the continuity of $\mathfrak{P}_G^k$ on $\mathsf{H}^m(G)$, we obtain the estimate

$$\mathcal{N}_G^m(U,b,k) \leqslant C_{b,m,k}\,\mathcal{N}_G^m(U,b), \tag{3.26a}$$

where the constant $C_{b,m,k}$ does not depend on U. On the other hand, the definition immediately implies the estimate

$$\mathcal{N}_G^m(U,b) \leqslant (b-k)^{-1}\mathcal{N}_G^m(U,b,k) \quad \text{if } b \neq k. \tag{3.26b}$$

From the last two inequalities it follows that for any fixed real number $\rho \in (0,\frac{1}{2}]$ we would obtain an equivalent norm to (3.25b) by defining B_ℓ as $(k_\ell-\rho, k_\ell+\rho)\cap(b_0,b_1)$ instead of $(k_\ell-\frac{1}{2}, k_\ell+\frac{1}{2})\cap(b_0,b_1)$.

3.3 Spaces defined by Mellin norms

The norms defined in (3.25b) suggest the introduction of a class of Sobolev spaces $\mathsf{N}^m_{\beta,\beta_0;\mathfrak{N}}$ with Mellin transforms meromorphic in a strip $\eta_0 < \operatorname{Re}\lambda < \eta$ and a fixed set of poles $\mathfrak{N}$.

Definition 3.10. Let $m \in \mathbb{N}$ and $\beta,\beta_0 \in \mathbb{R}$ be such that $\beta \leqslant \beta_0$, and let $\eta = -\beta - \frac{n}{2}$, $\eta_0 = -\beta_0 - \frac{n}{2}$. Let $\mathfrak{N}$ be a subset of $\mathbb{N}\cap[\eta_0,\eta]$.

- The functions $u \in \mathsf{N}^m_{\beta,\beta_0;\mathfrak{N}}(K)$ with support in $B(\mathbf{0},1)$ are the functions whose Mellin transform $\mathscr{M}[u]$ is holomorphic in the half-plane $\operatorname{Re}\lambda < \eta_0$ and has a meromorphic extension U to the half-plane $\operatorname{Re}\lambda < \eta$ satisfying the estimate
$$\mathcal{N}_G^m(U,[\eta_0,\eta],\mathfrak{N}) < \infty\,. \tag{3.27}$$
Let $\chi \in \mathscr{C}^\infty(\mathbb{R}^\nu)$ be a cut-off function with support in $B(\mathbf{0},1)$, equal to 1 in a neighborhood of the origin. The elements u of $\mathsf{N}^m_{\beta,\beta_0;\mathfrak{N}}(S)$ are defined by the two conditions that $\chi u \in \mathsf{N}^m_{\beta,\beta_0;\mathfrak{N}}(K)$ and $(1-\chi)u \in \mathsf{H}^m(S)$.
- In the case $\eta_0 = \min\{0,\eta\}$ and $\mathfrak{N} = \mathbb{N}\cap[\eta_0,\eta]$, the space $\mathsf{N}^m_{\beta,\beta_0;\mathfrak{N}}(K)$ will alternatively be denoted by $\mathsf{J}^m_{\max,\beta}(K)$.

Note that in this definition, the set of poles $\mathfrak{N}$ is contained in the interval $[\eta_0,\eta]$ determined by the weight exponents, but $\mathfrak{N}$ has no relation with the regularity order m. Thus, the residues at the poles which, according to Theorem 3.4, give an asymptotic expansion at the origin, can only be identified with the terms of a Taylor expansion in a generalized sense, in general, because the corresponding derivatives need not exist outside of the origin. With $m = 0$, for example, one gets weighted L^2 spaces with detached asymptotics.

For the maximal J-weighted Sobolev spaces $\mathsf{J}^m_{\max,\beta}$, the definition immediately yields the following properties.

Proposition 3.11. (a) *For all* $m \geqslant 0$, $\beta < \beta'$ *implies* $\mathsf{J}^m_{\max,\beta}(S) \subset \mathsf{J}^m_{\max,\beta'}(S)$.

(b) *For all* $\beta \in \mathbb{R}$, $0 \leqslant m' < m$ *implies* $\mathsf{J}^m_{\max,\beta}(S) \subset \mathsf{J}^{m'}_{\max,\beta}(S)$.

(c) $\partial^\alpha_{\mathbf{x}}$ *is continuous from* $\mathsf{J}^m_{\max,\beta}(S)$ *into* $\mathsf{J}^{m-|\alpha|}_{\max,\beta+|\alpha|}(S)$ *for any* $\alpha \in \mathbb{N}^n$, *any* $m \geqslant |\alpha|$, *and any* β.

(d) *The multiplication by* $\mathbf{x}^\alpha$ *is continuous from* $\mathsf{J}^m_{\max,\beta}(S)$ *into* $\mathsf{J}^m_{\max,\beta-|\alpha|}(S)$ *for any* $\alpha \in \mathbb{N}^n$.

From Definition 3.10 it follows that the poles of the Mellin transform of elements of $\mathsf{N}^m_{\beta,\beta_0;\mathfrak{N}}$ are associated with polynomials and that $\mathsf{N}^m_{\beta,\beta_0;\mathfrak{N}}$ can be split into a sum of a space with homogeneous norm and a space of polynomials.

Theorem 3.12. *Let* $m \in \mathbb{N}$ *and* $\beta, \beta_0 \in \mathbb{R}$ *be such that* $\beta \leqslant \beta_0$, *and let* $\eta = -\beta - \frac{n}{2}$, $\eta_0 = -\beta_0 - \frac{n}{2}$. *Let* $\mathfrak{N}$ *be a subset of* $\mathbb{N} \cap [\eta_0, \eta]$. *Let* $u \in \mathsf{N}^m_{\beta,\beta_0;\mathfrak{N}}(K)$ *with support in* $B(\mathbf{0},1)$, *and let* U *be its Mellin transform. Then for* $b \in (\eta_0, \eta] \setminus \mathfrak{N}$, *the inverse Mellin transform* u' *of* U *on the line* $\operatorname{Re}\lambda = b$ *belongs to* $\mathsf{K}^m_{-b-\frac{n}{2}}(K)$ *and*

$$u' - u = \sum_{k \in \mathfrak{N} \cap (\eta_0, b)} \operatorname{Res}_{\lambda = k}\{r^\lambda U(\lambda)\} \quad \textit{is a polynomial.} \tag{3.28}$$

The coefficients of the polynomial in (3.28) *depend continuously on* u *in the norm of* $\mathsf{N}^m_{\beta,\beta_0;\mathfrak{N}}(K)$.

Proof. Let $b \in (\eta_0, \eta] \setminus \mathfrak{N}$. By the definition of $\mathsf{N}^m_{\beta,\beta_0;\mathfrak{N}}(K)$, we have, in particular,

$$\mathcal{N}^m_G(U, b) = \left(\int_{\operatorname{Re}\lambda = b} \|U(\lambda)\|^2_{m;\,G;\,\lambda} \,\mathrm{d}\operatorname{Im}\lambda \right)^{\frac{1}{2}} < \infty.$$

Theorem 3.2 provides the existence of a function $u' \in \mathsf{K}^m_{-b-\frac{n}{2}}(K)$ such that

$$\mathscr{M}[u'](\lambda) = U(\lambda) \quad \forall \lambda,\ \operatorname{Re}\lambda = b,$$

and, according to Theorem 3.4, we have

$$u' - u = \frac{1}{2i\pi} \int_{\mathfrak{C}} r^\lambda U(\lambda)\,\mathrm{d}\lambda = \sum_{k \in \mathfrak{N} \cap (\eta_0, b)} \operatorname*{Res}_{\lambda = k} \{r^\lambda U(\lambda)\}.$$

Here, $\mathfrak{C}$ is a contour surrounding the poles of U in $\mathfrak{N} \cap [\eta_0, b]$.

It remains to show that the residual at $k \in \mathfrak{N} \cap (\eta_0, b)$ is a polynomial. From the finiteness of $\mathcal{N}^m_G(U, [\eta_0, \eta], \mathfrak{N})$ it follows, in particular, that

$$\sup_{|b-k|<1/2} \mathcal{N}^m_G(U, b, k) < \infty$$

and therefore that both $(\mathbb{I} - \mathfrak{P}^k_G)U(\lambda)$ and $(\lambda - k)\mathfrak{P}^k_G U(\lambda)$ are holomorphic at k. Hence

$$\operatorname*{Res}_{\lambda=k}\{r^\lambda U(\lambda)\} = \operatorname*{Res}_{\lambda=k}\{r^\lambda \mathfrak{P}^k_G U(\lambda)\} = r^k \mathfrak{P}^k_G \operatorname*{Res}_{\lambda=k} U(\lambda),$$

which is a polynomial in $\mathbf{x}$ of degree k. □

We can now complete the characterization of the K^m_β seminorm (cf. Lemma 3.6).

Theorem 3.13. *Let $\beta < \beta_0$ be two real numbers. We set $\eta = -\beta - \frac{n}{2}$, $\eta_0 = -\beta_0 - \frac{n}{2}$, and $\mathfrak{N}_m = \{0, \ldots, m-1\} \cap (\eta_0, \eta]$. Let $u \in \mathsf{K}^0_{\beta_0}(K)$ with support in $B(\mathbf{0}, 1)$. Let U be its Mellin transform.*

1. *The following two conditions are equivalent:*
   (a) *the seminorm $|u|_{\mathsf{K}^m_\beta(K)}$ is finite,*
   (b) *$u \in \mathsf{N}^m_{\beta,\beta_0;\mathfrak{N}_m}(K)$.*
2. *We have the equivalence of norms*

$$c\left(\|u\|_{\mathsf{K}^0_{\beta_0}(K)} + |u|_{\mathsf{K}^m_\beta(K)}\right) \leqslant \mathcal{N}^m_G(U, [\eta_0, \eta], \mathfrak{N}_m)$$
$$\leqslant C\left(\|u\|_{\mathsf{K}^0_{\beta_0}(K)} + |u|_{\mathsf{K}^m_\beta(K)}\right). \tag{3.29}$$

Proof. (a)⇒(b) and equivalence (3.29). Let $u \in \mathsf{K}^0_{\beta_0}(K)$ with finite $\mathsf{K}^m_\beta(K)$ seminorm and support in $B(\mathbf{0}, 1)$. According to Lemma 3.6, its Mellin transform is defined for $\operatorname{Re}\lambda \leqslant \eta_0$ and has a meromorphic extension U to the half-plane $\operatorname{Re}\lambda < \eta$ satisfying the estimates (3.15), i.e., the seminorm $|u|_{\mathsf{K}^m_\beta(K)}$ is equivalent to the norm

$$\sup_{b\in(\eta_0,\eta)} \left(\int_{\operatorname{Re}\lambda=b} |U(\lambda)|^2_{m;\,G;\,\mathscr{D}(\lambda)}\, \mathrm{d}\operatorname{Im}\lambda\right)^{\frac{1}{2}}. \tag{1}$$

But Lemma 3.8 reveals that the norm (1) is equivalent to

$$\mathcal{N}^m_G(U, [\eta_0, \eta], \overline{\mathfrak{N}}_m) \quad \text{with} \quad \overline{\mathfrak{N}}_m = \{0, \ldots, m-1\} \cap [\eta_0, \eta]. \tag{2}$$

Hence Lemma 3.6 yields the equivalence of the norm (2) with the seminorm $|u|_{\mathsf{K}^m_\beta(K)}$.

On the other hand, the norm $\|u\|_{\mathsf{K}^0_{\beta_0}(K)}$ is equivalent to $\mathcal{N}^0_G(U, \eta_0)$. Therefore, the norm

$$\|u\|_{\mathsf{K}^0_{\beta_0}(K)} + |u|_{\mathsf{K}^m_\beta(K)} \tag{3}$$

presented in (3.29) is equivalent to

$$\mathcal{N}^0_G(U, \eta_0) + \mathcal{N}^m_G(U, [\eta_0, \eta], \overline{\mathfrak{N}}_m). \tag{4}$$

It remains to prove that (4) is equivalent to $\mathcal{N}_G^m(U,[\eta_0,\eta],\mathfrak{N}_m)$.

• If $\mathfrak{N}_m = \overline{\mathfrak{N}}_m$ (this occurs if $\eta_0 \notin \{0,\dots,m-1\}$), we have the equality of norms $\mathcal{N}_G^m(U,[\eta_0,\eta],\mathfrak{N}_m) = \mathcal{N}_G^m(U,[\eta_0,\eta],\overline{\mathfrak{N}}_m)$. Moreover, $\mathcal{N}_G^0(U,\eta_0)$ is bounded by $\mathcal{N}_G^m(U,[\eta_0,\eta],\mathfrak{N}_m)$. Hence we obtain the desired equivalence.

• If $\mathfrak{N}_m \neq \overline{\mathfrak{N}}_m$, then $\eta_0 \in \{0,\dots,m-1\}$ and

$$\overline{\mathfrak{N}}_m = \mathfrak{N}_m \cup \{\eta_0\}. \tag{5}$$

Since all norms are equivalent on the range of $\mathfrak{P}_G^k$, we find the estimate

$$\mathcal{N}_G^m(U,\eta_0) \leqslant C\Big(\mathcal{N}_G^0(U,\eta_0) + \mathcal{N}_G^m(U,\eta_0,\eta_0)\Big) \leqslant C\cdot\text{norm (4)}. \tag{6}$$

Let us choose $b \in (\eta_0, \eta_0+\frac{1}{2})$. We have

$$\mathcal{N}_G^m(U,b) \leqslant C(b)\mathcal{N}_G^m(U,[\eta_0,\eta],\overline{\mathfrak{N}}_m), \tag{7}$$

where $C(b)$ means that this constant depends on b (and would blow up if b approaches η_0, cf. (3.26b)). The finiteness of $\mathcal{N}_G^m(U,\eta_0)$ and $\mathcal{N}_G^m(U,b)$ implies that $u \in \mathsf{K}_{\beta_0}^m(K) \cap \mathsf{K}_{-b-\frac{n}{2}}^m(K)$ and, by Theorem 3.3,

$$\mathcal{N}_G^m(U,[\eta_0,b],\varnothing) \leqslant C\Big(\mathcal{N}_G^m(U,\eta_0) + \mathcal{N}_G^m(U,b)\Big). \tag{8}$$

The estimates (6)–(8) yield that $\mathcal{N}_G^m(U,[\eta_0,b],\varnothing)$ is bounded by the norm (4), which, in association with (5), implies that $\mathcal{N}_G^m(U,[\eta_0,\eta],\mathfrak{N}_m)$ is bounded by (4). The converse estimate is obvious.

(b)⇒(a) Let $u \in \mathsf{K}_{\beta_0}^0(K) \cap \mathsf{N}_{\beta,\beta_0;\mathfrak{N}_m}^m(K)$ with support in $B(\mathbf{0},1)$. Since $\eta_0 \notin \mathfrak{N}_m$, we have, in particular (cf. (3.26b)),

$$\mathcal{N}_G^m(U,\eta_0) \leqslant C\,\mathcal{N}_G^m(U,[\eta_0,\eta],\mathfrak{N}_m).$$

Hence u belongs to $\mathsf{K}_{\beta_0}^m(K)$. Therefore, for all α, $|\alpha| = m$, the function $w_\alpha = r^m\partial_{\mathbf{x}}^\alpha u$ belongs to $\mathsf{K}_{\beta_0}^0(K)$. Let W_α be its Mellin transform. By construction, and thanks to Lemma 3.8, we find

$$\mathcal{N}_G^0(W_\alpha,[\eta_0,\eta],\varnothing) \leqslant C\,\mathcal{N}_G^m(U,[\eta_0,\eta],\mathfrak{N}_m).$$

Hence $w_\alpha \in \mathsf{K}_\beta^0(K)$, and therefore the $\mathsf{K}_\beta^m(K)$ seminorm of u is finite. □

3.4 Spaces defined by weighted seminorms

We have seen in Theorem 3.13 how a space defined by *two* weighted seminorms $|\cdot|_{\mathsf{K}_{\beta_0}^0}$ and $|\cdot|_{\mathsf{K}_\beta^m}$ has a Mellin characterization described by the space $\mathsf{N}_{\beta,\beta_0;\mathfrak{N}_m}^m$. We are now generalizing this to the case of spaces given by several weighted seminorms, and this will eventually lead to the Mellin characterization of the space J^m_β, which is defined by the $m+1$ seminorms $|\cdot|_{\mathsf{K}^\ell_{\beta+m-\ell}}$, $0 \leqslant \ell \leqslant m$ (cf. Definition 2.2).

Definition 3.14. Let $\mathfrak{L}$ be a subset of $\mathbb{N}$ that includes 0. For each $\ell \in \mathfrak{L}$ let β_ℓ be a weight exponent such that

$$\beta_\ell \text{ decreases as } \ell \text{ increases},$$

and denote $\mathfrak{B} = \{\beta_\ell : \ell \in \mathfrak{L}\}$. We define the associated norm

$$\|u\|_{\mathsf{J}^{\mathfrak{L}}_{\mathfrak{B}}(S)} = \Big(\sum_{\ell\in\mathfrak{L}} \sum_{|\alpha|=\ell} \|r^{\beta_\ell+\ell}\partial^\alpha_{\mathbf{x}} u\|^2_{0;S}\Big)^{\frac12} \equiv \Big(\sum_{\ell\in\mathfrak{L}} |u|^2_{\mathsf{K}^\ell_{\beta_\ell}(S)}\Big)^{\frac12}. \tag{3.30}$$

The Hilbert space defined by this norm is denoted by $\mathsf{J}^{\mathfrak{L}}_{\mathfrak{B}}(S)$.

This definition includes the weighted Sobolev spaces with homogeneous norms and those with nonhomogeneous norms as obvious special cases:

- We obtain the norm in K^m_β by choosing $\beta_\ell = \beta$ for all ℓ and $\mathfrak{L}$ any arbitrary subset contained in $\{0,\dots,m\}$ and containing 0 and m.
- According to Definition 2.2, we obtain the norm in J^m_β by choosing

$$\mathfrak{L} = \{0,\dots,m\}, \quad \beta_\ell = \beta + m - \ell.$$

- Finally, the space defined by the norm $\mathsf{K}^0_{\beta_0}$ and the seminorm K^m_β simply corresponds to $\mathfrak{L} = \{0, m\}$ and $\mathfrak{B} = \{\beta_0, \beta\}$.

We can use Theorem 3.13 to obtain a first Mellin characterization of the space $\mathsf{J}^{\mathfrak{L}}_{\mathfrak{B}}(S)$. We set, as usual, $\eta_\ell = -\beta_\ell - \frac{n}{2}$. Then we have

$$\mathsf{J}^{\mathfrak{L}}_{\mathfrak{B}}(S) = \bigcap_{\ell\in\mathfrak{L}} \mathsf{N}^\ell_{\beta_\ell,\beta_0;\mathfrak{N}_\ell}(S) \quad \text{with } \mathfrak{N}_\ell = \{0,\dots,\ell-1\} \cap (\eta_0, \eta_\ell]. \tag{3.31}$$

This can be simplified with the following result.

Lemma 3.15. *Let $\mathfrak{L}$, $\mathfrak{B}$ and $\mathfrak{N}_\ell$ be as above. Let $m = \max \mathfrak{L}$. Then there exists a unique subset $\mathfrak{N} \subset \mathfrak{N}_m = \{0,\dots,m-1\} \cap (\eta_0, \eta]$ such that there is a norm equivalence*

$$c\mathcal{N}^m_G(U, [\eta_0,\eta], \mathfrak{N}) \leqslant \max_{\ell\in\mathfrak{L}} \mathcal{N}^\ell_G(U, [\eta_0,\eta_\ell], \mathfrak{N}_\ell) \leqslant C\mathcal{N}^m_G(U, [\eta_0,\eta], \mathfrak{N}). \tag{3.32}$$

This set $\mathfrak{N}$ is given by

$$\mathfrak{N} = \mathfrak{N}_m \setminus \Big(\bigcup_{\ell\in\mathfrak{L}} [\ell, \eta_\ell]\Big). \tag{3.33}$$

Proof. Assume that $\mathfrak{N}$ is such that the norm equivalence (3.32) holds. Then any pole $k \in \mathfrak{N}$ that lies in an interval $(\eta_0, \eta_\ell]$ must appear as a pole in the corresponding set $\mathfrak{N}_\ell$ and vice versa. This means

$$\mathfrak{N} \cap (\eta_0, \eta_\ell] \subset \{0, \dots, \ell - 1\} \qquad \forall \ell \in \mathfrak{L}. \tag{3.34}$$

In other words, for $k \in \mathfrak{N}_m$ there holds $k \notin \mathfrak{N}$ if and only if there exists $\ell \in \mathfrak{L}$ such that

$$\ell \leqslant k \leqslant \eta_\ell.$$

This implies the formula (3.33) for $\mathfrak{N}$. Conversely, it is not hard to see that if we define the set $\mathfrak{N}$ by (3.33), then the norm equivalence (3.32) holds. □

Combining (3.31) with Lemma 3.15, we obtain the Mellin characterization of the space $\mathsf{J}^{\mathfrak{L}}_{\mathfrak{B}}(S)$.

Proposition 3.16. *Let* $\mathfrak{L}$ *and* $\mathfrak{B}$ *satisfy the conditions in Definition* 3.14. *Let* $m = \max \mathfrak{L}$, *and let* $\beta = \beta_m$. *Define* $\mathfrak{N}$ *by* (3.33). *Then*

$$\mathsf{J}^{\mathfrak{L}}_{\mathfrak{B}}(S) = \mathsf{N}^m_{\beta,\beta_0;\mathfrak{N}}(S)\,. \tag{3.35}$$

We have seen that with each set of seminorms given by $\mathfrak{L}$ and $\mathfrak{B}$ there is a unique associated set of poles $\mathfrak{N}$ that characterizes the space $\mathsf{N}^m_{\beta,\beta_0;\mathfrak{N}}(S)$ and therefore the space $\mathsf{J}^{\mathfrak{L}}_{\mathfrak{B}}(S)$. The converse is not always true, i.e., the spaces $\mathsf{N}^m_{\beta,\beta_0;\mathfrak{N}}$ cannot always be defined by a set of weighted Sobolev seminorms. A necessary condition is that $\mathfrak{N} \subset \mathfrak{N}_m$. But this is also sufficient:

For fixed m and β, let $\mathfrak{N}$ be a given subset of $\mathfrak{N}_m$. We can construct indices $\mathfrak{L}$ and weight exponents $\mathfrak{B}$ such that formula (3.33), and therefore the equality of spaces (3.35) in Proposition 3.16 holds. This can be done by setting

$$\mathfrak{L} = \{0\} \cup \left(\mathfrak{N}_m \setminus \mathfrak{N}\right) \cup \{m\}, \tag{3.36}$$

and for all $\ell \in \mathfrak{L}$, $\ell \neq 0, m$,

$$\beta_\ell = -\eta_\ell - \tfrac{n}{2} \quad \text{with} \quad \eta_\ell = \ell, \tag{3.37}$$

and $\eta_0 = 0$ if $0 \notin \mathfrak{N}$, $\eta_0 < 0$ arbitrary if $0 \in \mathfrak{N}$.

In this context, the counterexample (2.8), for instance, corresponds to the choice of $m \geqslant 2$, $\mathfrak{L} = \{0\} \cup \{2, \dots, m\}$, and $\eta_\ell = \eta - m + \ell$ with $m < \eta < m+1$, so that $\eta_\ell \in (\ell, \ell + 1)$. From these informations one obtains $\mathfrak{N} = \{1\}$.

From the equality (3.35) we conclude that the space $\mathsf{J}^{\mathfrak{L}}_{\mathfrak{B}}(S)$ depends only on m, $\beta = \beta_m$ and on the set of integers $\mathfrak{N}$. Several different choices of $\mathfrak{L}$ and $\mathfrak{B}$ can therefore lead to the same space. We have already seen this for the space K^m_β, where the choice of $\mathfrak{L}$ is arbitrary, as soon as it includes 0 and m. This observation expresses the fact that for the spaces with homogeneous norms, the intermediate seminorms are bounded by the two extreme seminorms. The set of poles $\mathfrak{N}$ is empty in this case.

Also for the space with nonhomogeneous norm J^m_β, several different choices of sets of seminorms are possible, as we will discuss now.

3.5 Mellin characterization of spaces with nonhomogeneous norms

For the weighted Sobolev space with nonhomogeneous norm J^m_β, several different choices of sets $\{\mathfrak{L}, \mathfrak{B}\}$ of seminorms are possible that lead to the same set of poles $\mathfrak{N}$ and define therefore, according to Proposition 3.16, the same space. The original definition of J^m_β corresponds to the choice $\mathfrak{L} = \{0, \dots, m\}$ and $\beta_\ell = \beta + m - \ell$, $\ell \in \mathfrak{L}$, which implies $\eta_{\ell+1} = \eta_\ell + 1$. From this information and formula (3.33) one easily deduces that $\mathfrak{N}$ is either empty or a set of consecutive integers starting with 0. It is nonempty if and only if $0 < \eta < m$, and in this case

$$\mathfrak{N} = \{0, \dots, m-1\} \cap (\eta - m, \eta] = \{0, \dots M\} \quad \text{with} \quad M = [\eta]\,. \tag{3.38}$$

Since $\mathfrak{N} = \varnothing$ corresponds to the space K^m_β and $\mathfrak{N} = \{0, \dots, [\eta]\}$ to the space $\mathsf{J}^m_{\max,\beta}$, we find the following classification of the space J^m_β.

Proposition 3.17. *Suppose that $m \in \mathbb{N}$, $\beta \in \mathbb{R}$, and $\eta = -\beta - \frac{n}{2}$.*
(a) *If $\eta < 0$, then $\mathsf{J}^m_\beta(S) = \mathsf{J}^m_{\max,\beta}(S) = \mathsf{K}^m_\beta(S)$.*
(b) *If $0 \leqslant \eta < m$, then $\mathsf{J}^m_\beta(S) = \mathsf{J}^m_{\max,\beta}(S)$.*
(c) *If $\eta \geqslant m$, then $\mathsf{J}^m_\beta(S) = \mathsf{K}^m_\beta(S)$.*

The set $\mathfrak{N}$ of integers (3.38) that characterizes J^m_β can also be obtained by other choices for the weight indices: We start again with $0 < \eta < m$ and $\mathfrak{L} = \{0, \dots, m\}$, but now we fix some integer ℓ_0 in the interval $(\eta, m]$. Then we define the weight indices β_ℓ in such a way that

$$\eta_\ell = \eta - \ell_0 + \ell \quad \text{for} \quad 0 \leqslant \ell \leqslant \ell_0 \qquad \text{and} \qquad \eta_\ell = \eta \quad \text{for} \quad \ell \geqslant \ell_0$$

Since $\eta_0 < 0$ and $\ell_0 - 1 \geqslant M$, we easily see that this set of weight indices defines the same set of degrees $\mathfrak{N} = \{0, \dots M\}$ as in (3.38). In this way, we prove the following "step-weighted" characterization of J^m_β.

Proposition 3.18. *Let $\beta \in \mathbb{R}$ and $m \in \mathbb{N}$ be such that $m > \eta = -\beta - \frac{n}{2}$. Let ρ be any real number in the interval $(-\frac{n}{2}, \beta + m]$. Then the norm in the space $\mathsf{J}^m_\beta(S)$ is equivalent to*

$$\Big(\sum_{|\alpha| \leqslant m} \| r^{\max\{\beta+|\alpha|,\,\rho\}} \partial^\alpha_{\mathbf{x}} u \|^2_{0;S} \Big)^{\frac{1}{2}}. \tag{3.39}$$

Corollary 3.19. *Let $\beta \in \mathbb{R}$. Set $\eta = -\beta - \frac{n}{2}$. Let m be a natural number, $m > \eta$. Then $\mathsf{J}^{m+1}_\beta(S) \subset \mathsf{J}^m_\beta(S)$.*

Proof. Using Proposition 3.18, we note that we can choose the same ρ for $\mathsf{J}^m_\beta(S)$ and $\mathsf{J}^{m+1}_\beta(S)$. The embedding $\mathsf{J}^{m+1}_\beta(S) \subset \mathsf{J}^m_\beta(S)$ follows. □

Still another choice giving the same result is possible. When $0 < \eta < m$, it suffices to take $\mathfrak{L} = \{0, m\}$, $\eta_m = \eta$, and any $\eta_0 < 0$. In this case,

$$\mathfrak{N} = \mathfrak{N}_m = \{0, \dots M\},$$

which corresponds to the identity $\mathsf{J}^m_\beta(S) = \mathsf{J}^m_{\max,\beta}(S)$. We obtain the corollary that when $0 < \eta < m$, the intermediate seminorms in the definition of $\mathsf{J}^m_\beta(S)$ are indeed bounded by the sum of the two extreme seminorms. This is not the case, as we have seen, if $\eta > m$.

Let us mention another identity that can be obtained from these purely combinatorial arguments, namely

$$\mathsf{K}^m_\beta(S) = \mathsf{J}^m_\beta(S) \cap \mathsf{K}^0_\beta(S).$$

This can be seen as follows. The intersection $\mathsf{J}^m_\beta(S) \cap \mathsf{K}^0_\beta(S)$ is included in the space $\mathsf{J}^{\mathfrak{L}}_{\mathfrak{B}}(S)$ with $\mathfrak{L} = \{0, m\}$ and $\beta_0 = \beta_m = \beta$. Then $\mathfrak{N} = \varnothing$, and we find that this latter space coincides with $\mathsf{K}^m_\beta(S)$.

Remark 3.20. Corollary 3.19 gives a partial response to the question of how to define spaces J^s_β with *noninteger* Sobolev index s. If $[s] > \eta$, the natural idea is to define the space of index s by Hilbert space interpolation between spaces with integer indices $[s]$ and $[s]+1$. The same possibility exists if $[s] + 1 \leqslant \eta$ since for $m + 1 \leqslant \eta$ the inclusion $\mathsf{J}^{m+1}_\beta(S) \subset \mathsf{J}^m_\beta(S)$ holds too because, according to Proposition 3.17 (c), the J-weighted spaces coincide with the K-weighted spaces in this range.

For fixed weight β both scales of spaces $\big(\mathsf{K}^m_\beta(S)\big)_{m\in\mathbb{N}}$ and $\big(\mathsf{J}^m_{\max,\beta}(S)\big)_{m\in\mathbb{N}}$ can be extended in a natural way by interpolation to scales with arbitrary real positive index. This definition, when extended by analogy to the $n-1$-dimensional conical manifold ∂K, is then also compatible with the trace operator, i.e., the trace space of $\mathsf{K}^m_\beta(K)$ is $\mathsf{K}^{m-\frac12}_{\beta+\frac12}(\partial K)$ and similarly for the $\mathsf{J}_{\max}$ scale.

There is, however, no natural definition of J^s_β for the remaining noninteger indices s for which $[s] \leqslant \eta < [s] + 1$. The problem is that if $m > \eta$, so that $\mathsf{J}^m_\beta(S) = \mathsf{J}^m_{\max,\beta}(S)$, then the trace space is also of the $\mathsf{J}_{\max}$ class because it contains nonzero constant functions. But if $m - \frac12 < \eta$, then the candidate for the trace space would be $\mathsf{J}^{m-\frac12}_{\beta+\frac12}$ and should be of the K class, which does not contain nonconstant functions.

As a further corollary of the Mellin description of the space J^m_β, we give an equivalent definition by derivatives in polar coordinates that is valid when $\eta < 1$ and will be useful later on:

Lemma 3.21. *Suppose that $\beta \in \mathbb{R}$, $\eta = -\beta - \frac{n}{2}$, and $m \in \mathbb{N}$, $m \geqslant 1$. We assume that $\eta < 1$. Then*

$$\Big\{ \sum_{1 \leqslant \ell + |\gamma| \leqslant m} \|r^\beta (r\partial_r)^\ell \partial^\gamma_\vartheta u\|^2_{0;S} + \|r^{\beta+1} u\|^2_{0,S} \Big\}^{\frac12} \tag{3.40}$$

defines a norm on $\mathsf{J}^m_\beta(S)$, equivalent to its natural norm.

Proof. If $\eta < 0$, the statement is clear because, in that case, $\mathsf{J}^m_\beta(S)$ coincides with $\mathsf{K}^m_\beta(S)$.

Let us suppose that $0 \leqslant \eta < 1$, and let u be such that its norm (3.40) is finite. Using a cut-off function, we can assume that u has the same regularity on K, with support in $\overline{S}$. Let $\mathscr{M}[u] =: U$ be the Mellin transform of u. By the Parseval identity, we have the equivalence

$$(1)\quad \sum_{1\leqslant \ell+k\leqslant m} \int_{\operatorname{Re}\lambda=\eta} |\lambda|^{2\ell} |U(\lambda)|^2_{k;\,G}\, \mathrm{d}\operatorname{Im}\lambda \simeq \sum_{1\leqslant \ell+|\gamma|\leqslant m} \|r^\beta (r\partial_r)^\ell \partial^\gamma_\vartheta u\|^2_{0;S}.$$

It is easy to see that we have the uniform estimate

$$\sum_{k=1}^m |U(\lambda)|_{k;\,G;\,\mathscr{D}(\lambda)} \leqslant C \sum_{1\leqslant \ell+k\leqslant m} |\lambda|^\ell |U(\lambda)|_{k;\,G}.$$

We deduce that $u \in \mathsf{J}^m_\beta(S)$.

Conversely, let $u \in \mathsf{J}^m_\beta(S)$. We apply Lemma 3.8. Outside a neighborhood of 0, we have the uniform estimate

$$(2)\qquad \|U(\lambda)\|_{m;\,G;\,\lambda} \leqslant C |U(\lambda)|_{m;\,G;\,\mathscr{D}(\lambda)}$$

and, in a bounded neighborhood of 0,

$$(3)\quad \|U(\lambda) - \mathfrak{P}^0 U(\lambda)\|_{m;\,G} + |\lambda| \|U(\lambda)\|_{m;\,G} \leqslant C |U(\lambda)|_{m;\,G;\,\mathscr{D}(\lambda)}.$$

Since $\mathfrak{P}^0 U(\lambda)$ is a constant, we have

$$(4)\qquad \sum_{k=1}^m |U(\lambda)|_{k;\,G} \leqslant C \|U(\lambda) - \mathfrak{P}^0 U(\lambda)\|_{m;\,G}.$$

We deduce from (2)–(4) that

$$\sum_{1\leqslant \ell+k\leqslant m} |\lambda|^\ell |U(\lambda)|_{k;\,G} \leqslant C |U(\lambda)|_{m;\,G;\,\mathscr{D}(\lambda)}.$$

The boundedness of norm (3.40) follows from (1) and Lemma 3.6. □

We can now collect the informations about the Mellin description of the space J^m_β. For this purpose, we introduce some notation concerning the Taylor expansion at the origin.

For $u \in \mathscr{C}^\infty(\overline{S})$ and $M \in \mathbb{N}$ we write $\mathfrak{T}^M u \in \mathbb{P}^M(S)$ for the *Taylor part* of u of degree M at $\mathbf{0}$:

$$\mathfrak{T}^M u = \sum_{|\alpha|\leqslant M} \partial^\alpha_{\mathbf{x}} u(\mathbf{0}) \frac{\mathbf{x}^\alpha}{\alpha!}. \tag{3.41}$$

By continuity, the coefficients of the Taylor expansion and therefore the *corner Taylor operator* $\mathfrak{T}^M$ can be defined on the space $\mathsf{N}^m_{\beta,\beta_0;\mathfrak{N}}(S)$, as soon as $\{0,\ldots,M\} \subset \mathfrak{N} \subset (\eta_0,\eta)$ (cf. Theorem 3.12).

The proofs of the following two theorems are contained in the results of the preceding section.

Theorem 3.22. *Let K be a regular cone in $\mathbb{R}^n$. Let $\beta \in \mathbb{R}$. We set, as usual*

$$\eta = -\beta - \tfrac{n}{2} \quad \text{and} \quad M = [\eta].$$

Let $\mathfrak{N} = \{0,\ldots,M\}$ if $M \geqslant 0$ and $m > \eta$, and $\mathfrak{N} = \varnothing$ in the other cases (either $M < 0$ or $m \leqslant \eta$). Let $u \in \mathsf{K}^0_{\beta+m}(K)$ with support in $B(\mathbf{0},1)$. Let U be its Mellin transform. Set $\eta_0 = \eta - m$ and $\beta_0 = \eta_0 - \frac{n}{2} = \beta + m$.

(a) *Then $u \in \mathsf{J}^m_\beta(K)$ if and only if $u \in \mathsf{N}^m_{\beta,\beta_0;\mathfrak{N}}(K)$. Moreover we have the equivalence of norms*

$$c\,\|u\|_{\mathsf{J}^m_\beta(K)} \;\leqslant\; \mathcal{N}^m_G(U,[\eta-m,\eta],\mathfrak{N}) \;\leqslant\; C\,\|u\|_{\mathsf{J}^m_\beta(K)}\,. \tag{3.42}$$

Furthermore, U is meromorphic in the half-plane $\operatorname{Re}\lambda < \eta$ with only possible poles on natural numbers and the residues of $r^\lambda U(\lambda)$ are polynomials.

(b) *Let $M^* = M$ if $\eta \neq M$ and $M^* = M-1$ if $\eta = M$. Let $b \in (M^*,\eta]$ with $b \neq \eta$ if $\eta = M$. Then the inverse Mellin transform u' of U on the line $\operatorname{Re}\lambda = b$ belongs to $\mathsf{K}^m_{-b-\frac{n}{2}}(K)$ and, with the notation (3.41),*

$$u' - u \;=\; \sum_{k=0}^{M^*} \operatorname*{Res}_{\lambda=k}\,\{r^\lambda\, U(\lambda)\} = -\mathfrak{T}^{M^*} u. \tag{3.43}$$

When $M < 0$ or $m \leqslant \eta$, the sum of residues collapses to 0, and u belongs to $\mathsf{K}^m_\beta(K)$.

We call the case $\eta \in \mathbb{N}$ *critical.* In the noncritical case, we can take $b = \eta$ in the previous result and obtain, therefore, the following relations between the space $\mathsf{J}^m_\beta(S)$ with nonhomogeneous norm and the space $\mathsf{K}^m_\beta(S)$ with homogeneous norm.

Theorem 3.23. *Let $K \subset \mathbb{R}^n$ be a cone, and let $S = K \cap B(\mathbf{0},1)$. Let $\beta \in \mathbb{R}$. We set*

$$\eta = -\beta - \tfrac{n}{2} \quad \text{and} \quad M = [\eta].$$

Let $m \in \mathbb{N}$. Then

(a) *if $\eta < 0$, the the spaces $\mathsf{J}^m_\beta(S)$ and $\mathsf{K}^m_\beta(S)$ coincide,*

(b) *if $\eta \geqslant 0$ and $m \leqslant \eta$, then the spaces $\mathsf{J}^m_\beta(S)$ and $\mathsf{K}^m_\beta(S)$ coincide,*

(c) *if $\eta \geqslant 0$ and $m > \eta$, then $\mathsf{J}^m_\beta(S)$ and $\mathsf{J}^m_{\max,\beta}(S)$ coincide and there are two cases:*

- ***noncritical case*** *$\eta \notin \mathbb{N}$: the corner Taylor operator $\mathfrak{T}^M$ defined in (3.41) is continuous from $\mathsf{J}^m_\beta(S)$ to $\mathbb{P}^M(S)$ and $\mathbb{I} - \mathfrak{T}^M$ is continuous from $\mathsf{J}^m_\beta(S)$ to $\mathsf{K}^m_\beta(S)$; the decomposition $u = (u - \mathfrak{T}^M u) + \mathfrak{T}^M u$ gives the direct* *sum*

$$\mathsf{J}^m_\beta(S) = \mathsf{K}^m_\beta(S) \oplus \mathbb{P}^M(S)\,; \tag{3.44}$$

• ***critical case*** $\eta \in \mathbb{N}$: *the operator* $\mathfrak{T}^{M-1}$ *is continuous on* $\mathsf{J}^m_\beta(S)$*, but* $\mathfrak{T}^M$ *is not; the space* $\mathsf{J}^m_\beta(S)$ *contains* $\mathsf{K}^m_\beta(S) \oplus \mathbb{P}^M(S)$ *as a strict subspace of infinite codimension.*

The structure of J^m_β in the critical case, and the generalization of the Taylor expansion in that case, is the subject of the following section.

4 Structure of Spaces with Nonhomogeneous Norms in the Critical Case

4.1 Weighted Sobolev spaces with analytic regularity

Using the monotonicity $\mathsf{K}^{m+1}_\beta(S) \subset \mathsf{K}^m_\beta(S)$ for all m and β and $\mathsf{J}^{m+1}_\beta(S) \subset \mathsf{J}^m_\beta(S)$ if $M > \eta = -\beta - \frac{n}{2}$, we introduce corresponding weighted spaces with infinite and with analytic regularity:

Definition 4.1. Let $\beta \in \mathbb{R}$ and $\eta = -\beta - \frac{n}{2}$.

- $\mathsf{K}^\infty_\beta(K) = \bigcap\limits_{m\in\mathbb{N}} \mathsf{K}^m_\beta(K)$.
- We denote by $\mathsf{A}_\beta(K)$ the subspace of the functions $u \in \mathsf{K}^\infty_\beta(K)$ satisfying the following analytic estimates for some $C > 0$

$$\exists C > 0 \quad \forall k \in \mathbb{N}, \quad |u|_{\mathsf{K}^k_\beta(K)} \leqslant C^{k+1} k!\,. \tag{4.1}$$

- $\mathsf{J}^\infty_\beta(S) = \bigcap\limits_{k\in\mathbb{N}\,,\ k>\eta} \mathsf{J}^k_\beta(S)$.
- The analytic weighted class $\mathsf{B}_\beta(S)$ with nonhomogeneous norm is the space of functions $u \in \mathsf{J}^\infty_\beta(S)$ such that there exists a constant $C > 0$ with

$$\forall k \in \mathbb{N} \quad \text{with} \quad k > \eta, \qquad |u|_{\mathsf{K}^k_\beta(K)} \leqslant C^{k+1} k!. \tag{4.2}$$

Note that in (4.2) the estimates are the same as in (4.1) but only for $k > \eta$. This suggests that for $\eta < 0$, we have $\mathsf{B}_\beta(S) = \mathsf{A}_\beta(S)$, which will be proved below.

For generalization of the Taylor expansion in the critical case, we develop the Mellin-domain analogue of an idea from [4], based on the splitting of u^M provided by the decomposition

$$U(\lambda) = (\mathbb{I} - \mathfrak{P}^M)U(\lambda) + \mathfrak{P}^M U(\lambda).$$

The first part is the Mellin transform of a function in $\mathsf{K}^m_\beta(K)$ and the second one has essentially a one dimensional structure – that is, the most important features of its structure are described by the behavior of functions of one variable – and it can be regularized in such a way that it splits again into two parts, one in the analytic class $\mathsf{B}_\beta(K)$, and the remaining part in $\mathsf{K}^m_\beta(K)$.

4.2 Mellin regularizing operator in one dimension

The main tool of the following analysis is a one-dimensional Mellin convolution operator:

Definition 4.2. We denote by $\mathfrak{K} : v \mapsto \mathfrak{K}v$ be the *Mellin convolution operator* defined by

$$\mathscr{M}[\mathfrak{K}v](\lambda) = \mathrm{e}^{\lambda^2}\mathscr{M}[v](\lambda). \tag{4.3}$$

Owing to the strong decay properties of the kernel e^{λ^2} in the imaginary direction, the operator $\mathfrak{K}$ has analytic regularizing properties in the scales K^m_β and J^m_β.

Proposition 4.3. *Let $\beta \in \mathbb{R}$ and $m \geqslant 1$. Then*

(a) *if $v \in \mathsf{K}^m_\beta(\mathbb{R}_+)$, then $\mathfrak{K}v \in \mathsf{A}_\beta(\mathbb{R}_+)$,*

(b) *if $v \in \mathsf{J}^m_{-\frac{1}{2}}(\mathbb{R}_+)$ with support in $I := [0,1]$, then $\mathfrak{K}v|_I$ belongs to the analytic class $\mathsf{B}_{-\frac{1}{2}}(I)$, and $v - \mathfrak{K}v \in \mathsf{K}^m_{-\frac{1}{2}}(\mathbb{R}_+)$,*

(c) *if $v \in \mathsf{J}^m_\beta(\mathbb{R}_+)$ with support in $I := [0,1]$, and if $\beta < -\frac{1}{2}$ so that v is continuous in 0, then $\mathfrak{K}v$ is continuous in 0 as well, and $\mathfrak{K}v(0) = v(0)$.*

The proof of this proposition is based on the following characterization of analytic classes by Mellin transformation.

Lemma 4.4. *Let $\beta \in \mathbb{R}$ and $\eta = -\beta - \frac{1}{2}$. We set $I = (0,1)$.*

(a) *Let $v \in \mathsf{K}^1_\beta(\mathbb{R}_+)$. Then v belongs to $\mathsf{A}_\beta(\mathbb{R}_+)$ if and only if $V := \mathscr{M}[v]$ satisfies*

$$\exists C > 0\ \forall k \geqslant 1, \quad \left\{ \int_{\operatorname{Re}\lambda=\eta} |\lambda|^{2k}\, |V(\lambda)|^2 \,\mathrm{d}\operatorname{Im}\lambda \right\}^{\frac{1}{2}} \leqslant C^{k+1}k! \tag{4.4}$$

(b) *Let $v \in \mathsf{J}^1_{-\frac{1}{2}}(\mathbb{R}_+)$. Then $v|_I$ belongs to $\mathsf{B}_{-\frac{1}{2}}(I)$ if (4.4) is satisfied with $\eta = 0$ and $V(\lambda) := \lambda^{-1}\mathscr{M}[r\partial_r v](\lambda)$.*

Proof. (a) According to Definition 4.1, $v \in \mathsf{A}_\beta(\mathbb{R}_+)$ if and only if

$$\exists C > 1\ \forall k \geqslant 0, \quad \|r^{\beta+k}\partial_r^k v\|_{0;\mathbb{R}_+} \leqslant C^{k+1}k!$$

Using (3.18), one can see that this is equivalent to

$$\exists C > 1 \; \forall k \geqslant 0, \quad \|r^{\beta}(r\partial_r)^k v\|_{0;\mathbb{R}_+} \leqslant C^{k+1} k!$$

Then (a) is a consequence of the Parseval equality.

(b) Let $v \in \mathsf{J}^1_{-\frac{1}{2}}(\mathbb{R}_+)$. With $V(\lambda) = \lambda^{-1}\mathscr{M}[r\partial_r v](\lambda)$, for any $k \geqslant 1$ the function $\lambda^k V(\lambda)$ is the Mellin transform of $(r\partial_r)^k v$ on the line $\operatorname{Re}\lambda = 0$. Thus, (4.4) with $\eta = 0$ implies the analytic estimates

$$\exists C > 0 \; \forall k \geqslant 1, \quad \|r^{-\frac{1}{2}}(r\partial_r)^k v\|_{0;\mathbb{R}_+} \leqslant C^{k+1} k!$$

Restricting this to I, and using Definition 4.1, we find that $v|_I \in \mathsf{B}_{-\frac{1}{2}}(I)$. □

Proof of Proposition 4.3. (a) Let $v \in \mathsf{K}^m_{\beta}(\mathbb{R}_+)$, and let V be the Mellin transform of v. It is defined for $\operatorname{Re}\lambda = \eta$ and, in particular, the norm

$$N_0 := \left\{ \int_{\operatorname{Re}\lambda=\eta} |V(\lambda)|^2 \, \mathrm{d}\operatorname{Im}\lambda \right\}^{\frac{1}{2}}$$

is finite. The Mellin transform of $\mathfrak{K}v$ is $\lambda \mapsto \mathrm{e}^{\lambda^2} V(\lambda)$. We have for any $k \geqslant 1$

$$\left\{ \int_{\operatorname{Re}\lambda=\eta} |\lambda|^{2k} \, |\mathrm{e}^{\lambda^2} V(\lambda)|^2 \, \mathrm{d}\operatorname{Im}\lambda \right\}^{\frac{1}{2}} \leqslant N_0 \sup_{\operatorname{Re}\lambda=\eta} |\lambda|^k |\mathrm{e}^{\lambda^2}|$$
$$\leqslant C(\eta) N_0 \sup_{\xi\geqslant 0} \xi^k \mathrm{e}^{-\xi^2} = C(\eta) N_0 \left(\frac{k}{2\mathrm{e}}\right)^{\frac{k}{2}}.$$

Therefore, the condition (4.4) is satisfied for the Mellin transform of $\mathfrak{K}v$. By Lemma 4.4 (a), $\mathfrak{K}v$ belongs to $\mathsf{A}_{\beta}(\mathbb{R}_+)$.

(b) Let $v \in \mathsf{J}^m_{-\frac{1}{2}}(\mathbb{R}_+)$ with support in I. By Corollary 3.19, $v \in \mathsf{J}^1_{-\frac{1}{2}}(\mathbb{R}_+)$. Now, V is defined as the Mellin transform of $r\partial_r v$ divided by λ. Thus, V coincides with $\mathscr{M}[v]$, where $\mathscr{M}[v]$ is well defined and the Mellin transform of $\mathfrak{K}v$ is given by $\mathrm{e}^{\lambda^2} V(\lambda)$. With the same arguments as above, we prove that $\mathfrak{K}v$ satisfies the assumptions of Lemma 4.4 (b). Hence $\mathfrak{K}v|_I \in \mathsf{B}_{-\frac{1}{2}}(I)$.

The Mellin transform of $v - \mathfrak{K}v$ is $(1-\mathrm{e}^{\lambda^2})V(\lambda)$. Since $r^{-\frac{1}{2}}(r\partial_r)^k v \in \mathsf{L}^2(\mathbb{R}_+)$ for $k = 1, \dots, m$, we have

$$\text{(1)} \qquad \sum_{k=1}^{m} \left\{ \int_{\operatorname{Re}\lambda=0} |\lambda|^{2k} \, |V(\lambda)|^2 \, \mathrm{d}\operatorname{Im}\lambda \right\}^{\frac{1}{2}} < \infty.$$

The function $\lambda \mapsto (1-\mathrm{e}^{\lambda^2})$ is bounded on the line $\operatorname{Re}\lambda = 0$ and has a double zero at $\lambda = 0$. Hence we deduce from (1) that

$$\sum_{k=0}^{m} \left\{ \int_{\operatorname{Re}\lambda=0} |\lambda|^{2k} \, |(1-\mathrm{e}^{\lambda^2})V(\lambda)|^2 \, \mathrm{d}\operatorname{Im}\lambda \right\}^{\frac{1}{2}} < \infty.$$

Therefore, $v - \mathfrak{K}v \in \mathsf{K}^m_{-\frac{1}{2}}(\mathbb{R}_+)$.

(c) Let $v \in \mathsf{J}^m_\beta(\mathbb{R}_+)$ with support in I, $\beta < -\frac{1}{2}$. It suffices to consider the case $m = 1$ and $-\frac{3}{2} < \beta < -\frac{1}{2}$. With $\eta = -\beta - \frac{1}{2}$ we then have $0 < \eta < 1$. Let V be the Mellin transform of v, and let $w = v - \mathfrak{K}v$. As above, we have

$$\mathscr{M}[w](\lambda) = (1 - \mathrm{e}^{\lambda^2})V(\lambda) = \frac{1 - \mathrm{e}^{\lambda^2}}{\lambda} U(\lambda), \quad \text{where } U = \mathscr{M}[r\partial_r v]\,.$$

Since the function $u = r\partial_r v$ belongs to $\mathsf{K}^0_\beta(\mathbb{R}^+)$ and has support in I, U is holomorphic for $\operatorname{Re}\lambda < \eta$, and $\mathscr{M}[w](\lambda)$ has the same property. It follows that $w \in \mathsf{K}^1_\beta(\mathbb{R}^+)$, which implies that w is continuous at 0 and $w(0) = 0$. □

4.3 Generalized Taylor expansions

We are now ready for the definition of the splitting which replaces the Taylor expansion in the critical case: Let us assume that the natural number M is critical. We are going to replace the homogeneous part

$$\mathfrak{T}^M u = \sum_{|\alpha|=M} \partial^\alpha_{\mathbf{x}} u(\mathbf{0})\, \frac{\mathbf{x}^\alpha}{\alpha!}\,,$$

of the corner Taylor expansion with a new operator $u \mapsto \mathfrak{K}^M u$ for which the point traces $\partial^\alpha_{\mathbf{x}} u(\mathbf{0})$ are replaced by moments defined thanks to the the dual basis (3.21) $(\varphi^M_\gamma)_{|\gamma|=M}$. Let us recall that:

$$\int_G \frac{\vartheta^\alpha}{\alpha!}\, \varphi^M_\gamma(\vartheta)\, \mathrm{d}\vartheta = \delta_{\alpha\gamma}\,, \quad |\alpha| = |\gamma| = M, \quad \vartheta^\alpha = r^{-M}\mathbf{x}^\alpha\,, \tag{4.5}$$

and this dual basis served to define the projection operator $\mathfrak{P}^M : \mathsf{L}^2(G) \to \mathsf{P}^M(G)$ as

$$\mathfrak{P}^M U = \sum_{|\alpha|=k} \langle U, \varphi^M_\alpha \rangle_G\, \frac{\vartheta^\alpha}{\alpha!}\,. \tag{4.6}$$

Definition 4.5. Let $M \in \mathbb{N}$. For $u \in \mathscr{C}^\infty(\overline{K})$, let $\mathfrak{T}^{M-1}u$ be its Taylor expansion at $\mathbf{0}$ of order $M - 1$, and let $u^M = u - \mathfrak{T}^{M-1}u$ be its Taylor remainder of order M, considered in polar coordinates (r, ϑ). With the dual basis (4.5), we define the moments of u^M:

$$\forall \alpha,\ |\alpha| = M, \quad d_\alpha(r) = \langle r^{-M}u^M(r, \cdot), \varphi^M_\alpha \rangle_G\,, \quad r > 0. \tag{4.7}$$

Let us fix a cut-off function $\chi \in \mathscr{C}^\infty_0((-1, 1))$, $\chi \equiv 1$ on $[-\frac{1}{2}, \frac{1}{2}]$. Then, using (4.3), the *regularizing operator* $\mathfrak{K}^M u$ is defined by

$$\mathfrak{K}^M u = \sum_{|\alpha|=M} \mathfrak{K}(\chi d_\alpha)\, \frac{\mathbf{x}^\alpha}{\alpha!}\,. \tag{4.8}$$

Remark 4.6. For $M = 0$, $d_\alpha \equiv d_0$ is the mean value of $u(r,\cdot)$ over G and $\mathfrak{K}^0 u = \mathfrak{K}(\chi d_0)$. In particular, if u is continuous in $\mathbf{0}$, then both d_0 and $\mathfrak{K}(\chi d_0)$ are continuous in 0, and $\mathfrak{K}^0 u(\mathbf{0}) = u(\mathbf{0})$ (cf. Proposition 4.3 (c)). More generally, for sufficiently smooth u

$$d_\alpha(0) = \mathfrak{K}(\chi d_\alpha)(0) = \partial_{\mathbf{x}}^\alpha u(\mathbf{0}).$$

The moments d_α are well defined in the critical case and have the following properties.

Proposition 4.7. *Let β be real such that $-\beta - \frac{n}{2}$ coincides with a nonnegative integer M. For $m > M$, let $u \in \mathsf{J}^m_\beta(K)$ with support in $B(\mathbf{0},1)$. Then the moments d_α as defined in* (4.7) *satisfy the following conditions*:

(a) *for all $|\alpha| = M$, $\chi d_\alpha \in \mathsf{J}^m_{-\frac{1}{2}}(\mathbb{R}_+)$,*

(b) *for all $\alpha| = M$, if $\chi d_\alpha \in \mathsf{K}^m_{-\frac{1}{2}}(\mathbb{R}_+)$, then $u - \mathfrak{T}^{M-1}u$ belongs to $\mathsf{K}^m_\beta(K)$.*

Proof. We can assume without restriction that $\chi u = u$. Let us set $v = u - \chi\mathfrak{T}^{M-1}u$. Then $\chi d_\alpha[u] = d_\alpha[v]$ and

$$v - \mathfrak{T}^{M-1}v = v \quad \text{and} \quad u - \mathfrak{T}^{M-1}u = v - (1-\chi)\mathfrak{T}^{M-1}u.$$

Since $(1-\chi)\mathfrak{T}^{M-1}u$ belongs to $\mathsf{K}^m_\beta(K)$, we can replace u with v and omit the cut-off function χ. We still denote v by u. The Mellin transform $U(\lambda)$ of u is *holomorphic* in the half-plane $\operatorname{Re}\lambda < M$ and, by Theorem 3.22, the norm

$$\sup_{b \in (M-\frac{1}{2}, M)} \Big\{ \int_{\substack{|\operatorname{Im}\lambda| \leqslant 1 \\ \operatorname{Re}\lambda = b}} \|(\mathbb{I} - \mathfrak{P}^M)U(\lambda)\|^2_{m;\,G} + |\lambda - M|^2 \|\mathfrak{P}^M U(\lambda)\|^2_{m;\,G} \,\mathrm{d}\operatorname{Im}\lambda$$

$$+ \int_{\substack{|\operatorname{Im}\lambda| \geqslant 1 \\ \operatorname{Re}\lambda = b}} \|U(\lambda)\|^2_{m;\,G;\,\lambda} \,\mathrm{d}\operatorname{Im}\lambda \Big\}^{\frac{1}{2}} \tag{4.9}$$

is bounded by $C\|u\|_{\mathsf{J}^m_\beta(K)}$.

As a mere consequence of the definition of $\mathfrak{P}^M$ (cf. (4.6)), we have the uniform inequality for $\operatorname{Re}\lambda < M$

$$\|\mathfrak{P}^M U(\lambda)\|_{m;\,G;\,\lambda} \leqslant C\|U(\lambda)\|_{m;\,G;\,\lambda}.$$

Hence we deduce from estimates (4.9) that

$$\sup_{b \in (M-\frac{1}{2}, M)} \int_{\operatorname{Re}\lambda = b} \|(\mathbb{I} - \mathfrak{P}^M)U(\lambda)\|^2_{m;\,G;\,\lambda} \,\mathrm{d}\operatorname{Im}\lambda \leqslant C\|u\|^2_{\mathsf{J}^m_\beta(K)}.$$

Thus, Theorem 3.2 yields that $\mathscr{M}^{-1}[(\mathbb{I} - \mathfrak{P}^M)U]$ belongs to $\mathsf{K}^m_\beta(K)$.

Let us set $D_\alpha = \mathscr{M}[d_\alpha]$. We have

$$D_\alpha(\lambda) = \mathscr{M}\Big[\langle r^{-M} u^M, \varphi_\alpha^M \rangle_G\Big](\lambda) = \mathscr{M}\Big[\langle u^M, \varphi_\alpha^M \rangle_G\Big](\lambda + M)$$
$$= \langle U(\lambda + M), \varphi_\alpha^M \rangle_G.$$

Therefore, by formulas (4.5) and (4.6), we find

$$D_\alpha(\lambda) = \langle \mathfrak{P}^M U(\lambda + M), \varphi_\alpha^M \rangle_G. \tag{1}$$

(a) We deduce from (1) and (4.9) that D_α is holomorphic in the half-plane $\operatorname{Re}\lambda < 0$ and

$$\sup_{b \in (-\frac{1}{2}, 0)} \left\{ \int_{\operatorname{Re}\lambda = b} \big(|\lambda|^2 + |\lambda|^{2m}\big)\, |D_\alpha(\lambda)|^2 \,\mathrm{d}\operatorname{Im}\lambda \right\}^{\frac{1}{2}}$$

is bounded. This allows us to prove that $d_\alpha \in \mathsf{J}^m_{-\frac{1}{2}}(\mathbb{R}_+)$.

(b) If $d_\alpha \in \mathsf{K}^m_{-\frac{1}{2}}(\mathbb{R}_+)$, then

$$\int_{\operatorname{Re}\lambda = 0} \big(1 + |\lambda|^{2m}\big)\, |D_\alpha(\lambda)|^2 \,\mathrm{d}\operatorname{Im}\lambda$$

is bounded. Since, by (1) and (4.5),

$$\mathfrak{P}^M U(\lambda + M) = \textstyle\sum_{|\alpha| = M} D_\alpha(\lambda)\, \frac{\vartheta^\alpha}{\alpha!},$$

we find

$$\int_{\operatorname{Re}\lambda = M} \big\| \mathfrak{P}^M U(\lambda) \big\|^2_{m;\, G;\, \lambda} \,\mathrm{d}\operatorname{Im}\lambda < \infty.$$

Hence $\mathscr{M}^{-1}[\mathfrak{P}^M U] \in \mathsf{K}^m_\beta(K)$. Since $\mathscr{M}^{-1}[(\mathbb{I} - \mathfrak{P}^M)U] \in \mathsf{K}^m_\beta(K)$, this ends the proof. □

We conclude this section with a result about the generalized Taylor expansion at the corner in the critical case. The homogeneous part of critical degree $\sum_{|\alpha|=M} \partial^\alpha_{\mathbf{x}} u(\mathbf{0}) \frac{\mathbf{x}^\alpha}{\alpha!}$ does not make sense because the Taylor coefficients $\partial^\alpha_{\mathbf{x}} u(\mathbf{0})$ are not bounded with respect to the J^m_β norm in this case. But one can replace the constants $\partial^\alpha_{\mathbf{x}} u(\mathbf{0})$ by "generalized constants", namely the analytic functions $\mathfrak{K}(\chi d_\alpha)$, which means that the homogeneous part of degree M of the Taylor expansion is replaced by $\mathfrak{K}^M u$, which is not a polynomial, but belongs to the analytic class $\mathsf{B}_\beta(S)$. The "Taylor remainder" then belongs to $\mathsf{K}^m_\beta(K)$.

Theorem 4.8. *Let β be such that $-\beta - \frac{n}{2} = M \in \mathbb{N}$, and let $u \in \mathsf{J}^m_\beta(K)$ with support in $B(\mathbf{0}, 1)$. Then*

$$u - \mathfrak{T}^{M-1} u - \mathfrak{K}^M u \in \mathsf{K}^m_\beta(S) \quad \text{and} \quad \mathfrak{K}^M u \in \mathsf{B}_\beta(S). \tag{4.10}$$

Proof. Let $u \in \mathsf{J}^m_\beta(K)$ with support in $B(\mathbf{0},1)$. By Proposition 4.7 (a), for all $|\alpha| = M$, χd_α belongs to $\mathsf{J}^m_{-\frac{1}{2}}(\mathbb{R}_+)$. By Proposition 4.3 (b), we deduce that $\mathfrak{K}(\chi d_\alpha)$ belongs to $\mathsf{B}_{-\frac{1}{2}}(I)$. Let us consider the function

$$v_\alpha : S \ni \mathbf{x} \mapsto \mathfrak{K}(\chi d_\alpha)(r)$$

and the class $\mathsf{B}_{-\frac{n}{2}}(S)$. This class is associated with $\eta = 0$. Therefore, using Lemma 3.21 we deduce that $v_\alpha \in \mathsf{B}_{-\frac{n}{2}}(S)$ as a direct consequence of the fact that $\mathfrak{K}(\chi d_\alpha) \in \mathsf{B}_{-\frac{1}{2}}(I)$. Multiplying by $\mathbf{x}^\alpha$, we find that $\mathbf{x} \mapsto \mathbf{x}^\alpha v_\alpha(\mathbf{x})$ belongs to $\mathsf{B}_{-\frac{n}{2}-M}(S) = \mathsf{B}_\beta(S)$. Finally $\mathfrak{K}^M u$ belongs to $\mathsf{B}_\beta(S)$.

Let $v = u - \mathfrak{T}^{M-1}u - \mathfrak{K}^M u$. It remains to show that $v \in \mathsf{K}^m_\beta(S)$. Denote by $d_\alpha[v]$ the moments of v defined like in (4.7). We note that

$$\chi d_\alpha[v] = \chi d_\alpha - \chi\mathfrak{K}(\chi d_\alpha).$$

But Proposition 4.7 (a) yields $\chi d_\alpha \in \mathsf{J}^m_{-\frac{1}{2}}(\mathbb{R}_+)$ and then, by Proposition 4.3 (b), we get $\chi d_\alpha - \mathfrak{K}(\chi d_\alpha) \in \mathsf{K}^m_{-\frac{1}{2}}(\mathbb{R}_+)$, hence $\chi d_\alpha[v] \in \mathsf{K}^m_{-\frac{1}{2}}(\mathbb{R}_+)$. The regularity $v \in \mathsf{K}^m_\beta(S)$ is then a consequence of Proposition 4.7 (b). □

Corollary 4.9. *Let β be such that $-\beta - \frac{n}{2} = M \in \mathbb{N}$ and $m > M$. Then the space $\mathsf{K}^m_\beta(S)$ is not closed in $\mathsf{J}^m_\beta(S)$ and the quotient $\mathsf{J}^m_\beta(S)/\mathsf{K}^m_\beta(S)$ is infinite dimensional.*

5 Conclusion

Theorems 3.22 and 4.8 can advantageously be used for the analysis of second order elliptic boundary value problems in domains Ω with corners. Let L be the interior operator, and let B be the operator on the boundary. L is supposed to be elliptic on $\overline{\Omega}$ and B to cover L on $\partial\Omega$. The order d of B is 0 or 1.

Theorem 3.22 fully characterizes the spaces J^m_β by Mellin transformation. This is an essential tool for stating necessary and sufficient conditions for (L, B) to define a Fredholm operator:

$$\mathsf{J}^m_\beta(\Omega) \longrightarrow \mathsf{J}^{m-2}_{\beta+2}(\Omega) \times \Gamma_{\partial\Omega}\mathsf{J}^{m-d}_{\beta+d}(\Omega),$$

where $\Gamma_{\partial\Omega}$ denotes the trace operator on $\partial\Omega$. When K^m_β spaces are involved instead, this condition is the absence of poles for the corner Mellin resolvents on certain lines $\{\operatorname{Re}\lambda = \text{const}\}$ (cf. [3]). Theorem 3.22 allows us to prove by Mellin transformation that the necessary and sufficient condition associated with spaces J^m_β is the injectivity modulo polynomials (cf. [2, 1]) on similar lines in the complex plane.

Theorem 4.8 allows us to prove an *analytic shift theorem* in J^m_β spaces for elliptic (L,B) with analytic coefficients: Roughly, this means that if a solution u belongs to $\mathsf{J}^2_\beta(\Omega)$ and is associated with a right-hand side in $\mathsf{RB}_\beta(\Omega) := \mathsf{B}_{\beta+2}(\Omega) \times \Gamma_{\partial\Omega}\mathsf{B}_{\beta+d}(\Omega)$, then u belongs to $\mathsf{B}_\beta(\Omega)$. This result relies on

1. The analytic shift theorem in the scale K^m_β: If $u \in \mathsf{K}^2_\beta(\Omega)$ and the right-hand side belongs to $\mathsf{RA}_\beta(\Omega)$, then $u \in \mathsf{A}_\beta(\Omega)$.
2. The splitting (4.10).

The analytic shift theorem in the scale K^m_β, that is with homogeneous norms, can be proved by a "standard" technique of dyadic refined partitions towards the corners combined with local analytic estimates in smooth regions. This technique cannot be directly applied to spaces with nonhomogeneous norms, hence the utility of the splitting (4.10) (cf. [1, Part II] for details)

References

1. Costabel, M., Dauge, M., Nicaise, S.: Corner Singularities and Analytic Regularity for Linear Elliptic Systems [In preparation]
2. Dauge, M.: Elliptic Boundary Value Problems in Corner Domains – Smoothness and Asymptotics of Solutions. Lect. Notes Math. **1341** Springer, Berlin (1988)
3. Kondrat'ev, V.A.: Boundary value problems for elliptic equations in domains with conical or angular points. Trans. Moscow Math. Soc. **16**, 227–313 (1967)
4. Kozlov, V.A., Maz'ya, V.G., Rossmann, J.: Elliptic Boundary Value Problems in Domains with Point Singularities. Am. Math. Soc., Providence, RI (1997)
5. Kozlov, V.A., Maz'ya, V.G., Rossmann, J.: Spectral Problems Associated with Corner Singularities of Solutions to Elliptic Equations. Am. Math. Soc., Providence, RI (2001)
6. Maz'ya, V.G., Plamenevskii, B.A.: Weighted spaces with nonhomogeneous norms and boundary value problems in domains with conical points. Transl., Ser. 2, Am. Math. Soc. **123**, 89–107 (1984)
7. Nazarov, S.A.: Vishik-Lyusternik method for elliptic boundary value problems in regions with conical points. I. The problem in a cone. Sib. Math. J. **22**, 594–611 (1981)
8. Nazarov, S.A., Plamenevskiĭ, B.A.: The Neumann problem for selfadjoint elliptic systems in a domain with a piecewise-smooth boundary. Am. Math. Soc. Transl. (2) **155**, 169–206 (1993)
9. Nazarov, S.A., Plamenevskii, B.A.: Elliptic Problems in Domains with Piecewise Smooth Boundaries. Walter de Gruyter, Berlin (1994)

Optimal Hardy–Sobolev–Maz'ya Inequalities with Multiple Interior Singularities

Stathis Filippas, Achilles Tertikas, and Jesper Tidblom

Dedicated to Professor Vladimir Maz'ya with esteem

Abstract We first establish a complete characterization of the Hardy inequalities in $\mathbb{R}^n$ involving distances to different codimension subspaces. In particular, the corresponding potentials have strong interior singularities. We then provide necessary and sufficient conditions for the validity of Hardy–Sobolev–Maz'ya inequalities with optimal Sobolev terms.

1 Introduction

For $n \geqslant 3$ we write $\mathbb{R}^n = \mathbb{R}^k \times \mathbb{R}^{n-k}$, $1 \leqslant k \leqslant n$. Introduce the affine subspace of codimension k:

$$S_k := \{x = (x_1, \dots x_k, \dots x_n) \in \mathbb{R}^n : \ x_1 = \dots = x_k = 0\}.$$

The Euclidean distance from a point $x \in \mathbb{R}^n$ to S_k is defined by the formula

Stathis Filippas
Department of Applied Mathematics, University of Crete, 71409 Heraklion and Institute of Applied and Computational Mathematics, FORTH, 71110 Heraklion, Greece
e-mail: `filippas@tem.uoc.gr`

Achilles Tertikas
Department of Mathematics, University of Crete, 71409 Heraklion and Institute of Applied and Computational Mathematics, FORTH, 71110 Heraklion, Greece
e-mail: `tertikas@math.uoc.gr`

Jesper Tidblom
The Erwin Schrödinger Institute, Boltzmanngasse 9, A-1090 Vienna, Austria
e-mail: `Jesper.Tidblom@esi.ac.at`

A. Laptev (ed.), *Around the Research of Vladimir Maz'ya I: Function Spaces*, International Mathematical Series 11, DOI 10.1007/978-1-4419-1341-8_5,

$$d(x) = d(x, S_k) = |\mathbf{X_k}|, \quad \mathbf{X_k} := (x_1, \dots, x_k, 0, \dots, 0).$$

The classical Hardy inequality in $\mathbb{R}^n$, where the distance is taken from S_k, reads

$$\int_{\mathbb{R}^n} |\nabla u|^2 dx \geqslant \left(\frac{k-2}{2}\right)^2 \int_{\mathbb{R}^n} \frac{u^2}{|\mathbf{X_k}|^2} dx, \quad u \in C_0^\infty(\mathbb{R}^n \setminus S_k), \tag{1.1}$$

where the constant $\frac{(k-2)^2}{4}$ is optimal. The Hardy inequality was improved and generalized in many different ways (cf., for example, [1, 2, 5, 8, 10, 11, 13, 14, 15, 8, 20, 27, 28] and the references therein).

On the other hand, the standard Sobolev inequality with critical exponent states that

$$\int_{\mathbb{R}^n} |\nabla u|^2 dx \geqslant S_n \left(\int_{\mathbb{R}^n} |u|^{\frac{2n}{n-2}} dx\right)^{\frac{n-2}{n}}, \quad u \in C_0^\infty(\mathbb{R}^n),$$

where $S_n = \pi n(n-2)\left(\frac{\Gamma(\frac{n}{2})}{\Gamma(n)}\right)^{2/n}$ is the best Sobolev constant [6, 25]. Versions of Sobolev inequalities involving subcritical exponents and weights can be found, for example, in [4, 7, 12].

Maz'ya [22, Section 2.1.6/3] combined both inequalities for $1 \leqslant k \leqslant n-1$ and establish that for any $u \in C_0^\infty(\mathbb{R}^n \setminus S_k)$

$$\int_{\mathbb{R}^n} |\nabla u|^2 dx \geqslant \left(\frac{k-2}{2}\right)^2 \int_{\mathbb{R}^n} \frac{u^2}{|\mathbf{X_k}|^2} dx + c_{k,Q} \left(\int_{\mathbb{R}^n} |\mathbf{X_k}|^{\frac{Q-2}{2}n-Q} |u|^Q dx\right)^{\frac{2}{Q}}, \tag{1.2}$$

with $2 < Q \leqslant 2^* = \frac{2n}{n-2}$. Concerning the best constant $c_{k,Q}$, Tertikas and Tintarev [26] proved that $c_{k,2^*} < S_n$ for $3 \leqslant k \leqslant n-1$, $n \geqslant 4$ or $k = 1$ and $n \geqslant 4$. In the case $k = 1$ and $n = 3$, Benguria, Frank, and Loss [9] (cf. also [21]) established that $c_{1,6} = S_3 = 3(\pi/2)^{4/3}$! Maz'ya and Shaposhnikova [23] recently computed the best constant in the case $k = 1$ and $Q = \frac{2(n+1)}{n-1}$. These are the only cases where the best constant $c_{k,Q}$ is known. For other type of Hardy–Sobolev inequalities cf. [16, 17, 24].

In the case $k = n$, i.e., when the distance is taken from the origin, the inequality (1.2) fails. Brezis and Vazquez [11] considered a bounded domain containing the origin and improved the Hardy inequality by adding a subcritical Sobolev term. It turns out that, in a bounded domain, one can have the critical Sobolev exponent at the expense of adding a logarithmic weight. More specifically, let

$$X(t) = (1 - \ln t)^{-1}, \quad 0 < t < 1.$$

Then an analogue of (1.2) in the case of a bounded domain Ω containing the origin for the critical exponent reads:

$$\int_\Omega |\nabla u|^2 dx - \left(\frac{n-2}{2}\right)^2 \int_\Omega \frac{u^2}{|x|^2} dx$$
$$\geqslant C_n(\Omega) \left(\int_\Omega X^{\frac{2(n-1)}{n-2}} \left(\frac{|x|}{D}\right) |u|^{\frac{2n}{n-2}} dx \right)^{\frac{n-2}{n}}, \quad u \in C_0^\infty(\Omega), \tag{1.3}$$

where $D = \sup_{x\in\Omega} |x|$ (cf [18]). The best constant in (1.3) was recently computed in [3] and has the form

$$C_n(\Omega) = (n-2)^{-\frac{2(n-1)}{n}} S_n.$$

Note that, in the case $n = 3$, once again $C_3(\Omega) = S_3 = 3(\pi/2)^{4/3}$!

In the recent work [19], we studied the Hardy–Sobolev–Maz'ya inequalities that involve distances taken from different codimension subspaces of the boundary. In particular, working in the upper half-space $\mathbb{R}^n_+ = \{x \in \mathbb{R}^n : x_1 > 0\}$ and taking distances from $S_k \subset \partial\mathbb{R}^n_+ \equiv S_1$, $k = 1, 2, \ldots, n$, we established that the following inequality holds for any $u \in C_0^\infty(\mathbb{R}^n_+)$

$$\int_{\mathbb{R}^n_+} |\nabla u|^2 dx \geqslant \int_{\mathbb{R}^n_+} \left(\frac{\beta_1}{x_1^2} + \frac{\beta_2}{|\mathbf{X_2}|^2} + \ldots + \frac{\beta_n}{|\mathbf{X_n}|^2} \right) u^2 dx \tag{1.4}$$

if and only if there exist nonpositive constants $\alpha_1, \ldots, \alpha_n$ such that

$$\beta_1 = -\alpha_1^2 + \frac{1}{4}, \quad \beta_m = -\alpha_m^2 + \left(\alpha_{m-1} - \frac{1}{2}\right)^2, \quad m = 2, 3, \ldots, n. \tag{1.5}$$

Moreover, if $\alpha_n < 0$, then one can add the critical Sobolev term on the right-hand side, thus obtaining the Hardy–Sobolev–Maz'ya inequality for any $u \in C_0^\infty(\mathbb{R}^n_+)$:

$$\int_{\mathbb{R}^n_+} |\nabla u|^2 dx \geqslant \int_{\mathbb{R}^n_+} \left(\frac{\beta_1}{x_1^2} + \frac{\beta_2}{|\mathbf{X_2}|^2} + \ldots + \frac{\beta_n}{|\mathbf{X_n}|^2} \right) u^2 dx$$
$$+ C \left(\int_{\mathbb{R}^n_+} |u|^{\frac{2n}{n-2}} dx \right)^{\frac{n-2}{n}}; \tag{1.6}$$

we refer to [19] for details.

In the present work, we consider the case where distances are again taken from different codimension subspaces $S_k \subset \mathbb{R}^n$, which, however, are now placed in the interior of the domain $\mathbb{R}^n$. We consider the cases $k = 3, \ldots, n$ since there is no positive Hardy constant if $k = 2$ (cf (1.1)) and the case $k = 1$ corresponds to the case studied in [19].

We formulate our first result .

Theorem A [improved Hardy inequality]. *Suppose the* $n \geqslant 3$.
(i) *Let* $\alpha_3, \alpha_4, \ldots, \alpha_n$ *be arbitrary real numbers, and let*

$$\beta_3 = -\alpha_3^2 + \frac{1}{4}, \quad \beta_m = -\alpha_m^2 + \left(\alpha_{m-1} - \frac{1}{2}\right)^2, \quad m = 4, \ldots, n.$$

Then for any $u \in C_0^\infty(\mathbb{R}^n)$

$$\int_{\mathbb{R}^n} |\nabla u|^2 dx \geqslant \int_{\mathbb{R}^n} \left(\frac{\beta_3}{|\mathbf{X_3}|^2} + \ldots + \frac{\beta_n}{|\mathbf{X_n}|^2} \right) u^2 dx.$$

(ii) *Suppose that for some real numbers* $\beta_3, \beta_4 \ldots, \beta_n$ *the following inequality holds:*

$$\int_{\mathbb{R}^n} |\nabla u|^2 dx \geqslant \int_{\mathbb{R}^n} \left(\frac{\beta_3}{|\mathbf{X_3}|^2} + \ldots + \frac{\beta_n}{|\mathbf{X_n}|^2} \right) u^2 dx$$

for any $u \in C_0^\infty(\mathbb{R}^n)$. *Then there exists nonpositive constants* $\alpha_3, \ldots, \alpha_n$ *such that*

$$\beta_3 = -\alpha_3^2 + \frac{1}{4}, \quad \beta_m = -\alpha_m^2 + \left(\alpha_{m-1} - \frac{1}{2}\right)^2, \quad m = 4, \ldots, n.$$

Note that the recursive formula for β in this theorem is the same as in (1.5). However, since the coefficients in Theorem A start from β_3 – and not from β_1 – the best constants in the case of interior singularities are different from the best constants when singularities of the same codimension are placed on the boundary. (cf., for example, Corollary 2.3 and [19, Corollary 2.4]).

To state our next results, we define

$$\beta_3 = -\alpha_3^2 + \frac{1}{4}, \quad \beta_m = -\alpha_m^2 + \left(\alpha_{m-1} - \frac{1}{2}\right)^2, \quad m = 4, \ldots, n. \tag{1.7}$$

The next theorem gives a complete answer as to when we can add a Sobolev term.

Theorem B [improved Hardy–Sobolev–Maz'ya inequality]. *Suppose that* $\alpha_3, \alpha_4, \ldots, \alpha_n$, $n \geqslant 3$, *are arbitrary nonpositive real numbers and* $\beta_3, \ldots, \beta_n$ *are given by* (1.7). *If* $\alpha_n < 0$, *then there exists a positive constant* C *such that for any* $u \in C_0^\infty(\mathbb{R}^n)$

$$\int_{\mathbb{R}^n} |\nabla u|^2 dx \geqslant \int_{\mathbb{R}^n} \left(\frac{\beta_3}{|\mathbf{X_3}|^2} + \ldots + \frac{\beta_n}{|\mathbf{X_n}|^2} \right) u^2 dx + C \left(\int_{\mathbb{R}^n} |\mathbf{X_2}|^{\frac{Q-2}{2}n - Q} |u|^Q dx \right)^{\frac{2}{Q}}, \tag{1.8}$$

for any $2 < Q \leqslant \frac{2n}{n-2}$. *If* $\alpha_n = 0$, *then there is no positive constant* C *such that (1.8) holds.*

The above result extends considerably the original inequality obtained by Maz'ya (1.2). First, by having at the same time all possible combinations of Hardy potentials involving the distances $|\mathbf{X_3}|, \ldots, |\mathbf{X_n}|$. Second, the weight in the Sobolev term is stronger than the weight used in (1.2).

We note that a similar result can be obtained in the setting of [19], where singularities are placed on the boundary $\partial\mathbb{R}^n_+$. More precisely, the following inequality holds for any $u \in C_0^\infty(\mathbb{R}^n_+)$:

$$\int_{\mathbb{R}^n_+} |\nabla u|^2 dx \geqslant \int_{\mathbb{R}^n_+} \left(\frac{\beta_1}{x_1^2} + \frac{\beta_2}{|\mathbf{X_2}|^2} \cdots + \frac{\beta_n}{|\mathbf{X_n}|^2} \right) u^2 dx + C \left(\int_{\mathbb{R}^n_+} x_1^{\frac{Q-2}{2}n-Q} |u|^Q dx \right)^{\frac{2}{Q}} \tag{1.9}$$

provided that $\alpha_n < 0$, where the constants β_i are given by (1.5) and $2 < Q \leqslant \frac{2n}{n-2}$. In this case, the weight on the right-hand side is even stronger than that in (1.8). In the light of (1.9), one may ask whether one can replace the weight $|\mathbf{X_2}|$ in (1.8) with $|x_1|$. It turns out that it is possible provided that we properly restrict the exponent Q. More precisely, the following assertion holds.

Theorem C. [improved Hardy–Sobolev–Maz'ya inequality]. *Suppose that $\alpha_3, \alpha_4, \ldots, \alpha_n$, $n \geqslant 3$, are arbitrary nonpositive real numbers and $\beta_3, \ldots, \beta_n$ are given by (1.7). If $\alpha_n < 0$, then there exists a positive constant C such that for any $u \in C_0^\infty(\mathbb{R}^n)$*

$$\int_{\mathbb{R}^n} |\nabla u|^2 dx \geqslant \int_{\mathbb{R}^n} \left(\frac{\beta_3}{|\mathbf{X_3}|^2} + \ldots + \frac{\beta_n}{|\mathbf{X_n}|^2} \right) u^2 dx + C \left(\int_{\mathbb{R}^n} |x_1|^{\frac{Q-2}{2}n-Q} |u|^Q dx \right)^{\frac{2}{Q}} \tag{1.10}$$

for any $\frac{2(n-1)}{n-2} < Q \leqslant \frac{2n}{n-2}$. If $\alpha_n = 0$ then there is no positive constant C such that (1.10) holds.

It is easy to see that the range of the exponent Q in Theorem C is optimal since otherwise the weight is not locally integrable. In the special case $\beta_3 = \ldots \beta_n = 0$, the corresponding weighted Sobolev inequality in (1.10) was proved by Maz'ya [22, Section 2.1.6/2].

An important role in our analysis is played by two weighted Sobolev inequalities, which are of independent interest (cf. Theorems 3.1 and 3.2).

The paper is organized as follows. In Section 2, we prove Theorem A. In Section 3, we prove Theorems B and C. The main ideas are similar to the ideas used in [19] to which we refer on various occasions. On the other hand, ideas or technical estimates that are different from [19] are presented in detail.

2 Improved Hardy Inequalities with Multiple Singularities

The following simple lemma may be found in [19].

Lemma 2.1. (i) *Let* $\mathbf{F} \in C^1(\Omega)$. *Then*

$$\int_\Omega |\nabla u|^2 dx = \int_\Omega \left(\mathrm{div}\mathbf{F} - |\mathbf{F}|^2\right) |u|^2 dx + \int_\Omega |\nabla u + \mathbf{F} u|^2 dx \quad \forall u \in C_0^\infty(\Omega). \tag{2.1}$$

(ii) *Let* $\phi > 0$, $\phi \in C^2(\Omega)$ *and* $u = \phi v$. *Then*

$$\int_\Omega |\nabla u|^2 dx = -\int_\Omega \frac{\Delta\phi}{\phi} u^2 dx + \int_\Omega \phi^2 |\nabla v|^2 dx \quad \forall u \in C_0^\infty(\Omega). \tag{2.2}$$

Proof. Expanding the square, we have

$$\int_\Omega |\nabla u + \mathbf{F} u|^2 dx = \int_\Omega |\nabla u|^2 dx + \int_\Omega |\mathbf{F}|^2 u^2 dx + \int_\Omega \mathbf{F} \cdot \nabla u^2 dx.$$

The identity (2.1) follows by integrating by parts the last term.

To prove (2.2), we apply (2.1) to $\mathbf{F} = -\frac{\nabla\phi}{\phi}$. An elementary calculation yields the result. □

Let us recall our notation

$$\mathbf{X_k} := (x_1, \ldots, x_k, 0, \ldots, 0) \quad \text{so that} \quad |\mathbf{X_k}|^2 = x_1^2 + \ldots + x_k^2.$$

In particular, $|\mathbf{X_n}| = |x|$. Now, we prove the first part of Theorem A.

Proof of Theorem A (i). Let $\gamma_3, \gamma_4, \ldots, \gamma_n$ be arbitrary real numbers. We set

$$\phi := |\mathbf{X_3}|^{-\gamma_3} |\mathbf{X_4}|^{-\gamma_2} \cdot \ldots \cdot |\mathbf{X_n}|^{-\gamma_n}$$

and

$$\mathbf{F} := -\frac{\nabla\phi}{\phi}.$$

An easy calculation shows that

$$\mathbf{F} = \sum_{m=3}^{n} \gamma_m \frac{\mathbf{X_m}}{|\mathbf{X_m}|^2}.$$

With this choice of $\mathbf{F}$, we get

$$\mathrm{div}\mathbf{F} = \sum_{m=3}^{n} \gamma_m \frac{(m-2)}{|\mathbf{X_m}|^2},$$

and

$$|\mathbf{F}|^2 = \sum_{m=3}^{n} \frac{\gamma_m^2}{|\mathbf{X_m}|^2} + 2\sum_{m=3}^{n}\sum_{j=1}^{m-1} \gamma_m\gamma_j \frac{\mathbf{X_m}}{|\mathbf{X_m}|^2}\frac{\mathbf{X_j}}{|\mathbf{X_j}|^2}$$

$$= \sum_{m=3}^{n} \frac{\gamma_m^2}{|\mathbf{X_m}|^2} + 2\sum_{m=3}^{n}\sum_{j=1}^{m-1} \frac{\gamma_m\gamma_j}{|\mathbf{X_j}|^2}.$$

Then

$$-\frac{\Delta\phi}{\phi} = \operatorname{div}\mathbf{F} - |\mathbf{F}|^2 = \sum_{m=3}^{n} \frac{\beta_m}{|\mathbf{X_m}|^2}, \tag{2.3}$$

where

$$\beta_3 = -\gamma_3(\gamma_3 - 1),$$

$$\beta_m = -\gamma_m\Big(2 - m + \gamma_m + 2\sum_{j=3}^{m-1} \gamma_j\Big), \quad m = 4, 5, \dots, n.$$

We set

$$\gamma_3 = \alpha_3 + \frac{1}{2},$$

$$\gamma_m = \alpha_m - \alpha_{m-1} + \frac{1}{2}, \quad m = 4, 5, \dots, n.$$

With this choice of γ, β are the same as in the statement of the theorem.

We use Lemma 2.1 with $\Omega = \mathbb{R}^n \setminus K_3$, where $K_3 := \{x \in \mathbb{R}^n : x_1 = x_2 = x_3 = 0\}$. We have

$$\int_{\mathbb{R}^n} |\nabla u|^2 dx \geqslant \int_{\mathbb{R}^n} (\operatorname{div}\mathbf{F} - |\mathbf{F}|^2)\, u^2 dx, \quad u \in C_0^\infty(\mathbb{R}^n \setminus K_3). \tag{2.4}$$

By a standard density argument, (2.4) is true even for $u \in C_0^\infty(\mathbb{R}^n)$. The result then follows from (2.3) and (2.4). □

Some interesting cases are presented in the following corollary.

Corollary 2.2. *Let k=3,...,n, $n \geqslant 3$, and let $u \in C_0^\infty(\mathbb{R}^n)$. Then*

$$\int_{\mathbb{R}^n} |\nabla u|^2 dx$$

$$\geqslant \int_{\mathbb{R}^n} \bigg(\Big(\frac{k-2}{2}\Big)^2 \frac{1}{|\mathbf{X_k}|^2} + \frac{1}{4}\frac{1}{|\mathbf{X_{k+1}}|^2} \dots + \frac{1}{4}\frac{1}{|\mathbf{X_n}|^2}\bigg) u^2 dx. \tag{2.5}$$

Moreover,

$$\int_{\mathbb{R}^n} |\nabla u|^2 dx \geqslant \left(\frac{k-2}{2}\right)^2 \int_{\mathbb{R}^n} \frac{u^2}{|\mathbf{X_k}|^2} dx + \left(\frac{n-k}{2}\right)^2 \int_{\mathbb{R}^n} \frac{u^2}{|x|^2} dx. \quad (2.6)$$

Proof. We first prove (2.5). In the case $k = 3$, we choose $\alpha_3 = \alpha_4 = \ldots = \alpha_n = 0$. Then all β_k are equal to $1/4$. In the general case $k > 3$, we choose $\alpha_m = -(m-2)/2$ if $m = 3, \ldots, k-1$ and $\alpha_m = 0$ if $m = k, \ldots, n$.

To prove (2.5), we choose $\alpha_m = -(m-2)/2$, when $m = 3, \ldots, k-1$, $a_k = 0$, $a_{k+l} = -\frac{l}{2}$, $l = 1, \ldots, n-k-1$, $a_n = 0$. □

Proof of Theorem A (ii). We first show that $\beta_3 \leqslant \frac{1}{4}$. Then $\beta_3 = -\alpha_3^2 + \frac{1}{4}$ for suitable $\alpha_3 \leqslant 0$. Then for such β_3 we prove that $\beta_4 \leqslant (\alpha_3 - \frac{1}{2})^2$. Therefore, $\beta_4 = -\alpha_4^2 + (\alpha_3 - \frac{1}{2})^2$ for suitable $\alpha_4 \leqslant 0$, and so on.

Step 1. Let us first prove the estimate for β_3. For this purpose, we set

$$Q_3[u] := \frac{\displaystyle\int_{\mathbb{R}^n} |\nabla u|^2 dx - \sum_{i=4}^{n} \beta_i \int_{\mathbb{R}^n} \frac{u^2}{(x_1^2 + x_2^2 + \ldots + x_i^2)} dx}{\displaystyle\int_{\mathbb{R}^n} \frac{u^2}{x_1^2 + x_2^2 + x_3^2} dx}. \quad (2.7)$$

It is clear that

$$\beta_3 \leqslant \inf_{u \in C_0^\infty(\mathbb{R}^n)} Q_3[u].$$

In the sequel, we show that

$$\inf_{u \in C_0^\infty(\mathbb{R}^n)} Q_3[u] \leqslant \frac{1}{4}. \quad (2.8)$$

Hence $\beta_3 \leqslant 1/4$.

Introduce a family of cut-off functions. For $j = 3, \ldots, n$ and $k_j > 0$ we set

$$\phi_j(t) = \begin{cases} 0, & t < 1/k_j^2, \\ 1 + \dfrac{\ln k_j t}{\ln k_j}, & 1/k_j^2 \leqslant t < 1/k_j, \\ 1, & t \geqslant 1/k_j \end{cases}$$

and

$$h_{k_j}(x) := \phi_j(r_j), \quad \text{where } r_j := |\mathbf{X_j}| = (x_1^2 + \ldots + x_j^2)^{\frac{1}{2}}.$$

Note that

$$|\nabla h_{k_j}(x)|^2 = \begin{cases} \dfrac{1}{\ln^2 k_j} \dfrac{1}{r_j^2}, & 1/k_j^2 \leqslant r_j \leqslant 1/k_j, \\ 0 & \text{otherwise.} \end{cases}$$

We also denote by $\phi(x)$ a radially symmetric $C_0^\infty(\mathbb{R}^n)$ function such that $\phi = 1$ for $|x| < 1/2$ and $\phi = 0$ for $|x| > 1$.

To prove (2.8), we consider the family of functions

$$u_{k_3}(x) = |\mathbf{X_3}|^{-\frac{1}{2}} h_{k_3}(x)\phi(x). \tag{2.9}$$

We show that

$$\frac{\displaystyle\int_{\mathbb{R}^n} |\nabla u_{k_3}|^2 dx - \sum_{i=4}^{n} \beta_i \int_{\mathbb{R}^n} \frac{u_{k_1}^2}{(x_1^2 + x_2^2 + \ldots + x_i^2)} dx}{\displaystyle\int_{\mathbb{R}^n} \frac{u_{k_3}^2}{x_1^2 + x_2^2 + x_3^2} dx}$$

$$= \frac{\displaystyle\int_{\mathbb{R}^n} |\nabla u_{k_3}|^2 dx}{\displaystyle\int_{\mathbb{R}^n} \frac{u_{k_3}^2}{x_1^2 + x_2^2 + x_3^2} dx} + o(1), \quad k_3 \to \infty. \tag{2.10}$$

To see this, let us first examine the behavior of the denominator. For large k_3 we compute

$$\begin{aligned}
&\int_{\mathbb{R}^n} |\mathbf{X_3}|^{-3} h_{k_3}^2 \phi^2 dx \\
&\geqslant C \int_{\frac{1}{k_3} < x_1^2 + x_2^2 + x_3^2 < \frac{1}{2}} (x_1^2 + x_2^2 + x_3^2)^{-\frac{3}{2}} dx_1 dx_2 dx_3 \\
&\geqslant C \int_0^{\pi} \int_{\frac{1}{k_3}}^{\frac{1}{2}} r^{-1} \sin\theta dr d\theta \geqslant C \ln k_3.
\end{aligned} \tag{2.11}$$

On the other hand, by the Lebesgue dominated theorem, the terms

$$\sum_{i=4}^{n} \beta_i \int_{\mathbb{R}_+^n} \frac{u_{k_3}^2}{(x_1^2 + x_2^2 + \ldots + x_i^2)} dx$$

are bounded as $k_3 \to \infty$. From this we conclude (2.10).

We now estimate the gradient term in (2.10). We have

$$\begin{aligned}
\int_{\mathbb{R}^n} |\nabla u_{k_3}|^2 dx &= \frac{1}{4} \int_{\mathbb{R}^n} |\mathbf{X_3}|^{-3} h_{k_3}^2 \phi^2 dx + \int_{\mathbb{R}^n} |\mathbf{X_3}|^{-1} |\nabla h_{k_3}|^2 \phi^2 \\
&\quad + \int_{\mathbb{R}^n} |\mathbf{X_3}|^{-1} h_{k_3}^2 |\nabla \phi|^2 + \text{ mixed terms.}
\end{aligned} \tag{2.12}$$

The first integral on the right-hand side of (2.12) behaves itself exactly as the denominator, i.e., goes to infinity like $O(\ln k_3)$. The last integral is bounded as $k_3 \to \infty$. For the middle integral we have

$$\int_{\mathbb{R}^n} |\mathbf{X_3}|^{-1}|\nabla h_{k_3}|^2\phi^2 \leqslant \frac{C}{\ln^2 k_3}\int_{\frac{1}{k_3^2}\leqslant(x_1^2+x_2^2+x_3^2)^{1/2}\leqslant\frac{1}{k_3}} |\mathbf{X_3}|^{-3}dx_1dx_2dx_3$$
$$\leqslant \frac{C}{\ln k_3}.$$

As a consequence of these estimates, we easily get that the mixed terms in (2.12) are of the order $o(\ln k_3)$ as $k_3 \to \infty$. Hence

$$\int_{\mathbb{R}^n} |\nabla u_{k_3}|^2 dx = \frac{1}{4}\int_{\mathbb{R}^n} |\mathbf{X_3}|^{-3}h_{k_3}^2\phi^2 dx + o(\ln k_3), \quad k_1 \to \infty. \tag{2.13}$$

From (2.10)–(2.13) we conclude that

$$Q_3[u_{k_3}] = \frac{1}{4} + o(1), \quad k_3 \to \infty.$$

Hence

$$\inf_{u\in C_0^\infty(\mathbb{R}^n)} Q_3[u] \leqslant \frac{1}{4}.$$

Consequently, $\beta_3 \leqslant 1/4$. Therefore, for a suitable nonnegative constant α_3 we have $\beta_3 = -\alpha_3^2 + 1/4$. We also set

$$\gamma_3 := \alpha_3 + \frac{1}{2}. \tag{2.14}$$

Step 2. We show that

$$\beta_4 \leqslant (\alpha_3 - \frac{1}{2})^2.$$

For this purpose, setting

$$Q_4[u] := \frac{1}{\displaystyle\int_{\mathbb{R}^n} \frac{u^2}{|\mathbf{X_4}|^2} dx}\Bigg[\int_{\mathbb{R}^n} |\nabla u|^2 dx - \Big(\frac{1}{4} - \alpha_3^2\Big)\int_{\mathbb{R}^n} \frac{u^2}{x_1^2+x_2^2+x_3^2} dx - \sum_{i=5}^n \beta_i \int_{\mathbb{R}^n} \frac{u^2}{|\mathbf{X_i}|^2} dx\Bigg] \tag{2.15}$$

will prove that

$$\inf_{u\in C_0^\infty(\mathbb{R}^n)} Q_4[u] \leqslant (\alpha_3 - 1/2)^2.$$

We now consider the family of functions

$$u_{k_3,k_4}(x) := |\mathbf{X_3}|^{-\gamma_3}|\mathbf{X_4}|^{\alpha_3-\frac{1}{2}}h_{k_3}(x)h_{k_4}(x)\phi(x)$$
$$=: |\mathbf{X_3}|^{-\gamma_3}v_{k_1,k_2}(x). \tag{2.16}$$

An a easy calculation shows that

$$Q_4[u_{k_3,k_4}] = \frac{1}{\displaystyle\int_{\mathbb{R}^n} |\mathbf{X_3}|^{-2\gamma_3} |\mathbf{X_4}|^{-2} v^2_{k_3,k_4} dx} \Bigg[\int_{\mathbb{R}^n} |\mathbf{X_3}|^{-2\gamma_3} |\nabla v_{k_3,k_4}|^2 dx$$

$$- \sum_{i=5}^{n} \beta_i \int_{\mathbb{R}^n_+} |\mathbf{X_3}|^{-2\gamma_3} |\mathbf{X_i}|^{-2} v^2_{k_3,k_4} dx \Bigg] \tag{2.17}$$

We next use the precise form of $v_{k_1,k_2}(x)$. Concerning the denominator of $Q_4[u_{k_3,k_4}]$, we have

$$\int_{\mathbb{R}^n} |\mathbf{X_3}|^{-2\gamma_3} |\mathbf{X_4}|^{-2} v^2_{k_3,k_4} dx$$

$$= \int_{\mathbb{R}^n} (x_1^2 + x_2^2 + x_3^2)^{-1/2-\alpha_3} (x_1^2 + x_2^2 + x_3^2 + x_4^2)^{\alpha_1 - \frac{3}{2}} h^2_{k_3} h^2_{k_4} \phi^2 dx.$$

Letting $k_3 \to \infty$, using the structure of the cut-off functions, and introducing polar coordinates, we get

$$\int_{\mathbb{R}^n} |\mathbf{X_3}|^{-2\gamma_3} |\mathbf{X_4}|^{-2} v^2_{\infty,k_4} dx$$

$$= \int_{\mathbb{R}^n} (x_1^2 + x_2^2 + x_3^2)^{-1/2-\alpha_3} (x_1^2 + x_2^2 + x_3^2 + x_4^2)^{\alpha_3 - \frac{3}{2}} h^2_{k_4} \phi^2 dx,$$

$$\geqslant C \int_{\frac{1}{k_4} < x_1^2 + x_2^2 + x_3^2 + x_4^2 < \frac{1}{2}} (x_1^2 + x_2^2 + x_3^2)^{-1/2-\alpha_3}$$

$$\times (x_1^2 + x_2^2 + x_3^2 + x_4^2)^{\alpha_3 - \frac{3}{2}} dx_1 dx_2 dx_3 dx_4$$

$$\geqslant C \int_{\frac{1}{k_4}}^{\frac{1}{2}} r^{-1} dr \geqslant C \ln k_4.$$

The terms in the numerator, multiplied by β_i, stay bounded as k_3 or k_4 go to infinity. We have

$$\int_{\mathbb{R}^n} |\mathbf{X_3}|^{-2\gamma_3} |\nabla v_{k_3,k_4}|^2 dx$$

$$= \Big(\alpha_3 - \frac{1}{2}\Big)^2 \int_{\mathbb{R}^n} |\mathbf{X_3}|^{-2\gamma_3} |\mathbf{X_4}|^{2\alpha_3 - 3} h^2_{k_3} h^2_{k_4} \phi^2 dx$$

$$+ \int_{\mathbb{R}^n} |\mathbf{X_3}|^{-2\gamma_3} |\mathbf{X_4}|^{2\alpha_3 - 1} |\nabla (h_{k_3} h_{k_4})|^2 \phi^2$$

$$+ \int_{\mathbb{R}^n} |\mathbf{X_3}|^{-2\gamma_3} |\mathbf{X_4}|^{2\alpha_3 - 1} h^2_{k_3} h^2_{k_4} |\nabla \phi|^2 + \text{ mixed terms.} \tag{2.18}$$

The first integral on the right hand side above is the same as the denominator of Q_4, and therefore is finite as $k_3 \to \infty$ and increases like $\ln k_4$ as $k_4 \to \infty$ (cf. (2.11)). The last integral is bounded, no matter how large k_3 and k_4 are. Concerning the middle term, we have

$$\begin{aligned} M[v_{k_3,k_4}] &:= \int_{\mathbb{R}^n} |\mathbf{X_3}|^{-2\gamma_3} |\mathbf{X_4}|^{2\alpha_3 - 1} |\nabla (h_{k_3} h_{k_4})|^2 \phi^2 dx \\ &= \int_{\mathbb{R}^n} |\mathbf{X_3}|^{-2\gamma_3} |\mathbf{X_4}|^{2\alpha_3 - 1} |\nabla h_{k_3}|^2 h_{k_4}^2 \phi^2 dx \\ &+ \int_{\mathbb{R}^n} |\mathbf{X_3}|^{-2\gamma_3} |\mathbf{X_4}|^{2\alpha_3 - 1} h_{k_3}^2 |\nabla h_{k_4}|^2 \phi^2 dx + \quad \text{mixed term} \\ &=: I_1 + I_2 + \quad \text{mixed term}. \end{aligned} \tag{2.19}$$

Since

$$|\mathbf{X_4}|^{2\alpha_3 - 1} h_{k_4}^2 = r_4^{2\alpha_3 - 1} \phi_4(r_4) \leqslant C_{k_4}, \quad 0 < r_4 < 1,$$

we easily get

$$I_1 \leqslant \frac{C}{(\ln k_3)^2} \int_{\frac{1}{k_3^2} < (x_1^2 + x_2^2 + x_3^2)^{1/2} < \frac{1}{k_3}} (x_1^2 + x_2^2 + x_3^2)^{-\alpha_3 - \frac{3}{2}} dx_1 dx_2 dx_3,$$

and therefore, since $\alpha_3 \leqslant 0$,

$$I_1 \leqslant \frac{C}{\ln k_3}, \quad k_3 \to \infty. \tag{2.20}$$

Moreover, since

$$|\mathbf{X_3}|^{-2\gamma_3} h_{k_3}^2 = r_3^{2\alpha_3 - 1} \phi_3(r_3) \leqslant C_{k_3}, \quad 0 < r_3 < 1,$$

we similarly get (for any k_3)

$$\begin{aligned} I_2 &\leqslant \frac{C}{(\ln k_4)^2} \int_{\frac{1}{k_4^2} < (x_1^2 + x_2^2 + x_3^2 + x_4^2)^{1/2} < \frac{1}{k_4}} \frac{dx_1 dx_2 dx_3 dx_4}{(x_1^2 + x_2^2 + x_3^2 + x_4^2)^{\frac{1}{2}}} \\ &\leqslant \frac{C}{\ln k_4}, \quad k_4 \to \infty. \end{aligned} \tag{2.21}$$

From (2.19)–(2.21) we find

$$M[v_{\infty,k_4}] = o(1), \quad k_4 \to \infty.$$

Returning to (2.18), we have

$$\int_{\mathbb{R}^n} |\mathbf{X_3}|^{-2\gamma_3} |\nabla v_{\infty,k_4}|^2 dx$$

$$= \int_{\mathbb{R}^n} |\mathbf{X_3}|^{-2\gamma_3} |\mathbf{X_4}|^{-2} v_{\infty,k_4}^2 dx + o(\ln k_4), \quad k_4 \to \infty. \tag{2.22}$$

Then

$$Q_4[u_{\infty,k_4}] = (\alpha_3 - 1/2)^2 + o(1), \quad k_4 \to \infty. \tag{2.23}$$

Consequently,

$$\beta_4 \leqslant (\alpha_3 - 1/2)^2,$$

and therefore

$$\beta_4 = -\alpha_4^2 + (\alpha_3 - 1/2)^2$$

for suitable $\alpha_4 \leqslant 0$. We also set

$$\gamma_4 = \alpha_4 - \alpha_3 + 1/2.$$

Step 3. The general case. At the $(q-1)$th step, $3 \leqslant q \leqslant n$, we have already established that

$$\begin{aligned} \beta_3 &= -\alpha_3^2 + 1/4, \\ \beta_m &= -\alpha_m^2 + (\alpha_{m-1} - 1/2)^2, \quad m = 4, 5, \dots, q-1, \end{aligned}$$

for suitable nonpositive constants a_i. Also, we have defined

$$\begin{aligned} \gamma_3 &= \alpha_3 + 1/2, \\ \gamma_m &= \alpha_m - \alpha_{m-1} + 1/2, \quad m = 4, 5, \dots, q-1. \end{aligned}$$

Our goal for the rest of the proof is to show that

$$\beta_q \leqslant (\alpha_{q-1} - 1/2)^2.$$

For this purpose, we consider the quotient

$$Q_q[u] := \frac{\displaystyle\int_{\mathbb{R}^n} |\nabla u|^2 dx - \sum_{q \neq i = 3}^{n} \beta_i \int_{\mathbb{R}^n} \frac{u^2}{|\mathbf{X_i}|^2} dx}{\displaystyle\int_{\mathbb{R}^n} \frac{u^2}{|\mathbf{X_q}|^2} dx}. \tag{2.24}$$

The test function is given by

$$\begin{aligned} &u_{k_3,k_q}(x) \\ &:= |\mathbf{X_3}|^{-\gamma_3} |\mathbf{X_4}|^{-\gamma_4} \dots |\mathbf{X_{q-1}}|^{-\gamma_{q-1}} |\mathbf{X_q}|^{\alpha_{q-1} - \frac{1}{2}} h_{k_4}(x) h_{k_q}(x) \phi(x) \\ &=: |\mathbf{X_3}|^{-\gamma_3} |\mathbf{X_4}|^{-\gamma_4} \dots |\mathbf{X_{q-1}}|^{-\gamma_{q-1}} . v_{k_q}(x). \end{aligned} \tag{2.25}$$

The proof is similar to that in the case $q = 4$ and goes along the lines of [19]. □

The following assertion is a direct consequence of the above theorem and shows that the constants obtained in Corollary 2.2 are sharp.

Corollary 2.3. *For* $3 \leqslant k \leqslant n$

$$\inf_{u \in C_0^\infty(\mathbb{R}^n)} \frac{\int_{\mathbb{R}^n} |\nabla u|^2 dx}{\int_{\mathbb{R}^n} \frac{|u|^2}{|\mathbf{X_k}|^2}} = \left(\frac{k-2}{2}\right)^2, \tag{2.26}$$

$$\inf_{u \in C_0^\infty(\mathbb{R}^n)} \frac{1}{\int_{\mathbb{R}^n} \frac{|u|^2}{|\mathbf{X_{m+1}}|^2} dx} \left[\int_{\mathbb{R}^n} |\nabla u|^2 dx - \left(\frac{k-2}{2}\right)^2 \int_{\mathbb{R}^n} \frac{|u|^2}{|\mathbf{X_k}|^2} dx \right.$$
$$\left. - \frac{1}{4} \int_{\mathbb{R}^n} \frac{|u|^2}{|\mathbf{X_{k+1}}|^2} dx - \ldots - \frac{1}{4} \int_{\mathbb{R}^n} \frac{|u|^2}{|\mathbf{X_m}|^2} dx \right] = \frac{1}{4} \tag{2.27}$$

for $k \leqslant m < n$. *Moreover,*

$$\inf_{u \in C_0^\infty(\mathbb{R}^n)} \frac{\int_{\mathbb{R}^n} |\nabla u|^2 dx - \left(\frac{k-2}{2}\right)^2 \int_{\mathbb{R}^n} \frac{|u|^2}{|\mathbf{X_k}|^2} dx}{\int_{\mathbb{R}^n} \frac{|u|^2}{|x|^2} dx} = \left(\frac{n-k}{2}\right)^2. \tag{2.28}$$

Proof. All the assertions are consequences of Theorem A. For (2.26) we take $\alpha_l = -\frac{l-2}{2}$, $l = 1, \ldots, k-1$. For (2.27) and (2.28) we take the a's of Corollary 2.2. □

3 Hardy–Sobolev–Maz'ya Inequalities

We first establish the following result that will be used for Theorem B.

Theorem 3.1 (weighted Sobolev inequality). *Let* $\sigma_2, \sigma_3, \ldots, \sigma_n$ *be real numbers,* $n \geqslant 2$. *We set* $c_l := \sigma_2 + \ldots + \sigma_l + l - 1$ *for* $2 \leqslant l \leqslant n$. *Assume that*

$$c_l > 0 \quad , \textit{whenever} \quad \sigma_l \neq 0,$$

for $l = 2, \ldots, n$. *Then there exists a positive constant* C *such that for any* $w \in C_0^\infty(\mathbb{R}^n)$

$$\int_{\mathbb{R}^n} |\mathbf{X_2}|^{\sigma_2} \ldots |\mathbf{X_n}|^{\sigma_n} |\nabla w| dx$$

$$\geqslant C\left(\int_{\mathbb{R}^n}\left(|\mathbf{X_2}|^b|\mathbf{X_3}|^{\sigma_3}\ldots|\mathbf{X_n}|^{\sigma_n}|w|\right)^q dx\right)^{\frac{1}{q}}, \tag{3.1}$$

where

$$b=\sigma_2-1+\frac{q-1}{q}n,\quad 1<q\leqslant\frac{n}{n-1}.$$

Proof. For $1<q\leqslant n/(n-1)$ and $b=\sigma_2-1+\frac{q-1}{q}n$ we easily obtain the following L^1 interpolation inequality:

$$\||\mathbf{X_2}|^b v\|_q\leqslant c_1\||\mathbf{X_2}|^{\sigma_2}v\|_{\frac{n}{n-1}}+c_2\||\mathbf{X_2}|^{\sigma_2-1}v\|_1.$$

Using the inequality

$$\left|\int_{\mathbb{R}^n}\operatorname{div}\mathbf{F}|v|dx\right|\leqslant\int_{\mathbb{R}^n}|\mathbf{F}||\nabla v|dx \tag{3.2}$$

with the vector field $\mathbf{F}=|\mathbf{X_2}|^{\sigma_2-1}\mathbf{X_2}$, we obtain

$$|\sigma_2+1|\int_{\mathbb{R}^n}|\mathbf{X_2}|^{\sigma_2-1}|v|dx\leqslant\int_{\mathbb{R}^n}|\mathbf{X_2}|^{\sigma_2}|\nabla v|dx.$$

Here, we restrict ourselves to $\sigma_2+1>0$ to ensure that $|\mathbf{X_2}|^{\sigma_2-1}\in L^1_{\text{loc}}(\mathbb{R}^n)$. Combining this inequality with the standard L^1 Sobolev inequality, we get

$$\||\mathbf{X_2}|^{\sigma_2}v\|_{\frac{n}{n-1}}\leqslant\||\mathbf{X_2}|^{\sigma_2}|\nabla v|\|_1.$$

Hence

$$\left(\int_{\mathbb{R}^n}(|\mathbf{X_2}|^b|v|)^q dx\right)^{1/q}\leqslant c\int_{\mathbb{R}^n}|\mathbf{X_2}|^{\sigma_2}|\nabla v|dx.$$

Setting $v=|\mathbf{X_3}|^{\sigma_3}w$ in the above inequality, we find

$$\begin{aligned}&\||\mathbf{X_2}|^b|\mathbf{X_3}|^{\sigma_3}|w|\|_q\\&\leqslant c\int_{\mathbb{R}^n}|\mathbf{X_2}|^{\sigma_2}|\mathbf{X_3}|^{\sigma_3}|\nabla w|dx+|\sigma_3|c\int_{\mathbb{R}^n}|\mathbf{X_2}|^{\sigma_2}|\mathbf{X_3}|^{\sigma_3-1}|w|dx.\end{aligned}$$

Taking $\mathbf{F}=|\mathbf{X_2}|^{\sigma_2}|\mathbf{X_3}|^{\sigma_3-1}\mathbf{X_3}$ in (3.2), we get

$$|\sigma_2+\sigma_3+2|\int_{\mathbb{R}^n}|\mathbf{X_2}|^{\sigma_2}|\mathbf{X_3}|^{\sigma_3-1}|w|dx\leqslant\int_{\mathbb{R}^n}|\mathbf{X_2}|^{\sigma_2}|\mathbf{X_3}|^{\sigma_3}|\nabla w|dx. \tag{3.3}$$

Here, we assumed that $\sigma_2+\sigma_3+2>0$ to guarantee that $|\mathbf{X_2}|^{\sigma_2}|\mathbf{X_3}|^{\sigma_3-1}\in L^1_{\text{loc}}(\mathbb{R}^n)$. The two previous estimates give

$$\||\mathbf{X_2}|^b|\mathbf{X_3}|^{\sigma_3}|w|\|_q\leqslant c\int_{\mathbb{R}^n}|\mathbf{X_2}|^{\sigma_2}|\mathbf{X_3}|^{\sigma_3}|\nabla w|dx.$$

If we would have $\sigma_3 = 0$, we obtain our result immediately and it is not necessary to check whether the constant $\sigma_2 + \sigma_3 + 2$ is positive or not.

We may repeat this procedure iteratively. At the lth step, we use the vector field

$$\mathbf{F} = |\mathbf{X_2}|^{\sigma_2}|\mathbf{X_3}|^{\sigma_3}\dots|\mathbf{X_l}|^{\sigma_l-1}\mathbf{X}_l$$

in (3.2) to get

$$|c_l|\ |||\mathbf{X_2}|^{\sigma_2}|\mathbf{X_3}|^{\sigma_3}\dots|\mathbf{X_l}|^{\sigma_l-1}w||_1 \leqslant \int_{\mathbb{R}^n}|\mathbf{X_2}|^{\sigma_2}|\mathbf{X_3}|^{\sigma_3}\dots|\mathbf{X_l}|^{\sigma_l}|\nabla w|dx.$$

As above, we do not need this inequality in the case $\sigma_l = 0$ and, if $\sigma_l \neq 0$, we assume $c_l = \sigma_2 + \ldots + \sigma_l + l - 1 > 0$ to ensure the integrability of the integrand on the left-hand side. Similarly, it follows that

$$c|||\mathbf{X_2}|^{b}|\mathbf{X_3}|^{\sigma_3}\dots|\mathbf{X_l}|^{\sigma_l}w||_q \leqslant \int_{\mathbb{R}^n}|\mathbf{X_2}|^{\sigma_2}|\mathbf{X_3}|^{\sigma_3}\dots|\mathbf{X_l}|^{\sigma_l}|\nabla w|dx,$$

which is (3.1). □

To prove Theorem C, we use the following variant of Theorem 3.1.

Theorem 3.2 (weighted Sobolev inequality). *Let $\sigma_1, \sigma_2, \ldots, \sigma_n$ be real numbers, $n \geqslant 2$. We set $\overline{c}_l := \sigma_1 + \ldots + \sigma_l + l - 1$ for $1 \leqslant l \leqslant n$ and assume that*

$$\overline{c}_l > 0, \quad \textit{whenever} \quad \sigma_l \neq 0,$$

for $l = 1, 2, \ldots, n$. Then there exists a positive constant C such that for any $w \in C_0^\infty(\mathbb{R}^n)$

$$\int_{\mathbb{R}^n}|x_1|^{\sigma_1}|\mathbf{X_2}|^{\sigma_2}\dots|\mathbf{X_n}|^{\sigma_n}|\nabla w|dx$$

$$\geqslant C\left(\int_{\mathbb{R}^n}\left(|x_1|^{b}|\mathbf{X_2}|^{\sigma_2}\dots|\mathbf{X_n}|^{\sigma_n}|w|\right)^q dx\right)^{\frac{1}{q}}, \tag{3.4}$$

where

$$b = \sigma_1 - 1 + \frac{q-1}{q}n, \quad < q \leqslant \frac{n}{n-1}.$$

Proof. Let

$$1 < q \leqslant n/(n-1) \quad \text{and} \quad b = \sigma_1 - 1 + \frac{q-1}{q}n.$$

We first consider the case $\sigma_1 > 0$. We use the following L^1 interpolation inequality

$$|||x_1|^{b}v||_q \leqslant c_1|||x_1|^{\sigma_2}v||_{\frac{n}{n-1}} + c_2|||x_1|^{\sigma_2-1}v||_1.$$

Arguing in the same way as in the proof of Theorem D, we find

$$\left(\int_{\mathbb{R}^n} (|x_1|^b |v|)^q dx\right)^{1/q} \leqslant c \int_{\mathbb{R}^n} |x_1|^{\sigma_1} |\nabla v| dx. \tag{3.5}$$

In the case $\sigma_1 = 0$, the inequality (3.5) is still valid (cf. [22, Section 2.1.6/1]).

The rest of the proof goes as in Theorem D, i.e., we apply (3.5) to $v = |\mathbf{X_2}|^{\sigma_3} w$ to get

$$|||x_1|^b |\mathbf{X_2}|^{\sigma_2} |w|||_q \leqslant c \int_{\mathbb{R}^n} |x_1|^{\sigma_1} |\mathbf{X_2}|^{\sigma_2} |\nabla w| dx + |\sigma_2| c \int_{\mathbb{R}^n} |x_1|^{\sigma_1} |\mathbf{X_2}|^{\sigma_2 - 1} |w| dx.$$

Setting $\mathbf{F} = |x_1|^{\sigma_1} |\mathbf{X_2}|^{\sigma_2 - 1} \mathbf{X_2}$ in (3.2), we get

$$|\sigma_1 + \sigma_2 + 1| \int_{\mathbb{R}^n} |x_1|^{\sigma_1} |\mathbf{X_2}|^{\sigma_2 - 1} |w| dx \leqslant \int_{\mathbb{R}^n} |x_1|^{\sigma_1} |\mathbf{X_2}|^{\sigma_2} |\nabla w| dx. \tag{3.6}$$

The condition $\overline{c}_2 = \sigma_1 + \sigma_2 + 1 > 0$ guarantees that $|x_1|^{\sigma_1} |\mathbf{X_2}|^{\sigma_2 - 1} \in L^1_{\text{loc}}(\mathbb{R}^n)$ and leads to

$$|||x_1|^b |\mathbf{X_2}|^{\sigma_2} |w|||_q \leqslant c \int_{\mathbb{R}^n} |x_1|^{\sigma_1} |\mathbf{X_2}|^{\sigma_2} |\nabla w| dx.$$

We omit further details. □

Now, we are ready to prove Theorem B.

Proof of Theorem B. As the first step, we establish that for any $v \in C_0^\infty(\mathbb{R}^n)$

$$\int_{\mathbb{R}^n} |\mathbf{X_2}|^{2\sigma_2 - \frac{2QB}{Q+2}} |\mathbf{X_3}|^{\frac{4\sigma_3}{Q+2}} \dots |\mathbf{X_n}|^{\frac{4\sigma_n}{Q+2}} |\nabla v|^2 dx$$

$$\geqslant C \left(\int_{\mathbb{R}^n} |\mathbf{X_2}|^{\frac{2QB}{Q+2}} |\mathbf{X_3}|^{\frac{2Q\sigma_3}{Q+2}} \dots |\mathbf{X_n}|^{\frac{2Q\sigma_n}{Q+2}} |v|^Q dx\right)^{\frac{2}{Q}} \tag{3.7}$$

provided that $c_l := \sigma_2 + \ldots + \sigma_l + (l-1) > 0$ if $\sigma_l \neq 0$, $2 \leqslant l \leqslant n$, where

$$B = \sigma_2 - 1 + \frac{Q-2}{2Q} n, \quad 2 < Q \leqslant \frac{2n}{n-2}.$$

To show (3.7), we apply Theorem 3.1 to the function $w = |v|^s$ with $s = \frac{Q+2}{2}$, $sq = Q$ and $b = B$. Trivial estimates give

$$C\left(\int_{\mathbb{R}^n} |\mathbf{X_2}|^{bq} |\mathbf{X_3}|^{\sigma_3 q} \dots |\mathbf{X_n}|^{\sigma_n q} |v|^{sq} dx\right)^{1/q}$$

$$\leqslant s \int_{\mathbb{R}^n} |\mathbf{X_2}|^{\sigma_3} |\mathbf{X_3}|^{\sigma_3} \cdot \ldots \cdot |\mathbf{X_n}|^{\sigma_n} |v|^{s-1} |\nabla v| dx.$$

Applying the Cauchy–Schwartz inequality to the right-hand side, we obtain the required result.

We use (3.7) with $\sigma_2 = \frac{1}{4}((Q-2)n - 2Q)$, so that $2\sigma_2 - \frac{2QB}{Q+2} = 0$. Note that the requirement

$$c_2 = \sigma_2 + 1 = \frac{1}{4}(Q-2)(n-2) > 0$$

is equivalent to $Q > 2$, and therefore is satisfied.

Then we use Lemma 2.1. Recall that for $\phi > 0$ and $u = \phi v$ with $v \in C_0^\infty(\mathbb{R}^n \setminus S_2)$

$$\int_{\mathbb{R}^n} |\nabla u|^2 dx + \int_{\mathbb{R}^n} \frac{\Delta\phi}{\phi} |u|^2 dx = \int_{\mathbb{R}^n} \phi^2 |\nabla v|^2 dx. \tag{3.8}$$

For ϕ we take

$$\begin{aligned} \phi(x) &= |\mathbf{X_3}|^{\frac{2\sigma_3}{Q+2}} |\mathbf{X_4}|^{\frac{2\sigma_4}{Q+2}} \ldots |\mathbf{X_n}|^{\frac{2\sigma_n}{Q+2}} \\ &= |\mathbf{X_3}|^{-\gamma_3} |\mathbf{X_4}|^{-\gamma_4} \cdot \ldots \cdot |\mathbf{X_n}|^{-\gamma_n}, \end{aligned} \tag{3.9}$$

where

$$\begin{aligned} \gamma_3 &= \alpha_3 + 1/2, \\ \gamma_m &= \alpha_m - \alpha_{m-1} + 1/2, \quad m = 3, \ldots, n. \end{aligned}$$

Therefore,

$$\sigma_m = -\frac{Q+2}{2} \gamma_m, \quad m = 3, \ldots, n.$$

Applying (3.7), we obtain

$$\int_{\mathbb{R}^n} \phi^2 |\nabla v|^2 dx \geqslant C \left(\int_{\mathbb{R}^n} |\mathbf{X_2}|^{\frac{Q-2}{2}n - Q} |\phi v|^Q dx \right)^{\frac{2}{Q}} \tag{3.10}$$

provided that for $3 \leqslant l \leqslant n$

$$c_l := \sigma_2 + \ldots + \sigma_l + l - 1 > 0, \quad \text{whenever} \quad \sigma_l \neq 0. \tag{3.11}$$

Combining (3.10) with (3.8), we get

$$\int_{\mathbb{R}^n} |\nabla u|^2 dx + \int_{\mathbb{R}^n} \frac{\Delta\phi}{\phi} |u|^2 dx \geqslant C \left(\int_{\mathbb{R}^n} |\mathbf{X_2}|^{\frac{Q-2}{2}n - Q} |u|^Q dx \right)^{\frac{2}{Q}}.$$

On the other hand, by Theorem A(i),

$$-\frac{\Delta\phi}{\phi} = \frac{\beta_3}{|\mathbf{X_3}|^2} + \ldots + \frac{\beta_n}{|\mathbf{X_n}|^2},$$

and the desired inequality follows.

It remains to check the condition (3.11). For $l = 2$ it was already checked.. For $3 \leqslant l \leqslant n$, after some calculations we find

$$\begin{aligned} c_l &= \sigma_2 + \ldots + \sigma_l + l - 1 \\ &= \frac{1}{4}(Q-2)(n-2) - \frac{Q+2}{2}(\gamma_3 + \ldots + \gamma_l) + l - 1 \\ &= \frac{Q+2}{2}\Big(-\alpha_l + \frac{(Q-2)(n-l)}{2(Q+2)}\Big). \end{aligned}$$

Recalling that $\alpha_l \leqslant 0$, we conclude that $l \leqslant n-1$ implies $c_l > 0$, whereas for $l = n$ we have $c_n > 0$ if and only if $\alpha_n < 0$. This proves (1.8) for $u \in C_0^\infty(\mathbb{R}^n \setminus S_2)$ and, by a density argument, the result holds for any $u \in C_0^\infty(\mathbb{R}^n)$.

In the rest of the proof, we show that (1.8) fails in the case $\alpha_n = 0$. For this purpose, we establish that

$$\inf_{u \in C_0^\infty(\mathbb{R}^n)} \frac{\displaystyle\int_{\mathbb{R}^n} |\nabla u|^2 dx - \beta_3 \int_{\mathbb{R}^n} \frac{|u|^2}{|\mathbf{X_3}|^2} dx - \ldots - \beta_n \int_{\mathbb{R}^n} \frac{|u|^2}{|\mathbf{X_n}|^2} dx}{\displaystyle\Big(\int_{\mathbb{R}^n} |\mathbf{X_2}|^{\frac{Q-2}{2}n - Q} |u|^Q dx \Big)^{\frac{2}{Q}}} = 0, \tag{3.12}$$

where $\beta_n = \big(\alpha_{n-1} - \frac{1}{2}\big)^2$. Let

$$u(x) = |\mathbf{X_3}|^{-\gamma_3} \ldots |\mathbf{X_{n-1}}|^{-\gamma_{n-1}} v(x).$$

A direct calculation, quite similar to that leading to (2.17), shows that the infimum in (3.12) is the same as the following infimum:

$$\inf_{v \in C_0^\infty(\mathbb{R}^n)} \frac{\displaystyle\int_{\mathbb{R}^n} \prod_{j=3}^{n-1} |\mathbf{X_j}|^{-2\gamma_j} |\nabla v|^2 dx - \beta_n \int_{\mathbb{R}^n} \prod_{j=3}^{n-1} |\mathbf{X_j}|^{-2\gamma_j} |\mathbf{X_n}|^{-2} v^2 dx}{\displaystyle\bigg(\int_{\mathbb{R}^n} \Big(|\mathbf{X_2}|^{\frac{Q-2}{2Q}n - 1} \prod_{j=3}^{n-1} |\mathbf{X_j}|^{-\gamma_j} \Big)^Q |v|^Q dx \bigg)^{\frac{2}{Q}}}. \tag{3.13}$$

We now choose the following test functions

$$v_{k_3, \varepsilon} = |\mathbf{X_n}|^{-\gamma_n + \varepsilon} h_{k_3}(x) \phi(x), \quad \varepsilon > 0, \tag{3.14}$$

where $h_{k_3}(x)$ and $\phi(x)$ are the same test functions as at the first step of the proof of Theorem A (ii). Under this choice, after straightforward calculations, quite similar to that used in the proof of Theorem A (ii), we obtain the following estimate for the numerator N in (3.13):

$$
\begin{aligned}
N[v_{\infty,\varepsilon}] &= \left(\left(\alpha_{n-1} - \frac{1}{2} + \varepsilon\right)^2 - \left(\alpha_{n-1} - \frac{1}{2}\right)^2\right) \\
&\quad \times \int_{\mathbb{R}^n} \prod_{j=3}^{n-1} |\mathbf{X_j}|^{-2\gamma_j} |\mathbf{X_n}|^{-2\gamma_n+2+\varepsilon} \phi^2(x)dx + O_\varepsilon(1) \\
&= C\varepsilon \int_{\mathbb{R}^n} r^{-1+2\varepsilon} \sin\theta_2 \prod_{j=3}^{n-1} (\sin\theta_j)^{1-2\alpha_j} \phi^2(r) d\theta_1 \ldots d\theta_{n-1} dr + O_\varepsilon(1) \\
&= C\varepsilon \int_0^1 r^{-1+\varepsilon} dr + O_\varepsilon(1).
\end{aligned}
$$

In the above calculations, we took the limit as $k_3 \to \infty$ and used polar coordinates in $(x_1, \ldots, x_n) \to (\theta_1, \ldots, \theta_{n-1}, r)$. We conclude that

$$
N[v_{\infty,\varepsilon}] < C, \quad \varepsilon \to 0. \tag{3.15}
$$

Similar calculations for the denominator D in (3.13) yields

$$
\begin{aligned}
D[v_{\infty,\varepsilon}] &= C\left(\int_{\mathbb{R}^n} r^{-1+\varepsilon Q} \prod_{j=2}^{n-1} (\sin\theta_j)^{j-n-1+Q(\frac{n-j}{2}-\alpha_j)} \phi^Q d\theta_1 \ldots d\theta_{n-1} dr\right)^{\frac{2}{Q}} \\
&\geqslant C\left(\int_0^{\frac{1}{2}} r^{-1+\varepsilon Q} dr\right)^{\frac{2}{Q}} = C\varepsilon^{-\frac{2}{Q}}.
\end{aligned}
$$

Then

$$
\frac{N[v_{\infty,\varepsilon}]}{D[v_{\infty,\varepsilon}]} \to 0 \quad \text{as} \quad \varepsilon \to 0.
$$

Therefore, the infimum in (3.13) or (3.12) vanishes. □

From Theorem B we obtain the following assertion.

Corollary 3.3. *Let* $3 \leqslant k < n$ *and* $2 < Q \leqslant \frac{2n}{n-2}$. *Then for any* $\beta_n < 1/4$ *there exists a positive constant* C *such that for all* $u \in C_0^\infty(\mathbb{R}^n)$

$$
\begin{aligned}
&\int_{\mathbb{R}^n} |\nabla u|^2 dx \\
&\geqslant \int_{\mathbb{R}^n} \left(\left(\frac{k-2}{2}\right)^2 \frac{1}{|\mathbf{X_k}|^2} + \frac{1}{4}\frac{1}{|\mathbf{X_{k-1}}|^2} + \ldots + \frac{1}{4}\frac{1}{|\mathbf{X_{n-1}}|^2} + \frac{\beta_n}{|\mathbf{X_n}|^2}\right) |u|^2 dx \\
&+ C\left(\int_{\mathbb{R}^n} |\mathbf{X_2}|^{\frac{Q-2}{2}n-Q} |u|^Q dx\right)^{\frac{2}{Q}}.
\end{aligned}
$$

If $\beta_n = 1/4$, *then the inequality fails.*

In the case $k = n$, for any $\beta_n < \frac{(n-2)^2}{4}$ there exists a positive constant C such that for all $u \in C_0^\infty(\mathbb{R}^n)$

$$\int_{\mathbb{R}^n} |\nabla u|^2 dx \geqslant \beta_n \int_{\mathbb{R}^n} \frac{u^2}{|x|^2} dx + C \left(\int_{\mathbb{R}^n} |\mathbf{X_2}|^{\frac{Q-2}{2}n - Q} |u|^Q dx \right)^{\frac{2}{Q}}.$$

The inequality fails for $\beta_n = \frac{(n-2)^2}{4}$.

Proof. In Theorem B, we make the following choice. In the case $k = 3$, we choose $\alpha_3 = \alpha_4 = \ldots = \alpha_{n-1} = 0$. In this case, $\beta_k = 1/4$, $k = 1, \ldots, n-1$. The condition $\alpha_n < 0$ is equivalent to $\beta_n < 1/4$. In the case $3 < k \leqslant n-1$, we choose $\alpha_m = -m/2$ if when $m = 1, 2, \ldots, k-1$ and $\alpha_m = 0$ if when $m = k, \ldots, n-1$. Finally, in the case $k = n$, we choose $\alpha_m = -(m-2)/2$ for $m = 3, 4, \ldots, n-1$. □

Proof of Theorem C. We first prove that the following inequality holds for any $v \in C_0^\infty(\mathbb{R}^n)$:

$$\int_{\mathbb{R}^n} |x_1|^{2\sigma_1 - \frac{2QB}{Q+2}} |\mathbf{X_2}|^{\frac{4\sigma_2}{Q+2}} \ldots |\mathbf{X_n}|^{\frac{4\sigma_n}{Q+2}} |\nabla v|^2 dx$$

$$\geqslant C \left(\int_{\mathbb{R}^n} |x_1|^{\frac{2QB}{Q+2}} |\mathbf{X_2}|^{\frac{2Q\sigma_2}{Q+2}} \ldots |\mathbf{X_n}|^{\frac{2Q\sigma_n}{Q+2}} |v|^Q dx \right)^{\frac{2}{Q}} \tag{3.16}$$

provided that $\overline{c}_l := \sigma_1 + \ldots + \sigma_l + (l-1) > 0$ if $\sigma_l \neq 0$, $1 \leqslant l \leqslant n$, where

$$B = \sigma_1 - 1 + \frac{Q-2}{2Q} n, \quad \frac{2(n-1)}{n-2} < Q \leqslant \frac{2n}{n-2}.$$

To show (3.16), we apply Theorem 3.2 to the function $w = |v|^s$ with $s = \frac{Q+2}{2}$, $sq = Q$ and $b = B$, and then use the Cauchy–Schwartz inequality.

We use (3.16) with $\sigma_1 = \frac{1}{4}((Q-2)n - 2Q)$ and $\sigma_2 = 0$. In this case, $2\sigma_1 - \frac{2QB}{Q+2} = 0$. The choice of ϕ is the same as in the proof of Theorem B. We find

$$\int_{\mathbb{R}^n} |\nabla u|^2 dx - \int_{\mathbb{R}^n} \left(\frac{\beta_3}{|\mathbf{X_3}|^2} + \ldots + \frac{\beta_n}{|\mathbf{X_n}|^2} \right) |u|^2 dx$$

$$\geqslant C \left(\int_{\mathbb{R}^n} |x_1|^{\frac{Q-2}{2}n - Q} |u|^Q dx \right)^{\frac{2}{Q}}$$

provided that $\overline{c}_l$ satisfy our assumptions of Theorem 3.2. However, it turns out that

$$\overline{c}_l = \frac{Q+2}{2} \left(-\alpha_l + \frac{(Q-2)(n-l)}{2(Q+2)} \right), \quad 1 \leqslant l \leqslant n,$$

and our assumptions are satisfied in the case $\alpha_n < 0$.

It remains to prove that (1.10) fails in the case $\alpha_n = 0$. For this purpose, we show that

$$\inf_{u \in C_0^\infty(\mathbb{R}^n)} \frac{\displaystyle\int_{\mathbb{R}^n} |\nabla u|^2 dx - \beta_3 \int_{\mathbb{R}^n} \frac{|u|^2}{|\mathbf{X_3}|^2} dx - \ldots - \beta_n \int_{\mathbb{R}^n} \frac{|u|^2}{|\mathbf{X_n}|^2} dx}{\left(\displaystyle\int_{\mathbb{R}^n} |x_1|^{\frac{Q-2}{2}n - Q} |u|^Q dx \right)^{\frac{2}{Q}}} = 0, \tag{3.17}$$

where $\beta_n = \left(\alpha_{n-1} - \frac{1}{2}\right)^2$. We can use the test functions as in the proof of Theorem B because they belong to the proper function space. The result follows by observing that, in this case, the weight is stronger than the weight in Theorem B. □

An easy consequence of the above theorem is the following assertion.

Corollary 3.4. *Let $3 \leqslant k < n$ and $\frac{2(n-1)}{n-2} < Q \leqslant \frac{2n}{n-2}$. Then, for any $\beta_n < 1/4$ there exists a positive constant C such that for all $u \in C_0^\infty(\mathbb{R}^n)$*

$$\begin{aligned} &\int_{\mathbb{R}^n} |\nabla u|^2 dx \\ &\geqslant \int_{\mathbb{R}^n} \Big(\Big(\frac{k-2}{2} \Big)^2 \frac{1}{|\mathbf{X_k}|^2} + \frac{1}{4} \frac{1}{|\mathbf{X_{k-1}}|^2} + \ldots + \frac{1}{4} \frac{1}{|\mathbf{X_{n-1}}|^2} + \frac{\beta_n}{|\mathbf{X_n}|^2} \Big) |u|^2 dx \\ &+ C \Big(\int_{\mathbb{R}^n} |x_1|^{\frac{Q-2}{2}n - Q} |u|^Q dx \Big)^{\frac{2}{Q}}. \end{aligned}$$

If $\beta_n = 1/4$, this inequality fails.

In the case $k = n$, for any $\beta_n < \frac{(n-2)^2}{4}$ there exists a positive constant C such that for all $u \in C_0^\infty(\mathbb{R}^n)$

$$\int_{\mathbb{R}^n} |\nabla u|^2 dx \geqslant \beta_n \int_{\mathbb{R}^n} \frac{u^2}{|x|^2} dx + C \Big(\int_{\mathbb{R}^n} |x_1|^{\frac{Q-2}{2}n - Q} |u|^Q dx \Big)^{\frac{2}{Q}}.$$

This inequality fails for $\beta_n = \frac{(n-2)^2}{4}$.

Acknowledgments. The third author is thanking the Departments of Mathematics and Applied Mathematics of University of Crete for the invitation as well as the warm hospitality.

References

1. Adimurthi, Chaudhuri, N., Ramaswamy, M.: An improved Hardy–Sobolev inequality and its application. Proc. Am. Math. Soc. **130**, no. 2, 489–505 (2002)

2. Adimurthi, Esteban, M.J.: An improved Hardy–Sobolev inequality in $W^{1,p}$ and its application to Schrödinger operators. Nonlinear Differ. Equations Appl. **12**, no. 2, 243–263 (2005)
3. Adimurthi, Filippas, S., Tertikas, A.: On the best constant of Hardy–Sobolev inequalities. Nonlinear Anal. T.M.A. **70**, no. 8 2826–2833 (2009)
4. Alvino, A., Ferone, V., Trombetti, G.: On the best constant in a Hardy–Sobolev inequality. Appl. Anal. **85**, no 1–3, 171–180 (2006)
5. Ancona, A.: On strong barriers and an inequality of Hardy for domains in $\mathbb{R}^n$. J. London Math. Soc. 34 **2**, 274–290 (1986)
6. Aubin, T.: Probléme isopérimétric et espace de Sobolev. J. Differ. Geom. **11**, 573–598 (1976)
7. Badiale, M., Tarantello, G.: A Sobolev–Hardy inequality with applications to a nonlinear elliptic equation arising in astrophysics. Arch. Ration. Mech. Anal. **163**, no. 4, 259–293 (2002)
8. Barbatis, G., Filippas, S., Tertikas, A.: A unified approach to improved L^p Hardy inequalities with best constants. Trans. Am. Math. Soc. **356**, 2169–2196 (2004)
9. Benguria, R.D., Frank, R.L., Loss, M.: The sharp constant in the Hardy–Sobolev–Maz'ya inequality in the three dimensional upper half space. Math. Res. Lett. **15**, no. 4, 613–622 (2008)
10. Brezis, H., Marcus, M.: Hardy's inequalities revisited Dedicated to Ennio De Giorgi. Ann. Scuola Norm. Sup. Pisa Cl. Sci. 4 **25**, 217–237 (1997)
11. Brezis, H., Vázquez, J.L.: Blow-up solutions of some nonlinear elliptic problems. Rev. Mat. Univ. Comp. Madrid **10**, 443–469 (1997)
12. Caffarelli, L., Kohn, R.V., Nirenberg, L.: First order interpolation inequalities with weights. Compos. Math. **53**, no. 3, 259–275 (1984)
13. Cianchi, A., Ferone, A.: Hardy inequalities with non-standard remainder terms. Ann. Inst. H. Poincaré Anal. Non Lineaire **25**, no. 5, 889–906 (2008)
14. Davies, E.B.: The Hardy constant. Quart. J. Math. 2 **46**, 417–431 (1995)
15. Dolbeault, J., Esteban, M.J., Loss, M., Vega, L.: An analytical proof of Hardy-like inequalities related to the Dirac operator. J. Funct. Anal. **216**, no. 1, 1–21 (2004)
16. Filippas, S., Maz'ya, V., Tertikas, A.: Sharp Hardy–Sobolev inequalities. C. R., Math., Acad. Sci. Paris **339**, no. 7, 483–486 (2004)
17. Filippas, S., Maz'ya, V., Tertikas, A.: Critical Hardy–Sobolev Inequalities. J. Math. Pures Appl. (9) **87**, 37–56 (2007)
18. Filippas, S., Tertikas, A.: Optimizing Improved Hardy Inequalities. J. Funct. Anal. **192**, 186–233 (2002); Corrigendum. ibid. **255**, 2095 (2008)
19. Filippas, S., Tertikas, A., Tidblom, J.: On the structure of Hardy–Sobolev–Maz'ya inequalities. J. Eur. Math. Soc. [To appear]
20. Hoffmann-Ostenhof, M., Hoffmann-Ostenhof, T., Laptev, A.: A geometrical version of Hardy's inequality. J. Func. Anal. **189**, 539–548 (2002)
21. Mancini, G., Sandeep, K.: On a semilinear elliptic equation in $\mathbb{H}^n$. Ann. Sc. Norm. Super. Pisa Cl. Sci. (5) **7** no. 4, 635–671 (2008)
22. Maz'ya, V.G.: Sobolev Spaces. Springer, Berlin etc. (1985)
23. Maz'ya, V., Shaposhnikova, T.: A collection of sharp dilation invariant inequalities for differentiable functions. In: Maz'ya, V. (ed.), Sobolev Spaces in Mathematics. I: Sobolev Type Inequalities. Springer, New York; Tamara Rozhkovskaya Publisher, Novosibirsk. International Mathematical Series **8**, 223–247 (2009)

24. Pinchover, Y., Tintarev, K.: On the Hardy–Sobolev–Maz'ya inequality and its generalizations. In: Maz'ya, V. (ed.), Sobolev Spaces in Mathematics. I: Sobolev Type Inequalities. Springer, New York; Tamara Rozhkovskaya Publisher, Novosibirsk. International Mathematical Series **8**, 281–297 (2009)
25. Talenti, G.: Best constant in Sobolev inequality. Ann. Mat. Pura Appl. (4) **110**, 353–372 (1976)
26. Tertikas, A., Tintarev, K.: On existence of minimizers for the Hardy–Sobolev–Maz'ya inequality. Ann. Mat. Pura Appl. **186**, 645–662 (2007)
27. Tidblom, J.: A geometrical version of Hardy's inequality for $\overset{\circ}{W}{}^{1,p}(\Omega)$. Proc. Am . Math. Soc. 8 **132**, 2265–2271 (2004)
28. Tidblom, J.: A Hardy inequality in the half-space. J. Func. Anal. 2 **221**, 482–495 (2005)

Sharp Fractional Hardy Inequalities in Half-Spaces

Rupert L. Frank and Robert Seiringer

Dedicated to V.G. Maz'ya

Abstract We determine the sharp constant in the Hardy inequality for fractional Sobolev spaces on half-spaces. Our proof relies on a nonlinear and nonlocal version of the ground state representation.

1 Introduction and Main Results

This short note is motivated by the paper [4] concerning Hardy inequalities in the half-space $\mathbb{R}_+^N := \{(x', x_N) : x' \in \mathbb{R}^{N-1}, x_N > 0\}$. The fractional Hardy inequality states that for $0 < s < 1$ and $1 \leqslant p < \infty$ with $ps \neq 1$ there is a positive constant $\mathcal{D}_{N,p,s}$ such that

$$\iint_{\mathbb{R}_+^N \times \mathbb{R}_+^N} \frac{|u(x) - u(y)|^p}{|x - y|^{N+ps}}\, dx\, dy \geqslant \mathcal{D}_{N,p,s} \int_{\mathbb{R}_+^N} \frac{|u(x)|^p}{x_N^{ps}}\, dx \tag{1.1}$$

for all $u \in C_0^\infty(\overline{\mathbb{R}_+^N})$ if $ps < 1$ and for all $u \in C_0^\infty(\mathbb{R}_+^N)$ if $ps > 1$. In [4], the sharp (i.e., the largest possible) value of the constant $\mathcal{D}_{N,2,s}$ for $p = 2$ is calculated. Our goal in this note is to determine the sharp constant $\mathcal{D}_{N,p,s}$ for *arbitrary* p.

Rupert L. Frank
Department of Mathematics, Princeton University, Washington Road, Princeton, NJ 08544, USA
e-mail: rlfrank@math.princeton.edu

Robert Seiringer
Department of Physics, Princeton University, P. O. Box 708, Princeton, NJ 08544, USA
e-mail: rseiring@princeton.edu

A. Laptev (ed.), *Around the Research of Vladimir Maz'ya I: Function Spaces*, International Mathematical Series 11, DOI 10.1007/978-1-4419-1341-8_6,

Indeed, we shall see that the sharp inequality (1.1) follows by a minor modification of the approach introduced in [7]. In that note, we calculated the sharp constant $\mathcal{C}_{N,p,s}$ in the inequality

$$\iint_{\mathbb{R}^N\times\mathbb{R}^N} \frac{|u(x)-u(y)|^p}{|x-y|^{N+ps}}\,dx\,dy \geqslant \mathcal{C}_{N,p,s}\int_{\mathbb{R}^N}\frac{|u(x)|^p}{|x|^{ps}}\,dx \tag{1.2}$$

for all $u\in C_0^\infty(\mathbb{R}^N)$ if $1\leqslant p<N/s$ and for all $u\in C_0^\infty(\mathbb{R}^N\setminus\{0\})$ if $p>N/s$. A (nonsharp) version of (1.2) was used by Maz'ya and Shaposhnikova [10] in order to simplify and extend considerably a result of Bourgain, Brezis, and Mironescu [5] on the norm of the embedding $\dot{W}_p^s(\mathbb{R}^N)\subset L_{Np/(N-ps)}(\mathbb{R}^N)$. Our proof of (1.2) relied on a *ground state substitution*, i.e., on writing $u(x)=\omega(x)v(x)$, where $\omega(x)=|x|^{-(N-ps)/p}$ is a solution of the Euler–Lagrange equation corresponding to (1.2). In this note, we prove (1.1) using that $\omega(x)=x_N^{-(1-ps)/p}$ satisfies the Euler–Lagrange equation corresponding to (1.1).

We refer to [4, 6, 8] and the references therein for motivations and applications of fractional Hardy inequalities.

In order to state our main result, let $1\leqslant p<\infty$ and $0<s<1$ with $ps\neq 1$ and denote by $\mathcal{W}_p^s(\mathbb{R}_+^N)$ the completion of $C_0^\infty(\mathbb{R}_+^N)$ with respect to the left-hand side of (1.1). It is a consequence of the Hardy inequality that this completion is a space of functions. Moreover, it is well known that for $ps<1$, $\mathcal{W}_p^s(\mathbb{R}_+^N)$ coincides with the completion of $C_0^\infty(\overline{\mathbb{R}_+^N})$.

Theorem 1.1 (sharp fractional Hardy inequality). *Let $N\geqslant 1$, $1\leqslant p<\infty$, and $0<s<1$ with $ps\neq 1$. Then for all $u\in\mathcal{W}_p^s(\mathbb{R}_+^N)$*

$$\iint_{\mathbb{R}_+^N\times\mathbb{R}_+^N} \frac{|u(x)-u(y)|^p}{|x-y|^{N+ps}}\,dx\,dy \geqslant \mathcal{D}_{N,p,s}\int_{\mathbb{R}_+^N}\frac{|u(x)|^p}{x_N^{ps}}\,dx \tag{1.3}$$

with

$$\mathcal{D}_{N,p,s} := 2\pi^{(N-1)/2}\frac{\Gamma((1+ps)/2)}{\Gamma((N+ps)/2)}\int_0^1 \left|1-r^{(ps-1)/p}\right|^p \frac{dr}{(1-r)^{1+ps}}\,. \tag{1.4}$$

The constant $\mathcal{D}_{N,p,s}$ is optimal. If $p=1$ and $N=1$, equality holds if and only if u is proportional to a nonincreasing function. If $p>1$ or if $p=1$ and $N\geqslant 2$, the inequality is strict for any function $0\not\equiv u\in\mathcal{W}_p^s(\mathbb{R}_+^N)$.

For $p\geqslant 2$ the inequality (1.3) holds even with a remainder term.

Theorem 1.2 (sharp Hardy inequality with remainder). *Let $N\geqslant 1$, $2\leqslant p<\infty$ and $0<s<1$ with $ps\neq 1$. Then for all $u\in\mathcal{W}_p^s(\mathbb{R}_+^N)$ and $v:=x_N^{(1-ps)/p}u$,*

$$\iint_{\mathbb{R}_+^N\times\mathbb{R}_+^N} \frac{|u(x)-u(y)|^p}{|x-y|^{N+ps}}\,dx\,dy - \mathcal{D}_{N,p,s}\int_{\mathbb{R}_+^N}\frac{|u(x)|^p}{x_N^{ps}}\,dx$$

$$\geqslant c_p \iint_{\mathbb{R}^N_+ \times \mathbb{R}^N_+} \frac{|v(x)-v(y)|^p}{|x-y|^{N+ps}} \frac{dx}{x_N^{(1-ps)/2}} \frac{dy}{y_N^{(1-ps)/2}}, \tag{1.5}$$

where $\mathcal{D}_{N,p,s}$ *is given by* (1.4) *and* $0 < c_p \leqslant 1$ *is given by*

$$c_p := \min_{0<\tau<1/2} \left((1-\tau)^p - \tau^p + p\tau^{p-1} \right). \tag{1.6}$$

If $p = 2$, *then* (1.5) *is an equality with* $c_2 = 1$.

We conclude this section by mentioning an open problem concerning fractional Hardy–Sobolev–Maz'ya inequalities. If $p \geqslant 2$ and $0 < s < 1$ with $1 < ps < N$, is it true that the left-hand side of (1.5) is bounded from below by a positive constant times

$$\left(\int_{\mathbb{R}^N_+} |u|^q \, dx \right)^{p/q}, \qquad q = Np/(N-ps)\,?$$

The analogous estimate for $s = 1$,

$$\int_{\mathbb{R}^N_+} |\nabla u|^p \, dx - \left(\frac{p-1}{p} \right)^p \int_{\mathbb{R}^N_+} \frac{|u(x)|^p}{x_N^p} \, dx \geqslant \sigma_{N,p} \left(\int_{\mathbb{R}^N_+} |u|^q \, dx \right)^{p/q}, \tag{1.7}$$

where $q = Np/(N-p)$, is due to Maz'ya (for $p = 2$) [9] and Barbatis–Filippas–Tertikas (for $2 < p < N$) [2]; see also [3] for the sharp value of $\sigma_{3,2}$. The proof of (1.7) is based on the analogue of (1.5),

$$\int_{\mathbb{R}^N_+} |\nabla u|^p \, dx - \left(\frac{p-1}{p} \right)^p \int_{\mathbb{R}^N_+} \frac{|u(x)|^p}{x_N^p} \, dx \geqslant c_p \int_{\mathbb{R}^N_+} |\nabla v|^p x_N^{p-1} \, dx \,,$$

where $u = x_N^{(p-1)/p} v$.

2 Proofs

2.1 General Hardy inequalities

This subsection is a quick reminder of the results in [7]. Throughout we fix $N \geqslant 1$, $p \geqslant 1$ and an open set $\Omega \subset \mathbb{R}^N$. Let k be a nonnegative measurable function on $\Omega \times \Omega$ satisfying $k(x,y) = k(y,x)$ for all $x, y \in \Omega$ and define

$$E[u] := \iint_{\Omega \times \Omega} |u(x) - u(y)|^p k(x,y) \, dx \, dy.$$

Our key assumption for proving a Hardy inequality for the functional E is the following.

Assumption 2.1. Let ω be an a.e. positive, measurable function on Ω. There exists a family of measurable functions k_ϵ, $\epsilon > 0$, on $\Omega \times \Omega$ satisfying $k_\epsilon(x,y) = k_\epsilon(y,x)$, $0 \leqslant k_\epsilon(x,y) \leqslant k(x,y)$, and

$$\lim_{\epsilon \to 0} k_\epsilon(x,y) = k(x,y) \tag{2.1}$$

for a.e. $x, y \in \Omega$. Moreover, the integrals

$$V_\epsilon(x) := 2\,\omega(x)^{-p+1} \int_\Omega (\omega(x) - \omega(y))\, |\omega(x) - \omega(y)|^{p-2}\, k_\epsilon(x,y)\, dy \tag{2.2}$$

are absolutely convergent for a.e. x, belong to $L_{1,\mathrm{loc}}(\Omega)$ and $V := \lim_{\epsilon \to 0} V_\epsilon$ exists weakly in $L_{1,\mathrm{loc}}(\Omega)$, i.e.,

$$\int V_\epsilon g\, dx \to \int V g\, dx$$

for any bounded g with compact support in Ω.

The following abstract Hardy inequality was proved in [7] in the special case $\Omega = \mathbb{R}^N$. The general case considered here is proved by exactly the same arguments.

Proposition 2.2. *Under Assumption* 2.1*, for any u with compact support in Ω and $E[u]$ and $\int V_+ |u|^p\, dx$ finite one has*

$$E[u] \geqslant \int_\Omega V(x) |u(x)|^p\, dx\ . \tag{2.3}$$

For $p \geqslant 2$ a stronger version of (2.3) is valid which includes a remainder term.

Proposition 2.3. *Let $p \geqslant 2$. Under Assumption* 2.1*, for any u with compact support in Ω write $u = \omega v$ and assume that $E[u]$, $\int V_+ |u|^p\, dx$, and*

$$E_\omega[v] := \iint_{\Omega \times \Omega} |v(x) - v(y)|^p\, \omega(x)^{\frac{p}{2}} k(x,y) \omega(x)^{\frac{p}{2}}\, dx\, dy$$

are finite. Then

$$E[u] - \int_\Omega V(x) |u(x)|^p\, dx \geqslant c_p\, E_\omega[v] \tag{2.4}$$

with c_p from (1.6)*. If $p = 2$, then* (2.4) *is an equality with $c_2 = 1$.*

2.2 Proof of Theorem 1.1

Throughout this subsection, we fix $N \geqslant 1$, $0 < s < 1$, and $p \neq 1/s$ and we abbreviate

$$\alpha := (1 - ps)/p\,.$$

We will deduce the sharp Hardy inequality (1.3) using the general approach in the previous subsection with the choice

$$\omega(x) = x_N^{-\alpha}\,, \quad k(x,y) = |x-y|^{-N-ps}\,, \quad V(x) = \mathcal{D}_{N,p,s} x_N^{-ps}\,. \tag{2.5}$$

The key observation is

Lemma 2.4. *One has uniformly for x from compacts in $\mathbb{R}^N_+$*

$$\begin{aligned} &2 \lim_{\epsilon \to 0} \int_{y \in \mathbb{R}^N_+\,, |x_N - y_N| > \epsilon} (\omega(x_N) - \omega(y_N))\, |\omega(x_N) - \omega(y_N)|^{p-2}\, k(x,y)\, dy \\ &\quad = \frac{\mathcal{D}_{N,p,s}}{|x|_N^{ps}}\, \omega(x)^{p-1} \end{aligned} \tag{2.6}$$

with $\mathcal{D}_{N,p,s}$ from (1.4).

Proof. First, let $N = 1$. Then it follows from [7, Lemma 3.1] that

$$2 \lim_{\epsilon \to 0} \int_{y>0\,, |x-y|>\epsilon} (\omega(x) - \omega(y))\, |\omega(x) - \omega(y)|^{p-2}\, k(x,y)\, dy = \frac{\mathcal{D}_{1,p,s}}{x^{ps}}\, \omega(x)^{p-1}$$

uniformly for x from compacts in $(0, \infty)$. To be more precise, in [7, Lemma 3.1] the y-integral was extended over the whole axis. Therefore, the difference between the constant $\mathcal{C}_{1,s,p}$ in [7, (3.2)] and our $\mathcal{D}_{1,p,s}$ here comes from the absolutely convergent integral

$$2 \int_{-\infty}^{0} (\omega(x) - \omega(|y|))\, |\omega(x) - \omega(|y|)|^{p-2}\, \frac{dy}{(x-y)^{1+ps}}\,.$$

This proves the assertion for $N = 1$. In order to extend the assertion to higher dimensions, we use the fact (cf. [1, (6.2.1)]) that

$$\begin{aligned} &\int_{\mathbb{R}^{N-1}} \frac{dy'}{(|x'-y'|^2 + m^2)^{(N+ps)/2}} \\ &= |\mathbb{S}^{N-2}| m^{-1-ps} \int_0^\infty \frac{r^{N-2}\, dr}{(r^2+1)^{(N+ps)/2}} \\ &= \frac{1}{2} |\mathbb{S}^{N-2}| m^{-1-ps} \frac{\Gamma((N-1)/2)\; \Gamma((1+ps)/2)}{\Gamma((N+ps)/2)} \end{aligned} \tag{2.7}$$

for $N \geqslant 2$. Recalling

$$|\mathbb{S}^{N-2}| = 2\pi^{(N-1)/2}/\Gamma((N-1)/2)$$

concludes the proof. □

Proof of Theorem 1.1. According to Lemma 2.4, Assumption 2.1 is satisfied with kernel

$$k_\epsilon(x,y) = |x-y|^{-N-ps}\chi_{\{|x_N-y_N|>\epsilon\}}.$$

Hence the inequality (1.3) for $u \in C_0^\infty(\mathbb{R}_+^N)$ follows from Proposition 2.2. By density, it holds for all $u \in \mathcal{W}_p^s(\mathbb{R}_+^N)$. Strictness for $p > 1$ follows by the same argument as in [7]. In order to discuss equality in (1.3) for $p = 1$, we first note that for equality it is necessary that u is proportional to a nonnegative function, which we assume henceforth. From [7, (2.18)] we see that equality holds if and only if for a.e. x and y with $\omega(x_N) > \omega(y_N)$ (i.e., $x_N < y_N$) one has

$$|\omega(x_N)v(x) - \omega(y_N)v(y)| - (\omega(x_N)v(x) - \omega(y_N)v(y)) = 0$$

for $v(x) := \omega(x_N)^{-1}u(x)$. Since for numbers $a, b \geqslant 0$ the equality $|a-b| - (a-b) = 0$ holds if and only if $b \leqslant a$, we conclude that for a.e. x and y with $x_N < y_N$ one has $\omega(y_N)v(y) \leqslant \omega(x_N)v(x)$, i.e., $u(y) \leqslant u(x)$. If $N = 1$, this means that u is nonincreasing. If $N \geqslant 2$, one sees that for a function u with this property the integral $\int_{\mathbb{R}_+^N} |u| x_N^{-s}\,dx$ is infinite, unless $u \equiv 0$. This proves the strictness assertion in Theorem 1.1.

The fact that the constant is sharp for $N = 1$ was shown in [7] (with $\mathbb{R}_+$ replaced by $\mathbb{R}$, but this only leads to trivial modifications). In order to prove sharpness in higher dimensions, we consider functions of the form $u_n(x) = \chi_n(x')\phi(x_N)$, where

$$\chi_n(x') = \begin{cases} 1 & \text{if } |x'| \leqslant n\,, \\ n+1-|x'| & \text{if } n < |x'| < n+1\,, \\ 0 & \text{if } |x'| \geqslant n+1\,. \end{cases}$$

An easy calculation using (2.7) shows that

$$\frac{\displaystyle\iint_{\mathbb{R}_+^N\times\mathbb{R}_+^N} \frac{|u_n(x)-u_n(y)|^p}{|x-y|^{N+ps}}\,dx\,dy}{\displaystyle\int_{\mathbb{R}_+^N} \frac{|u_n(x)|^p}{x_N^{ps}}\,dx} \to A\,\frac{\displaystyle\iint_{\mathbb{R}_+\times\mathbb{R}_+} \frac{|\phi(x_N)-\phi(y_N)|^p}{|x_N-y_N|^{1+ps}}\,dx_N\,dy_N}{\displaystyle\int_{\mathbb{R}_+} \frac{|\phi(x)|^p}{x_N^{ps}}\,dx_N}$$

as $n \to \infty$ with $A := \frac{1}{2}|\mathbb{S}^{N-2}|\Gamma((N-1)/2)\ \Gamma((1+ps)/2)/\Gamma((N+ps)/2)$. Since $A = \mathcal{D}_{N,p,s}/\mathcal{D}_{1,p,s}$, the sharpness of $\mathcal{D}_{N,p,s}$ for $N \geqslant 2$ follows from the sharpness of $\mathcal{D}_{1,p,s}$ for $N = 1$. □

Proof of Theorem 1.2. The inequality (1.2) follows immediately from Proposition 2.3. □

Acknowledgments. Support through DFG grant FR 2664/1-1 (R.F.) and U.S. NSF grants PHY-0652854 (R.F.) and PHY-0652356 (R.S.) is gratefully acknowledged.

References

1. Abramowitz, M., Stegun, I.A.: Handbook of Mathematical Functions with Formulas, Graphs, and Mathematical Tables. Dover Publications, New York (1992)
2. Barbatis, S., Filippas, S., Tertikas, A.: A unified approach to improved L^p Hardy inequalities with best constants. Trans. Am. Math. Soc. **356**, no. 6, 2169–2196 (2004)
3. Benguria, R.D., Frank, R.L., Loss, M.: The sharp constant in the Hardy–Sobolev–Maz'ya inequality in the three dimensional upper half-space. Math. Res. Lett. **15**, no. 4, 613–622 (2008)
4. Bogdan, K., Dyda, B.: The best constant in a fractional Hardy inequality. Preprint (2008); arXiv:0807.1825v1
5. Bourgain, J., Brezis, H., Mironescu, P.: Limiting embedding theorems for $W^{s,p}$ when $s \uparrow 1$ and applications. J. Anal. Math. **87**, 77–101 (2002)
6. Dyda, B.: A fractional order Hardy inequality. Illinois J. Math. **48**, no. 2, 575–588 (2004)
7. Frank, R.L., Seiringer, R.: Non-linear ground state representations and sharp Hardy inequalities. J. Funct. Anal. **255**, 3407–3430 (2008)
8. Krugljak, N., Maligranda, L., Persson, L.E.: On an elementary approach to the fractional Hardy inequality. Proc. Am. Math. Soc. **128**, no. 3, 727–734 (2000)
9. Maz'ya, V.G.: Sobolev Spaces. Springer, Berlin etc. (1985)
10. Maz'ya, V., Shaposhnikova, T.: On the Bourgain, Brezis, and Mironescu theorem concerning limiting embeddings of fractional Sobolev spaces. J. Funct. Anal. **195**, no. 2, 230–238 (2002); Erratum: ibid **201**, no. 1, 298–300 (2003)

Proof [illegible] The inequality (1.2) follows immediately from Propo-sition [illegible]

Acknowledgements [illegible] through [illegible] and U.S. NSF [illegible] gratefully acknowledged.

References

1. [illegible] Functions with [illegible], and Mathematical [illegible], New York (1972)
2. [illegible] Hardy inequalities with [illegible] 356, no. 6, 2469–2496 (2004)
3. [illegible] in the Hardy [illegible] upper half space [illegible] 15 [illegible]
4. [illegible] The [illegible] constant [illegible] Functional [illegible]
5. [illegible]
6. [illegible] 1277–1286 (2011)
7. [illegible] and [illegible] Third [illegible]
8. [illegible] Mathematical [illegible] 128, no. 3 [illegible] (1999)
9. [illegible]
10. [illegible] 208 [illegible]

Collapsing Riemannian Metrics to Sub-Riemannian and the Geometry of Hypersurfaces in Carnot Groups

Nicola Garofalo and Christina Selby

Dedicated to Professor Maz'ya
with affection and admiration

Abstract Given a Carnot group with step two $\mathbb{G}$, we study the limit as $\epsilon \to 0$ of a family of left-invariant Riemannian metrics g_ϵ on $\mathbb{G}$. In such metrics, neither the Ricci tensor nor the sectional curvatures are bounded from below. Nonetheless, our main result shows that the Riemannian first and second variation formulas in the g_ϵ-metrics converge to the corresponding sub-Riemannian ones. This testifies of an intrinsic stability of some delicate cancellation processes and makes the method of collapsing Riemannian metrics a possible tool for attacking various fundamental open questions in sub-Riemannian geometry. Finally, we mention that the restriction to groups of step two is purely for ease of exposition and that the same method can be applied to Carnot groups of arbitrary step.

1 Introduction

There has recently been growing interest in first and second variation formulas for the horizontal perimeter measure of smooth (C^2) hypersurfaces in Carnot groups. As in the Riemannian case, such formulas play a pervasive

Nicola Garofalo
Department of Mathematics, Purdue University, West Lafayette, IN 47907, USA and Università di Padova, 35131 Padova, Italy
e-mail: garofalo@math.purdue.edu,garofalo@dmsa.unipd.it

Christina Selby
The Johns Hopkins University, Physics Laboratory, MD, USA
e-mail: Christina.Selby@jhuapl.edu

A. Laptev (ed.), *Around the Research of Vladimir Maz'ya I: Function Spaces*, International Mathematical Series 11, DOI 10.1007/978-1-4419-1341-8_7,

role in basic problems such as the Bernstein and the isoperimetric problems. In all works on the subject, such formulas have been derived by directly developing sub-Riemannian analogs of various fundamental tools from Riemannian geometry, such for instance the notions of horizontal connection, horizontal perimeter etc.

The celebrated isoperimetric inequality states that for any measurable set with locally finite perimeter $E \subset \mathbb{R}^n$ (a Caccioppoli set)

$$|E|^{\frac{n-1}{n}} \leqslant n^{-1}\omega^{-\frac{1}{n}} P(E, \mathbb{R}^n),$$

where ω_n is the n-dimensional measure of the unit ball in $\mathbb{R}^n$ and $P(E)$ is the variational perimeter of E according to De Giorgi. A fundamental result, proved independently by Maz'ya [25] and by Fleming and Rishel [15], states that the above inequality is, in fact, equivalent to the following sharp form of the Gagliardo–Nirenberg inequality, also known as the geometric Sobolev embeddiing:

$$\left(\int_{\mathbb{R}^n} |f|^{\frac{n}{n-1}} dx\right)^{\frac{n-1}{n}} \leqslant n^{-1}\omega^{-\frac{1}{n}} \operatorname{Var}(f; \mathbb{R}^n), \qquad f \in BV(\mathbb{R}^n),$$

where $\operatorname{Var}(f; \mathbb{R}^n)$ means the total variation of f and $BV(\mathbb{R}^n)$ denotes the space of functions of bounded variation in $\mathbb{R}^n$. The extremal sets in the isoperimetric inequality are balls (up to measure zero). According to A.D. Alexandrov's soap bubble theorem, these are the only smooth compact sets whose boundary has positive constant mean curvature. If instead one asks for Caccioppoli sets of zero mean curvature, one is led to the Bernstein problem.

A central tool in the calculus of variations and geometry are the first and second variation formulas of the area (cf., for example, [24, 35, 9, 3]. Besides their intrinsic interest, such formulas play a fundamental role in the study of various basic problems, such as, for instance, the above-mentioned isoperimetric and the Bernstein problems. Over the past few years there was an explosion of interest in the study of sub-Riemannian analogues of such problems and there presently exists a rapidly growing literature. It is not surprising that, in these developments, appropriate first and second variation formulas for the sub-Riemannian perimeter have occupied a central position. For instance, the second variation formula has recently led in [11] to the discovery of the role of stability in the sub-Riemannian Bernstein problem in the Heisenberg group $\mathbb{H}^1$, and to its complete solution in [13, 14] (cf. also [1] for an analogous result for intrinsic graphs).

Sub-Riemannian first and second variation formulas were found in [7, 10, 2, 18, 19, 1, 29, 33]. Although these papers are based on different approaches, they all have in common the fact that the authors work directly with the sub-Riemannian geometry induced by the ambient space on the submanifold. The following works contain various applications of these formulas: [8, 11, 13, 12,

34, 31]. While in such formulas the horizontal perimeter and mean curvature are the leading characters, they also entail several other entities whose deeper geometric meaning is hard to grasp since they appear a posteriori, often after long and complex calculations.

This paper was motivated by the desire of understanding the first and second variation of the sub-Riemannian area from a Riemannian viewpoint. Given a Carnot group $\mathbb{G}$, roughly speaking the method employed in this paper is based on defining a one-parameter family $\{g_\epsilon\}_\epsilon$ of left-invariant Riemannian metrics in $\mathbb{G}$, applying results from Riemannian geometry, and then taking the limit as ϵ converges to zero. More precisely, given a C^2 hypersuface $M \subset \mathbb{G}$, our main objective was the study of the limiting behavior of the induced Riemannian metrics on M with the purpose of unraveling the sub-Riemannian geometry of M in the limit as $\epsilon \to 0$. This program has been successful and, as a by-product, we have been able to obtain first and second variation formulas which, when specialized to the Heisenberg group $\mathbb{H}^1$, give back those first obtained in [7] using CR geometry and in [10] with sub-Riemannian calculus. Such formulas also coincide with the more general ones independently obtained in [27] and [19] by exploiting the Lagrangian framework of the Poincaré–Cartan differential forms. One notable aspect is that, despite the fact that the Ricci tensor or the sectional curvatures are not bounded from below, the Riemannian second fundamental form of M in the ϵ-metric converges to the sub-Riemannian second fundamental form introduced in [10] and [18].

The present work focuses on Carnot groups of step 2, but the same technique could very well be applied to more general groups. The method employed in this paper has already been explored with different purposes by several other people (cf., for example, [21, 22, 30, 32, 8, 6, 4]. Of course, there is a large literature in Riemannian geometry on collapsing Riemannian metrics (cf., for example, [16, 17] and [26]), but such a powerful method is used in that context in a perspective which is somewhat different from that of the present paper. For instance, in our context, the Ricci tensor is not bounded from below, nor are such the sectional curvatures of the approximating Riemannian metrics. It seems plausible that a thorough investigation of the method used in this paper could have far-reaching applications in sub-Riemannian geometry. For instance, one could use it to investigate the fundamental open question of estimates of Alexandrov–Bakelman–Pucci type for equations with rough coefficients in non variational form. In this perspective, we hope that the geometric information contained in the present paper will prove useful.

A brief description of the organization of the present paper is as follows. The first section provides the necessary background on Carnot groups of step 2. In the following section, an orthonormal basis with respect to the metric which makes $\{X_1, \ldots, X_m\, T_{1,\epsilon}, \ldots, T_{k,\epsilon}\}$ an orthonormal basis is given. Here, $T_{s,\epsilon} = \sqrt{\epsilon} T_s$. This basis is what is used extensively in completing the geometric calculations of Section 5. In this section, the reader find computations of the ϵ-tangential divergence, ϵ-mean curvature, and the ϵ-second fundamental form. Applying the Riemannian divergence theorem, some interesting integration-by-parts formulas are derived. The geometric quantities are all of the ingredients needed to prove the first and second variation formula found in Section 6. The second variation formula is proved for the Heisenberg group and for variations in the horizontal normal direction. This restriction is for computational ease. It should be clear to the reader that our technique could very well be used for more general variations.

In the model setting of the Heisenberg group $\mathbb{H}^1$, for an arbitrary (compactly supported) deformation of a surface first and second variation formulas were obtained in [7] and in [10]. In [11], the notion of stable minimal surface was introduced on the basis of the second variation formula in [10]. In [10], stability inequalities were obtained. Note that such inequalities have recently found application to the solution of the Bernstein problem in the Heisenberg group $\mathbb{H}^1$ (cf. [13, 14, 1]). A Bernstein type theorem is also found in [7], and results related to those in [14] were recently obtained in [20]. The first variation formula for H-perimeter implies that any minimizer of H-perimeter must be an H-minimal surface. These surfaces are also studied in [8, 33, 34]. In particular, a solution to the isoperimetric problem for the first Heisenberg group is found in [34] (cf. also [12] for a partial result). First and second variation formulas for more general groups were computed in [29] and [19], concurrently to this work. These papers exploit the Lagrangian framework of the Poincaré–Cartan differential forms.

2 Hypersurfaces in Carnot Groups of Step 2

A Carnot group $\mathbb{G}$ of step 2 is a real Lie group whose Lie algebra $\mathfrak{g}$ admits a stratification nilpotent of step 2. This means that there exist vector spaces $V_1, V_2 \subset \mathfrak{g}$ such that $\mathfrak{g} = V_1 \oplus V_2$, $[V_1, V_1] = V_2$, and $[V_1, V_2] = \{0\}$. The vector space V_1 is called the *horizontal layer*. The exponential mapping $\exp : \mathfrak{g} \to \mathbb{G}$ is an analytic diffeomorphism of $\mathfrak{g}$ onto $\mathbb{G}$. In general, the Baker–Campbell–Hausdorff formula [36] states that

$$\exp(\xi)\exp(\eta) = \exp\left(\xi + \eta + \frac{1}{2}[\xi,\eta] + \frac{1}{12}\{[\xi,[\xi,\eta]] - [\eta,[\xi,\eta]]\} + \dots\right), \quad (2.1)$$

where the dots indicate commutators or order four and higher. For a group of step 2 commutators of order greater than 2 are zero. Using (2.1), one can determine the group law of $\mathbb{G}$ through knowledge of the bracket relationships of the vector spaces V_1 and V_2.

An example of paramount importance of a Carnot group of step 2 is the Heisenberg group, $\mathbb{H}^n$. In this case, $V_1 = \mathbb{R}^{2n} \times \{0\}$, $V_2 = \{0\} \times \mathbb{R}$, and $\mathbb{H}^n \simeq \mathbb{R}^{2n+1}$. For x, $y \in \mathbb{R}^n$, let $g = (x, y, t)$, $g' = (x', y', t')$. The noncommutative group law in $\mathbb{H}^n$ is given by

$$g \cdot g' = \left(x + x', y + y', t + t' + \frac{1}{2} \left(\langle x, y' \rangle - \langle x', y \rangle \right) \right).$$

Throughout the paper, $m = \dim V_1$, $k = \dim V_2$, and [$n = m + k$ is the topological dimension of $\mathbb{G}$. Let $\{e_1, \ldots, e_m\}$ denote an orthonormal basis for V_1 and $\{\epsilon_1, \ldots, \epsilon_k\}$ denote an orthonormal basis for V_2. The notation $\langle \,, \rangle_{\mathfrak{g}}$ means the inner product on $\mathfrak{g}$ for which the preceding bases are orthonormal. We also use the notation $\langle \,, \rangle$ when the dependence on $\mathfrak{g}$ is understood. Using the exponential map, we define the *exponential coordinates* of $g \in \mathbb{G}$ as follows:

$$x_i(g) = \langle \exp^{-1}(g), e_i \rangle_{\mathfrak{g}}, \quad i = 1, \ldots, m, \tag{2.2}$$

$$t_s(g) = \langle \exp^{-1}(g), \epsilon_s \rangle_{\mathfrak{g}}, \quad s = 1, \ldots, k. \tag{2.3}$$

Using these coordinates, we routinely identify an element $g \in \mathbb{G}$ with the point $(x_1, \ldots, x_m, t_1, \ldots, t_k) \in \mathbb{R}^{m+k}$, i.e.,

$$g = (x_1, \ldots, x_m, t_1, \ldots, t_k).$$

The group constants b^s_{ij} are defined by

$$b^s_{ij} = \langle [e_i, e_j], \epsilon_s \rangle_{\mathfrak{g}}, \quad i, j = 1, \ldots, m, \quad s = 1, \ldots, k. \tag{2.4}$$

Therefore,

$$[e_i, e_j] = \sum_{s=1}^{k} b^s_{ij} \epsilon_s, \quad i, j = 1, \ldots, m. \tag{2.5}$$

The operators of left and right translation by an element $g \in \mathbb{G}$ are denoted L_g and R_g respectively. In particular,

$$L_g(g') = g \cdot g', \qquad R_g(g') = g' \cdot g.$$

The following left-invariant vector fields are defined on $\mathbb{G}$:

$$X_i(g) = (L_g)_* (e_i), \quad i = 1, \ldots, m, \tag{2.6}$$

$$T_s(g) = (L_g)_* (\epsilon_s), \quad s = 1, \ldots, k. \tag{2.7}$$

Using the Baker–Campbell–Hausdorff formula, one obtains the following lemma.

Lemma 2.1. *For each $i = 1, \ldots, m$*

$$X_i = \frac{\partial}{\partial x_i} + \frac{1}{2} \sum_{s=1}^{k} \sum_{j=1}^{m} b^s_{ji} x_j \frac{\partial}{\partial t_s}. \tag{2.8}$$

For $s = 1, \ldots, k$

$$T_s = \frac{\partial}{\partial t_s}. \tag{2.9}$$

We notice explicitly that, as a consequence of (2.8) and (2.9), one has

$$\frac{\partial}{\partial x_i} = X_i - \frac{1}{2}\sum_{s=1}^{k}\sum_{j=1}^{m} b_{ji}^s x_j T_s. \tag{2.10}$$

A vector field ζ on $\mathbb{G}$ is said to be *horizontal* if

$$\zeta = \sum_{i=1}^{m} a_i X_i,$$

for suitable $a_i \in C^\infty(\mathbb{G})$. Let $\mathbb{G}$ be a Carnot group of step 2. For an oriented hypersurface $M \subset \mathbb{G}$ we denote by N a Riemannian nonunit normal to M and by $\nu = N/|N|$ the Riemannian Gauss map.

Definition 2.2. A *horizontal normal* $N_H : M \to H\mathbb{G}$ is defined by

$$N_H = \sum_{i=1}^{m} \langle \nu, X_j \rangle X_j,$$

and the horizontal unit normal or *horizontal Gauss map*, denoted ν_H, is defined by

$$\nu_H = \frac{N_H}{|N_H|},$$

where it exists.

The function $W = |N_H|$ is called the *angle function*. It is clear that $W = 0$ if and only if $\langle \nu, X_j \rangle = 0$ for $j = 1, \ldots, m$.

Definition 2.3. The set of points in M where ν_H does not exist is called the *characteristic set* of M and is denoted by $\Sigma(M)$ or Σ if the dependence on M is understood. It is clear that

$$\Sigma = \{g \in M \mid \langle \nu(g), X_j(g) \rangle = 0, j = 1, \ldots, m\} = \{g \in M \mid W(g) = 0\}$$

is a closed subset of M.

Definition 2.4. The *horizontal divergence* of $\phi = \sum_{i=1}^m \phi_i X_i$ is defined as

$$\mathrm{div}_H \phi = \sum_{i=1}^{m} X_i \phi_i,$$

and the *horizontal gradient* of $u \in C^1(\mathbb{G})$ is defined as

$$\nabla_H u = \sum_{i=1}^{m} X_i u X_i.$$

Note that $\nabla_H u$ is the projection of the Riemannian gradient of u onto the horizontal bundle $H\mathbb{G}$.

Definition 2.5. The *horizontal mean curvature* at a point $p \in M \backslash \Sigma(M)$ is given by

$$\mathcal{H} = div_H \nu_H.$$

Let Ω be an open set in $\mathbb{G}$. The *H-variation* of $u \in L^1_{loc}(\Omega)$ with respect to Ω is defined as

$$\mathrm{Var}_H(u;\Omega) = \sup\Big\{\int_\Omega u\, \mathrm{div}_H \phi\, dh : \phi \in C^1_0(\Omega; H\mathbb{G}), |\phi| \leqslant 1\Big\}.$$

A function u is said to belong to $BV_H(\Omega)$ if $\mathrm{Var}_H(u;\Omega) < \infty$. The *$H$-perimeter measure* of a measurable set $E \subset \mathbb{G}$ with respect to an open set $\Omega \subset \mathbb{G}$ is defined as $|\partial E|_H(\Omega) = \mathrm{Var}_H(\chi_E;\Omega)$. The set E has *finite H-perimeter* if $|\partial E|_H(\mathbb{G}) < \infty$. By the Riesz representation theorem, $|\partial E|_H$ is a Radon measure on $\mathbb{G}$ and there exists a measurable section ν_E of $H\mathbb{G}$ such that $|\nu_E| = 1$ and

$$\int_E \mathrm{div}_H \phi\, dh = -\int_{\mathbb{G}} \langle \phi, \nu_E\rangle\, d|\partial E|_H$$

for any $\phi \in C^1_0(\mathbb{G}, H\mathbb{G})$. The section ν_E is called the *generalized inward normal* to E. The following assertion was proved in [5].

Proposition 2.6. *If E has C^1 boundary and finite H-perimeter in Ω, then*

$$|\partial E|_H(\Omega) = \int_{\partial E\cap\Omega} \Big(\sum_{i=1}^{m} \langle \nu, X_i\rangle^2\Big)^{1/2} d\mathcal{H}^{n-1} = \int_{\partial E\cap\Omega} |N_H| d\mathcal{H}^{n-1},$$

where $\mathcal{H}^{n-1}$ is the $(n-1)$-dimensional Hausdorff measure and ν is the unit outward normal to ∂E (recall that $n = m + k$).

By this formula, it is clear that the measure on ∂E, defined by

$$\sigma_H(\partial E \cap \Omega) \stackrel{def}{=} P_H(E;\Omega)$$

on the open sets of ∂E, is absolutely continuous with respect to σ and its density is represented by the angle function W of ∂E. We formalize this observation in the following definition.

Definition 2.7. For a bounded domain $E \subset \mathbb{G}$ of class C^1 we denote by

$$d\sigma_H \;=\; |N_H| d\mathcal{H}^{n-1} = W d\mathcal{H}^{n-1}, \tag{2.11}$$

the H-*perimeter measure* supported on ∂E (here, W is the angle function of ∂E).

Note that if Ω can be described as $\Omega = \{\Phi < 0\}$ for some defining function Φ which is C^1 in a neighborhood U of Ω and with non vanishing gradient $\nabla\Phi$ in U, then it can be shown that on $M = \partial\Omega$ the horizontal normal is given by

$$\nu_H = -\frac{\nabla_H \Phi}{|\nabla_H \Phi|}$$

away from the characteristic set $\Sigma(M)$.

3 The Limit as $\epsilon \to 0$ of the Rescaled ϵ-Volume Forms on M

Hereafter, for a Carnot group $\mathbb{G}$ with step 2 we denote by $\langle \cdot, \cdot \rangle$ the left-invariant Riemannian metric which makes $\{X_1, \ldots, X_m, T_1, \ldots, T_k\}$ an orthonormal basis of $T\mathbb{G}$. If we consider the rescaled basis of the vertical layer defined by

$$T_{s,\epsilon} = \sqrt{\epsilon}\, T_s, \qquad s = 1, \ldots k,$$

then from (2.10) we obtain

$$\frac{\partial}{\partial x_i} = X_i - \frac{1}{2\sqrt{\epsilon}} \sum_{s=1}^{k} \sum_{j=1}^{m} b_{ji}^s x_j T_{s,\epsilon}. \tag{3.1}$$

Lemma 3.1. *Let g^ϵ be the Riemannian metric which makes $\{X_1, \ldots, X_m, T_{1,\epsilon}, \ldots, T_{k,\epsilon}\}$ an orthonormal basis. Denote its matrix entries g_{ij}^ϵ. Denote by $g^{\epsilon,ij}$ the matrix entries of its inverse. Then*

$$g_{ij}^\epsilon = \left\langle \frac{\partial}{\partial x_i}, \frac{\partial}{\partial x_j} \right\rangle_\epsilon = \delta_{ij} + \frac{1}{4\epsilon} b_{mi}^s b_{kj}^s x_m x_k$$

for $i, j = 1, \ldots, m$,

$$g_{m+r,m+s}^\epsilon = \left\langle \frac{\partial}{\partial t_r}, \frac{\partial}{\partial t_s} \right\rangle_\epsilon = \frac{1}{\epsilon} \delta_{rs}$$

for $s, r = 1, \ldots, k$, and

$$g_{i,m+s}^\epsilon = \left\langle \frac{\partial}{\partial x_i}, \frac{\partial}{\partial t_s} \right\rangle_\epsilon = -\frac{1}{2\epsilon} b_{mj}^s x_m$$

for $i = 1, \ldots, m$, $s = 1, \ldots, k$. Further,

$$g^{\epsilon,ij} = \delta_{ij}$$

for $i, j = 1, \ldots, m$,

$$g^{\epsilon,m+r,m+s} = \epsilon\delta_{ij} + \frac{1}{4} b^r_{lq} b^s_{hq} x_l x_h$$

for $s, r = 1, \ldots, k$, *and*

$$g^{\epsilon,i,m+s} = \frac{1}{2} b^s_{qi} x_q$$

for $i = 1, \ldots, m$, $s = 1, \ldots, k$.

Proof. The formulas for g^ϵ_{ij} follow directly from the definition of X_i. The formula for the inverse is verified in this proof. In general, to prove that a matrix N is the inverse of a matrix M, one needs to show that $N_{ik}M_{kj} = \delta_{ij}$. For $v > m$ we set $r = v - m$. For $i, j = 1, \ldots, m$

$$g^\epsilon_{iv} g^{\epsilon,vj} = \sum_{v=1}^{m} \Big(\delta_{iv} + \frac{1}{4\epsilon} b^w_{ui} b^w_{zv} x_u x_z\Big)(\delta_{vj}) + \sum_{r=1}^{k} \Big(-\frac{1}{2\epsilon} b^r_{li} x_l\Big)\Big(\frac{1}{2} b^r_{hj} x_h\Big) = \delta_{ij}.$$

For $s = 1, \ldots, k$, $j = 1, \ldots, m$

$$g^\epsilon_{m+s,v} g^{\epsilon,vj} = \sum_{v=1}^{m} \Big(-\frac{1}{2\epsilon} b^s_{lv} x_l\Big)(\delta_{vj}) + \sum_{r=1}^{k} \Big(\frac{1}{\epsilon}\delta_{rs}\Big)\Big(\frac{1}{2} b^r_{hj} x_h\Big) = 0 = \delta_{m+s,j}.$$

For $i = 1, \ldots, m$, $s = 1, \ldots, k$

$$\begin{aligned} g^\epsilon_{iv} g^{\epsilon,v,m+s} &= \sum_{v=1}^{m} \Big(\delta_{iv} + \frac{1}{4\epsilon} b^r_{li} b^r_{hv} x_l x_h\Big)\Big(\frac{1}{2} b^s_{tv} x_t\Big) \\ &+ \sum_{r=1}^{k} \Big(-\frac{1}{2\epsilon} b^r_{li} x_l\Big)\Big(\epsilon\delta_{rs} + \frac{1}{4} b^s_{tq} b^r_{hq} x_t x_h\Big) = 0 = \delta_{i,m+s}. \end{aligned}$$

Finally, for $v > m$ we set $t = v - m$. Then for $r, s = 1, \ldots, k$

$$\begin{aligned} g^\epsilon_{m+r,v} g^{\epsilon,v,m+s} &= \sum_{v=1}^{m} \Big(-\frac{1}{2\epsilon} b^r_{lv} x_l\Big)\Big(\frac{1}{2} b^s_{hv} x_h\Big) \\ &+ \sum_{t=1}^{k} \Big(\frac{1}{\epsilon}\delta_{rt}\Big)\Big(\epsilon\delta_{ts} + \frac{1}{4\epsilon} b^t_{lq} b^s_{hq} x_l x_h\Big) = \delta_{rs} = \delta_{m+r,m+s}. \end{aligned}$$

The proof is complete. □

In the case of $\mathbb{H}^1$, by Lemma 3.1, we have

$$g^\epsilon = \begin{pmatrix} 1+\frac{y^2}{4\epsilon} & -\frac{xy}{4\epsilon} & \frac{y}{2\epsilon} \\ -\frac{xy}{4\epsilon} & 1+\frac{x^2}{4\epsilon} & -\frac{x}{2\epsilon} \\ \frac{y}{2\epsilon} & -\frac{x}{2\epsilon} & \frac{1}{\epsilon} \end{pmatrix}, \tag{3.2}$$

$$(g^\epsilon)^{-1} = \begin{pmatrix} 1 & 0 & -\frac{y}{2} \\ 0 & 1 & \frac{x}{2} \\ -\frac{y}{2} & \frac{x}{2} & \epsilon + \frac{x^2+y^2}{4} \end{pmatrix}. \tag{3.3}$$

The following assertion, which can be found in [10] for $\mathbb{H}^n$, is crucial to the results of this section.

Proposition 3.2. *In $\mathbb{G}$, we consider the left-invariant Riemannian metric $g^\epsilon = [g^\epsilon_{ij}]$ which makes $\{X_1, \dots, X_m, T_{1,\epsilon}, \dots, T_{k,\epsilon}\}$ an orthonormal basis. Given a C^2 codimension one submanifold $M \subset \mathbb{G}$, denote by $\sigma^\epsilon(M)$ the corresponding surface area of M. Then the sub-Riemannian area of M is given by*

$$\sigma_H(M) = \lim_{\epsilon \to 0} \sqrt{\epsilon}\sigma^\epsilon(M).$$

Proof. Consider a bounded open chart $U \subset M$. By employing a partition of unity argument, it suffices to prove the statement for U. Now, let $\Omega \subset \mathbb{R}^{n-1}$ be such that U is represented by $F : \Omega \to U$, with $F \in C^2(\Omega)$. Then

$$\sigma^\epsilon(U) = \int_\Omega |N^\epsilon|_\epsilon \, dp.$$

One may assume that

$$U = \{(x,t) \in \mathbb{R}^{m+k} : \Phi(x,t) = 0\}.$$

Further, Φ can be chosen so that

$$|N_\epsilon|_\epsilon = \frac{1}{\sqrt{\epsilon}} |\nabla^\epsilon \Phi|_\epsilon.$$

Using the formula for $g^{\epsilon,ij}$ in Lemma 3.1, one has (re-labelling summation indices as needed)

$$\nabla^\epsilon \Phi = g^{\epsilon,ij} \frac{\partial \Phi}{\partial x_i} \frac{\partial}{\partial x_j} = \left(\frac{\partial \Phi}{\partial x_j} + \frac{1}{2} b^s_{hj} x_h \frac{\partial \Phi}{\partial t_s} \right) \frac{\partial}{\partial x_j}$$
$$+ \frac{1}{2} b^s_{hi} x_h \frac{\partial \Phi}{\partial x_i} \frac{\partial}{\partial t_s} + \epsilon \frac{\partial \Phi}{\partial t_s} \frac{\partial}{\partial t_s} + \frac{1}{4} b^r_{lq} b^s_{hq} x_l x_h \frac{\partial \Phi}{\partial t_r} \frac{\partial}{\partial t_s}$$

$$= X_j\Phi\frac{\partial}{\partial x_j} + X_q\Phi\left(\frac{1}{2}b^s_{hq}x_h\frac{\partial}{\partial t_s}\right) + \epsilon\frac{\partial\Phi}{\partial t_s}\frac{\partial}{\partial t_s}$$

$$= X_j\Phi X_j + \epsilon\frac{\partial\Phi}{\partial t_s}\frac{\partial}{\partial t_s} = \nabla_H\Phi + \sqrt{\epsilon}\sum_{s=1}^{k} T_s\Phi T_{s,\epsilon}.$$

Note that $|\nabla^\epsilon\Phi|_\epsilon$ converges uniformly to $|\nabla_H\Phi|$ since $T_s\Phi$ is bounded in view of the regularity of U. Therefore,

$$\begin{aligned}\lim_{\epsilon\to 0}\sqrt{\epsilon}\sigma^\epsilon(U) &= \lim_{\epsilon\to 0}\int_\Omega \sqrt{\epsilon}|N_\epsilon|_\epsilon\, dp \\ &= \lim_{\epsilon\to 0}\int_\Omega |\nabla^\epsilon\Phi|_\epsilon\, dp = \lim_{\epsilon\to 0}\int_\Omega \sqrt{|N_H|^2 + \epsilon\sum_{s=1}^{k}(T_s\Phi)^2 d\sigma_H} \\ &= \int_\Omega |N_H|\, dp = \int_U d\sigma_H\end{aligned}$$

by Definition 2.7. □

4 Orthonormal Basis

Our first objective in this section is to produce an explicit basis for the so-called horizontal tangent bundle to an hypersuface $M \subset \mathbb{G}$ (cf. Definition 4.1 below). Such a basis then be completed to one of the full tangent bundle TM. We denote by N a Riemannian nonunit normal to M. It is clear that

$$N = \sum_{i=1}^{m} p_i X_i + \sum_{j=1}^{k} \omega_j T_j, \tag{4.1}$$

where $p_i = \langle N, X_i\rangle$, $\omega_j = \langle N, T_j\rangle$. We introduce some quantities which will appear in the computations:

$$\begin{aligned} W &= \left(\sum_{i=1}^{m} p_i^2\right)^{1/2}, \\ \gamma_i &= \left(\sum_{j=1}^{i} p_j^2\right)^{1/2}, \quad i = 1, \ldots, m, \\ \tau_j &= \left(W^2 + \sum_{r=1}^{j-1}\omega_r^2\right)^{1/2}, \quad j = 1, \ldots, k+1. \end{aligned} \tag{4.2}$$

As we have already observed in the previous section, W vanishes precisely on the characteristic set Σ of M. Away from Σ, we define

$$\overline{p}_j = \frac{p_j}{W}, \quad j = 1, \ldots, m, \tag{4.3}$$

$$\overline{\omega}_s = \frac{\omega_s}{W}, \quad s = 1, \ldots, k. \tag{4.4}$$

The horizontal Gauss map on $M \setminus \Sigma$ is defined as

$$\nu_H = \sum_{j=1}^{m} \overline{p}_j X_j. \tag{4.5}$$

Definition 4.1. The *horizontal tangent bundle* of M is the intersection of the tangent bundle of M with the horizontal subbundle $H\mathbb{G}$. We denote it as

$$HTM = TM \cap H\mathbb{G}.$$

The subbundle HTM plays an important role in various geometric definitions such as *horizontal second fundamental form, horizontal mean curvature*, and others to be discussed in the following section.

Let $x \in M$ be a noncharacteristic point. Note that $x \notin \Sigma$ guarantees $W(x) \neq 0$, and therefore $\tau_j(x) \neq 0$ for every $j = 1, \ldots, k+1$. Without loss of generality, we assume that $p_1(x) \neq 0$. Note that this assumption guarantees, in particular, that $\gamma_i(x) \neq 0$ for every $i = 1, \ldots, m$.

Proposition 4.2. *Define*

$$E_i = \frac{-p_{i+1} \sum_{j=1}^{i} p_j X_j + \gamma_i^2 X_{i+1}}{\gamma_i \gamma_{i+1}} \tag{4.6}$$

for $i = 1, \ldots, m-1$ *and*

$$F_s = \frac{-\omega_s \sum_{j=1}^{m} p_j X_j - \sum_{r=1}^{s-1} \omega_r \omega_s T_r + \tau_s^2 T_s}{\tau_s \tau_{s+1}} \tag{4.7}$$

for $s = 1, \ldots, k$. *If we set*

$$\mathcal{E}_i = \begin{cases} E_i, & i = 1, \ldots, m-1, \\ F_{i-m+1}, & i = m, \ldots, m+k-1 = n-1, \end{cases}$$

then $\{\mathcal{E}_1, \ldots, \mathcal{E}_{n-1}\}$ *is an orthonormal basis for* $T_x M$.

Proof. We begin by showing that $\{\mathcal{E}_1, \ldots, \mathcal{E}_{n-1}\} \in T_x M$. For $i = 1, \ldots, m-1$ from (4.1) and (4.6) we have

$$\langle E_i, N\rangle = \frac{-p_{i+1}\sum_{j=1}^{i-1} p_j^2 + \gamma_i^2 p_{i+1}}{\gamma_i\gamma_{i+1}} = 0.$$

Similarly, for $s = 1, \ldots, k$ from (4.1) and (4.7) we have

$$\langle F_s, N\rangle = \frac{-\omega_s \sum_{i=1}^{m} p_i^2 - \omega_s \sum_{r=1}^{s-1} \omega_r^2 + \omega_s\tau_s^2}{\tau_s\tau_{s+1}} = \frac{-\omega_s\tau_s^2 + \omega_s\tau_s^2}{\tau_s\tau_{s+1}} = 0.$$

Hence the vectors $\mathcal{E}_1, \ldots, \mathcal{E}_{n-1}$ belong to T_xM. We next show that they constitute an orthonormal basis for T_xM, i.e., $\langle \mathcal{E}_i, \mathcal{E}_j\rangle = \delta_{ij}$ for $i, j = 1, \ldots, n-1$. For $i < j$

$$\langle E_i, E_j\rangle = \frac{(p_{i+1})(p_{j+1})\sum_{k=1}^{i} p_k^2 + (-p_{j+1})(p_{i+1})\gamma_i^2}{\gamma_i\gamma_{i+1}\gamma_j\gamma_{j+1}} = 0$$

and

$$\langle E_i, E_i\rangle = \frac{(p_{i+1})^2\sum_{j=1}^{i} p_j^2 + \gamma_i^4}{\gamma_i^2\gamma_{i+1}^2} = \frac{\gamma_i^2\gamma_{i+1}^2}{\gamma_i^2\gamma_{i+1}^2} = 1.$$

Next, for $r < s$,

$$\langle F_s, F_r\rangle = \frac{\omega_s\omega_r W^2 + \omega_r\omega_s \sum_{k=1}^{s-1} \omega_k^2 - \tau_s^2\omega_r\omega_s}{\tau_s\tau_{s+1}\tau_r\tau_{r+1}} = 0$$

and

$$\langle F_s, F_s\rangle = \frac{\omega_s^2 W^2 + \tau_s^4}{\tau_s^2\tau_{s+1}^2} = \frac{\tau_s^2\tau_{s+1}^2}{\tau_s^2\tau_{s+1}^2} = 1.$$

Finally,

$$\langle E_i, F_s\rangle = \frac{\omega_s p_{i+1}\sum_{j=1}^{i} p_j^2 - \omega_s p_{i+1}\gamma_i^2}{\gamma_i\gamma_{i+1}\tau_s\tau_{s+1}} = 0.$$

The proof is complete. □

Remark 4.3. In the proof of Proposition 4.2, we see that the vectors $E_1, \ldots, E_{m-1}$ constitute an orthonormal subset for T_xM. Since, by definition, $E_1, \ldots, E_{m-1} \in H_x\mathbb{G}$, we infer that (at a point where $p_1(x) \neq 0$) the vectors $\{E_1, \ldots, E_{m-1}\}$ provide an orthonormal basis of HT_xM.

Although the following identity is already contained in the statement in Remark 4.3, for future reference let us observe explicitly that for $i = 1, \ldots, m$

$$\langle E_i, \nu_H\rangle = \frac{1}{|N_H|}\langle E_i, N_H\rangle = \frac{1}{|N_H|}\langle E_i, N\rangle = 0;$$

furthermore,

$$\langle E_i, T \rangle = 0.$$

We denote

$$E_{i,j} = \langle E_i, X_j \rangle.$$

Henceforth, we adopt the summation convention over repeated indices. Recall the definition of the group constants b_{ij}^s from (2.4). We introduce the quantities

$$\xi_{ij}^s = b_{ml}^s E_{i,l} E_{j,m}, \quad i, j = 1, \dots m-1, \ s = 1, \dots k. \tag{4.8}$$

We also define the vector fields,

$$Z^s = b_{ij}^s \overline{p}_j X_i, \quad s = 1, \dots, k, \tag{4.9}$$

$$Y = \nu_H + \sum_{s=1}^{k} \overline{\omega}_s T_s, \tag{4.10}$$

and

$$Y^{\perp}{}_s = -\overline{\omega}_s \nu_H + T_s, \quad s = 1, \dots k. \tag{4.11}$$

Consider an oriented hypersurface $M \subset \mathbb{G}$ and N_ϵ as above. It is easy to see that for the unit normal ν_ϵ to M with respect to $\langle \, , \, \rangle_\epsilon$

$$\lim_{\epsilon \to 0} \nu_\epsilon = \nu_H \tag{4.12}$$

and this limit is uniform in ϵ for points in a relatively compact subset of M. Some analogous quantities to W and τ_j are defined. Let

$$W_\epsilon = \left(\sum_{i=1}^{m} p_i^2 + \epsilon \sum_{j=1}^{k} \omega_j^2 \right)^{1/2}, \tag{4.13}$$

$$\tau_{j,\epsilon} = \left(W^2 + \epsilon \sum_{r=1}^{j-1} \omega_r^2 \right)^{1/2}. \tag{4.14}$$

One can observe that

$$\langle \nu_\epsilon, X_i \rangle_\epsilon = \langle \nu_\epsilon, X_i \rangle = \frac{W}{W_\epsilon} \overline{p}_i, \tag{4.15}$$

$$\langle \nu_\epsilon, T_{s,\epsilon} \rangle_\epsilon = \sqrt{\epsilon} \frac{W}{W_\epsilon} \overline{\omega}_s. \tag{4.16}$$

In general, for a vector field V

$$\langle V, X_i \rangle_\epsilon = \langle V, X_i \rangle$$

and for $V = \langle V, X_i \rangle X_i + \langle V, T_s \rangle T_s$

$$\langle V, T_{s,\epsilon} \rangle_\epsilon = \frac{1}{\sqrt{\epsilon}} \langle V, T_s \rangle.$$

The following lemma will be useful.

Lemma 4.4. *Let W_ϵ be the same as previously defined in the context of $\mathbb{H}^n$. Then*

$$\frac{W^2}{W_\epsilon^2} = 1 - \epsilon \frac{W}{W_\epsilon} \sum_{j=1}^{k} \varpi_j^2, \tag{4.17}$$

$$\frac{1}{\epsilon}\left(\frac{W}{W_\epsilon}\right)^4 = \frac{1}{\epsilon} - 2\sum_{j=1}^{k} \varpi_j^2 \left(\frac{W}{W_\epsilon}\right)^2. \tag{4.18}$$

Proof. By definition,

$$W_\epsilon^2 = \sum_{i=1}^{m} p_i^2 + \epsilon \sum_{j=1}^{k} \omega_j^2 = W^2 + \epsilon \sum_{j=1}^{k} \omega_j^2.$$

Hence

$$\frac{W_\epsilon^2}{W^2} = \sum_{i=1}^{m} \overline{p}_i^2 + \epsilon \sum_{j=1}^{k} \varpi_j^2 = 1 + \epsilon \sum_{j=1}^{k} \varpi_j^2$$

and

$$\frac{W^2}{W_\epsilon^2} = 1 - \epsilon \frac{W}{W_\epsilon} \sum_{j=1}^{k} \varpi_j^2.$$

The rest of the proof follows similarly from definitions. □

An orthonormal basis for TM_x with respect to the metric $\langle \, , \rangle_\epsilon$ can be defined in a similar way as Proposition 4.2.

Proposition 4.5. *Define*

$$E_i = \frac{-p_{i+1} \sum_{j=1}^{i} p_j X_j + \gamma_i^2 X_{i+1}}{\gamma_i \gamma_{i+1}} \tag{4.19}$$

for $i = 1, \ldots, m-1$ and

$$F_s^\epsilon = \frac{-\sqrt{\epsilon}\omega_s \sum_{i=1}^m p_i X_i - \epsilon\omega_s \sum_{r=1}^{s-1} \omega_r T_{r,\epsilon} + \tau_{s,\epsilon}^2 T_{s,\epsilon}}{\tau_{s,\epsilon}\tau_{s+1,\epsilon}} \tag{4.20}$$

for $s = 1, \dots, k$. *Then* $\{\mathcal{E}_1^\epsilon, \dots, \mathcal{E}_{n-1}^\epsilon\}$ *is an orthonormal basis for* TM_x *with respect to* $\langle \, , \, \rangle_\epsilon$, *where*

$$\mathcal{E}_i^\epsilon = E_i, \quad i = 1, \dots, m-1,$$

and

$$\mathcal{E}_s^\epsilon = F_{s-m+1}^\epsilon, \quad s = m, \dots, n-1.$$

The proof is analogous to that of Proposition 4.2 and will be omitted.

In the setting of the Heisenberg group, $\mathbb{G} = \mathbb{H}^n$, one can make the following observation:

$$F^\epsilon = \sqrt{\epsilon}\frac{W}{W_\epsilon} Y^\perp, \tag{4.21}$$

where $Y^\perp$ is defined by (4.11).

5 Geometric Guantities with respect to the Collapsing Metrics

In this section, several geometric quantities with respect to the metric $\langle \, , \, \rangle_\epsilon$ are computed. These quantities are used, along with Proposition 3.2, to derive several integration by parts formulas with respect to the horizontal perimeter measure. Other quantities are computed for use in the following section. Many quantities are computed for general Carnot groups of step 2, but for simplifying the notation some quantities are computed only for the Heisenberg group.

The Levi–Civita connection ∇ on $T\mathbb{G}$ is defined by the metric $\langle \, , \, \rangle$. One can prove the following lemma as in [10].

Lemma 5.1. *Let* ∇ *be the Levi-Civita connection on* $T\mathbb{G}$ *defined by the metric* $\langle \, , \, \rangle$. *Then*

$$\nabla_{X_i} X_j = \frac{1}{2}\sum_{s=1}^k b_{ij}^s T_s,$$

$$\nabla_{T_s} T_p = 0,$$

$$\nabla_{X_i} T_s = \nabla_{T_s} X_i = -\frac{1}{2}\sum_{j=1}^m b_{ij}^s X_j.$$

Similarly, the Levi–Civita connection ∇^ϵ is defined by the metric $\langle\, , \rangle_\epsilon$.

The proof of the following lemma is similar to that of the previous lemma.

Lemma 5.2. *Let ∇^ϵ be the Levi-Civita connection on $T\mathbb{G}$ defined by the metric $\langle\, , \,\rangle_\epsilon$. Then*

$$\nabla^\epsilon{}_{X_i} X_j = \frac{1}{2\sqrt{\epsilon}} \sum_{s=1}^{k} b_{ij}^s T_{s,\epsilon}, \quad i, j = 1, \dots m,$$

$$\nabla^\epsilon{}_{T_{s,\epsilon}} T_{p,\epsilon} = 0, \quad s, p = 1, \dots k,$$

$$\nabla^\epsilon{}_{X_i} T_{s,\epsilon} = \nabla^\epsilon{}_{T_{s,\epsilon}} X_i = -\frac{1}{2\sqrt{\epsilon}} \sum_{j=1}^{m} b_{ij}^s X_j, \quad i = 1, \dots m,\ s = 1, \dots k.$$

The above information will be useful in computing the geometric quantities found in the sequel. Recall a definition of the Riemannian curvature tensor (cf., for example, [23] for more details).

Definition 5.3. Let X, Y, Z, and W be vectors in TM, where M is a Riemannian manifold. Let ∇ be the Levi–Civita connection on TM with respect to the metric $\langle\, , \,\rangle$. Then

$$\mathrm{Rm}(X, Y, Z, W) = \langle R(X, Y)Z, W\rangle,$$

where $R(X, Y)Z = \nabla_X \nabla_Y Z - \nabla_Y \nabla_X Z - \nabla_{[X,Y]} Z$.

Recall the symmetries of the Riemannian curvature tensor.

Proposition 5.4. *The Riemannian curvature tensor has the following symmetries for any vector fields X, Y, Z, and W:*

(a) $\mathrm{Rm}(W, X, Y, Z) = -\mathrm{Rm}(X, W, Y, Z)$,

(b) $\mathrm{Rm}(W, X, Y, Z) = -\mathrm{Rm}(W, X, Z, Y)$,

(c) $\mathrm{Rm}(W, X, Y, Z) = \mathrm{Rm}(Y, Z, W, X)$.

Lemma 5.1 allows one to compute the Riemannian curvature tensor with respect to the metric $\langle\, , \,\rangle$.

Lemma 5.5. *Let $\langle\, , \,\rangle$ be the metric which makes $\{X_1, \dots, X_m, T_1, \dots, T_k\}$ an orthonormal basis. Then the following are the only nonzero quantities of the Riemannian curvature tensor:*

$$\mathrm{Rm}(X_i, X_m, X_j, X_l) = \frac{1}{4}(2b_{im}^q b_{jl}^q + b_{ij}^q b_{ml}^q - b_{mj}^q b_{il}^q),$$

$$\mathrm{Rm}(X_i, X_m, T_r, T_s) = \frac{1}{4}(b_{il}^r b_{ml}^s - b_{mk}^r b_{ik}^s),$$

$$\mathrm{Rm}(T_r, X_m, X_j, T_s) = \frac{1}{4} b_{jl}^r b_{ml}^s.$$

In particular, in $\mathbb{H}^n$, for $i = 1, \ldots, n$

$$\begin{aligned}
\mathrm{Rm}(X_i, X_{n+i}, X_{n+i}, X_i) &= -\frac{3}{4}, \\
\mathrm{Rm}(X_i, T, T, X_i) &= \frac{1}{4}, \\
\mathrm{Rm}(X_{n+i}, T, T, X_{n+i}) &= \frac{1}{4}.
\end{aligned}$$

Lemma 5.2 allows one to compute the Riemannian curvature tensor with respect to the metric $\langle\ ,\ \rangle_\epsilon$.

Lemma 5.6. *Let g_ϵ be the metric which makes $\{X_1, \ldots, X_m, T_{1,\epsilon}, \ldots, T_{k,\epsilon}\}$ an orthonormal basis. Then the following are the only nonzero quantities of the Riemannian curvature tensor:*

$$\begin{aligned}
\mathrm{Rm}^\epsilon(X_i, X_m, X_j, X_l) &= \frac{1}{4\epsilon}(2b^q_{im}b^q_{jl} + b^q_{ij}b^q_{ml} - b^q_{mj}b^q_{il}), \\
\mathrm{Rm}^\epsilon(X_i, X_m, T_{r,\epsilon}, T_{s,\epsilon}) &= \frac{1}{4\epsilon}(b^r_{il}b^s_{ml} - b^r_{mk}b^s_{ik}), \\
\mathrm{Rm}^\epsilon(T_{r,\epsilon}, X_m, X_j, T_{s,\epsilon}) &= \frac{1}{4\epsilon}b^r_{jl}b^s_{ml}.
\end{aligned}$$

In particular, in $\mathbb{H}^n$, for $i = 1, \ldots, n$

$$\begin{aligned}
\mathrm{Rm}^\epsilon(X_i, X_{n+i}, X_{n+i}, X_i) &= -\frac{3}{4\epsilon}, \\
\mathrm{Rm}^\epsilon(X_i, T_\epsilon, T_\epsilon, X_i) &= \frac{1}{4\epsilon}, \\
\mathrm{Rm}^\epsilon(X_{n+i}, T_\epsilon, T_\epsilon, X_{n+i}) &= \frac{1}{4\epsilon}.
\end{aligned}$$

The following lemma will be used in the proof of a second variation formula for the H-perimeter in $\mathbb{H}^n$.

Lemma 5.7. *Let E_i, $i = 1, \ldots, 2n-1$, F^ϵ be as previously defined for the Heisenberg group $\mathbb{H}^n$. Then*

$$\begin{aligned}
\sum_{i=1}^{2n-1} \mathrm{Rm}^\epsilon(\nu_\epsilon, E_i, E_i, \nu_\epsilon) &= \overline{\omega}^2\frac{2n-1}{4} + \frac{3}{4}\overline{\omega}^2\left(\frac{W}{W_\epsilon}\right)^2 \\
&\quad - \frac{3}{4\epsilon} - \epsilon\overline{\omega}^4\left(\frac{W}{W_\epsilon}\right)^2\frac{2n-1}{4} \qquad (5.1)
\end{aligned}$$

and

$$\mathrm{Rm}^\epsilon(\nu_\epsilon, F^\epsilon, F^\epsilon, \nu_\epsilon) = \frac{1}{4\epsilon} - \frac{1}{2}\overline{\omega}^2\left(\frac{W}{W_\epsilon}\right)^2 + \frac{1}{2}\overline{\omega}^2\left(\frac{W}{W_\epsilon}\right)^4 + \epsilon\frac{1}{4}\overline{\omega}^4\left(\frac{W}{W_\epsilon}\right)^4. \tag{5.2}$$

Proof. The assertion follows directly by definition and Lemma 5.6. We only prove (5.2) as an illustration.

$$\begin{aligned}
&\sum_{i=1}^{2n-1} \mathrm{Rm}^\epsilon(\nu_\epsilon, F^\epsilon, F^\epsilon, \nu_\epsilon)\\
&= \epsilon\left(\frac{W}{W_\epsilon}\right)^2 \mathrm{Rm}^\epsilon(\nu_\epsilon, Y^\perp, Y^\perp, \nu_\epsilon)\\
&= \epsilon\left(\frac{W}{W_\epsilon}\right)^2 (\langle \nu_\epsilon, T_\epsilon\rangle_\epsilon{}^2 \langle Y^\perp, X_r\rangle^2 \,\mathrm{Rm}^\epsilon(T_\epsilon, X_r, X_r, T_\epsilon)\\
&+ \langle \nu_\epsilon, T_\epsilon\rangle_\epsilon{}^2 \langle Y^\perp, X_{n+r}\rangle^2 \,\mathrm{Rm}^\epsilon(T_\epsilon, X_{n+r}, X_{n+r}, T_\epsilon)\\
&+ \langle \nu_\epsilon, X_r\rangle^2 \langle Y^\perp, T_\epsilon\rangle_\epsilon{}^2 \,\mathrm{Rm}^\epsilon(X_r, T_\epsilon, T_\epsilon, X_r)\\
&+ \langle \nu_\epsilon, X_r\rangle^2 \langle Y^\perp, X_{n+r}\rangle^2 \,\mathrm{Rm}^\epsilon(X_r, X_{n+r}, X_{n+r}, X_r)\\
&+ \langle \nu_\epsilon, X_{n+r}\rangle^2 \langle Y^\perp, X_r\rangle^2 \,\mathrm{Rm}^\epsilon(X_{n+r}, X_r, X_r, X_{n+r})\\
&+ 2\langle \nu_\epsilon, T_\epsilon\rangle_\epsilon \langle \nu_\epsilon, X_r\rangle \langle Y^\perp, T_\epsilon\rangle_\epsilon \langle Y^\perp, X_r\rangle \,\mathrm{Rm}^\epsilon(T_\epsilon, X_r, T_\epsilon, X_r)\\
&+ 2\langle \nu_\epsilon, T_\epsilon\rangle_\epsilon \langle \nu_\epsilon, X_{n+r} > \langle Y^\perp, T_\epsilon\rangle_\epsilon \langle Y^\perp, X_{n+r}\rangle \,\mathrm{Rm}^\epsilon(T_\epsilon, X_{n+r}, T_\epsilon, X_{n+r})\\
&+ 2\langle \nu_\epsilon, X_r\rangle \langle \nu_\epsilon, X_{n+r}\rangle \langle Y^\perp, X_r\rangle \langle Y^\perp, X_{n+r}\rangle \,\mathrm{Rm}^\epsilon(X_r, X_{n+r}, X_r, X_{n+r}))\\
&= \left(\frac{W}{W_\epsilon}\right)^4 \frac{1}{4\epsilon} + \frac{1}{2}\overline{\omega}^2\left(\frac{W}{W_\epsilon}\right)^4 + \frac{\epsilon}{4}\overline{\omega}^4\left(\frac{W}{W_\epsilon}\right)^4\\
&= \frac{1}{4\epsilon} - \frac{1}{2}\overline{\omega}^2\left(\frac{W}{W_\epsilon}\right)^2 + \frac{1}{2}\overline{\omega}^2\left(\frac{W}{W_\epsilon}\right)^4 + \frac{\epsilon}{4}\overline{\omega}^4\left(\frac{W}{W_\epsilon}\right)^4
\end{aligned}$$

by Lemma 4.4. □

Consider the decomposition of a vector V into the ϵ-tangential and ϵ-normal components. Let $V \in C_0^1(\mathcal{U}\backslash\Sigma)$. Define

$$\begin{aligned}
V_\epsilon = V &= \langle V, \mathcal{E}_i^\epsilon\rangle_\epsilon \mathcal{E}_i^\epsilon + \langle V, \nu_\epsilon\rangle_\epsilon \nu_\epsilon\\
&= \sum_{i=1}^{m-1} \langle V, E_i\rangle E_i + \langle V, F_s^\epsilon\rangle_\epsilon F_s^\epsilon + \langle V, \nu_\epsilon\rangle_\epsilon \nu_\epsilon.
\end{aligned}$$

Proposition 5.8. *Let V_ϵ be defined as above. Then*

$$V = \lim_{\epsilon\to 0} V_\epsilon = \langle V, E_i\rangle E_i + \langle V, T_s\rangle Y_s^\perp + \langle V, Y\rangle \nu_H.$$

The convergence is uniform in ϵ on relatively compact subsets of $\mathcal{U}\backslash\Sigma$.

Proof. First, consider the term $\langle V, \nu_\epsilon\rangle_\epsilon \nu_\epsilon$:

$$\begin{aligned}\langle V, \nu_\epsilon\rangle_\epsilon \nu_\epsilon &= \left(\langle V, X_i\rangle\langle \nu_\epsilon, X_i\rangle + \langle V, T_{s,\epsilon}\rangle_\epsilon \langle T_{s,\epsilon}, \nu_\epsilon\rangle_\epsilon\right)\nu_\epsilon \\ &= \frac{W}{W_\epsilon}\left(\langle V, \nu_H\rangle + \langle V, T_s\rangle \overline{\omega}_s\right)\nu_\epsilon.\end{aligned}$$

Therefore,

$$\lim_{\epsilon\to 0}\langle V, \nu_\epsilon\rangle_\epsilon \nu_\epsilon = \lim_{\epsilon\to 0}\left(\frac{W}{W_\epsilon}\langle V, \nu_H\rangle + \frac{W}{W_\epsilon}\overline{\omega}_s\langle V, T_s\rangle\right)\nu_\epsilon = \langle V, Y\rangle \nu_H.$$

(Note that $\frac{W}{W_\epsilon}$ converges to 1 uniformly as $\epsilon \to 0$ on compact subsets of $\mathcal{U}\backslash\Sigma$.) Consider $\langle V, F_s^\epsilon\rangle_\epsilon F_s^\epsilon$:

$$\begin{aligned}&\langle V, F_s^\epsilon\rangle F_s^\epsilon = \left(\langle V, X_i\rangle\langle X_i, F_s^\epsilon\rangle + \langle V, T_{r,\epsilon}\rangle_\epsilon \langle T_{r,\epsilon}, F_s^\epsilon\rangle_\epsilon\right)F_s^\epsilon \\ &= \left(-\sqrt{\epsilon}\frac{\omega_s\, p_i}{\tau_{s,\epsilon}\tau_{s+1,\epsilon}}\langle V, X_i\rangle - \sqrt{\epsilon}\sum_{r\langle s}\langle V, T_r\rangle \frac{\omega_r\omega_s}{\tau_{s,\epsilon}\tau_{s+1,\epsilon}} + \frac{1}{\sqrt{\epsilon}}\langle V, T_s\rangle \frac{\tau_{s,\epsilon}}{\tau_{s+1,\epsilon}}\right)F_s^\epsilon;\end{aligned}$$

F_s^ϵ has no $1/\epsilon^r$, $r > 0$ terms in it. Thus, one may ignore the terms in the parentheses that converge to zero since only the limit is being considered. However, it should be noted that these terms do converge uniformly on relatively compact subsets of $\mathcal{U}\backslash\Sigma$. What remains is

$$\frac{1}{\sqrt{\epsilon}}\frac{\langle V, T_s\rangle}{(\tau_{s+1,\epsilon})^2}\Big(-\sqrt{\epsilon}\omega_s p_i X_i - (\epsilon)^{\frac{3}{2}}\omega_s\sum_{r<s}\omega_r T_r + \sqrt{\epsilon}\,(\tau_{s,\epsilon})^2\, T_s\Big).$$

Therefore,

$$\lim_{\epsilon\to 0}\langle V, F_s^\epsilon\rangle_\epsilon F_s^\epsilon = -\langle V, T_s\rangle \overline{\omega}_s \nu_H + \langle V, T_s\rangle T_s = \langle V, T_s\rangle Y_s^\perp,$$

and the limit is uniform on relatively compact subsets of $\mathcal{U}\backslash\Sigma$. □

The limits of other geometric quantities can also be computed.

Definition 5.9. Let M be an n-dimensional oriented hypersurface in a Riemannian manifold $\mathbb{G}$, and let $\{\varepsilon_1, \ldots, \varepsilon_n\}$ be an orthonormal basis for TM in a neighborhood of a point $x \in M$. Let ∇ be the Levi–Civita connection on $T\mathbb{G}$. Then the *tangential divergence* at x of a C^1 vector field V on M is given by

$$\operatorname{div}_M V = \sum_{i=1}^{n}\langle \nabla_{\varepsilon_i} V, \varepsilon_i\rangle.$$

Definition 5.10. Let M, $\mathbb{G}$, ∇ be as in Definition 5.9,. Assume that M is C^2. Let ν be the unit normal to M at x. Then the mean curvature of M at x is given by

$$H = \mathrm{div}_M \nu.$$

We now introduce the sub-Riemannian counterpart of the various geometric quantities so far discussed.

Definition 5.11. Let E_i, $i = 1, \ldots, m-1$, be an orthonormal basis for HT_xM. The *horizontal tangential divergence* of a C^1 vector field V is given by

$$\mathrm{div}_{H,M} V = \sum_{i=1}^{m-1} \langle \nabla_{E_i} V, E_i \rangle,$$

where ∇ is the Levi–Civita connection with respect to the metric $\langle \, , \, \rangle$ which makes $\{X_1, \ldots, X_m, T_1, \ldots, T_k\}$ an orthonormal basis for $\mathbb{G}$.

Definition 5.12. Let M be a C^2 hypersurface in $\mathbb{G}$. Then the *horizontal mean curvature* of M at the point $x \in M \backslash \Sigma$ is given by

$$\mathcal{H} = \mathrm{div}_{H,M} \nu_H.$$

Theorem 5.13. *Let*

$$V = V^i X_i + \frac{1}{\sqrt{\epsilon}} \langle V, T_s \rangle T_{s,\epsilon},$$

where $V^i = \langle V, X_i \rangle_\epsilon$, $V \in C_0^1(\mathcal{U} \backslash \Sigma)$. *Then*

$$\lim_{\epsilon \to 0} \mathrm{div}_{\epsilon,M} V = \mathrm{div}_{H,M} V + Y_s^{\perp} \left(\langle V, T_s \rangle \right) + \overline{\omega}_s \langle Z^s, V \rangle,$$

and the convergence is uniform in ϵ *on relatively compact subsets of* $\mathcal{U} \backslash \Sigma$.

Proof. By definition,

$$\mathrm{div}_{\epsilon,M} V_\epsilon = \sum_{i=1}^{m-1} \langle \nabla^\epsilon{}_{E_i} V, E_i \rangle_\epsilon + \sum_{s=1}^{k} \langle \nabla^\epsilon{}_{F_s^\epsilon} V, F_s^\epsilon \rangle_\epsilon = I_\epsilon + II_\epsilon.$$

Now, compute

$$I_\epsilon = \sum_{i=1}^{m-1} E_i(V^k) E_{i,k} + E_i \Big(\frac{1}{\sqrt{\epsilon}} \langle V, T_s \rangle \Big) \langle T_{s,\epsilon}, E_i \rangle_\epsilon$$

$$+ V^k \langle \nabla^\epsilon{}_{E_i} X_k, E_i \rangle_\epsilon + \frac{1}{\sqrt{\epsilon}} \langle V, T_s \rangle \langle \nabla^\epsilon{}_{E_i} T_{s,\epsilon}, E_i \rangle_\epsilon$$

$$= \sum_{i=1}^{m-1} E_i(V^k)E_{i,k} + \frac{1}{\sqrt{\epsilon}}\langle V, T_s\rangle E_{i,r}E_{i,q}\langle \nabla^\epsilon{}_{X_r} T_{s,\epsilon}, X_q\rangle_\epsilon$$
$$= \langle \nabla_{E_i} V, E_i\rangle + \frac{1}{\sqrt{\epsilon}}\langle V, T_s\rangle E_{i,r}E_{i,q}\Big(-\frac{1}{2\sqrt{\epsilon}} b^s_{rq}\Big)$$
$$= \langle \nabla_{E_i} V, E_i\rangle \text{ (by antisymmetry of } b^s_{rq})$$
$$= \mathrm{div}_{H,M} V.$$

One may also compute

$$II_\epsilon = \sum_{s=1}^{k} \langle \nabla^\epsilon{}_{F_s^\epsilon} V, F_s^\epsilon\rangle_\epsilon$$
$$= F_s^\epsilon(V_i)\langle X_i, F_s^\epsilon\rangle_\epsilon + F_s^\epsilon\Big(\frac{1}{\sqrt{\epsilon}}\langle V, T_r\rangle\Big)\langle T_{r,\epsilon}, F_s^\epsilon\rangle_\epsilon$$
$$+ V^i\langle \nabla^\epsilon{}_{F_s^\epsilon} X_i, F_s^\epsilon\rangle_\epsilon + \frac{1}{\sqrt{\epsilon}}\langle V, T_r\rangle\langle \nabla^\epsilon{}_{F_s^\epsilon} T_{r,\epsilon}, F_s^\epsilon\rangle_\epsilon$$
$$= A_\epsilon + B_\epsilon + C_\epsilon + D_\epsilon.$$

One can easily see that $A_\epsilon \to 0$. Consider B_ϵ:

$$B_\epsilon = \frac{1}{\sqrt{\epsilon}} F_s^\epsilon\left(\langle V, T_r\rangle\right)\langle T_{r,\epsilon}, F_s^\epsilon\rangle_\epsilon$$
$$= \left(-\omega_s p_i X_i\left(\langle V, T_r\rangle\right) - \epsilon\omega_q\omega_s T_q\left(\langle V, T_r\rangle\right) + \tau_{s,\epsilon}^2 T_s\left(\langle V, T_r\rangle\right)\right)\frac{\langle T_{r,\epsilon}, F_s^\epsilon\rangle_\epsilon}{\tau_{s,\epsilon}\tau_{s+1,\epsilon}}.$$

By the definition of F_s^ϵ, the above quantity converges to zero uniformly on relatively compact subsets of $\mathcal{U}\backslash\Sigma$ unless $r = s$. In that case,

$$\langle T_{s,\epsilon}, F_{s,\epsilon}\rangle_\epsilon = \frac{\tau_{s,\epsilon}}{\tau_{s+1,\epsilon}}$$

and

$$\lim_{\epsilon\to 0} B_\epsilon = -\overline{\omega}_s \nu_H\left(\langle V, T_s\rangle\right) + T_s\left(\langle V, T_s\rangle\right) = Y_s^\perp\left(\langle V, T_s\rangle\right).$$

Next,

$$C_\epsilon = V^i\langle F_s^\epsilon, X_m\rangle\langle F_s^\epsilon, X_l\rangle\langle \nabla^\epsilon{}_{X_m} X_i, X_l\rangle$$
$$+ V^i\langle F_s^\epsilon, T_{q,\epsilon}\rangle_\epsilon\langle F_s^\epsilon, T_{r,\epsilon}\rangle_\epsilon\langle \nabla^\epsilon{}_{T_{q,\epsilon}} X_i, T_{r,\epsilon}\rangle_\epsilon$$
$$+ V^i\langle F_s^\epsilon, X_m\rangle\langle F_s^\epsilon, T_{q,\epsilon}\rangle_\epsilon\langle \nabla^\epsilon{}_{X_m} X_i, T_{q,\epsilon}\rangle_\epsilon$$
$$+ V^i\langle F_s^\epsilon, T_{q,\epsilon}\rangle_\epsilon\langle F_s^\epsilon, X_m\rangle\langle \nabla^\epsilon{}_{T_{q,\epsilon}} X_i, X_m\rangle_\epsilon$$

$$= \frac{V^i}{\sqrt{\epsilon}} \langle F_s^\epsilon, X_m \rangle \langle F_s^\epsilon, T_{q,\epsilon} \rangle_\epsilon b_{mi}^q$$

$$= -V^i \frac{\omega_s p_m \tau_{s,\epsilon}^2 b_{mi}^s}{\tau_{s,\epsilon}^2 \tau_{s+1,\epsilon}^2} + \sum_{q=1}^{s-1} \epsilon \frac{\omega_s \omega_q \omega_s p_m b_{mi}^q}{\tau_{s,\epsilon} \tau_{s+1,\epsilon}}$$

Thus,

$$\lim_{\epsilon \to 0} C_\epsilon = V^i \overline{\omega}_s \overline{p}_m b_{im}^s = \overline{\omega}_s \langle Z^s, V \rangle,$$

and the convergence is uniform in ϵ on relatively compact subsets of $\mathcal{U} \backslash \Sigma$. Finally,

$$D_\epsilon = \frac{1}{\sqrt{\epsilon}} \langle V, T_r \rangle \langle F_s^\epsilon, X_m \rangle \langle F_s^\epsilon, X_l \rangle \langle \nabla^\epsilon{}_{X_l} T_{r,\epsilon}, X_m \rangle_\epsilon$$

$$= \frac{1}{\sqrt{\epsilon}} \langle V, T_r \rangle \left(\frac{\sqrt{\epsilon}\omega_s p_m}{\tau_{s,\epsilon}\tau_{s+1,\epsilon}} \right) \left(\frac{\sqrt{\epsilon}\omega_s p_l}{\tau_{s,\epsilon}\tau_{s+1,\epsilon}} \right) \left(\frac{1}{2\sqrt{\epsilon}} b_{lm}^r \right)$$

$$\equiv 0 \quad \text{[by the antisymmetry of } b_{lm}^r\text{]}.$$

This concludes the proof. □

Proposition 5.14. *Let H_ϵ be the Riemannian mean curvature of $M \subset \mathbb{G}$ with respect to the metric $\langle\ ,\ \rangle_\epsilon$. Then*

$$\lim_{\epsilon \to 0} H_\epsilon = \mathcal{H},$$

and the convergence is uniform for relatively compact subsets of $\mathcal{U} \backslash \Sigma$.

Proof. Take $V = \nu_\epsilon$ in the proof of Theorem 5.13. Then

$$\frac{1}{\sqrt{\epsilon}} \langle \nu_\epsilon, T_s \rangle = \sqrt{\epsilon}\overline{\omega}_s.$$

The proof proceeds as the proof of Theorem 5.13. □

The following "horizontal divergence theorem" can be obtained from Theorem 5.13 and Proposition 5.8:

Corollary 5.15. *Let $M \subset \mathcal{U} \subset \mathbb{G}$. Then if $V \in C_0(\mathcal{U} \backslash \Sigma)$,*

$$\int_M \operatorname{div}_{H,M} V + Y^\perp (\langle V, T_s \rangle) + \overline{\omega}_s \langle Z^s, V \rangle \, d\sigma_H = \int_M \mathcal{H} \langle V, Y \rangle \, d\sigma_H. \tag{5.3}$$

Proof. The Riemannian divergence theorem states

$$\int_M \operatorname{div}_{\epsilon,M} V \, d\sigma^\epsilon = \int_M H_\epsilon \langle V, \nu_\epsilon \rangle_\epsilon \, d\sigma^\epsilon .$$

The conclusion follows from Theorem 5.13 and Propositions 5.14 and 3.2. □

Definition 5.16. The *horizontal Laplace–Beltrami operator* is defined by

$$\Delta_{H,M} f = \operatorname{div}_{H,M} \nabla_H^M f = E_i \left(E_i f \right).$$

One can also obtain the following integration by parts formulas:

Corollary 5.17. *Let $f \in C_0^1(\mathcal{U} \backslash \Sigma(M))$. Then*

$$\int_M E_i(f) + f \sum_{r=1}^{m-1} \langle \nabla_{E_r} E_i, E_r \rangle \, d\sigma_H = - \int_M \overline{\omega}_s f \langle Z^s, E_i \rangle \, d\sigma_H, \tag{5.4}$$

$$\int_M Y_s^\perp f \, d\sigma_H = \int_M \overline{\omega}_s f \mathcal{H} \, d\sigma_H, \tag{5.5}$$

$$\int_M Z^s f \, d\sigma_H = - \int_M \overline{\omega}_s f |Z^s| \, d\sigma_H - \int_M f \operatorname{div}_{H,M} Z^s \, d\sigma_H. \tag{5.6}$$

In particular, if $g \in C_0^2(\mathcal{U} \backslash \Sigma(M))$,

$$\int_M \Delta_{H,M} g \, d\sigma_H = - \int \overline{\omega}_s Z^s g \, d\sigma_H - \int_M E_i(f) \sum_{r=1}^{m-1} \langle \nabla_{E_r} E_i, E_r \rangle \, d\sigma_H. \tag{5.7}$$

Proof. First, one may recognize that

$$\operatorname{div}_{H,M} V = E_i \left(\langle V, E_i \rangle \right) + \langle V, E_m \rangle \langle \nabla_{E_i} E_m, E_i \rangle + \mathcal{H} \langle V, \nu_H \rangle. \tag{5.8}$$

This can be seen in the following calculation which uses Lemma 5.1:

$$\begin{aligned} \operatorname{div}_{H,M} V &= \langle \nabla_{E_i} V, E_i \rangle \\ &= E_i \left(\langle V, E_i \rangle \right) + \langle V, E_m \rangle \langle \nabla_{E_i} E_m, E_i \rangle \\ &\quad + E_i \left(\langle V, T_r \rangle \right) \langle T_r, E_i \rangle + \langle V, T_r \rangle \langle \nabla_{E_i} T_r, E_i \rangle \\ &\quad + E_i \left(\langle V, \nu_H \rangle \right) \langle \nu_H, E_i \rangle + \langle V, \nu_H \rangle \langle \nabla_{E_i} \nu_H, E_i \rangle \\ &= E_i \left(\langle V, E_i \rangle \right) + \langle V, E_m \rangle \langle \nabla_{E_i} E_m, E_i \rangle + \mathcal{H} \langle V, \nu_H \rangle. \end{aligned}$$

(This used the fact that $\{E_1, \ldots E_{m-1}, \nu_H, T_1, \ldots, T_k\}$ is an orthonormal basis.) Plugging this information into equation (5.3), one obtains

$$\int_M E_i \left(\langle V, E_i \rangle \right) + \langle V, E_m \rangle \langle \nabla_{E_i} E_m, E_i \rangle$$

$$+Y_s^{\perp}(\langle V,T_s\rangle)+\overline{\omega}_s\langle Z^s,V\rangle\,d\sigma_H=\int_M\overline{\omega}_s\mathcal{H}\langle V,T_s\rangle\,d\sigma_H. \tag{5.9}$$

Since $Z^s\in HTM$, we have $Z^s=\langle Z^s,E_i\rangle E_i$. Setting $V=fE_k$, we obtain (5.4). To obtain (5.5), let $V=fT_s$. For (5.6) we set $V=fZ^s$. Finally, for 5.7 we set $f=E_ig$ in (5.4). □

Remark 5.18. It is trivial to see that for $\mathbb{H}^1$, $\mathrm{div}_{H,M}Z=0$. Further, for the Heisenberg group, $\mathbb{H}^n$, $|Z|=1$.

Next, recall the definition of the second fundamental form.

Definition 5.19. Let M be an n-dimensional oriented hypersurface of a Riemannian manifold $\mathbb{G}$ and ε_i, $i=1,\ldots,n$ be an orthonormal basis for TM at a point $x\in M$. Let ∇ be the Levi–Civita connection on $T\mathbb{G}$, and let ν be the unit normal. Then the *second fundamental form* is given by the $n\times n$ symmetric matrix

$$A_{i,j}=\langle\nabla_{\varepsilon_i}\nu,\varepsilon_j\rangle.$$

The following notion of a horizontal second fundamental form was given in [18]. It is equivalent to previous definitions found in works such as [10].

Definition 5.20. Let M be an oriented hypersurface in $\mathbb{G}$. Let E_i, $i=1,\ldots,m-1$, be an orthonormal basis for HT_xM. Then the *horizontal second fundamental form* is the $(m+k-1)\times(m+k-1)$ matrix

$$A^H_{ij}=\langle\nabla_{E_i}\nu_H,E_j\rangle.$$

Note that the horizontal second fundamental form is not, in general, symmetric. Note that

$$\langle\nabla_{E_i}\nu_H,E_j\rangle-\langle\nabla_{E_j}\nu_H,E_i\rangle=\langle\,[E_i,E_j],\nu_H\rangle.$$

The quantity $[E_i,E_j]$ is a tangent vector, but it is not necessarily horizontal. Therefore, one cannot conclude that $\langle[E_i,E_j],\nu_H\rangle=0$. This will be discussed in more detail later.

The following quantities will be very useful in following calculations.

Lemma 5.21. *The following statements hold:*

i. $\langle\nabla^{\epsilon}{}_{X_j}\nu_\epsilon,X_k\rangle_\epsilon=X_j\left(\overline{p}_k\dfrac{W}{W_\epsilon}\right)+\dfrac{1}{2}\dfrac{W}{W_\epsilon}\overline{\omega}_r b^r_{kj}$,

ii. $\langle\nabla^{\epsilon}{}_{T_{s,\epsilon}}\nu_\epsilon,X_k\rangle_\epsilon=\sqrt{\epsilon}T_s\left(\overline{p}_k\dfrac{W}{W_\epsilon}\right)+\dfrac{1}{2\sqrt{\epsilon}}\dfrac{W}{W_\epsilon}\overline{p}_l b^s_{kl}$,

iii. $\langle\nabla^{\epsilon}{}_{X_j}\nu_\epsilon,T_{r,\epsilon}\rangle_\epsilon=\sqrt{\epsilon}X_j\left(\overline{\omega}_r\dfrac{W}{W_\epsilon}\right)+\dfrac{1}{2\sqrt{\epsilon}}\dfrac{W}{W_\epsilon}b^r_{jk}$,

iv. $\langle \nabla^{\epsilon}{}_{T_{s,\epsilon}} \nu_\epsilon, T_{r,\epsilon} \rangle_\epsilon = \epsilon T_s \left(\overline{\omega}_r \frac{W}{W_\epsilon} \right)$.

Proof. These statements follow directly from the definitions. We only prove assertion i:

$$\begin{aligned}\langle \nabla^{\epsilon}{}_{X_j} \nu_\epsilon, X_k \rangle_\epsilon &= X_j \left(\overline{p}_k \frac{W}{W_\epsilon} \right) + \overline{p}_m \frac{W}{W_\epsilon} \langle \nabla^{\epsilon}{}_{X_j} X_m, X_k \rangle_\epsilon \\ &\quad + X_j \left(\sqrt{\epsilon}\, \overline{\omega}_s \frac{W}{W_\epsilon} \right) \langle T_{s,\epsilon}, X_k \rangle_\epsilon + \sqrt{\epsilon}\, \overline{\omega}_r \frac{W}{W_\epsilon} \langle \nabla^{\epsilon}{}_{X_j} T_{r,\epsilon}, X_k \rangle_\epsilon \\ &= X_j \left(\overline{p}_k \frac{W}{W_\epsilon} \right) + \frac{1}{2} \frac{W}{W_\epsilon} \overline{\omega}_r b^r_{kj}.\end{aligned}$$

The proof is complete. □

Introduce the notation which will be useful in the further calculations. Consider J_ϵ. It is written as

$$J_\epsilon = (\epsilon)^s J^s,$$

where J^s is independent of ϵ. For example, if

$$J_\epsilon = \epsilon(2+x) + \sqrt{\epsilon} \left(\frac{2}{\sqrt{1-x^2}} \right),$$

then,

$$J^1 = 2 + x, \quad J^{1/2} = \frac{2}{\sqrt{1-x^2}}.$$

Now, the second fundamental form will be computed with respect to the ϵ-metric. To simplify the notation, we focus on the model case $\mathbb{H}^n$. However, the calculations can be generalized to Carnot groups of step 2.

Proposition 5.22. *Let M be a C^2 oriented hypersurface in $\mathbb{H}^n$. Then*

$$A^\epsilon_{i,j} = \begin{cases} \dfrac{W}{W_\epsilon} A^0_{i,j}, & i,j = 1, \dots 2n-1, \\ \sqrt{\epsilon} A^{1/2}_{i,j} + \frac{1}{\sqrt{\epsilon}} A^{-1/2}_{i,j} + o\left(\epsilon^{3/2}\right), & i = 1, \dots 2n-1, j = 2n, \\ \epsilon \dfrac{W}{W_\epsilon} A^1_{i,j} + o\left(\epsilon^2\right), & i,j = 2n, \end{cases}$$

where

$$A^0_{i,j} = A^H_{i,j} + \frac{1}{2} \overline{\omega} \xi_{ij}$$

for $i,j = 1, \dots 2n-1$, and for $i = 1, \dots 2n-1$

$$A^{1/2}_{i,2n} = E_i\left(\overline{\omega}\right) = A^{1/2}_{2n,i} = Y^{\perp}\left(\overline{p}_j\right) E_{i,j} - \overline{\omega}^2 \langle Z, E_i\rangle,$$

$$A^{-1/2}_{i,2n} = \frac{1}{2}\langle Z, E_i\rangle,$$

and

$$A^{1}_{2n,2n} = Y^{\perp}\left(\overline{\omega}\right) = \langle \nabla_{Y^{\perp}} Y, T\rangle.$$

Proof. Lemma 5.21 is used throughout the proof. First take $i, j = 1, \dots m-1$:

$$\begin{aligned}
A^{\epsilon}_{i,j} &= \langle \nabla^{\epsilon}{}_{E_i}\nu_{\epsilon}, E_j\rangle_{\epsilon} = E_{i,k}E_{j,l}\langle \nabla^{\epsilon}{}_{X_k}\nu_{\epsilon}, X_l\rangle_{\epsilon} \\
&= E_{i,k}E_{j,l}\Big(X_k\Big(\overline{p}_l\frac{W}{W_{\epsilon}}\Big) + \frac{1}{2}\frac{W}{W_{\epsilon}}\overline{\omega} b_{lk}\Big) \\
&= E_{j,l}E_i\Big(\overline{p}_l\frac{W}{W_{\epsilon}}\Big) + \frac{1}{2}\frac{W}{W_{\epsilon}}\overline{\omega}\xi_{ij} = \frac{W}{W_{\epsilon}}\Big(E_{j,l}E_i\Big(\overline{p}_l\Big) + \frac{1}{2}\overline{\omega}\xi_{ij}\Big) \\
&= \frac{W}{W_{\epsilon}}\Big(\langle \nabla_{E_i}\nu_H, E_j\rangle + \frac{1}{2}\overline{\omega}\xi_{ij}\Big) = \frac{W}{W_{\epsilon}}\Big(A^H_{ij} + \frac{1}{2}\overline{\omega}\xi_{ij}\Big).
\end{aligned}$$

Next, let $i = 1, \dots m-1$. Then

$$\begin{aligned}
A^{\epsilon}_{i,2n} &= \langle \nabla^{\epsilon}{}_{E_i}\nu_{\epsilon}, F^{\epsilon}\rangle_{\epsilon} \\
&= \sqrt{\epsilon}\frac{W}{W_{\epsilon}}\Big(E_{i,k}\langle Y^{\perp}, X_l\rangle\langle \nabla^{\epsilon}{}_{X_k}\nu_{\epsilon}, X_l\rangle_{\epsilon} + E_{i,k}\langle Y^{\perp}, T_{\epsilon}\rangle_{\epsilon}\langle \nabla^{\epsilon}{}_{X_k}\nu_{\epsilon}, T_{\epsilon}\rangle_{\epsilon}\Big) \\
&= \sqrt{\epsilon}\frac{W}{W_{\epsilon}}\Big(-\overline{\omega}E_i\Big(\frac{W}{W_{\epsilon}}\Big) - \overline{\omega}\frac{W}{W_{\epsilon}}\overline{p}_j E_i(\overline{p}_j) + \frac{1}{2}\frac{W}{W_{\epsilon}}\overline{\omega}^2\langle Z, E_i\rangle \\
&\qquad \overline{\omega}E_i\Big(\frac{W}{W_{\epsilon}}\Big) + \frac{W}{W_{\epsilon}}E_i(\overline{\omega}) + \frac{1}{2\epsilon}\frac{W}{W_{\epsilon}}\langle Z, E_i\rangle\Big) \\
&= \Big(\frac{W}{W_{\epsilon}}\Big)^2\Big(\sqrt{\epsilon}\Big(E_i(\overline{\omega}) + \frac{1}{2}\overline{\omega}^2\langle Z, E_i\rangle\Big) + \frac{1}{\sqrt{\epsilon}}\Big(\frac{1}{2}\langle Z, E_i\rangle\Big)\Big) \\
&= \Big(1 - \epsilon\overline{\omega}^2\Big(\frac{W}{W_{\epsilon}}\Big)^2\Big)\Big(\sqrt{\epsilon}\Big(E_i(\overline{\omega}) + \frac{1}{2}\overline{\omega}^2\langle Z, E_i\rangle\Big) + \frac{1}{\sqrt{\epsilon}}\Big(\frac{1}{2}\langle Z, E_i\rangle\Big)\Big) \\
&= \sqrt{\epsilon}E_i\Big(\overline{\omega}\Big) + \frac{1}{\sqrt{\epsilon}}\Big(\frac{1}{2}\langle Z, E_i\rangle\Big) \\
&\quad + \epsilon^{3/2}\Big(\frac{W}{W_{\epsilon}}\Big)^2\Big(- \overline{\omega}^2\Big(E_i(\overline{\omega}) + \frac{1}{2}\overline{\omega}^2\langle Z, E_i\rangle\Big) + \frac{1}{2}\overline{\omega}^4\langle Z, E_i\rangle\Big).
\end{aligned}$$

The second equality is obtained by a similar calculation for $\langle \nabla^{\epsilon}{}_{F^{\epsilon}}\nu_{\epsilon}, E_i\rangle_{\epsilon}$. Finally,

$$\begin{aligned}
A^{\epsilon}_{2n,2n} &= \langle \nabla^{\epsilon}{}_{F^{\epsilon}}\nu_{\epsilon}, F^{\epsilon}\rangle_{\epsilon} = \epsilon\Big(\frac{W}{W_{\epsilon}}\Big)^2 \langle \nabla^{\epsilon}{}_{Y^{\perp}}\nu_{\epsilon}, Y^{\perp}\rangle_{\epsilon} \\
&= \epsilon\Big(\frac{W}{W_{\epsilon}}\Big)^2 \Big(-\overline{\omega}\overline{p}_m Y^{\perp}\Big(\overline{p}_m \frac{W}{W_{\epsilon}}\Big) + Y^{\perp}\Big(\overline{\omega}\frac{W}{W_{\epsilon}}\Big) \\
&\quad + \frac{W}{W_{\epsilon}}\overline{p}_m \langle \nabla^{\epsilon}{}_{Y^{\perp}}X_m, Y^{\perp}\rangle_{\epsilon} + \sqrt{\epsilon}\overline{\omega}\frac{W}{W_{\epsilon}}\langle \nabla^{\epsilon}{}_{Y^{\perp}}T_{\epsilon}, Y^{\perp}\rangle_{\epsilon}\Big) \\
&= \epsilon\Big(\frac{W}{W_{\epsilon}}\Big)^2 \Big(\frac{W}{W_{\epsilon}} Y^{\perp}\big(\overline{\omega}\big) - \frac{1}{\epsilon}\frac{W}{W_{\epsilon}}\overline{p}_m\overline{p}_l\overline{\omega}b_{lm} + \frac{1}{2}\frac{W}{W_{\epsilon}}\overline{\omega}^2\overline{p}_m\overline{p}_l b_{ml}\Big) \\
&= \epsilon\Big(\frac{W}{W_{\epsilon}}\Big)^3 Y^{\perp}(\overline{\omega}) = \epsilon\frac{W}{W_{\epsilon}}\Big(1 - \epsilon\overline{\omega}^2\Big(\frac{W}{W_{\epsilon}}\Big)\Big)Y^{\perp}\big(\overline{\omega}\big) \\
&= \epsilon\frac{W}{W_{\epsilon}}Y^{\perp}\big(\overline{\omega}\big) - \epsilon^2\overline{\omega}^2\Big(\frac{W}{W_{\epsilon}}\Big)^2 Y^{\perp}\big(\overline{\omega}\big).
\end{aligned}$$

The proof is complete. □

Let us make some remarks concerning the horizontal second fundamental form. For $i, j = 1, \ldots, 2n-1$

$$A^{\epsilon}_{i,j} = \frac{W}{W_{\epsilon}}\left(A^H_{ij} + \frac{1}{2}\overline{\omega}_s\xi^s_{ij}\right)$$

is symmetric for every ϵ, and therefore the limit must be symmetric. By definition, $\xi_{ij} = -\xi_{ji}$. Therefore,

$$\lim_{\epsilon\to 0}\frac{1}{2}\left(A^{\epsilon}_{i,j} + A^{\epsilon}_{j,i}\right) = A^H_{ij} + \frac{1}{2}\overline{\omega}_s\xi^s_{ij} = \frac{1}{2}\left(A^H_{ij} + A^H_{ji}\right).$$

This leads to the following definition.

Definition 5.23. Let $A^H_{ij} = \langle \nabla_{E_i}\nu_H, E_j\rangle$. Then the *symmetrized horizontal second fundamental form* is defined as

$$A^{H,s}_{ij} = \frac{1}{2}\left(A^H_{ij} + A^H_{ji}\right) = A^H_{ij} + \frac{1}{2}\overline{\omega}_s\xi^s_{ij}.$$

Remark 5.24. Note that for $i, j = 1, \ldots, m-1$

$$\langle \nabla_{E_i}\nu_H, E_j\rangle = -\langle \nabla_{E_i}E_j, \nu_H\rangle.$$

Then

$$
\begin{aligned}
0 &= \left(A_{ji}^H + \frac{1}{2}\overline{\omega}_s \xi_{ji}\right) - \left(A_{ij}^H + \frac{1}{2}\overline{\omega}_s \xi_{ij}^s\right) \\
&= \left(-\langle \nabla_{E_j} E_i, \nu_H\rangle + \frac{1}{2}\overline{\omega}_s \xi_{ji}^s\right) - \left(-\langle \nabla_{E_i} E_j, \nu_H\rangle + \frac{1}{2}\overline{\omega}_s \xi_{ij}^s\right) \\
&= \langle\, [E_i, E_j], \nu_H\rangle + \overline{\omega}_s \xi_{ji}^s.
\end{aligned}
$$

Therefore,

$$
\langle\, [E_i, E_j], \nu_H\rangle = \overline{\omega}_s \xi_{ij}^s.
$$

We close this section with some useful observations. First, we note that

$$
\langle V, F_s^\epsilon\rangle_\epsilon = \sqrt{\epsilon}\Big(\rho_s^\epsilon \langle V, \nu_H\rangle + \rho_s^\epsilon \sum_{q<s} \overline{\omega}_q \langle V, T_q\rangle\Big) + \frac{1}{\sqrt{\epsilon}}\left(\lambda_s^\epsilon \langle V, T_s\rangle\right).
$$

Denote

$$
\mathcal{F}_s^{(1/2,\epsilon)} = \rho_s^\epsilon \langle V, \nu_H\rangle + \rho_s^\epsilon \sum_{q<s} \overline{\omega}_q \langle V, T_q\rangle \tag{5.10}
$$

and

$$
\mathcal{F}_s^{(-1/2,\epsilon)} = \lambda_s^\epsilon \langle V, T_s\rangle. \tag{5.11}
$$

In particular, in $\mathbb{H}^n$,

$$
\begin{aligned}
\langle V, F^\epsilon\rangle_\epsilon &= \sqrt{\epsilon}\frac{W}{W_\epsilon}\left(-\overline{\omega}\langle V, \nu_H\rangle\right) + \frac{1}{\sqrt{\epsilon}}\frac{W}{W_\epsilon}\left(\langle V, T\rangle\right) \\
&= \sqrt{\epsilon}\frac{W}{W_\epsilon}\mathcal{F}^{1/2} + \frac{1}{\sqrt{\epsilon}}\frac{W}{W_\epsilon}\mathcal{F}^{-1/2},
\end{aligned}
$$

where

$$
\mathcal{F}^{1/2} = -\overline{\omega}\langle V, \nu_H\rangle, \quad \mathcal{F}^{-1/2} = \langle V, T\rangle.
$$

6 First and Second Variation Formulas for H-Perimeter

Using the geometric tools developed in Section 5, we can now derive first and second variation formulas for the H-perimeter. Throughout the section, M denotes an oriented C^2 hypersurface in $\mathbb{G}$. Note that the variations are taken in neighborhoods away from characteristic points. This is necessary because the horizontal Gauss map is not even defined on the characteristic set. We note however that, using appropriate cut-offs, it is possible to extend these formulas to some situations where the deformation is performed near the characteristic set. It should also be mentioned that such formulas are particularly useful in the study of the Bernstein problem and the latter is concerned with noncharacteristic entire graphs.

Lemma 6.1. *Let $M_s = M + sV$, where $V \in C_0^2(\mathcal{U}\backslash\Sigma)$. Then*

$$\sqrt{\epsilon}\frac{d}{ds}\sigma^{\epsilon}(M_s)$$

converges uniformly as $\epsilon \to 0$ *for* $s \in [-\delta, \delta]$, *where* $\delta > 0$ *is sufficiently small.*

Proof. By a partition of unity argument, it suffices to establish the desired conclusion when V is supported in a relatively compact open chart $U_s \subset M_s$. As in the proof of Proposition 3.2, one can write

$$\sqrt{\epsilon}\sigma^{\epsilon}(U_s) = \int_{\Omega} \sqrt{\sum_{i=1}^{m} (p_i^s)^2 + \epsilon \sum_{l=1}^{k} (\omega_l^s)^2}\, d\omega, \tag{6.1}$$

where

$$N_s = \sum_{i=1}^{m} p_i^s X_i + \sum_{l=1}^{k} \omega_l^s T_l$$

is the normal vector to M_s. Let M be given by a parametrization $\theta : \Omega \to \mathbb{G}$ and V by a parametrization $\gamma : \Omega \to \mathbb{G}$. Then M_s is given by the parametrization $\theta + s\gamma$. Further, N_s is the determinant of a matrix with entries that are the derivatives of $\theta + s\gamma$ with respect to variables in Ω. Therefore, the components of N_s are polynomials in s. By the regularity of M_s, p_i^s and ω_l^s are bounded on relatively compact subsets of $\mathcal{U}\backslash\Sigma$. One can compute,

$$\sqrt{\epsilon}\frac{d}{ds}\sigma^{\epsilon}(U_s) = \int_{\Omega} \frac{p_i^s \frac{d}{ds} p_i^s + \epsilon \omega_s^l \frac{d}{ds}\omega_l^s}{\sqrt{\sum_{i=1}^{m} (p_i^s)^2 + \epsilon \sum_{l=1}^{k} (\omega_l^s)^2}}\, d\omega$$

and conclude that the term

$$\frac{\epsilon \omega_s \frac{d}{ds}\omega_s}{\sqrt{\sum_{i=1}^{m} (p_i^s)^2 + \epsilon \sum_{l=1}^{k} (\omega_l^s)^2}}$$

converges to zero uniformly since

$$\frac{\omega_l^s \frac{d}{ds}\omega_l^s}{\sqrt{\sum_{i=1}^{m} (p_i^s)^2 + \epsilon \sum_{l=1}^{k} (\omega_l^s)^2}}$$

is uniformly bounded for points in $\overline{\Omega}$ and $s \in [-\delta, \delta]$ by the preceding comments. Note that

$$\sum_{i=1}^{m} (p_i^s)^2 \geqslant J,$$

for some $J > 0$ since the derivative of $\epsilon\sigma^{\epsilon}(U_s)$ is zero except in the support of V. The terms

$$\frac{p_i^s \frac{d}{ds} p_i^s}{\sqrt{\sum_{i=1}^m (p_i^s)^2 + \epsilon \sum_{l=1}^k (\omega_l^s)^2}}, \quad i = 1, \dots, m,$$

also converge uniformly as $\epsilon \to 0$, by the Dini convergence theorem. In particular, for each point in U_s

$$\frac{p_i^s \frac{d}{ds} p_i^s}{\sqrt{\sum_{i=1}^m (p_i^s)^2 + \epsilon \sum_{l=1}^k (\omega_l^s)^2}}$$

is monotone increasing as $\epsilon \to 0$, and clearly the quantity converges pointwise for $s \in [-\delta, \delta]$, $(u, v) \in \overline{\Omega}$. This completes the proof. □

Let M be an oriented Riemannian submanifold of the Riemannian manifold N. Let V be a compactly supported vector field, and let

$$M_t = M + tV.$$

The first variation of the Riemannian volume can be found in [9].

Proposition 6.2 (first variation for Riemannian submanifolds).

$$\frac{d}{dt} \sigma(M_t)|_{t=0} = \int_M H g(V, \nu)\, d\sigma,$$

where g is the metric on the ambient submanifold N, ν is the unit normal to M, and $d\sigma$ is the standard Riemannian surface measure on M.

Theorem 6.3 (first variation formula for the H-perimeter). *Let $M \subset \mathbb{G}$, a Carnot group of step 2, where M_s is as in lemma* 6.1. *Then*

$$\frac{d}{ds} \sigma_H(M_s)|_{s=0} = \int_M \mathcal{H} \langle V, Y \rangle \, d\sigma_H.$$

Proof. Writing the Riemannian first variation formula in the present context yields

$$\frac{d}{ds} \sigma^\epsilon(M_s)|_{s=0} = \int_M H_\epsilon \langle V, \nu_\epsilon \rangle_\epsilon \, d\sigma^\epsilon.$$

The limit of the right-hand side follows directly from Corollary 5.14 and Propositions 5.8 and 3.2. By Lemma 6.1 and Proposition 3.2, the left-hand side converges to

$$\frac{d}{ds} \sigma_H(M_s)|_{s=0}.$$

The proof is complete. □

We explicitly observe that the formula in Theorem 6.3 matches that found in works such as [10] and [18]. Following a similar procedure, we next want to derive a second variation formula for the horizontal perimeter. We recall the basic facts from Riemannian geometry. The formula stated below can be derived from the results of [9].

Proposition 6.4. *Let M, ν, and V be the same as in Proposition* 6.2. *Let* $\{\mathcal{E}_1,\dots,\mathcal{E}_n\}$ *be an orthonormal basis for* TM. *Then*

$$\frac{d^2}{ds^2}\sigma(M_s)|_{s=0} = \int_M (\mathrm{div}_M V)^2 + \sum_{i=1}^{n} (\langle \nabla_{\mathcal{E}_i} V, \nu\rangle)^2 - \sum_{i,j=1}^{n} \langle \nabla_{\mathcal{E}_i} V, \mathcal{E}_j\rangle \langle \nabla_{\mathcal{E}_j} V, \mathcal{E}_i\rangle - \sum_{i=1}^{n} \mathrm{Rm}(V,\mathcal{E}_i,\mathcal{E}_i,V)\, d\sigma.$$

In order to use this formula, we need to compute some pointwise identities of some of the terms in the integrand.

Lemma 6.5. *Let $\mathcal{E}_i$ be as in Proposition* 6.4. *Then*

$$\langle \nabla_{\mathcal{E}_i} V, \nu\rangle = E_i\left(\langle V,\nu\rangle\right) - \langle V, E_m\rangle A_{im}, \tag{6.2}$$

$$\langle \nabla_{\mathcal{E}_i} V, \mathcal{E}_j\rangle = \mathcal{E}_i\left(\langle V,\mathcal{E}_j\rangle\right) + \langle V,\mathcal{E}_m\rangle \Gamma^j_{im} + \langle V,\nu\rangle A_{ij}, \tag{6.3}$$

$$\mathrm{div}_M V = \mathcal{E}_i\left(\langle V,\mathcal{E}_i\rangle\right) + H\langle V,\nu\rangle. \tag{6.4}$$

Proof. One can write $V = \langle V,\mathcal{E}_i\rangle \mathcal{E}_i + \langle V,\nu\rangle\nu$, and use the fact that

$$\nabla_X(fY) = f\nabla_X Y + XfY$$

to obtain the first two statements. For the final statement,

$$\begin{aligned}\mathrm{div}_M V &= \mathcal{E}_i\left(\langle V,\mathcal{E}_i\rangle\right) + \langle V,\mathcal{E}_m\rangle\langle \nabla_{\mathcal{E}_i} V,\mathcal{E}_i\rangle + \langle V,\nu\rangle\langle \nabla_{\mathcal{E}_i}\nu,\mathcal{E}_i\rangle \\ &= \mathcal{E}_i\left(\langle V,\mathcal{E}_i\rangle\right) + \langle V,\mathcal{E}_m\rangle \Gamma^i_{im} + H\langle V,\nu\rangle.\end{aligned}$$

Since this is a pointwise identity, one may assume that $\mathcal{E}_i$ is a coordinate frame. Therefore, by symmetries of the Christoffel symbols,

$$\Gamma^i_{im} = \Gamma^i_{mi} = -\Gamma^i_{mi} = 0.$$

The proof is complete. □

We can now compute the second variation for the case of deformations in the normal direction. The variation that will be used is $V = h\nu_\epsilon$, where $h \in C^2_0(\mathcal{U}\backslash\Sigma)$. Therefore, V depends on ϵ and Lemma 6.1 cannot be directly applied because the components of N^ϵ_s (to be denoted $p^{s,\epsilon}_i$, $i = 1,\dots,2n$, and $\omega s,\epsilon$) may depend on ϵ. Therefore, one must explicitly calculate N^ϵ_s in order to observe the effect of this dependence on ϵ.

Lemma 6.6. *Let* $M_{s,\epsilon} = M + sh\nu_\epsilon$, $h \in C^2_0(\mathcal{U}\backslash\Sigma)$, $s \in [-\delta,\delta]$. *Then*

$$\lim_{\epsilon\to 0}\sqrt{\epsilon}\sigma^{\epsilon}(M_{s,\epsilon}) = \sigma_H(M_s).$$

Further,

$$\frac{d}{ds}\sqrt{\epsilon}\sigma^{\epsilon}(M_{s,\epsilon})$$

and

$$\frac{d^2}{ds^2}\sqrt{\epsilon}\sigma^{\epsilon}(M_{s,\epsilon})$$

converge uniformly as $\epsilon \to 0$ *for* $s \in [-\delta, \delta]$.

Proof. For $u \in \Omega \subset \mathbb{R}^{2n}$ a point on M is denoted by

$$g(u) = \sum_{i=1}^{2n} g^i(u)\frac{\partial}{\partial x_i} + g^{2n+1}(u)\frac{\partial}{\partial t}$$

and a point on $M_{s,\epsilon}$ is denoted by

$$\begin{aligned} g_s^{\epsilon}(u) &= g(u) + sh(g(u))\nu_{\epsilon}(g(u)) \\ &= \left(g^i + sh\nu_{\epsilon}^i\right)\frac{\partial}{\partial x_i} + \Big(g^{2n+1} + sh\nu_{\epsilon}^{2n+1} \\ &\quad - \frac{1}{2}\sum_{i=1}^{n} sh\nu_{\epsilon}^i(g^{n+i} + sh\nu_{\epsilon}^{n+i}) + \frac{1}{2}\sum_{i=1}^{n} sh\nu_{\epsilon}^{n+i}(g^i + sh\nu_{\epsilon}^i)\Big)\frac{\partial}{\partial t}, \end{aligned}$$

where

$$\nu_{\epsilon}^i = \langle \nu_{\epsilon}, X_i\rangle, \quad \nu_{\epsilon}^{2n+1} = \langle \nu_{\epsilon}, T_{\epsilon}\rangle_{\epsilon}.$$

Denote

$$x_i^{s,\epsilon}(u) = g^i(u) + sh\nu_{\epsilon}^i, \quad i = 1, \ldots, 2n,$$

and

$$\begin{aligned} t^{s,\epsilon}(u) = g^{2n+1} + sh\nu_{\epsilon}^{2n+1} - \frac{1}{2}\sum_{i=1}^{n} sh\nu_{\epsilon}^i\left(g^{n+i} + sh\nu_{\epsilon}^{n+i}\right) \\ + \frac{1}{2}\sum_{i=1}^{n} sh\nu_{\epsilon}^{n+i}\left(g^i + sh\nu_{\epsilon}^i\right) \end{aligned}$$

so that

$$g_s^{\epsilon} = x_i^{s,\epsilon}\frac{\partial}{\partial x_i} + t^{s,\epsilon}\frac{\partial}{\partial t}.$$

Recall that

$$\nu_{\epsilon} = \frac{W}{W_{\epsilon}}\nu_H + \sqrt{\epsilon}\frac{W}{W_{\epsilon}}\varpi T_{\epsilon}.$$

The normal of $M_{s,\epsilon}$ is the determinant of the $(2n+1)\times(2n+1)$ matrix with first row $X_1,\dots,X_{2n},T_\epsilon$ and remaining components consisting of $\langle (g_s^\epsilon)_{u_j}, X_i\rangle$ or $\langle (g_s^\epsilon)_{u_j}, T_\epsilon\rangle_\epsilon$ for $i,j=1,\dots,2n$. Then

$$
\begin{aligned}
(g_s^\epsilon)_{u_j} &= (x_i^{s,\epsilon})_{u_j}\frac{\partial}{\partial x_i} + (t^{s,\epsilon})_{u_j}\frac{\partial}{\partial t}\\
&= (x_i^{s,\epsilon})_{u_j} X_i(g_s^\epsilon) + \frac{1}{\sqrt{\epsilon}}\Big((t^{s,\epsilon})_{u_j} + \frac{1}{2}(x_{n+i}^{s,\epsilon}(x_i)_{u_j} - x_i^{s,\epsilon}(x_{n+i}^{s,\epsilon})_{u_j})\Big)T_\epsilon.
\end{aligned}
$$

It is easy to check that the u_j derivatives of any component of ν_ϵ converge uniformly as $\epsilon\to 0$ by the same arguments as in Lemma 6.1. Since the components of N_s^ϵ are multiples of $\langle (g_\epsilon^s)_{u_j}, X_i\rangle$, $\langle (g_\epsilon^s)_{u_j}, T_\epsilon\rangle_\epsilon$ and are bounded, they must converge uniformly as $\epsilon\to 0$. Recall that, by the proof of Lemma 6.1, the components of N_s^ϵ are polynomials in s. Recalling that

$$
\sqrt{\epsilon}\sigma^\epsilon(M_{s,\epsilon}) = \int_\Omega \sqrt{\sum_{i=1}^{2n}(p_i^{s,\epsilon})^2 + \epsilon\,(\omega^{s,\epsilon})^2}\, d\omega,
$$

one can reach the conclusions from plugging in the derived formulas for $p_i^{s,\epsilon}$ and $\omega^{s,\epsilon}$. □

We can finally prove the main result of this section.

Theorem 6.7 (second variation formula for the H-perimeter). *Let $M\subset\mathbb{H}^n$ be a C^2 hypersurface, and let $M_s = M + sV$, where $V = h\nu_H$ and $h\in C_0^1(\mathcal{U}\setminus\Sigma(M))$. Then*

$$
\begin{aligned}
\frac{d^2}{ds^2}\sigma_H(M_s)|_{s=0} &= \int_M \sum_{i=1}^{2n-1}|E_i(\langle V,\nu_H\rangle)|^2 + \langle V,\nu_H\rangle^2(\mathcal{H}^2 - \sum_{i,j=1}^{2n-1}|A_{ij}^{H,s}|^2\\
&\quad - \langle Z, X_m\rangle Y^\perp(\overline{p}_m) - Z(\overline{\omega}) - \frac{2n-2}{4}\overline{\omega}^2)\, d\sigma_H\\
&= \int_M \sum_{i=1}^{2n-1}|E_i(\langle V,\nu_H\rangle)|^2\\
&\quad + \langle V,\nu_H\rangle^2\left(\mathcal{H}^2 - \sum_{i,j=1}^{2n-1}|A_{ij}^{H,s}|^2 - 2Z\Big(\overline{\omega}\right) - \frac{2n+2}{4}\overline{\omega}^2\Big)\, d\sigma_H.
\end{aligned}
$$

Proof. We use the above notation. For the sake of simplicity, terms converging to zero uniformly are ignored. For example, the expression

$$
\left(\sqrt{\epsilon}A^{1/2} + \frac{1}{\sqrt{\epsilon}}A^{-1/2}\right)^2
$$

is written as

$$\frac{1}{\epsilon}(A^{-1/2})^2 + 2A^{-1/2}A^{1/2}.$$

Let $M_{s,\epsilon}$ be as in Lemma 6.6. Denote $V_\epsilon = h\nu_\epsilon$ and $V = h\nu_H$. By Proposition 5.8, $V_\epsilon \to V$. Note that the Riemannian second variation formula for the variation V_ϵ yields

$$\begin{aligned}\frac{d^2}{ds^2}\sqrt{\epsilon}\sigma^\epsilon(M_{s,\epsilon})|_{s=0} &= \sqrt{\epsilon}\int_M H_\epsilon^2 \langle V, \nu_\epsilon\rangle_\epsilon^{\,2} + \sum_{i=1}^{2n} |\mathcal{E}_i^\epsilon(\langle V, \nu_\epsilon\rangle_\epsilon)|^2 \\ &\quad - \langle V, \nu_\epsilon\rangle_\epsilon^{\,2} \sum_{i,j=1}^{2n} \left(A_{i,j}^\epsilon\right)^2 - \sum_{i=1}^{2n} \mathrm{Rm}^\epsilon(V, \mathcal{E}_i^\epsilon, \mathcal{E}_i^\epsilon, V)\, d\sigma^\epsilon \\ &= \sqrt{\epsilon}\int_M I_\epsilon + II_\epsilon + III_\epsilon + IV_\epsilon\, d\sigma^\epsilon.\end{aligned}$$

By Lemma 6.6, the left-hand side converges to

$$\frac{d^2}{ds^2}\sigma_H(M_s)|_{s=0}.$$

Next, the right-hand side is computed. First, I_ϵ:

$$\begin{aligned}I_\epsilon &= \sum_{i=1}^{2n} |\mathcal{E}_i(\langle V, \nu_\epsilon\rangle_\epsilon)|^2 = \sum_{i=1}^{2n-1} |E_i(\langle V, \nu_\epsilon\rangle)|^2 + |F^\epsilon(\langle V, \nu_\epsilon\rangle)|^2 \\ &= \sum_{i=1}^{2n-1} |E_i(\langle V, \nu_\epsilon\rangle)|^2 + |\sqrt{\epsilon}\frac{W}{W_\epsilon} Y^\perp(\langle V, \nu_\epsilon\rangle)|^2.\end{aligned}$$

Thus,

$$\lim_{\epsilon\to 0} I_\epsilon = \sum_{i=1}^{2n-1} |E_i(\langle V, \nu_H\rangle)|^2.$$

Using Corollary 5.14 and Proposition 5.8, we find

$$\lim_{\epsilon\to 0} H_\epsilon^2 \langle V, \nu_\epsilon\rangle_\epsilon^{\,2} = \mathcal{H}^2 \langle V, \nu_H\rangle^2.$$

Next, III_ϵ is computed:

$$\begin{aligned}III_\epsilon = -\langle V, \nu_\epsilon\rangle_\epsilon^{\,2}\Bigg(&\sum_{i,j=1}^{2n-1}\left(\frac{W}{W_\epsilon}A_{ij}^{H,s}\right)^2 + \sum_{i=1}^{2n-1}\left(\sqrt{\epsilon}A_{i,2n}^{1/2} + \frac{1}{\sqrt{\epsilon}}A_{i,2n}^{-1/2}\right)^2 \\ &+ \sum_{j=1}^{2n-1}\left(\sqrt{\epsilon}A_{2n,j}^{1/2} + = \frac{1}{\sqrt{\epsilon}}A_{2n,j}^{-1/2}\right)^2 + \left(\epsilon\frac{W}{W_\epsilon}A_{2n,2n}^1\right)^2\Bigg)\end{aligned}$$

$$
\begin{aligned}
&= -\langle V, \nu_\epsilon\rangle_\epsilon{}^2 \Big(\sum_{i,j=1}^{2n-1} \Big(\frac{W}{W_\epsilon} A_{ij}^{H,s} \Big)^2 + 2A_{i,2n}^{-1/2} A_{i,2n}^{1/2} + 2A_{2n,j}^{-1/2} A_{2n,j}^{1/2} \Big) \\
&\quad + \frac{1}{\epsilon} \Big(-2\langle V, \nu_\epsilon\rangle_\epsilon{}^2 \sum_{i=1}^{2n-1} (A_{i,2n}^{-1/2})^2 \Big) \\
&= -\langle V, \nu_\epsilon\rangle_\epsilon{}^2 \Big(\sum_{i,j=1}^{2n-1} \Big(\frac{W}{W_\epsilon} A_{ij}^{H,s} \Big)^2 + Z(\overline{\omega}) + \langle Z, X_l\rangle Y^\perp(\overline{p}_l) - 2\overline{\omega}^2 \Big) \\
&\quad + \frac{1}{\epsilon} \Big(-\frac{1}{2} \langle V, \nu_\epsilon\rangle_\epsilon{}^2 \Big).
\end{aligned}
$$

By the symmetry of A_{ij}. this expression can be written as

$$
III_\epsilon = -\langle V, \nu_\epsilon\rangle_\epsilon{}^2 \Big(\sum_{i,j=1}^{2n-1} \Big(\frac{W}{W_\epsilon} A_{ij}^{H,s} \Big)^2 + 2Z(\overline{\omega}) \Big) + \frac{1}{\epsilon} \Big(-\frac{1}{2} \langle V, \nu_\epsilon\rangle_\epsilon{}^2 \Big).
$$

Next, Lemma 4.2 is used to calculate IV_ϵ:

$$
\begin{aligned}
VI_\epsilon &= -\langle V, \nu_\epsilon\rangle_\epsilon{}^2 \Big(\sum_{i=1}^{2n-1} \mathrm{Rm}^\epsilon(\nu_\epsilon, E_i, E_i, \nu_\epsilon) + \mathrm{Rm}^\epsilon(\nu_\epsilon, F^\epsilon, F^\epsilon, \nu_\epsilon) \Big) \\
&= -\langle V, \nu_\epsilon\rangle_\epsilon{}^2 \Big(\overline{\omega}^2 \frac{2n-1}{4} + \frac{3}{4} \overline{\omega}^2 \Big(\frac{W}{W_\epsilon} \Big)^2 - \frac{1}{\epsilon} \frac{3}{4} \Big) \\
&\quad - \langle V, \nu_\epsilon\rangle_\epsilon{}^2 \Big(\frac{1}{\epsilon} \frac{1}{4} - \frac{1}{2} \overline{\omega}^2 \Big(\frac{W}{W_\epsilon} \Big)^2 + \frac{1}{2} \overline{\omega}^2 \Big(\frac{W}{W_\epsilon} \Big)^4 \Big) \\
&= \frac{1}{\epsilon} \Big(\frac{1}{2} \langle V, \nu_\epsilon\rangle_\epsilon{}^2 \Big) \\
&\quad - \langle V, \nu_\epsilon\rangle_\epsilon{}^2 \overline{\omega}^2 \Big(\frac{2n-1}{4} + \frac{3}{4} \Big(\frac{W}{W_\epsilon} \Big)^2 - \frac{1}{2} \Big(\frac{W}{W_\epsilon} \Big)^2 + \frac{1}{2} \Big(\frac{W}{W_\epsilon} \Big)^4 \Big).
\end{aligned}
$$

Note that the $1/\epsilon$ terms from II_ϵ and IV_ϵ cancel. Therefore,

$$
\begin{aligned}
\lim_{\epsilon\to 0} III_\epsilon + IV_\epsilon = -\langle V, \nu_H\rangle^2 \Big(\sum_{i,j=1}^{2n-1} \Big(A_{ij}^{H,s} \Big)^2 + Z(\overline{\omega}) + \langle Z, X_l\rangle Y^\perp(\overline{p}_l) \Big) \\
- \langle V, \nu_H\rangle^2 \overline{\omega}^2 \Big(\frac{2n-2}{4} \Big),
\end{aligned}
$$

and alternatively,

$$\lim_{\epsilon\to 0} III_\epsilon + IV_\epsilon = -\langle V, \nu_H\rangle^2 \Big(\sum_{i,j=1}^{2n-1} \left(A_{ij}^{H,s}\right)^2 + 2Z(\overline{\omega}) + \frac{2n+2}{4}\overline{\omega}^2 \Big).$$

Combining these quantities with the limits of I_ϵ, II_ϵ and applying Propositions 3.2, 5.8, we obtain the limit of the right-hand side. Thus, we obtain the required conclusion (note that the convergence in ϵ is uniform by the same type arguments previously stated). □

Acknowledgments. The second author was supported in part by NSF (grant DMS-0701001).

References

1. Adesi, V.B., Cassano, S., Vittone, D.: The Bernstein problem for intrinsic graphs in the Heisenberg group and calibrations. Calc. Var. Partial Differ. Equ. **30**, no. 1, 17–49 (2007)
2. Bonk, M., Capogna, L.: Horizontal Mean Curvature Flow in the Heisenberg Group. Preprint (2005)
3. Bryant, R., Griffiths, P., Grossmann, D.: Exterior Differential Systems and Euler-Lagrange Partial Differential Equations. Univ. Chicago Press (2003)
4. Capogna, L, Citti, G., Manfredini, M.: Regularity of minimal surfaces in the one-dimensional Heisenberg group In: "Bruno Pini" Mathematical Analysis Seminar, University of Bologna Department of Mathematics: Academic Year 2006/2007 (Italian), pp. 147–162. Tecnoprint, Bologna (2008)
5. Capogna, L, Danielli, D., Garofalo, N.: The geometric Sobolev embedding for vector fields and the isoperimetric inequality. Commun. Anal. Geom. **2**, no. 2, 203–215 (1994)
6. Capogna, L, Pauls, S.D., Tyson, J.T.: Convexity and Horizontal Second Fundamental Forms for Hypersurfaces in Carnot Groups. Trans. Am. Math. Soc. [To appear]
7. Cheng, J.H., Hwang, J.F., Malchiodi, A., Yang, P.: Minimal surfaces in pseudohermitian geometry and the Bernstein problem in the Heisenberg group. Ann. Sc. Norm. Sup. Pisa **1**, 129-177 (2005)
8. Cheng, J.H., Hwang, J.F., Yang, P.: Existence and uniqueness for p-area minimizers in the Heisenberg group. Math. Ann. **337**, no. 2, 253–293 (2007)
9. Colding, T.H., Minicozzi II, W.P.: Minimal Surfaces, Courant Lecture Notes in Mathematics.**4**. University Courant Institute of Mathematical Sciences, New York (1999)
10. Danielli, D., Garofalo, N., Nhieu, D.-M.: Sub-Riemannian calculus on hypersurfaces in Carnot groups. Adv. Math. **215**, no. 1, 292–378 (2007)
11. Danielli, D., Garofalo, N., Nhieu, D.-M.: A notable family of entire intrinsic minimal graphs in the Heisenberg group which are not perimeter minimizing. Am. J. Math. **130**, no. 2, 317–339 (2008)
12. Danielli, D., Garofalo, N., Nhieu, D.-M.: A partial solution of the isoperimetric problem for the Heisenberg group. Forum Math. **20**, no. 1, 99–143 (2008)

13. Danielli, D., Garofalo, N., Nhieu, D.-M., Pauls, S.: Instability of Graphical Strips and a Positive Answer to the Bernstein Problem in the Heisenberg Group $\mathbb{H}^1$. J. Differ. Geom. **81**, no. 2, 251–295 (2009)
14. Danielli, D., Garofalo, N., Nhieu, D.-M., Pauls, S.: Stable Complete Embedded Minimal Surfaces in $\mathbb{H}^1$ with Empty Characteristic Locus are Vertical Planes. Preprint (2008)
15. Fleming, W.H., Rishel, R.: An integral formula for total gradient variation. Arch. Math. **11**, 218-222 (1960)
16. Gromov, M.: Carnot-Carathéodory spaces seen from within. In Sub-Riemannian Geometry. Birkhäuser (1996)
17. Gromov, M.: Metric Structures for Riemannian and Non-Riemannian Spaces. Birkhäuser (1998)
18. Hladky, R., Pauls, S.: Constant mean curvature surfaces in sub-Riemannian geometry. J. Differ. Geom. **79**, no. 1, 111–139 (2008)
19. Hladky, R., Pauls, S.: Variation of Perimeter Measure in Sub-Riemannian Geometry. Preprint (2007)
20. A. Hurtado, M. Ritoré & C. Rosales, The classification of complete stable area-stationary surfaces in the Heisenberg group $\mathbb{H}^1$. Preprint (2008)
21. Korányi, A.: Geometric aspects of analysis on the Heisenberg group. In: Topics in Modern Harmonic Analysis I, II (Turin/Milan, 1982), pp. 209–258. Ist. Naz. Alta Mat. Francesco Severi, Rome (1983)
22. Korányi, A.: Horizontal normal vectors and conformal capacity of spherical rings in the Heisenberg group. Bull. Sc. Math. **111**, 3-21 (1987)
23. Lee, J.M.: Riemannian Manifolds. Springer, New York (1997)
24. Massari, U., Miranda, M.: Minimal Surfaces of Codimension One. North-Holland (1984)
25. Maz'ya, V.G.: Sobolev Spaces. Springer, Berlin etc. (1985)
26. Montefalcone, F.: Hypersurfaces and variational formulas in sub-Riemannian Carnot groups. J. Math. Pures Appl. (9) **87**, no. 5, 453–494 (2007)
27. Monti, R.: Heisenberg isoperimetric problem. The axial case. Adv. Calc. Var. **1**, no. 1, 93–121 (2008)
28. Monti, R., Rickly, M.: Convex Isoperimetric Sets in the Heisenberg Group. Preprint (2006)
29. Montgomery, R.: A Tour of Subriemannian Geometries. Their Geodesics and Applications. Am. Math. Soc., Providence, RI (2002)
30. Pansu, P.: Une inégalité isopérimétrique sur le groupe de Heisenberg. C. R. Acad. Sci. Paris I Math. **295**, no. 2, 127-130 (1982)
31. Ritorè, M.: A Proof by Calibration of an Isoperimetric Inequality in the Heisenberg Group $\mathbb{H}^n$. Preprint (2008)
32. Pauls, S.D.: Minimal surfaces in the Heisenberg group. Geom. Dedicata **104**, 201–231 (2004)
33. Ritorè, M., Rosales, C.: Rotationally Invariant Hypersurfaces with Constant Mean Curvature in the Heisenberg Group $\mathbb{H}^n$. Preprint (2005)
34. Ritorè, M., Rosales, C.: Area Stationary Surfaces in the Heisenberg Group $\mathbb{H}^1$. Preprint (2005)
35. Simon, L.: Lectures on Geometric Measure Theory. Australian Univ. (1983)
36. Varadarajan, V.S.: Lie Groups, Lie Algebras, and Their Representations. Springer, New York etc. (1974)

Sobolev Homeomorphisms and Composition Operators

Vladimir Gol'dshtein and Alexander Ukhlov

Abstract We study the invertibility of bounded composition operators of Sobolev spaces. We prove that if a homeomorphism φ of Euclidean domains D and D' generates, by the composition rule $\varphi^* f = f \circ \varphi$, a bounded composition operator of the Sobolev spaces $\varphi^* : L^1_\infty(D') \to L^1_p(D)$, $p > n-1$, has finite distortion and the Luzin N-property, then the inverse φ^{-1} generates the bounded composition operator from $L^1_{p'}(D)$, $p' = p/(p-n+1)$, into $L^1_1(D')$.

1 Introduction

We consider homeomorphisms $\varphi : D \to D'$ of the Euclidean domains $D, D' \subset \mathbb{R}^n$, $n \geqslant 2$, generating bounded composition operators on Sobolev spaces. We say that a homeomorphism $\varphi : D \to D'$ possesses the (p,q)*-composition property*, $1 \leqslant q \leqslant p \leqslant \infty$, if it generates a bounded composition operator $\varphi^* : L^1_p(D') \to L^1_q(D)$ by the chain rule $\varphi^*(f) := f \circ \varphi$.

Homeomorphisms generating bounded composition operators of Sobolev spaces $L^1_1(D')$ and $L^1_1(D)$ (possess the $(1,1)$-composition property) were introduced by V. G. Maz'ya [14] as a class of sub-areal mappings. His pioneering work established a connection between geometrical properties of homeomorphisms and the corresponding Sobolev spaces. This study was extended in the monograph [15], where the theory of multipliers was applied to the change of variable problem in Sobolev spaces.

Vladimir Gol'dshtein
Ben Gurion University of the Negev, P.O.B. 653, Beer Sheva 84105, Israel
e-mail: vladimir@bgu.ac.il

Alexander Ukhlov
Ben Gurion University of the Negev, P.O.B. 653, Beer Sheva 84105, Israel
e-mail: ukhlov@math.bgu.ac.il

A. Laptev (ed.), *Around the Research of Vladimir Maz'ya I: Function Spaces*, International Mathematical Series 11, DOI 10.1007/978-1-4419-1341-8_8,

Homeomorphisms possessing the (n,n)-composition property also admit a direct geometric description. As was proved in [22], the class of such homeomorphisms coincides with the well-known class of quasiconformal mappings. It is known that the mappings that are inverse to the quasiconformal homeomorphisms are also quasiconformal and have the same (n,n)-composition property. But this invertibility property is correct only for $p = n$. If we assume that the inverse homeomorphism possesses the same (p,p)-composition property as an original homeomorphism for $p \neq n$, then we have a bi-Lipschitz homeomorphism (in terms of interior geodesic metrics). This was proved for $p > n$ in [23], for $n-1 < p < n$ in [5], and for $1 \leqslant p < n$ in [12]. A geometric descriptions of homeomorphisms possessing the (p,p)-composition property has been done in [4] only for $p > n-1$.

New effects arise when we study composition operators in Sobolev spaces that decrease the integrability of weak first order derivatives. This problem was studied in a connection with the theory of mappings of finite distortion in [20] and in a connection with the embedding theorems in [3]. It was proved [20] that if a homeomorphism $\varphi : D \to D'$ possesses the (p,q)-composition property with $n-1 < q \leqslant p < \infty$, then the inverse homeomorphism $\varphi^{-1} : D' \to D$ possesses the (q',p')-composition property with $q' = q/(q-n+1)$, $p' = p/(p-n+1)$.

In this paper, we study the invertibility of a composition operator in the limit case $p = \infty$.

Theorem A. *Assume that a homeomorphism $\varphi : D \to D'$ has finite distortion, the Luzin N-property (the image of a set measure zero is a set measure zero) and generates a bounded composition operator*

$$\varphi^* : L^1_\infty(D') \to L^1_q(D), \quad q > n-1.$$

Then the inverse mapping $\varphi^{-1} : D' \to D$ generates the bounded composition operator

$$(\varphi^{-1})^* : L^1_{q'}(D) \to L^1_1(D'), \quad q' = q/(q-n+1).$$

Any homeomorphism with the (p,q)-composition property belongs to the Sobolev class $W^1_{1,\mathrm{loc}}$ (cf., for example [20]). Therefore, the "invertibility problem" for composition operators in Sobolev spaces is connected with the regularity of Sobolev homeomorphisms, which was studied by many authors. In particular, as is proved in [16], if a mapping $\varphi \in W^1_{n,\mathrm{loc}}(D)$ and $J(x,\varphi) > 0$ for almost all $x \in D$, then φ^{-1} belongs to $W^1_{1,\mathrm{loc}}(D')$.

The local regularity of plane homeomorphisms in the Sobolev space $W^1_1(D)$ was studied in [9]. Recently [10], in the case $\mathbb{R}^n$, $n \geqslant 3$, it was proved that if the norm of the derivative $|D\varphi|$ belongs to the Lorentz space $L^{n-1,1}(D)$ and a mapping $\varphi : D \to D'$ has finite distortion, then the inverse mapping belongs to the Sobolev space $W^1_{1,\mathrm{loc}}(D')$ and has finite distortion.

2 Composition Operators in Sobolev Spaces

We define the Sobolev space $L^1_p(D)$, $1 \leqslant p \leqslant \infty$, equipped with the seminorm, as a space of locally summable, weakly differentiable functions $f : D \to \mathbb{R}$ such that

$$\|f \mid L^1_p(D)\| = \|\nabla f \mid L_p(D)\|, \quad 1 \leqslant p \leqslant \infty.$$

A function f belongs to $L^1_{p,\mathrm{loc}}(D)$ if $f \in L^1_p(K)$ for every compact subset $K \subset D$. The Sobolev space $\overset{\circ}{L}{}^1_p(D)$ is the closure of $C_0^\infty(D)$ in $L^1_p(D)$. As usual, $C_0^\infty(D)$ denotes the space of infinitely smooth functions with compact support.

Note that smooth functions are dense in $L^1_p(D)$, $1 \leqslant p < \infty$ (cf., for example [13, 1]). If $p = \infty$, we can only assert that for arbitrary function $f \in L^1_p(D)$ there exists a sequence of smooth functions $\{f_k\}$ converges locally uniformly to f and $\|f_k \mid L^1_\infty(D)\| \to \|f \mid L^1_\infty(D)\|$ (cf. [1]).

A mapping $\varphi : D \to \mathbb{R}^n$ belongs to $L^1_p(D)$, $1 \leqslant p \leqslant \infty$, if its coordinate functions φ_j belong to $L^1_p(D)$, $j = 1, \dots, n$. In this case, the formal Jacobi matrix $D\varphi(x) = \big(\frac{\partial \varphi_i}{\partial x_j}(x)\big)$, $i, j = 1, \dots, n$, and the Jacobian $J(x, \varphi) = \det D\varphi(x)$ are well defined at almost all points $x \in D$. The norm $|D\varphi(x)|$ of $D\varphi(x)$ is the norm of the corresponding linear operator $D\varphi(x) : \mathbb{R}^n \to \mathbb{R}^n$ defined by the matrix $D\varphi(x)$. We use the same notation for the matrix and the corresponding linear operator.

In the theory of mappings with bounded mean distortion, additive set functions play an important role. Recall that a nonnegative mapping Φ defined on open subsets of D is a *finitely quasiadditive* set function [25] if

1) for any $x \in D$ there exists δ, $0 < \delta < \operatorname{dist}(x, \partial D)$, such that $0 \leqslant \Phi(B(x, \delta)) < \infty$ (hereinafter, $B(x, \delta) = \{y \in \mathbb{R}^n : |y - x| < \delta\}$);

2) for any finite collection $U_i \subset U \subset D$, $i = 1, \dots, k$, of mutually disjoint open sets $\sum\limits_{i=1}^{k} \Phi(U_i) \leqslant \Phi(U)$.

It is obvious that the last inequality can be extended to a countable collection of mutually disjoint open sets in D, so that a finitely quasiadditive set function is also *countable quasiadditive.*

If, instead of the second condition, we suppose that for any finite collection $U_i \subset D$, $i = 1, \dots, k$, of mutually disjoint open subsets of D

$$\sum_{i=1}^{k} \Phi(U_i) = \Phi(U),$$

then such a set function is said to be *finitely additive.* If the last equality can be extended to a countable collection of mutually disjoint open subsets of D, then such a set function is said to be countable additive.

A nonnegative mapping Φ defined on open subsets of D is called a monotone set function [25] if $\Phi(U_1) \leqslant \Phi(U_2)$, where $U_1 \subset U_2 \subset D$ are open sets.

Note that a monotone (countable) additive set function is the (countable) quasiadditive set function.

We formulate an auxiliary result from [25] in the following convenient way.

Proposition 2.1. *Let a monotone finitely additive set function Φ be defined on open subsets of a domain $D \subset \mathbb{R}^n$. Then for almost all $x \in D$ the volume derivative*

$$\Phi'(x) = \lim_{\delta \to 0, B_\delta \ni x} \frac{\Phi(B_\delta)}{|B_\delta|}$$

is finite and for any open set $U \subset D$

$$\int_U \Phi'(x)\, dx \leqslant \Phi(U).$$

We say that a mapping $\varphi : D \to D'$ generates a *bounded composition operator*

$$\varphi^* : L^1_p(D') \to L^1_q(D), \quad 1 \leqslant q \leqslant p \leqslant \infty,$$

if for every $f \in L^1_p(D')$ we have $f \circ \varphi \in L^1_q(D)$ and

$$\|\varphi^* f \mid L^1_q(D)\| \leqslant K \|f \mid L^1_p(D')\|.$$

Theorem 2.1. *A homeomorphism $\varphi : D \to D'$ between two domains $D, D' \subset \mathbb{R}^n$ generates a bounded composition operator*

$$\varphi^* : L^1_\infty(D') \to L^1_q(D), \quad 1 < q < +\infty,$$

if and only if φ belongs to the Sobolev space $L^1_q(D)$.

Proof. Necessity. Substituting the test functions $f_j(y) = y_j \in L^1_\infty(D')$, $j = 1, \dots, n$, in the inequality

$$\|\varphi^* f \mid L^1_q(D)\| \leqslant K \|f \mid L^1_\infty(D')\|,$$

we see that φ belongs to $L^1_q(D)$.

Sufficiency. Let $f \in L^1_\infty(D') \cap C^\infty(D')$. Then

$$\|\varphi^* f \mid L^1_q(D)\| = \Big(\int_D |\nabla (f \circ \varphi)|^q \, dx \Big)^{\frac{1}{q}} \leqslant \Big(\int_D |D\varphi|^q |\nabla f|^q (\varphi(x))\, dx \Big)^{\frac{1}{q}}$$

$$\leqslant \Big(\int\limits_D |D\varphi)|^q\,dx\Big)^{\frac{1}{q}} \|f \mid L^1_\infty(D')\| = \|\varphi \mid L^1_q(D)\| \cdot \|f \mid L^1_\infty(D')\|.$$

For an arbitrary function $f \in L^1_\infty(D')$ we consider a sequence of smooth functions $f_k \in L^1_\infty(D')$ such that

$$\lim_{k\to\infty} \|f_k \mid L^1_\infty(D')\| = \|f \mid L^1_\infty(D')\|$$

and f_k converges locally uniformly to f in D'. Then the sequence $\varphi^* f_k$ converges locally uniformly to $\varphi^* f$ in D and is a bounded sequence in $L^1_q(D)$. Because of the reflexivity of $L^1_q(D)$, $1 < q < \infty$, there exists a subsequence $f_{k_l} \in L^1_q(D)$ weakly converging to $f \in L^1_q(D)$; moreover,

$$\|\varphi^* f \mid L^1_q(D)\| \leqslant \liminf_{l\to\infty} \|\varphi^* f_{k_l} \mid L^1_q(D)\|.$$

Passing to limit as $l \to +\infty$ in the inequality

$$\|\varphi^* f_{k_l} \mid L^1_q(D)\| \leqslant K \|f_{k_l} \mid L^1_\infty(D')\|,$$

we obtain

$$\|\varphi^* f \mid L^1_q(D)\| \leqslant K \|f \mid L^1_\infty(D')\|.$$

The proof is complete. □

The following theorem establishes the "localization" property of the composition operator in spaces of functions with compact support and/or its closure in L^1_∞.

Theorem 2.2. *Suppose that a homeomorphism $\varphi : D \to D'$ between two domains $D, D' \subset \mathbb{R}^n$ generates a bounded composition operator*

$$\varphi^* : L^1_\infty(D') \to L^1_q(D),\ 1 \leqslant q < +\infty.$$

Then there exists a bounded monotone countable additive function $\Phi(A')$ defined on open bounded subsets of D' such that for every $f \in \overset{\circ}{L}{}^1_\infty(A')$

$$\int\limits_{\varphi^{-1}(A)} |\nabla(f \circ \varphi)|^q\,dx \leqslant \Phi(A') \operatorname{ess\,sup}_{y\in A'} |\nabla f|^q(y).$$

Proof. Define $\Phi(A')$ as follows (cf. [20, 24])

$$\Phi(A') = \sup_{f\in \overset{\circ}{L}{}^1_\infty(A')} \left(\frac{\|\varphi^* f \mid L^1_q(D)\|}{\|f \mid \overset{\circ}{L}{}^1_\infty(A')\|} \right)^q.$$

Let $A_1' \subset A_2'$ be bounded open subsets of D'. Extending functions of the space $\overset{\circ}{L}{}^1_\infty(A_1')$ by zero to the set A_2', we obtain the inclusion

$$\overset{\circ}{L}{}^1_\infty(A_1') \subset \overset{\circ}{L}{}^1_\infty(A_2').$$

It is obvious that

$$\|f \mid \overset{\circ}{L}{}^1_\infty(A_1')\| = \|f \mid \overset{\circ}{L}{}^1_\infty(A_2')\|$$

for every $f \in \overset{\circ}{L}{}^1_\infty(A_1')$. Since

$$\begin{aligned}\Phi(A_1') = \sup_{f\in \overset{\circ}{L}{}^1_\infty(A_1')} \left(\frac{\|\varphi^* f \mid L^1_q(D)\|}{\|f \mid \overset{\circ}{L}{}^1_\infty(A_1')\|}\right)^q &= \sup_{f\in \overset{\circ}{L}{}^1_\infty(A_1')} \left(\frac{\|\varphi^* f \mid L^1_q(D)\|}{\|f \mid \overset{\circ}{L}{}^1_\infty(A_2')\|}\right)^q \\ &\leqslant \sup_{f\in \overset{\circ}{L}{}^1_\infty(A_2')} \left(\frac{\|\varphi^* f \mid L^1_q(D)\|}{\|f \mid \overset{\circ}{L}{}^1_\infty(A_2')\|}\right)^q = \Phi(A_2'),\end{aligned}$$

the set function Φ is monotone.

Consider open disjoint subsets A_i', $i \in \mathbb{N}$, of the domain D' such that $A_0' = \bigcup\limits_{i=1}^{\infty} A_i'$. Choose arbitrary functions $f_i \in \overset{\circ}{L}{}^1_\infty(A_i')$ such that

$$\|\varphi^* f_i \mid L^1_q(D)\| \geqslant \left(\Phi(A_i')\left(1-\frac{\varepsilon}{2^i}\right)\right)^{\frac{1}{q}} \|f_i \mid \overset{\circ}{L}{}^1_\infty(A_i')\|$$

and

$$\|f_i \mid \overset{\circ}{L}{}^1_\infty(A_i')\| = 1,$$

where $i \in \mathbb{N}$ and $\varepsilon \in (0,1)$ is a fixed number. Setting $g_N = \sum\limits_{i=1}^{N} f_i$, we find

$$\begin{aligned}\|\varphi^* g_N \mid L^1_q(D)\| &\geqslant \left(\sum_{i=1}^{N} \left(\Phi(A_i')\left(1-\frac{\varepsilon}{2^i}\right)\right) \|f_i \mid \overset{\circ}{L}{}^1_\infty(A_i')\|^q\right)^{1/q} \\ &= \left(\sum_{i=1}^{N} \Phi(A_i')\left(1-\frac{\varepsilon}{2^i}\right)\right)^{\frac{1}{q}} \left\|g_N \mid \overset{\circ}{L}{}^1_\infty\left(\bigcup_{i=1}^{N} A_i'\right)\right\| \\ &\geqslant \left(\sum_{i=1}^{N} \Phi(A_i') - \varepsilon\Phi(A_0')\right)^{\frac{1}{q}} \left\|g_N \mid \overset{\circ}{L}{}^1_\infty\left(\bigcup_{i=1}^{N} A_i'\right)\right\|\end{aligned}$$

since the sets where $\nabla\varphi^* f_i$ do not vanish are disjoint. By the last inequality, we have

$$\Phi(A_0')^{\frac{1}{q}} \geqslant \sup \frac{\|\varphi^* g_N \mid L^1_q(D)\|}{\| g_N \mid \overset{\circ}{L}{}^1_\infty\big(\bigcup\limits_{i=1}^{N} A_i'\big)\|} \geqslant \Big(\sum_{i=1}^{N} \Phi(A_i') - \varepsilon\Phi(A_0')\Big)^{\frac{1}{q}},$$

where the upper bound is taken over all above functions

$$g_N \in \overset{\circ}{L}{}^1_\infty\Big(\bigcup_{i=1}^{N} A_i'\Big).$$

Since both N and ε are arbitrary, we have

$$\sum_{i=1}^{\infty} \Phi(A_i') \leqslant \Phi\Big(\bigcup_{i=1}^{\infty} A_i'\Big).$$

The inverse inequality can be proved directly. Indeed, choose functions $f_i \in \overset{\circ}{L}{}^1_\infty(A_i')$ such that $\|f_i \mid \overset{\circ}{L}{}^1_\infty(A_i')\| = 1$. Setting $g = \sum\limits_{i=1}^{\infty} f_i$, we find

$$\begin{aligned}\|\varphi^* g \mid L^1_q(D)\| &\leqslant \Big(\sum_{i=1}^{\infty} \Phi(A_i')\|f_i \mid \overset{\circ}{L}{}^1_\infty(A_i')\|^q\Big)^{1/q} \\ &= \Big(\sum_{i=1}^{\infty} \Phi(A_i')\Big)^{\frac{1}{q}} \Big\|g_N \mid \overset{\circ}{L}{}^1_\infty\Big(\bigcup_{i=1}^{\infty} A_i'\Big)\Big\|\end{aligned}$$

since the sets where $\nabla\varphi^* f_i$ do not vanish are disjoint. From the last inequality we have

$$\Phi\Big(\bigcup_{i=1}^{\infty} A_i'\Big)^{\frac{1}{q}} \leqslant \sup \frac{\|\varphi^* g \mid L^1_q(D)\|}{\| g \mid \overset{\circ}{L}{}^1_\infty\big(\bigcup\limits_{i=1}^{\infty} A_i'\big)\|} \leqslant \Big(\sum_{i=1}^{\infty} \Phi(A_i')\Big)^{\frac{1}{q}},$$

where the upper bound is taken over all functions $g \in \overset{\circ}{L}{}^1_\infty\big(\bigcup\limits_{i=1}^{\infty} A_i'\big)$.

By the definition of Φ,

$$\|\varphi^* f \mid L^1_q(D)\|^p \leqslant \Phi(A')\|f \mid \overset{\circ}{L}{}^1_\infty(A')\|^q$$

Since the support of $f \circ \varphi$ is contained in $\varphi^{-1}(A')$, we have

$$\int\limits_{\varphi^{-1}(A)} |\nabla(f \circ \varphi)|^q \, dx \leqslant \Phi(A')\operatorname{ess\,sup}_{y\in A'} |\nabla f|^q(y).$$

The theorem is proved. □

Recall some basic facts about p-capacity. Let $G \subset \mathbb{R}^n$ be an open set, and let $E \subset G$ be a compact set. For $1 \leqslant p \leqslant \infty$ the *p-capacity* of the ring (E, G) is defined as

$$\operatorname{cap}_p(E,G) = \inf\left\{ \int_G |\nabla u|^p : u \in L^1_p(G) \cap C_0^\infty(G),\ u \geqslant 1 \text{ on } E \right\}.$$

Functions $u \in L^1_p(G) \cap C_0^\infty(G)$, $u \geqslant 1$ on E, are called *admissible* for the ring (E, G).

We need the following estimate of the p-capacity [11].

Lemma 2.1. *Let E be a connected closed subset of an open bounded set $G \subset \mathbb{R}^n$, $n \geqslant 2$, and let $n-1 < p < \infty$. Then*

$$\operatorname{cap}_p^{n-1}(E,G) \geqslant c \frac{(\operatorname{diam} E)^p}{|G|^{p-n+1}},$$

where c is a constant depending only on n and p.

We define the class BVL of mappings with finite variation. We say that a mapping $\varphi : D \to \mathbb{R}^n$ belongs to the class $\mathrm{BVL}(D)$ (i.e., has finite variation on almost all straight lines) if it has finite variation on almost all straight lines l parallel to the coordinate axis: for any finite number of points $t_1, \dots, t_k$ in such a line l

$$\sum_{i=0}^{k-1} |\varphi(t_{i+1}) - \varphi(t_i)| < +\infty.$$

For a mapping φ with finite variation on almost all straight lines the partial derivatives $\partial \varphi_i / \partial x_j$, $i, j = 1, \dots, n$, exists almost everywhere in D.

The following assertion was announced in [21].

Theorem 2.3. *Suppose that a homeomorphism $\varphi : D \to D'$ generates a bounded composition operator $\varphi^* : L^1_\infty(D') \to L^1_q(D)$, $q > n-1$. Then the inverse homeomorphism $\varphi^{-1} : D' \to D$ belongs to the class* $\mathrm{BVL}(D')$.

Proof. Take an arbitrary n-dimensional open parallelepiped P such that $\overline{P} \subset D'$ and the edges of P are parallel to the coordinate axes. Let us show that φ^{-1} has finite variation on almost all intersections of P with lines parallel to the x_n-axis.

Let P_0 be the projection of P onto the subspace $x_n = 0$, and let I be the projection of P onto the x_n-axis. Then $P = P_0 \times I$. The monotone countable-additive function Φ determines a monotone countable additive function of open sets $A \subset P_0$ by the rule $\Phi(A, P_0) = \Phi(A \times I)$. For almost all $z \in P_0$ the expression

$$\overline{\Phi'}(z, P_0) = \overline{\lim_{r\to 0}} \left[\frac{\Phi(B^{n-1}(z,r), P_0)}{r^{n-1}} \right]$$

is finite [18] (here, $B^{n-1}(z,r)$ is the $(n-1)$-dimensional ball of radius $r>0$ centered at the point z).

The n-dimensional Lebesgue measure $\Psi(U) = |\varphi^{-1}(U)|$, where U is an open set in D', is a monotone countable additive function. Therefore, it also determines a monotone countable additive function $\Psi(A, P_0) = \Psi(A \times I)$ defined on open sets $A \subset P_0$. Hence $\overline{\Psi'}(z, P_0)$ is finite for almost all points $z \in P_0$.

Choose an arbitrary point $z \in P_0$, where $\overline{\Phi'}(z, P_0) < +\infty$ and $\overline{\Psi'}(z, P_0) < +\infty$. On the section $I_z = \{z\} \times I$ of the parallelepiped P, we take arbitrary mutually disjoint closed intervals $\Delta_1, \dots, \Delta_k$ of length $b_1, \dots, b_k$ respectively. Let R_i denote the open set of points the distance from which to Δ_i is smaller than a given number $r>0$:

$$R_i = \{x \in G : \operatorname{dist}(x, \Delta_i) < r\}.$$

Consider the ring (Δ_i, R_i). Let $r>0$ be such that $r < cb_i$ for $i = 1, \dots, k$, where c is a sufficiently small constant. Then the function $u_i(x) = \operatorname{dist}(x, \Delta_i)/r$ is admissible for the ring (Δ_i, R_i).

By Theorem 2.2,

$$\|\varphi^* u_i \mid L^1_q(D)\|^q \leqslant \Phi(A') \|u_i \mid \overset{\circ}{L}{}^1_\infty(A')\|^q$$

for every u_i, $i = 1, \dots, k$.

Hence for every ring (Δ_i, R_i), $i = 1, \dots, k$,

$$\operatorname{cap}_q^{\frac{1}{q}}(\varphi^{-1}(\Delta_i), \varphi^{-1}(R_i)) \leqslant \Phi(R_i)^{\frac{1}{q}} \operatorname{cap}_\infty(\Delta_i, R_i).$$

The function $u_i(x) = \operatorname{dist}(x, \Delta_i)/r$ is admissible for the ring (Δ_i, R_i), and we obtain the upper estimate

$$\operatorname{cap}_\infty(\Delta_i, R_i) \leqslant |\nabla u_i| = 1/r.$$

Using the lower bound for the capacity of the ring (Lemma 2.1), we obtain the inequality

$$\left(\frac{(\operatorname{diam} \varphi^{-1}(\Delta_i))^{q/(n-1)}}{|\varphi^{-1}(R_i)|^{(q-n+1)/(n-1)}} \right)^{\frac{1}{q}} \leqslant c_1 \Phi(R_i)^{\frac{1}{q}} \frac{1}{r}$$

which implies

$$\operatorname{diam} \varphi^{-1}(\Delta_i) \leqslant c_2 \left(\frac{|\varphi^{-1}(R_i)|}{r^{n-1}} \right)^{\frac{q-n+1}{q}} \left(\frac{\Phi(R_i)}{r^{n-1}} \right)^{\frac{n-1}{q}}.$$

Taking the sum with respect to $i = 1, \dots, k$, we find

$$\sum_{i=1}^{k} \operatorname{diam} \varphi^{-1}(\Delta_i) \leqslant c_2 \sum_{i=1}^{k} \left(\frac{|\varphi^{-1}(R_i)|}{r^{n-1}} \right)^{\frac{q-n+1}{q}} \left(\frac{\Phi(R_i)}{r^{n-1}} \right)^{\frac{n-1}{q}}.$$

Hence

$$\sum_{i=1}^{k} \operatorname{diam} \varphi^{-1}(\Delta_i) \leqslant c_2 \left(\sum_{i=1}^{k} \frac{|\varphi^{-1}(R_i)|}{r^{n-1}} \right)^{\frac{q-n+1}{q}} \left(\sum_{i=1}^{k} \frac{\Phi(R_i)}{r^{n-1}} \right)^{\frac{n-1}{q}}.$$

Using the Bezikovich type theorem [7] for the estimate of Φ in terms of the multiplicity of a cover, we obtain

$$\sum_{i=1}^{k} \operatorname{diam} \varphi^{-1}(\Delta_i) \leqslant c_3 \left(\frac{|\varphi^{-1}(\bigcup_{i=1}^{k} R_i)|}{r^{n-1}} \right)^{\frac{q-n+1}{q}} \left(\frac{\Phi(\bigcup_{i=1}^{k} R_i)}{r^{n-1}} \right)^{\frac{n-1}{q}}.$$

Hence

$$\sum_{i=1}^{k} \operatorname{diam} \varphi^{-1}(\Delta_i) \leqslant c_3 \left(\frac{|\varphi^{-1}(B^{n-1}(z,r), P_0)|}{r^{n-1}} \right)^{\frac{q-n+1}{q}} \left(\frac{\Phi(B^{n-1}(z,r), P_0)}{r^{n-1}} \right)^{\frac{n-1}{q}}.$$

Since $\overline{\Phi'}(z, P_0) < +\infty$ and $\overline{\Psi'}(z, P_0) < +\infty$, we find

$$\sum_{i=1}^{k} \operatorname{diam} \varphi^{-1}(\Delta_i) < +\infty.$$

Therefore, $\varphi^{-1} \in \mathrm{BVL}(D')$. □

3 Proof of the Main Result

Recall the change of variable formula for the Lebesgue integral [8]. Let a mapping $\varphi : D \to \mathbb{R}^n$ belong to $W^1_{1,\mathrm{loc}}(D)$. Then there exists a measurable set $S \subset D$, $|S| = 0$, such that the mapping $\varphi : D \setminus S \to \mathbb{R}^n$ possesses the Luzin N-property and the change of variable formula

$$\int_E f \circ \varphi(x) |J(x,\varphi)| \, dx = \int_{\mathbb{R}^n \setminus \varphi(S)} f(y) N_f(E,y) \, dy$$

holds for any measurable set $E \subset D$ and nonnegative Borel measurable function $f : \mathbb{R}^n \to \mathbb{R}$. Here, $N_f(y,E)$ is the multiplicity function defined as the number of preimages of y under the mapping f in E.

If a mapping φ possesses the Luzin N-property (the image of a set of measure zero has measure zero), then $|\varphi(S)| = 0$ and the second integral can be written as the integral over $\mathbb{R}^n$. If a homeomorphism $\varphi : D \to D'$ belongs to the Sobolev space $W^1_{n,\mathrm{loc}}(D)$, then φ possesses the Luzin N-property and the change of variable formula holds [19].

As in [17] (cf. also [10]), we introduce the measurable function

$$\mu(y) = \begin{cases} \left(\dfrac{|\operatorname{adj} D\varphi|(x)}{|J(x,\varphi)|}\right)_{x=\varphi^{-1}(y)} & \text{if } x \in D \setminus S \quad \text{and} \quad J(x,\varphi) \neq 0, \\ 0 & \text{otherwise.} \end{cases}$$

Since the homeomorphism φ has finite distortion, the function $\mu(y)$ is well defined almost everywhere in D'.

The following lemma was proved (but not formulated) in [10] under an additional assumption that $|D\varphi|$ belongs to the Lorentz space $L^{n-1,n}(D)$.

Lemma 3.1. *Let a homeomorphism $\varphi : D \to D'$, $\varphi(D) = D'$ belong to the Sobolev space $L^1_q(D)$ for some $q > n-1$. Then the function μ is locally integrable in D'.*

Proof. Using the change of variable formula for the Lebesgue integral [8] and the Luzin N-property of φ, we obtain the equality

$$\int_{D'} \mu(y)\, dy = \int_{D'\setminus\varphi(S)} \mu(y)\, dy = \int_{D\setminus S} |\mu(\varphi(x))|J(x,\varphi)|\, dx = \int_{D} |\operatorname{adj} D\varphi|(x)\, dx.$$

Applying the Hölder inequality, for every compact subset $F' \subset D'$ we have

$$\int_{F'} \mu(y)\, dy \leqslant \int_{F} |\operatorname{adj} D\varphi|(x)\, dx \leqslant C \int_{F} |D\varphi|^{n-1}(x)\, dx,$$

where $F' = \varphi(F)$. Therefore, μ belongs to $L_{1,\mathrm{loc}}(D')$ since φ belongs to $L^1_q(D)$, $q > n-1$. Hence $\varphi \in L^1_{n-1,\mathrm{loc}}(D)$. □

Proof of Theorem A. We show that $\varphi^{-1} \in \mathrm{ACL}(D')$. Since the absolute continuity is a local property, it suffices to show that φ^{-1} belongs to ACL on every compact subset of D'. Consider an arbitrary cube $Q' \Subset D'$, $\overline{Q'} \Subset D$, with edges parallel to the coordinate axes. We set $Q = \varphi^{-1}(Q')$. For $i = 1, \dots n$ we denote $Y_i = (x_1, \dots, x_{i-1}, x_{i+1}, \dots, x_n)$,

$$F_i(x) = (\varphi_1(x), \dots, \varphi_{i-1}(x), \varphi_{i+1}(x), \dots, \varphi_n(x))$$

and Q'_i is the intersection of Q' with $Y_i = \mathrm{const}$.

Using the change of variable formula and the Fubini theorem [2], we obtain the following estimate

$$\int\limits_{F_i(Q)} H^{n-1}(dY_i) \int\limits_{Q_i'} \mu(y)\, H^1(dy) = \int\limits_{Q'} \mu(y)\, dy = \int\limits_{Q} |\operatorname{adj} D\varphi|(x)\, dx < +\infty.$$

Hence for almost all $Y_i \in F_i(Q)$

$$\int\limits_{Q_i'} \mu(y)\, H^1(dy) < +\infty.$$

Let $\operatorname{ap} J\varphi(x)$ be an approximate Jacobian of the trace of φ on the set $\varphi^{-1}(Q_i')$ [2]. Consider a point $x \in Q$ at which there exists a nondegenerated approximate differential $\operatorname{ap} Df(x)$ of the mapping $\varphi : D \to D'$. Let $L : \mathbb{R}^n \to \mathbb{R}^n$ be a linear mapping induced by this approximate differential $\operatorname{ap} Df(x)$. Denote by P the image of the unit cube Q_0 under the linear mapping L and by P_i the intersection of P with the image of the line $x_i = 0$. Denote by d_i the length of P_i. Then

$$d_i \cdot |\operatorname{adj} DF_i|(x) = |Q_0| = |J(x,\varphi)|.$$

So, since $d_i = \operatorname{ap} J\varphi(x)$, for almost all $x \in Q \setminus Z$, $Z = \{x \in D : J(x,\varphi) = 0\}$, we have

$$\operatorname{ap} J\varphi(x) = \frac{|J(x,\varphi)|}{|\operatorname{adj} DF_i|(x)}.$$

So, for an arbitrary compact set $A' \subset Q_i'$ and almost all $Y_i \subset F_i(Q)$ the following inequality holds:

$$\begin{aligned} H^1(\varphi^{-1}(A')) &\leqslant \int\limits_{\varphi^{-1}(A')} \frac{|\operatorname{adj} D\varphi|(x)}{|\operatorname{adj} DF_i|(x)}\, H^1(dx) \\ &= \int\limits_{\varphi^{-1}(A')} \frac{|\operatorname{adj} D\varphi|(x)}{|J(x,\varphi)|} \cdot \frac{|J(x,\varphi)|}{|\operatorname{adj} DF_i|(x)}\, H^1(dx) \\ &= \int\limits_{\varphi^{-1}(A')} \mu(\varphi(x))\, \operatorname{ap} J\varphi(x)\, H^1(dx). \end{aligned}$$

Using the change of variable formula for the Lebesgue integral [2], we find

$$H^1(f^{-1}(A')) \leqslant \int\limits_{A'} \mu(y)\, H^1(dy) < +\infty.$$

Therefore, the mapping φ^{-1} is absolutely continuous on almost all lines in $D'$and is weakly differentiable.

Since the homeomorphism φ possesses the Luzin N-property, the preimage of a set of positive measure is a set of positive measure. Hence, for the volume derivative of the inverse mapping we have

$$J_{\varphi^{-1}}(y) = \lim_{r\to 0} \frac{|\varphi^{-1}(B(y,r))|}{|B(y,r)|} > 0$$

almost everywhere in D'. So $J(y,\varphi^{-1}) \neq 0$ for almost all points $y \in D$. The integrability of the q'-distortion follows from the inequality

$$|D\varphi^{-1}|(y) \leqslant |D\varphi(x)|^{n-1}/|J(x,\varphi)|$$

which holds for almost all points $y = \varphi(x) \in D'$.

Indeed, with the help of the change of variable formula, we have

$$\begin{aligned}
\int\limits_{D'} \left(\frac{|D\varphi^{-1}(y)|^{q'}}{|J(y,\varphi^{-1})|}\right)^{\frac{1}{q'-1}} dy &= \int\limits_{D'} \left(\frac{|D\varphi^{-1}(y)|}{|J(y,\varphi^{-1})|}\right)^{\frac{q'}{q'-1}} |J(y,\varphi^{-1})|\, dy \\
&\leqslant \int\limits_{D} \left(\frac{|D\varphi^{-1}(\varphi(x))|}{|J(\varphi(x),\varphi^{-1})|}\right)^{\frac{q'}{q'-1}} dx \\
&\leqslant \int\limits_{D} |D\varphi(x)|^q\, dx < +\infty
\end{aligned}$$

since φ belongs to $L^1_q(D)$ in view of Theorem 2.1.

The boundedness of the composition operator follows from the integrability of p'-distortion [20]. □

Acknowledgement. This work was partially supported by Israel Scientific Foundation (grant 1033/07).

The authors thank Professor Jan Malý for useful discussions.

References

1. Burenkov, V.I.: Sobolev Spaces on Domains. Teubner-Texter zur Mathematik, Stuttgart (1998)
2. Federer, H.: Geometric Measure Theory. Springer, Berlin (1969)
3. Gol'dshtein, V., Gurov, L.: Applications of change of variable operators for exact embedding theorems. Integral Equat. Operator Theory **19**, no. 1, 1–24 (1994)
4. Gol'dshtein, V., Gurov, L., Romanov, A.S.: Homeomorphisms that induce monomorphisms of Sobolev spaces. Israel J. Math. **91**, 31–60 (1995)
5. Gol'dshtein, V., Romanov, A.S.: Transformations that preserve Sobolev spaces (Russian). Sib. Mat. Zh. **25**, no. 3, 55–61 (1984); English transl.: Sib. Math. J. **25**, 382–399 (1984)

6. Gol'dshtein, V., Ukhlov, A.: Weighted Sobolev spaces and embedding theorems. Trans. Am. Math. Soc. **361**, no. 7, 3829–3850 (2009)
7. Gusman, M.: Differentiation of integrals in $\mathbb{R}^n$. Springer, Berlin etc. (1975)
8. Hajlasz, P.: Change of variable formula under minimal assumptions. Colloq. Math. **64**, no. 1, 93–101 (1993)
9. Hencl, S., Koskela, P.: Regularity of the inverse of a planar Sobolev homeomorphism. Arch. Ration. Mech. Anal. **180**, no. 1, 75–95 (2006)
10. Hencl, S., Koskela, P., Malý, J.: Regularity of the inverse of a Sobolev homeomorphism in space. Proc. Roy. Soc. Edinburgh Sect. A. **136**, no. 6, 1267–1285 (2006)
11. Kruglikov, V.I.: Capacities of condensers and spatial mappings quasiconformal in the mean (Russian). Mat. Sb. **130**, no. 2, 185–206 (1986)
12. Markina, I.G.: A change of variable that preserves the differential properties of functions. Sib. Math. J. **31**, no. 3, 73–84 (1990); English transl.: Sib. Math. J. **31**, no. 3, 422–432 (1990)
13. Maz'ya, V.G.: Sobolev Spaces. Springer, Berlin etc. (1985)
14. Maz'ya, V.G.: Weak solutions of the Dirichlet and Neumann problems (Russian). Tr. Mosk. Mat. O-va. **20**, 137–172 (1969)
15. Maz'ya, V.G., Shaposhnikova, T.O.: Theory of Multipliers in Spaces of Differentiable Functions. Pitman, Boston etc. (1985) Russian edition: Leningrad. Univ. Press, Leningrad (1986)
16. Muller, S., Tang, Q., Yan, B.S.: On a new class of elastic deformations not allowing for cavitation. Ann. Inst. H. Poincaré. Anal. Non-Lineaire. **11**, no. 2, 217–243 (1994)
17. Peshkichev, Yu.A.: Inverse mappings for homeomorphisms of the class BL (Russian). Mat. Zametki **53**, no. 5, 98–101 (1993); English transl.: Math. Notes **53**, no. 5, 520–522 (1993)
18. Rado, T., Reichelderfer, P.V.: Continuous Transformations in Analysis. Springer, Berlin (1955)
19. Reshetnyak, Yu.G.: Some geometrical properties of functions and mappings with generalized derivatives (Russian). Sib. Mat. Zh. **7**, 886–919 (1966); English transl.: Sib. Math. J. **7**, 704–732 (1967)
20. Ukhlov, A.D.: On mappings generating the embeddings of Sobolev spaces (Russian). Sib. Mat. Zh. **34**, no. 1, 185–192 (1993); English transl.: Sib. Math. J. **34**, no. 1, 165–171 (1993)
21. Ukhlov, A.: Differential and geometrical properties of Sobolev mappings (Russian). Mat. Zametki **75**, no. 2, 317-320 (2004); English transl.: Math. Notes **75**, no. 2, 291-294 (2004)
22. Vodop'yanov, S.K., Gol'dshtein, V.M.: Structure isomorphisms of spaces W_n^1 and quasiconformal mappings (Russian). Sib. Mat. Zh. **16**, 224–246 (1975); English transl.: Sib. Math. J. **16**, no. 2, 174–189 (1975)
23. Vodop'yanov, S.K., Gol'dshtein, V.M.: Functional characteristics of quasiisometric mappings (Russian). Sib. Mat. Zh. **17**, 768–773 (1976); Englsih transl.: Sib. Math. J. **17**, no. 4, 580–584 (1976)
24. Vodop'yanov, S.K., Ukhlov, A.D.: Superposition operators in Sobolev spaces (Russian). Izv. Vyssh. Uchebn. Zaved., Mat. no. 10, 11-33 (2002); English transl.: Russ. Math. **46**, no. 10, 9-31 (2002)
25. Vodop'yanov, S.K.; Ukhlov, A.D. Set functions and their applications in the theory of Lebesgue and Sobolev spaces. I (Russian). Mat. Tr. **6**, no. 2, 14-65 (2003); English transl.: Sib. Adv. Math. **14**, no. 4, 78-125 (2004)

Extended L^p Dirichlet Spaces

Niels Jacob and René L. Schilling

Abstract In order to develop a nonlinear L^p potential theory for generators of Markov processes, we propose a definition of extended Dirichlet spaces in an L^p setting. We study these spaces in the context of generalized Bessel potential spaces which arise from the Γ-transform of a given L^p sub-Markovian semigroup. First considerations on related variational capacities are presented.

1 Introduction

Maz'ya's research showed that capacities are an essential and powerful tool in the study of Sobolev and related function spaces (cf. the important monograph [18] by Maz'ya). A further seminal contribution of his was the initiation of nonlinear potential theory (cf. the survey by Maz'ya and Khavin [19] and the references therein).

Our interest in both topics grew out of investigations on Dirichlet forms. Dirichlet spaces were originally introduced by Beurling and Deny [1, 2] as an approach to potential theory based on the notion of energy. In the hands of Fukushima [6] (cf. also [7, 10]) and Silverstein [22], Dirichlet forms became central to stochastic analysis. The fact that with a regular Dirichlet space we can associate a Hunt process and then base a stochastic calculus on Hilbert space methods, led to unforseen progress in the 1970s and 1980s.

Niels Jacob
Department of Mathematics, Swansea University, Singleton Park, Swansea SA2 8PP, UK
e-mail: n.jacob@swansea.ac.uk

René L. Schilling
Technische Universität Dresden, Institut für Stochastik, D-01062 Dresden, Germany
e-mail: rene.schilling@tu-dresden.de

A. Laptev (ed.), *Around the Research of Vladimir Maz'ya I: Function Spaces*, International Mathematical Series 11, DOI 10.1007/978-1-4419-1341-8_9,

The construction of a transition function when starting with a regular Dirichlet form depends heavily on the notion of sets of capacity zero and the existence of quasicontinuous modifications of elements in the given Dirichlet space $(\mathcal{E}, \mathcal{F})$. The "usual" capacity is the variational capacity $\text{cap}_{1,2}(\cdot)$ defined with the help of the bilinear form $\mathcal{E}_1(u,v) = \mathcal{E}(u,v) + \langle u,v\rangle_{L^2}$. Despite its enormous success and its generality, this approach has one obvious shortcoming: the existence of exceptional sets and, therefore, the lack of pointwise notions.

The original approach of Beurling and Deny uses the functional space generated in a certain sense by the form $\mathcal{E}$, whereas nowadays we start with $\mathcal{E}$ on the domain of $\mathcal{E}_1 := \mathcal{E} + \langle\cdot,\cdot\rangle_{L^2}$ which is always a subspace of L^2. In some cases, especially in the case of transient Dirichlet forms, starting with $(\mathcal{E}, \mathcal{F})$, $\mathcal{F} = D(\mathcal{E}_1)$, the corresponding extended Dirichlet space $\mathcal{F}_e$ is studied and is related to the original construction of Beurling and Deny.

In a series of papers, the authors, partially in co-authorship with Farkas or Hoh, started to investigate systematically the L^p-theory corresponding to Dirichlet forms (cf. [4, 5, 11, 14, 15]). Our investigations were stimulated by pioneering ideas of Malliavin [17], and, in particular, by Fukushima and Kaneko [9, 16, 8], to get a better control on exceptional sets using (r,p)-capacities. In this paper, we continue this line of research; we are especially interested in L^p versions of extended Dirichlet spaces. In order to achieve a certain self-contained presentation, we discuss in Section 2 why an L^p approach should lead to more regularity and a better control of exceptional sets. In Section 3, we collect some results obtained so far and we include some minor extensions. In Section 4, we give a definition of an extended L^p Dirichlet space and investigate some of its properties.

Our notation will be fairly standard or self-explanatory. In all other cases, our basic reference text is the treatise [12].

2 The Case for an L^p Theory for Dirichlet Forms

Let $(T_t)_{t\geqslant 0}$ be a symmetric sub-Markovian semigroup on $L^2(X,m)$, where X is a locally compact Hausdorff space and m is a Borel measure with full support. We tacitly assume that all semigroups are strongly continuous and contractive on the respective Banach spaces. We write $(A, D(A))$ for the infinitesimal generator of the semigroup and $(\mathcal{E}, \mathcal{F})$ for the associated Dirichlet form. As usual, $\mathcal{F} = D(\mathcal{E}_1)$, where $\mathcal{E}_1(u,v) = \mathcal{E}(u,v) + \langle u,v\rangle_{L^2}$. Note that $T_t|_{L^2\cap L^p}$ has for every $p \in (0,\infty)$ an extension to L^p such that $(T_t)_{t\geqslant 0}$ is a sub-Markovian semigroup on $L^p(X,m)$. Moreover, $(T_t)_{t\geqslant 0}$ is, on each L^p, an analytic semigroup. A proof of the first result is given by Davies [3], and the second result is due to Stein [23]. Combining both results yields

$$T_t\left(L^p(X,m)\right) \subset \bigcap_{k \geqslant 1} D((-A)^k), \tag{2.1}$$

where $(A, D(A))$ denotes the generator of $(T_t)_{t\geqslant 0}$ as semigroup in L^p. It is important to observe that (2.1) reduces statements about $T_t f$, $f \in L^p(X,m)$, to statements about elements from $\bigcap_{k\geqslant 1} D((-A)^k)$. In particular, if we happen to know that we have an embedding $D((-A)^{k_0}) \hookrightarrow C_b(X)$ for some $k_0 \in \mathbb{N}$, we find that $f \in L^p(X,m)$ implies $T_t f \in C_b(X)$. For any Borel set $B \in \mathcal{B}(X)$ with $m(B) < \infty$, we conclude that the equivalence class of $T_t \mathbf{1}_B \in L^p(X,m)$ contains a unique, continuous representative; this means that we may define a transition function

$$p_t(x,B) := T_t \mathbf{1}_B(x) \quad \text{for } all \quad x \in X. \tag{2.2}$$

This enables us to construct a corresponding Markov process *without any exceptional set.*

If $X = \mathbb{R}^n$ and m is Lebesgue measure, we know that a large class of (symmetric) pseudo-differential operators $-q(x,D)$ generates sub-Markovian semigroups. Quite often we have

$$D\big(q(x,D)^k\big) = H^{\psi,2k}(\mathbb{R}^n), \tag{2.3}$$

where $\psi : \mathbb{R}^n \to \mathbb{R}$ is a fixed continuous negative definite function. If $\psi(\xi) \geqslant c_0\,|\xi|^{\rho_0}$ for some constants $c_0, \rho_0 > 0$ and sufficiently large values of $|\xi| \geqslant R_0$, Sobolev's imbedding theorem implies that $H^{\psi,2k}(\mathbb{R}^n) \subset C_\infty(\mathbb{R}^n)$ if k is large enough. Recall that $\psi : \mathbb{R}^n \to \mathbb{R}$ is a continuous negative definite function if $\psi(0) \geqslant 0$ and $\xi \mapsto e^{-t\psi(\xi)}$ is, for each $t > 0$, positive definite in the usual sense. Moreover, u is in the anisotropic Bessel potential space $H^{\psi,s}(\mathbb{R}^n)$, $s \geqslant 0$, if $u \in L^2(\mathbb{R}^n)$ and

$$\|u\|^2_{H^{\psi,s}} := \int_{\mathbb{R}^n} (1+\psi(\xi))^2\, |\widehat{u}(\xi)|^2\, d\xi < \infty.$$

The inclusion (2.3) requires some regularity for the symbol $x \mapsto -q(x,\xi)$ of the generator $-q(x,D)$; even in the case of differential operators this regularity is not guaranteed and, indeed, it is of interest to reduce the regularity of the "coefficients." Nevertheless, in many cases where a second order elliptic differential operator $-L(x,D)$ in divergence form extends to a generator A of an analytic L^p sub-Markovian semigroup, we find that $\bigcap_{k\geqslant 1} D((-A)^k) \subset W^{1,p}(\mathbb{R}^n)$. If we work on a smooth domain G and we subject $-L(x,D)$ to certain boundary conditions, we can get $\bigcap_{k\geqslant 1} D((-A)^k) \subset W^{1,p}(G)$. Again, a Sobolev type imbedding theorem yields for large enough p, i.e., $p > n$, $W^{1,p}(\mathbb{R}^n) \subset C_b(\mathbb{R}^n)$ or $W^{1,p}(G) \subset C_b(G)$. Similar results hold for more general, nonlocal operators as is easily seen by combining subordination in the sense of Bochner with complex interpolation results [13].

It is clear that not all cases are covered by such an approach. In the next section, we discuss how (r,p)-capacities can be used to improve the situation.

3 Bessel Potential Spaces and (r,p)-Capacities

We always denote by $(T_t)_{t\geqslant 0}$ a strongly continuous, sub-Markovian contraction semigroup on $L^p(X,m)$ for some fixed $p\in(1,\infty)$. Its infinitesimal generator is denoted by $(A,D(A))$, and we assume that the dual semigroup $(T_t^*)_{t\geqslant 0}$ is again sub-Markovian on the space $L^{p'}(X,m)$ with $p'=\frac{p}{p-1}$. For the generators we clearly have

$$\int_X (Au)\cdot v\,dm = \int_X u\cdot A^*v\,dm$$

for all $u\in D(A)$ and $v\in D(A^*)\subset L^{p'}(X,m)$. The *$\Gamma$-transform* of $(T_t)_{t\geqslant 0}$ is defined by the Bochner integral

$$V_r u := \frac{1}{\Gamma(\frac{r}{2})}\int_0^\infty t^{\frac{r}{2}-1}\,e^{-t}\,T_t u\,dt,\quad u\in L^p(X,m). \tag{3.1}$$

One way to read (3.1) is to see $(V_r)_{r>0}$ as subordinate (in the sense of Bochner) to $(T_t)_{t\geqslant 0}$ (cf., for example, [20] or [4]). This makes it obvious that $(V_r)_{r>0}$ is a sub-Markovian semigroup on $L^p(X,m)$. Since the V_r are injective, we can define Bessel potential spaces by

$$\mathcal{F}_{r,p} := V_r(L^p),\quad \|f\|_{\mathcal{F}_{r,p}} = \|u\|_{L^p},\quad \text{where}\quad f = V_r u. \tag{3.2}$$

From [4] we know that

$$\mathcal{F}_{r,p} = D((1-A)^{r/2}),\quad V_r = (1-A)^{-r/2},\quad \|f\|_{\mathcal{F}_{r,p}} = \|(1-A)^{r/2}f\|_{L^p},$$

and we can show (cf. [5]) that for $r\in(0,1]$ and $f\in D(A)$

$$\frac{1}{3}\big(\|(-A)^r f\|_{L^p} + \|f\|_{L^p}\big) \leqslant \|(1-A)^r f\|_{L^p} \leqslant \big(\|(-A)^r f\|_{L^p} + \|f\|_{L^p}\big). \tag{3.3}$$

We now modify (3.1) by introducing a parameter $\lambda\in(0,1]$ so that

$$V_{r,\lambda} u := \frac{1}{\Gamma(\frac{r}{2})}\int_0^\infty t^{\frac{r}{2}-1}\,e^{-\lambda t}\,T_t u\,dt,\quad u\in L^p(X,m). \tag{3.4}$$

Using $\lambda - A = \lambda(1-\frac{1}{\lambda}A)$ together with (3.3), we yield

$$\frac{1}{3}\big(\|(-A)^r f\|_{L^p} + \lambda^r \|f\|_{L^p}\big) \leqslant \|(\lambda - A)^r f\|_{L^p}$$
$$\leqslant \big(\|(-A)^r f\|_{L^p} + \lambda^r \|f\|_{L^p}\big) \tag{3.5}$$

and deduce that the norms

$$\|(\lambda - A)^r f\|_{L^p}, \quad \|(1 - A)^r f\|_{L^p}, \quad \|(-A)^r f\|_{L^p} + \|f\|_{L^p}$$

are equivalent for all $\lambda > 0$. It is clear that the comparison constants depend on λ. Consequently,

$$\mathcal{F}_{r,p;\lambda} := V_{r,\lambda}(L^p) = \mathcal{F}_{r,p}, \quad V_{r,\lambda} = (\lambda - A)^{-r/2}. \tag{3.6}$$

Later on, we are interested in the limiting case as $\lambda \to 0$.

Let us introduce the corresponding L^p *energy forms*

$$\mathcal{E}_\lambda^{(r,p)}(f, g) := \int_X J_p((\lambda - A)^{r/2} f) \cdot (\lambda - A)^{r/2} g \, dm, \quad 0 \leqslant \lambda \leqslant 1, \tag{3.7}$$

where $J_p : L^p \to L^{p'}$, $\frac{1}{p} + \frac{1}{p'} = 1$, is the duality map induced by $j_p(x) = x\,|x|^{p-2}$. We use the shorthand $\mathcal{E}_\lambda^{(r,p)}(f) = \mathcal{E}_\lambda^{(r,p)}(f,f)$ for the diagonal terms. Note that $\mathcal{E}_\lambda^{(r,p)}$ is for $\lambda > 0$ and $(r,p) \neq (1,2)$ *different from* $\mathcal{E}^{(r,p)} + \lambda\langle\cdot,\cdot\rangle_{L^p}$, but (3.5) guarantees that $\mathcal{E}_\lambda^{(r,p)}(f)$ and $\mathcal{E}^{(r,p)}(f) + \lambda\|f\|_{L^p}^2$ are always comparable. As domain of $\mathcal{E}_\lambda^{(r,p)}$ we choose $\mathcal{F}_{r,p}$. Because of (3.5), $\mathcal{E}_0^{(r,p)}$ is indeed defined on $\mathcal{F}_{r,p}$ but it could well be possible to define $\mathcal{E}_0^{(r,p)}$ for elements f with $f \notin L^p(X, m)$, hence $f \notin \mathcal{F}_{r,p}$.

For $\lambda > 0$ the functional $\mathcal{E}_\lambda^{(r,p)}$ is strictly convex, coercive, and Gâteaux differentiable. The Gâteaux derivative at $f \in \mathcal{F}_{r,p}$ is given by

$$\mathcal{A}_{r,\lambda}^{(p)} f = (\lambda - A^*)^{r/2} \left(|(\lambda - A)^{r/2} f|^{p-2} \cdot (\lambda - A)^{r/2} f \right) \tag{3.8}$$

(cf. [11]). As in [11], one can show that $\mathcal{A}_{r,\lambda}^{(p)} : \mathcal{F}_{r,p} \to \mathcal{F}_{r,p}^*$ is uniformly monotone and for every unbounded set coercive. However, the constant in the uniform monotonicity estimate depends on λ. More precisely,

$$\left\langle \mathcal{A}_{r,\lambda}^{(p)} f - \mathcal{A}_{r,\lambda}^{(p)} g,\, f - g \right\rangle_{L^2} \geqslant 2^{2-p} \int_X |(\lambda - A)^{r/2}(f - g)|^p \, dm$$
$$\geqslant c(\lambda)\, \|f - g\|_{\mathcal{F}_{r,p}}^p.$$

In addition, $\mathcal{A}_{r,\lambda}^{(p)}$ is hemicontinuous.

We are mainly interested in the case where p is large. Without loss of generality, we will, from now on, assume that $p \geqslant 2$. For open sets $G \subset X$ we define

$$\Gamma_{r,p}(G) := \{f \in \mathcal{F}_{r,p} \,:\, f \geqslant \mathbf{1}_G \;\; m\text{-a.e.}\} \tag{3.9}$$

and the capacity

$$\mathrm{cap}_{r,p,\lambda}(G) := \inf\{\mathcal{E}_\lambda^{r,p}(f) \,:\, f \in \Gamma_{r,p}(G)\}. \tag{3.10}$$

For arbitrary $A \subset X$

$$\mathrm{cap}_{r,p,\lambda}(A) := \inf\{\mathrm{cap}_{r,p,\lambda}(G) \,:\, G \;\text{ open and }\; A \subset G\}$$

then defines a Choquet capacity. Using standard arguments [9], one proves

Theorem 3.1. *For $2 \leqslant p < \infty$ and every open set $G \subset X$ with $\mathrm{cap}_{r,p,\lambda}(G) < \infty$ there exists a unique equilibrium potential $f_{G,\lambda} \in \mathcal{F}_{r,p}$ such that $f_{G,\lambda} \geqslant \mathbf{1}_G$ m-a.e. and*

$$\mathrm{cap}_{r,p,\lambda}(G) = \|f_{G,\lambda}\|_{\mathcal{F}_{r,p}}.$$

Moreover, there exists a unique $u_{G,\lambda} \in L^p(X,m)$, $u_{G,\lambda} \geqslant 0$ m-a.e., with $f_{G,\lambda} = V_{r,\lambda} u_{G,\lambda}$.

The equilibrium potential is characterized (cf. [11]) by the following inequalities

$$\left\langle \mathcal{A}_{r,\lambda}^{(p)} f_{G,\lambda},\, \phi - f_{G,\lambda} \right\rangle_{L^2} \geqslant 0 \;\text{ for all }\; \phi \in \Gamma_{r,p}(G)$$

or, equivalently,

$$\mathcal{E}_\lambda^{r,p}(f_{G,\lambda}, \phi - f_{G,\lambda}) \geqslant 0 \;\text{ for all }\; \phi \in \Gamma_{r,p}(G).$$

Following [19], we introduce the nonlinear potential operator

$$\mathcal{U}_{r,\lambda}^{(p)} := (\mathcal{A}_{r,\lambda}^{(p)})^{-1} : \mathcal{F}_{r,p}^* \to \mathcal{F}_{r,p}$$

which is given by

$$\mathcal{U}_{r,\lambda}^{(p)} u := V_{r,\lambda}\big(|V_{r,\lambda}^* u|^{p'-2} \cdot V_{r,\lambda}^* u\big). \tag{3.11}$$

It follows (cf. [11]) that $f_{G,\lambda} = \mathcal{U}_{r,\lambda}^{(p)} u_{G,\lambda}$.

Definition 3.1. We say that $\mathcal{F}_{r,p}$ has the *truncation property* if all Lipschitz continuous functions Φ with Lipschitz constant 1 operate on $\mathcal{F}_{r,p}$ and do not increase the $\mathcal{F}_{r,p}$ norm, i.e., if for all $f \in \mathcal{F}_{r,p}$

$$\Phi(f) \in \mathcal{F}_{r,p} \quad\text{and}\quad \mathcal{E}^{(r,p)}(\Phi(f)) + \|\Phi(f)\|_{L^p}^2 \leqslant \mathcal{E}^{(r,p)}(f) + \|f\|_{L^p}^2.$$

Note that $\mathcal{F}_{r,p}$ always has the truncation property for $r \leqslant 1$. As in [11], we deduce that

$$\operatorname{cap}_{r,p,\lambda}(G) = \mathcal{E}_\lambda^{(r,p)}(f_{G,\lambda}, \phi) \quad \text{for all } \phi \in \mathcal{F}_{r,p} \text{ with } \phi|_G = 1 \ m\text{-a.e.}$$

With the techniques from [14] it is possible to investigate the relation of $\operatorname{cap}_{r,p,\lambda}$ to other capacities associated with $(T_t)_{t\geqslant 0}$ or $\mathcal{U}_{r,\lambda}^{(p)}$.

The following monotonicity result was (with $\lambda = 1$) proved by Fukushima and Kaneko [9]

$$\operatorname{cap}_{r,p,\lambda}(A) \leqslant \operatorname{cap}_{s,q,\lambda}(A) \quad \text{for all } r \leqslant s \text{ and } p \leqslant q. \tag{3.12}$$

Kaneko [16] (cf. also [12, Vol. 3]) succeeded to associate Hunt processes with certain L^p sub-Markovian semigroups. In this construction, exceptional sets are measured in terms of (r,p)-capacities. For large values of p it may happen that $\operatorname{cap}_{r,p}(A) = 0$ already implies that $A = \varnothing$. This means that *no* exceptional sets enter at all in the construction of a process. Again we see that the L^p approach leads to higher regularity results and allows us to avoid exceptional sets.

4 Extended L^p-Dirichlet Spaces

For $f, g \in \mathcal{F}_{r,p}$ we know that

$$\lim_{\lambda \to 0+} \mathcal{E}_\lambda^{(r,p)}(f,g) = \mathcal{E}_0^{(r,p)}(f,g). \tag{4.1}$$

Note that $\mathcal{E}_0^{(r,p)}(f,g)$ need not be a definite form and it might be happen that (for some extension)

$$(-A)^{r/2} f \in L^p(X,m) \quad \text{while} \quad f \notin L^p(X,m).$$

In fact, any homogeneous second order elliptic differential operator on $\mathbb{R}^n$ with constant coefficients has a kernel that contains all constant and all linear functions on $\mathbb{R}^n$ which are both not in $L^p(\mathbb{R}^n, dx)$. For $p = 2$, and any symmetric bilinear form on $L^2(X,m)$ given by a Beurling–Deny representation without killing and local terms,

$$\mathcal{E}(f,g) = \iint_{X\times X} \big(f(x+y) - f(x)\big)\big(g(x+y) - g(x)\big)\, J(dx,dy), \tag{4.2}$$

we find, that $\mathcal{E}$ is well defined and zero for all constant functions—but these are not in $L^2(X,m)$ unless $m(X) < \infty$. Thus, it is a reasonable question to find and to investigate $\mathcal{E}_0^{(r,p)}$ on its *natural* domain. Whenever $\mathcal{F}_{r,p}$ is contained in an "exterior world," this becomes a bit easier. For example, consider all measurable functions u for which $\mathcal{E}_0^{(r,p)}(u,u)$ is finite. In the case of translation invariant forms on $L^2(\mathbb{R}^n, dx)$ which are uniquely determined by some negative definite ψ, natural candidates would be all $u \in \mathcal{S}'(\mathbb{R}^n)$ such that $\sqrt{|\psi|}\,\widehat{u} \in L^2(\mathbb{R}^n, dx)$.

In this section, we want to construct the L^p analogue of an extended Dirichlet space using the classical approach as in [7] or [10]. This means we start with elements in $\mathcal{F}_{r,p}$ and construct a suitable space of measurable functions onto which we extend $\mathcal{E}_0^{(r,p)}$.

Let $(T_t)_{t\geqslant 0}$ and $(A, D(A))$ be as in Section 3. Assume that $r \in (0,1]$ and $2 \leqslant p < \infty$. Note that $-(-A)^r$ generates for every $r \in (0,1]$ again a sub-Markovian semigroup, and we write $\left(\beta+(-A)^{r/2}\right)^{-1}$, $\beta > 0$, for the resolvent operators of the square roots $-(-A)^{r/2}$. These operators coincide with the Γ-transforms $V_{r,\beta}$ of the semigroup generated by $-(-A)^r$ (cf. (3.4)). It is well known that $\left(\beta + (-A)^{r/2}\right)^{-1}$ can be represented as

$$\left(\beta + (-A)^{r/2}\right)^{-1} u(x) = \int_X u(y)\, \rho_{\beta,r}(x, dy)$$

with a sub-Markovian kernel $\beta\rho_{\beta,r}(x,dy)$. In particular, $\left(\beta + (-A)^{r/2}\right)^{-1}$ extends to $L^\infty(X,m)$ (cf., for example, [21]). Consider the Yosida approximation of $(-A)^{r/2}$,

$$(-A)^{r/2}_\beta := \beta(-A)^{r/2}\left(\beta + (-A)^{r/2}\right)^{-1} = \beta - \beta^2\left(\beta + (-A)^{r/2}\right)^{-1}$$

which we can rewrite in the form

$$(-A)^{r/2}_\beta f(x) = \beta \int_X \big(f(x) - f(y)\big)\, \beta\rho_{\beta,r}(x,dy) + \beta\big(1 - \beta\rho_{\beta,r}(x,X)\big) f(x). \tag{4.3}$$

This shows, in particular, that we can define $(-A)^{r/2}_\beta$ on $L^\infty(X,m)$ and not only on $L^p(X,m)$. As in [10, (1.3.16)], we introduce on $\mathcal{F}_{r,p}$

$$\mathcal{E}^{(r,p),(\beta)}(f,g) := \int J_p\big((-A)^{r/2}_\beta f\big) \cdot (-A)^{r/2}_\beta g\, dm, \quad f,g \in \mathcal{F}_{r,p}. \tag{4.4}$$

In particular, we have

$$\begin{aligned}
\mathcal{E}^{(r,p),(\beta)}(f) &:= \mathcal{E}^{(r,p),(\beta)}(f,f) \\
&= \int_X \left|(-A)^{r/2}_\beta f\right|^p dm \\
&= \int_X \left|\beta \int_X \big(f(x) - f(y)\big)\, \beta\rho_{\beta,r}(x,dy)\right|^p m(dx) \\
&= \left\|(-A)^{r/2}_\beta f\right\|^p_{L^p}.
\end{aligned}$$

Lemma 4.1. *For $f \in \mathcal{F}_{r,p}$ and all $0 < \beta \leqslant \alpha < \infty$ we have*

$$\mathcal{E}^{(r,p),(\beta)}(f) \leqslant \mathcal{E}^{(r,p),(\alpha)}(f).$$

In particular,

$$\mathcal{E}^{(r,p)}(f) = \lim_{\lambda \to 0} \mathcal{E}_\lambda^{(r,p)}(f) = \sup_{\beta > 0} \mathcal{E}^{(r,p),(\beta)}(f) \in [0, \infty] \tag{4.5}$$

is well defined.

Proof. On $\mathcal{F}_{r,p}$, we have

$$(-A)_\beta^{r/2} = \beta\big(\beta + (-A)^{r/2}\big)^{-1}(-A)^{r/2}.$$

Since $\beta\big(\beta+(-A)^{r/2}\big)^{-1}$ is an L^p contraction, we conclude that $\mathcal{E}^{(r,p),(\beta)}(f) \leqslant \mathcal{E}^{(r,p)}(f) < \infty$.

For $\alpha \geqslant \beta$ we find

$$\begin{aligned}(-A)_\beta^{r/2} &= \beta(-A)^{r/2}\big(\beta + (-A)^{r/2}\big)^{-1}\\ &= \frac{\beta}{\alpha} \cdot \big(\alpha + (-A)^{r/2}\big)\big(\beta + (-A)^{r/2}\big)^{-1}(-A)_\alpha^{r/2}.\end{aligned}$$

Moreover, we have for the L^p-L^p operator norm

$$\begin{aligned}\left\|\left\|\frac{\beta}{\alpha} \cdot \big(\alpha + (-A)^{r/2}\big)\big(\beta + (-A)^{r/2}\big)^{-1}\right\|\right\| &= \frac{\beta}{\alpha}\left\|\left\|(\alpha - \beta)\big(\beta + (-A)^{r/2}\big)^{-1} + 1\right\|\right\|\\ &\leqslant \frac{\beta}{\alpha}\left((\alpha - \beta)\frac{1}{\beta} + 1\right) = 1,\end{aligned}$$

which proves the monotonicity of $\beta \mapsto \mathcal{E}^{(r,p),(\beta)}(f)$.

Since $(-A)_\beta^{r/2}$ is the Yosida approximation of the operator $(-A)^{r/2}$, we know that

$$L^p\text{-}\lim_\beta (-A)_\beta^{r/2} f = (-A)^{r/2} f$$

for all $f \in \mathcal{F}_{r,p}$. This proves (4.5). □

Since we can extend $(-A)_\beta^{r/2}$ to $L^\infty(X, m)$, we can define

$$\mathcal{D}_{r,p} := \left\{u \in L^\infty(X, m) \,:\, \sup_{\beta > 0} \mathcal{E}^{(r,p),(\beta)}(u) < \infty\right\}$$

and extend $\mathcal{E}^{(r,p)}$ from $\mathcal{F}_{r,p} \cap L^\infty(X, m)$ onto $\mathcal{D}_{r,p}$ by

$$\mathcal{E}^{(r,p)}(u) := \sup_{\beta > 0} \mathcal{E}^{(r,p),(\beta)}(u), \quad u \in \mathcal{D}_{r,p}.$$

In order to proceed, we need the notion of bounded pointwise convergence, *bp-convergence.* We say that a sequence $(f_j)_{j \in \mathbb{N}}$ of functions $f_j : X \to \mathbb{R}$

converges boundedly pointwise to a function $f : X \to \mathbb{R}$ if $\sup_{j\in\mathbb{N}} |f_j(x)| \leqslant c < \infty$ for every $x \in X$ and $\lim_{j\to\infty} f_j = f$ m-a.e. We use $f_j \xrightarrow[j\to\infty]{\text{bp}} f$ as a shorthand.

Definition 4.1. The space $\mathcal{F}^{\text{bp}}_{r,p}$ consists of all measurable functions $f : X \to \mathbb{R}$ with the property that there is a sequence $(f_j)_{j\in\mathbb{N}} \subset \mathcal{F}_{r,p}$ which is an $\mathcal{E}^{(r,p)}$ Cauchy sequence and which converges in bp-sense to f, i.e.,

$$\lim_{j,k\to\infty} \mathcal{E}^{(r,p)}(f_j - f_k) = 0 \quad \text{and} \quad f_j \xrightarrow[j\to\infty]{\text{bp}} f.$$

Any such sequence $(f_j)_{j\in\mathbb{N}}$ is called an *approximating sequence* of $f \in \mathcal{F}^{\text{bp}}_{r,p}$.

We can now define a first extension of $(\mathcal{E}^{(r,p)}, \mathcal{F}_{r,p})$. Since all elements of $\mathcal{F}^{\text{bp}}_{r,p}$ are bounded functions, this is yet an intermediate step.

Theorem 4.1. *Let $f \in \mathcal{F}^{\text{bp}}_{r,p}$, and let $(f_j)_{j\in\mathbb{N}} \subset \mathcal{F}_{r,p}$ be any approximating sequence. Then*

$$\lim_{j\to\infty} \mathcal{E}^{(r,p)}(f_j) \tag{4.6}$$

exists and does not depend on the choice of the approximating sequence.

In particular, $\mathcal{F}^{\text{bp}}_{r,p} \subset \mathcal{D}_{r,p}$ and (4.6) defines an extension of $\mathcal{E}^{(r,p)}$ onto $\mathcal{F}^{\text{bp}}_{r,p}$ which again is denoted by $\mathcal{E}^{(r,p)}$.

Proof. Let $f \in \mathcal{F}^{\text{bp}}_{r,p}$, and let $(f_j)_{j\in\mathbb{N}}$ be an approximating sequence. Then we have

$$\begin{aligned}\left|\mathcal{E}^{(r,p)}(f_j)^{1/p} - \mathcal{E}^{(r,p)}(f_k)^{1/p}\right| &= \left|\left\|(-A)^{r/2} f_j\right\|_{L^p} - \left\|(-A)^{r/2} f_k\right\|_{L^p}\right| \\ &\leqslant \left\|(-A)^{r/2}(f_j - f_k)\right\|_{L^p} \\ &= \mathcal{E}^{(r,p)}(f_j - f_k)^{1/p}.\end{aligned}$$

Since $(f_j)_{j\in\mathbb{N}}$ is an $\mathcal{E}^{(r,p)}$ Cauchy sequence, we conclude that $(\mathcal{E}^{(r,p)}(f_j))_{j\in\mathbb{N}}$ is a Cauchy sequence in $\mathbb{R}$ and $\left((-A)^{r/2} f_j\right)_{j\in\mathbb{N}}$ is a Cauchy sequence in $L^p(X, m)$. This shows that both limits

$$\lim_{j\to\infty} \mathcal{E}^{(r,p)}(f_j) \quad \text{and} \quad L^p\text{-}\lim_{j\to\infty} (-A)^{r/2} f_j$$

exist. Using the definition of bp-convergence we can adapt the argument of [10, pp. 35–6]. By dominated convergence, we get

$$\left(\lim_{k\to\infty} f_k\right) = \lim_{k\to\infty} \left(\beta + (-A)^{r/2}\right)^{-1} f_k,$$

which implies

$$(-A)^{r/2}_\beta \left(\lim_{k\to\infty} f_k\right) = \lim_{k\to\infty} (-A)^{r/2}_\beta f_k.$$

Let $K \subset X$ be compact. Consider $L^p(K) := L^p(X, \mathbf{1}_K \cdot m)$ as a subspace of $L^p(X) = L^p(X, m)$. Using the Fatou lemma, we get

$$\begin{aligned}
\big\|(-A)_\beta^{r/2}(f_j - f)\big\|_{L^p(K)} &= \big\|(-A)_\beta^{r/2}(f_j - \lim_{k\to\infty} f_k)\big\|_{L^p(K)} \\
&= \big\|\lim_{k\to\infty} (-A)_\beta^{r/2}(f_j - f_k)\big\|_{L^p(K)} \\
&\leqslant \liminf_{k\to\infty} \big\|(-A)_\beta^{r/2}(f_j - f_k)\big\|_{L^p(X)} \\
&\leqslant \liminf_{k\to\infty} \big\|(-A)^{r/2}(f_j - f_k)\big\|_{L^p(X)}.
\end{aligned}$$

For the last estimate we use that $f_j, f_k \in \mathcal{F}_{r,p}$ and $\beta\big(\beta + (-A)^{r/2}\big)^{-1}$ is a contraction on $L^p(X)$. This shows that for every $\epsilon > 0$ we can find some $N_\epsilon \in \mathbb{N}$ such that

$$\big\|(-A)_\beta^{r/2}(f_j - f)\big\|_{L^p(K)} \leqslant \epsilon \quad \text{for all} \quad j \geqslant N_\epsilon.$$

Since N_ϵ does not depend on β or K, we can use the monotone convergence theorem to arrive at

$$\big\|(-A)_\beta^{r/2}(f_j - f)\big\|_{L^p(X)} = \sup_{K\subset X} \big\|(-A)_\beta^{r/2}(f_j - f)\big\|_{L^p(K)} \leqslant \epsilon$$

for all $\beta > 0$, $j \geqslant N_\epsilon$. This, in turn, shows that

$$\sup_{\beta>0} \big\|(-A)_\beta^{r/2}(f_j - f)\big\|_{L^p(X)} \leqslant \epsilon \quad \text{for all} \quad j \geqslant N_\epsilon,$$

and we conclude that $f_j, f \in \mathcal{D}_{r,p}$. By the definition of $\mathcal{D}_{r,p}$, we find for the extension of $\mathcal{E}^{(r,p)}$ to $\mathcal{D}_{r,p}$ that

$$\mathcal{E}^{(r,p)}(f_j - f) = \lim_{\beta\to\infty} \big\|(-A)_\beta^{r/2}(f_j - f)\big\|_{L^p(K)} \leqslant \epsilon \quad \text{for all} j \geqslant N_\epsilon,$$

implying that $\lim_{j\to\infty} \mathcal{E}^{(r,p)}(f_j - f) = 0$, i.e., $\lim_{j\to\infty} \mathcal{E}^{(r,p)}(f_j)$ does not depend on the approximating sequence and (4.6) defines an extension of $\mathcal{E}^{(r,p)}$ to $\mathcal{F}_{r,p}^{\mathsf{bp}}$. □

Corollary 4.1. $\mathcal{F}_{r,p} \cap L^\infty(X, m) = \mathcal{F}_{r,p}^{\mathsf{bp}} \cap L^p(X, m)$.

Proof. The inclusion $\mathcal{F}_{r,p} \cap L^\infty(X, m) \subset \mathcal{F}_{r,p}^{\mathsf{bp}} \cap L^p(X, m)$ is obvious.

Conversely, we pick $f \in \mathcal{F}_{r,p}^{\mathsf{bp}} \cap L^p(X, m)$. Let $(f_j)_{j\in\mathbb{N}}$ be an approximating sequence. Since $(f_j)_{j\in\mathbb{N}}$ is an $\mathcal{E}^{(r,p)}$ Cauchy sequence, we see from the very definition that $\lim_{j\to\infty}(-A)^{r/2} f_j = w$ in $L^p(X, m)$. Since $\beta\big(\beta + (-A)^{r/2}\big)^{-1}$ is a bounded operator on $L^p(X, m)$, we get

$$L^p\text{-}\lim_{j\to\infty} (-A)^{r/2} \beta\big(\beta + (-A)^{r/2}\big)^{-1} f_j = \beta\big(\beta + (-A)^{r/2}\big)^{-1} w.$$

On the other hand, we can use the integral representation (4.3) of $(-A)^{r/2}_{\beta} = (-A)^{r/2}\beta\big(\beta+(-A)^{r/2}\big)^{-1}$ and find from bp-convergence and the dominated convergence theorem that

$$\begin{aligned}
&\lim_{j\to\infty}(-A)^{r/2}\beta\big(\beta+(-A)^{r/2}\big)^{-1}f_j(x)\\
&\quad=\lim_{j\to\infty}\left[\beta\int_X\big(f_j(x)-f_j(y)\big)\,\beta\rho_{\beta,r}(x,dy)+\beta\big(1-\rho_{\beta,r}(x,X)\big)f_j(x)\right]\\
&\quad=\beta\int_X\big(f(x)-f(y)\big)\,\beta\rho_{\beta,r}(x,dy)+\beta\big(1-\rho_{\beta,r}(x,X)\big)f(x)\\
&\quad=(-A)^{r/2}\beta\big(\beta+(-A)^{r/2}\big)^{-1}f(x).
\end{aligned}$$

Since $f\in L^p(X,m)$, the last expression makes sense and we get $\beta\big(\beta+(-A)^{r/2}\big)^{-1}w = (-A)^{r/2}\beta\big(\beta+(-A)^{r/2}\big)^{-1}f$ almost everywhere. Consequently,

$$\lim_{\beta\to\infty}(-A)^{r/2}\beta\big(\beta+(-A)^{r/2}\big)^{-1}f=\lim_{\beta\to\infty}\beta\big(\beta+(-A)^{r/2}\big)^{-1}w=w$$

in $L^p(X,m)$, and our claim follows from the closedness of the operator $(-A)^{r/2}$ and the simple observation that $f\in D\big((-A)^{r/2}\big)=\mathcal{F}_{r,p}$ if and only if $(-A)^{r/2}_{\beta}f=(-A)^{r/2}\big(\beta+(-A)^{r/2}\big)^{-1}f$ converges in L^p as $\beta\to\infty$. □

The following theorem contains a further characterization of $\mathcal{F}^{\mathsf{bp}}_{r,p}$.

Theorem 4.2.

$$\mathcal{F}^{\mathsf{bp}}_{r,p}=\Big\{f:X\to\mathbb{R}: f \textit{ is measurable},\\ \exists\,(f_j)_{j\in\mathbb{N}}\subset\mathcal{F}_{r,p},\ \sup_{j\in\mathbb{N}}\mathcal{E}^{(r,p)}(f_j)<\infty,\ f_j\xrightarrow[j\to\infty]{\mathrm{bp}}f\Big\}.$$

Proof. Since an $\mathcal{E}^{(r,p)}$ Cauchy sequence is bounded, $\mathcal{F}^{\mathsf{bp}}_{r,p}$ is clearly included in the set on the right-hand side. Conversely, let $(f_j)_{j\in\mathbb{N}}\subset\mathcal{F}_{r,p}$ be a sequence such that

$$\sup_{j\in\mathbb{N}}\mathcal{E}^{(r,p)}(f_j)=c<\infty\quad\text{and}\quad f_j\xrightarrow{\mathrm{bp}}f.$$

By the definition of bp-convergence, we have $\sup_j\|f_j\|_{L^\infty}\leqslant\varkappa$. Since $\mathcal{E}^{(r,p)}(f_j)=\|v_j\|^p$, where $v_j=(-A)^{r/2}f_j$, the Banach–Alaoglu theorem shows that there is a subsequence $(v_{n(k)})_{k\in\mathbb{N}}$ converging weakly in $L^p(X,m)$ to some $v\in L^p(X,m)$. By the Banach–Saks theorem, we find that the Cesàro means

$$u_{n(k)}:=\frac{1}{k}\sum_{j=1}^{k}v_{n(j)}\xrightarrow{k\to\infty}v.$$

converge strongly to v. With $g_k := \frac{1}{k}\sum_{j=1}^k f_{n(j)}$ we find $(-A)^{r/2}g_k = u_{n(k)}$ as well as

$$\begin{aligned}\mathcal{E}^{(r,p)}(g_k)^{1/p} &= \left\|(-A)^{r/2}g_k\right\|_{L^p} \leqslant \frac{1}{k}\sum_{j=1}^k \left\|(-A)^{r/2}f_{n(j)}\right\|_{L^p}\\ &= \frac{1}{k}\sum_{j=1}^k \mathcal{E}^{(r,p)}(f_{n(j)})^{1/p} \leqslant \frac{1}{k}\sum_{j=1}^k c^{1/p} = c^{1/p}.\end{aligned}$$

By construction, $g_k \in \mathcal{F}_{r,p}$ and

$$\mathcal{E}^{(r,p)}(g_k - g_\ell) = \|u_{n(k)} - u_{n(\ell)}\|_{L^p}^p \xrightarrow{k,\ell\to\infty} 0.$$

It is obvious that the Cesàro means g_k inherit the bp-convergence from the sequence $(f_j)_{j\in\mathbb{N}}$,

$$|g_k| = \left|\frac{1}{k}\sum_{j=1}^k f_j(x)\right| \leqslant \frac{1}{k}\sum_{1}^k \|f_j\|_{L^\infty} \leqslant \varkappa$$

and

$$g_k(x) = \frac{1}{k}\sum_{j=1}^k f_j(x) \xrightarrow{k\to\infty} f(x).$$

Thus, $f \in \mathcal{F}_{r,p}$ with approximating sequence $(g_k)_{k\in\mathbb{N}}$. □

Remark 4.1. If we write $\mathcal{F}_\mathsf{e}$ for the "usual" extended L^2 Dirichlet space as in [10], then it is not hard to see that $\mathcal{F}_\mathsf{e} \cap L^\infty(X,m) = \mathcal{F}_{1,2}^{\mathsf{bp}}$. This means that our extension is compatible with the traditional setup.

So far, all functions $f \in \mathcal{F}_{r,p}^{\mathsf{bp}}$ are necessarily bounded. In order to abandon this technical condition, we use a *truncation* technique. For this we need to assume that $\mathcal{F}_{r,p}$ has the truncation property (cf. Definition 3.1). This is always the case where $r \leqslant 1$. We define for any measurable function f and all $c > 0$ the truncation operator Θ_c by

$$\Theta_c f(x) := (-c) \vee f(x) \wedge c. \tag{4.7}$$

Definition 4.2. The *extended Dirichlet space* of $(\mathcal{E}^{(r,p)}, \mathcal{F}_{r,p})$ is given by

$$\begin{aligned}\mathcal{F}_{r,p}^{\mathsf{e}} := \Big\{ f : X \to \mathbb{R} : f \text{ is measurable},\\ \Theta_j f \in \mathcal{F}_{r,p}^{\mathsf{bp}},\ j\in\mathbb{N}, \text{ and } \sup_{j\in\mathbb{N}} \mathcal{E}^{(r,p)}(\Theta_j f) < \infty \Big\}.\end{aligned}$$

Note that $\mathcal{E}^{(r,p)}$ is defined on $\mathcal{F}_{r,p}^{\mathsf{bp}}$. Since $\Theta_j\Theta_{j+1}f = \Theta_j f$, we find

$$\mathcal{E}^{(r,p)}(\Theta_j f) = \mathcal{E}^{(r,p)}(\Theta_j \Theta_{j+1} f) \leqslant \mathcal{E}^{(r,p)}(\Theta_{j+1} f),$$

which allows us to extend $\mathcal{E}^{(r,p)}$ onto $\mathcal{F}_{r,p}^{\mathsf{e}}$ by

$$\mathcal{E}^{(r,p)}(f) := \sup_{j\in\mathbb{N}} \mathcal{E}^{(r,p)}(\Theta_j f) = \lim_{j\to\infty} \mathcal{E}^{(r,p)}(\Theta_j f). \tag{4.8}$$

The next result states that our construction coincides with the "usual" definition of an extended Dirichlet space in L^2.

Theorem 4.3. *The extended Dirichlet space $\mathcal{F}_{r,p}^{\mathsf{e}}$ consists of all measurable functions $f : X \to \mathbb{R}$ which admit an approximating sequence $(f_j)_{j\in\mathbb{N}} \subset \mathcal{F}_{r,p}$ such that*

$$f_j \xrightarrow{j\to\infty} f \quad \text{a.e. and} \quad \sup_{j\in\mathbb{N}} \mathcal{E}^{(r,p)}(f_j) < \infty. \tag{4.9}$$

Proof. Assume that f satisfies the conditions stated in the theorem for some approximating sequence $(f_j)_{j\in\mathbb{N}} \subset \mathcal{F}_{r,p}$. By assumption, $\Theta_k f_j \in \mathcal{F}_{r,p}$ and, since $\lim_{j\to\infty} \Theta_k f_j = \Theta_k f$ a.e. and $|\Theta_k f_j| \leqslant k$, we know that $\Theta_k f_j \xrightarrow[j\to\infty]{\mathsf{bp}} \Theta_k f$. Because of the truncation property, $\mathcal{E}^{(r,p)}(\Theta_k f_j) \leqslant \mathcal{E}^{(r,p)}(f_j)$ for all k. Thus, $\Theta_k f \in \mathcal{F}_{r,p}^{\mathsf{bp}}$. Moreover, $\lim_{k\to\infty} \Theta_k f = f$ a.e. and

$$\mathcal{E}^{(r,p)}(\Theta_k f) = \lim_{j\to\infty} \mathcal{E}^{(r,p)}(\Theta_k f_j) \leqslant \liminf_{j\to\infty} \mathcal{E}^{(r,p)}(f_j) \leqslant \sup_{j\in\mathbb{N}} \mathcal{E}^{(r,p)}(f_j) < \infty,$$

where we used (4.9) for the last estimate. This shows that $f \in \mathcal{F}_{r,p}^{\mathsf{e}}$.

Conversely, let $f \in \mathcal{F}_{r,p}^{\mathsf{e}}$. By definition, $\Theta_k f \in \mathcal{F}_{r,p}^{\mathsf{bp}}$ for all $k \in \mathbb{N}$. Therefore, we can find for each $k \in \mathbb{N}$ a sequence $(g_{k,j})_{j\in\mathbb{N}} \subset \mathcal{F}_{r,p}$ such that

$$\lim_{j\to\infty} g_{k,j} = \Theta_k f, \quad \lim_{j\to\infty} \mathcal{E}^{(r,p)}(g_{k,j}) = \mathcal{E}^{(r,p)}(\Theta_k f).$$

For $k \in \mathbb{N}$ fixed we can select subsequences $(g_{k,j(n)})_{n\in\mathbb{N}} \subset (g_{k,j})_{j\in\mathbb{N}}$ such that

$$\left| \mathcal{E}^{(r,p)}(g_{k,j(n)}) - \mathcal{E}^{(r,p)}(\Theta_k f) \right| \leqslant 2^{-k} \quad \text{for all } n \geqslant k.$$

The diagonal sequence $(g_{k,j(k)})_{k\in\mathbb{N}}$ satisfies $g_{k,j(k)} \in \mathcal{F}_{r,p}$ and $\lim_{k\to\infty} g_{k,j(k)} = f$ a.e. Since $f \in \mathcal{F}_{r,p}^{\mathsf{e}}$, we find for every $k \in \mathbb{N}$

$$\begin{aligned} \mathcal{E}^{(r,p)}(g_{k,j(k)}) &\leqslant \sup_{n\geqslant k} \mathcal{E}^{(r,p)}(g_{k,j(n)}) \\ &\leqslant \sup_{n\geqslant k} \left| \mathcal{E}^{(r,p)}(g_{k,j(n)}) - \mathcal{E}^{(r,p)}(\Theta_k f) \right| + \mathcal{E}^{(r,p)}(\Theta_k f) \\ &\leqslant 2^{-k} + \sup_{k\in\mathbb{N}} \mathcal{E}^{(r,p)}(\Theta_k f) < \infty. \end{aligned}$$

This shows that f has an approximating sequence which satisfies the conditions stated in (4.9). □

Remark 4.2. The Banach–Saks argument used in the proof of Theorem 4.2 shows that we may replace (4.9) in Theorem 4.3 by the seemingly stronger assertion

$$f_j \xrightarrow{j\to\infty} f \text{ a.e. and } \lim_{j,k\to\infty} \mathcal{E}^{(r,p)}(f_j - f_k) = 0. \tag{4.9$'$}$$

Corollary 4.2. *We have* $\mathcal{F}^{\mathsf{e}}_{r,p} \cap L^p(X,m) = \mathcal{F}_{r,p}$.

Proof. The inclusion $\mathcal{F}_{r,p} \subset \mathcal{F}^{\mathsf{e}}_{r,p} \cap L^p(X,m)$ is obvious. Assume that $f \in \mathcal{F}^{\mathsf{e}}_{r,p}$ as well as $f \in L^p(X,m)$. By the definition of the extended Dirichlet space (cf. Definition 4.2), we know that $\Theta_j f \in \mathcal{F}^{\mathsf{bp}}_{r,p}$ and $\sup_{j\in\mathbb{N}} \mathcal{E}^{(r,p)}(\Theta_j f) < \infty$. The Banach-Saks argument used in the proof of Theorem 4.2 shows that the Cesàro means of some subsequence $(\Theta_{j(k)} f)_{k\in\mathbb{N}} \subset (\Theta_j f_j)_{j\in\mathbb{N}}$,

$$\Sigma_n f := \frac{1}{n} \sum_{k=1}^{n} \Theta_{j(k)} f,$$

are an $\mathcal{E}^{(r,p)}$ Cauchy sequence.

On the other hand, $f \in L^p(X,m)$ implies that both $\Theta_j f$ and $\Sigma_n f$ converge in $L^p(X,m)$ to f. Moreover, as $\Theta_j f \in L^p(X,m)$, we get $\Theta_j f \in \mathcal{F}^{\mathsf{bp}}_{r,p} \cap L^p(X,m)$ and, by Corollary 4.1, $\Theta_j f \in \mathcal{F}_{r,p}$.

Thus, $\Sigma_n f \in \mathcal{F}_{r,p}$ which tells us that

$$\lim_{\ell,n\to\infty} \left\|(-A)^{r/2}(\Sigma_\ell f - \Sigma_n f)\right\|_{L^p} = \lim_{\ell,n\to\infty} \mathcal{E}^{(r,p)}(\Sigma_\ell f - \Sigma_n f) = 0.$$

Since L^p-$\lim_{n\to\infty} \Sigma_n f = f$ and $(-A)^{r/2}$ is a closed operator, we finally see that $f \in D\big((-A)^{r/2}\big) = \mathcal{F}_{r,p}$. □

5 The Limiting Case $\lambda \to 0$ and Transience

For $\lambda > 0$ and $f \in \mathcal{F}_{r,p}$ we find from [5, Proposition 1.4.9 and Remark 1.4.11] that

$$\begin{aligned}\left|\mathcal{E}^{(r,p)}_\lambda(f)^{1/p} - \mathcal{E}^{(r,p)}(f)^{1/p}\right| &= \left|\left\|(\lambda - A)^{r/2} f\right\|_{L^p} - \left\|(-A)^{r/2} f\right\|_{L^p}\right| \\ &\leqslant \left\|\left[(\lambda - A)^{r/2} - (-A)^{r/2}\right] f\right\|_{L^p} \\ &\leqslant \lambda^{r/2} \|f\|_{L^p},\end{aligned}$$

and therefore $\lim_{\lambda\to 0} \mathcal{E}^{(r,p)}_\lambda(f) = \mathcal{E}^{(r,p)}(f)$ for all $f \in \mathcal{F}_{r,p}$.

Moreover, it is easy to see that for all f from the domain of $(\lambda - A)^{r/2}$ and $\mathcal{A}^{(p)}_{r,\lambda}$ respectively

$$(\lambda - A)^{r/2} f \xrightarrow{\lambda \to 0} (-A)^{r/2} f, \quad \mathcal{A}_{r,\lambda}^{(p)} f \xrightarrow{\lambda \to 0} \mathcal{A}_r^{(p)} f.$$

Here, $\mathcal{A}_r^{(p)}$ is defined on $\mathcal{F}_{r,p}$ by

$$\mathcal{A}_r^{(p)} f = (-A^*)^{r/2} \circ J_p \circ (-A)^{r/2} f = (-A^*)^{r/2} \big(\big|(-A)^{r/2} f\big|^{p-2} \cdot (-A)^{r/2} f \big).$$

A close inspection of the proof of Lemma 2.1 in [11] reveals that the operator $\mathcal{A}_r^{(p)}$ is on $\mathcal{F}_{r,p}$ the Gâteaux derivative of the functional

$$f \mapsto \frac{1}{p} \int_X \big|(-A)^{r/2} f\big|^p \, dm.$$

If we switch to the extension $\mathcal{F}_{r,p}^{\mathsf{e}}$ and try to extend the considerations made for $(\mathcal{E}_\lambda^{(r,p)}, \mathcal{F}_{r,p})$ in Section 3 to $(\mathcal{E}^{(r,p)}, \mathcal{F}_{r,p}^{\mathsf{e}})$, we run into a problem: $f \mapsto \big(\mathcal{E}^{(r,p)}(f)\big)^{1/p}$ need not be definite for $f \in \mathcal{F}_{r,p}^{\mathsf{e}}$. To overcome this problem, we recall the notion of transience [15].

Definition 5.1. The form $(\mathcal{E}^{(r,p)}, \mathcal{F}_{r,p})$ is said to be *transient* if there exists some $G \in L^1(X, m)$ such that $G > 0$ m-a.e. and

$$\left[\int_X |f| \cdot G \, dm \right]^p \leqslant \mathcal{E}^{(r,p)}(f) \quad \text{for all } \ f \in \mathcal{F}_{r,p}.$$

For a transient form $(\mathcal{E}^{(r,p)}, \mathcal{F}_{r,p})$ and $f, g \in \mathcal{F}_{r,p}$ we get

$$\Big\langle \mathcal{A}^{(r,p)} f - \mathcal{A}^{(r,p)} g, f - g \Big\rangle \geqslant 2^{2-p} \, \mathcal{E}^{(r,p)}(f-g) \geqslant 2^{2-p} \langle |f-g|, G \rangle^p. \quad (5.1)$$

This shows that $\mathcal{A}^{(r,p)} : \mathcal{F}_{r,p} \to \mathcal{F}_{r,p}^*$ is uniformly monotone with respect to $\mathcal{E}^{(r,p)}$, coercive on unbounded sets, and convex.

Because of transience, $f \mapsto \mathcal{E}^{(r,p)}(f)$ is a norm on $\mathcal{F}_{r,p}^{\mathsf{e}}$ and $(\mathcal{F}_{r,p}^{\mathsf{e}}, \mathcal{E}^{(r,p)})$ is a Banach space. Therefore, all properties of $\mathcal{A}^{(r,p)}$ on $\mathcal{F}_{r,p}$ still hold on $\mathcal{F}_{r,p}^{\mathsf{e}}$. In particular, the minimization problem

Find for an open set $G \subset \mathbb{R}^n$

$$\operatorname{cap}_{r,p}^{\mathsf{e}}(G) = \inf \big\{ \mathcal{E}^{(r,p)}(f) \ : \ f \in \Gamma_{r,p}^{\mathsf{e}}(G) \big\}, \quad (5.2)$$

where $\Gamma_{r,p}^{\mathsf{e}}(G) = \big\{ u \in \mathcal{F}_{r,p}^{\mathsf{e}} \ : \ u \geqslant \mathbf{1}_G \ \ m\text{-a.e.} \big\}$

can now be treated with variational methods in the $\mathcal{F}_{r,p}^{\mathsf{e}}$-context in exactly the same way as we could solve (3.9), (3.10) for $\mathcal{F}_{r,p}$. Since $\Gamma_{r,p}(G) \subset \Gamma_{r,p}^{\mathsf{e}}(G)$, we have

$$\operatorname{cap}_{r,p}^{\mathsf{e}}(G) \leqslant \liminf_{\lambda \to 0} \operatorname{cap}_{r,p,\lambda}(G). \quad (5.3)$$

At the moment, however, it is an open problem to find conditions which allow us to extend Fukushima's construction of a Markov process associated with a regular Dirichlet form to the case $(\mathcal{E}^{(r,p)}, \mathcal{F}^{\mathrm{e}}_{r,p})$ with $\mathrm{cap}^{\mathrm{e}}_{r,p}$ as related capacity.

References

1. Beurling, A., Deny, J.: Espaces de Dirichlet. I. Le cas élémentaire. Acta Math. **99**, 203–224 (1958)
2. Beurling, A., Deny, J.: Dirichlet Spaces. Proc. Natl. Acad. Sci. U.S.A. **45**, 208–215 (1959)
3. Davies, E.B.: Heat Kernels and Spectral Theory. Cambridge Univ. Press, Cambridge (1989)
4. Farkas, W., Jacob, N., Schilling, R.L.: Feller semigroups, L^p-sub-Markovian semigroups, and applications to pseudo-differential operators with negative definite symbols. Forum Math. **13**, 51–90 (2001)
5. Farkas, W., Jacob, N., Schilling, R.L.: Function spaces related to continuous negative definite functions: ψ-Bessel potential spaces. Diss. Math. CCCXCIII, 1–62 (2001)
6. Fukushima, M.: Dirichlet spaces and strong Markov processes. Trans. Am. Math. Soc. **162**, 185–224 (1971)
7. Fukushima, M.: Dirichlet Forms and Markov Processes. Kodansha and North-Holland, Tokyo and Amsterdam (1980)
8. Fukushima, M.: Two topics related to Dirichlet forms: quasi-everywhere convergence and additive functionals. In: Dell'Antonio, G., Mosco, U. (eds.), Dirichlet Forms. Lect. Notes Math. **1563**, 21–53 (1993)
9. Fukushima, M., Kaneko, H.: On (r,p)-capacities for general Markovian semigroups. In: Albeverio, S. (ed.), Infinite Dimensional Analysis and Stochastic Processes, pp. 41–47. Pitman, Boston (1985)
10. Fukushima, M., Oshima, Y., Takeda, M.: Dirichlet Forms and Symmetric Markov Processes. Walter de Gruyter, Berlin (1994)
11. Hoh, W., Jacob, N.: Towards an L^p potential theory for sub-Markovian semigroups: variational inequalities and balayage theory. J. Evol. Equ. **4**, 297–312 (2004)
12. Jacob, N.: Pseudo Differential Operators and Markov Processes (3 vols.). Imperial College Press, London (2001, 2002, 2005)
13. Jacob, N., Schilling, R.L.: Some Dirichlet spaces obtained by subordinate reflected diffusions, Rev. Mat. Iberoam. **15**, 59–91 (1999)
14. Jacob, N., Schilling, R.L.: Towards an L^p-potential theory for sub-Markovian semigroups: kernels and capacities. Acta Math. Sinica **22**, 1227–1250 (2006)
15. Jacob, N., Schilling, R.L.: On a Poincaré type inequality for energy forms in L^p. Mediterr. J. Math. **4**, 33–44 (2007)
16. Kaneko, H.: On (r,p)-capacities for Markov processes. Osaka J. Math. **23**, 325–336 (1986)
17. Malliavin, P.: Implicit functions in finite corank on the Wiener space. In: Proceedings Taniguchi Symp. Stoch. Anal. Katata and Kyoto 1982, pp. 369–386. North-Holland, Amsterdam (1984)
18. Maz'ya, V.G.: Sobolev Spaces. Springer, Berlin etc. (1985)

19. Maz'ya, V.G., Khavin, V.P.: Nonlinear potential theory (Russian). Usp. Mat. Nauk **27**, 67–138 (1972); English transl.: Russ. Math. Surv. **27**, 71–148 (1973)
20. Schilling, R.L.: Subordination in the sense of Bochner and a related functional calculus, J. Aust. Math. Soc. Ser. A **64**, 368–396 (1998)
21. Schilling, R.L.: A note on invariant sets. Probab. Math. Stat. **24**, 47–66 (2004)
22. Silverstein, M.L.: Symmetric Markov Processes. Springer, Lect. Notes Math. vol. **426**, Berlin (1974)
23. Stein, E.M.: Topics in Harmonic Analysis Related to the Littlewood–Paley Theory. Princeton Univ. Press, Princeton, NJ (1970)

Characterizations for the Hardy Inequality

Juha Kinnunen and Riikka Korte

Abstract Necessary and sufficient conditions for the validity of a multidimensional version of the Hardy inequality are discussed. A characterization through a boundary Poincaré inequality is considered.

1 Introduction

We discuss necessary and sufficient conditions for the validity of the following multidimensional version of the Hardy inequality. Let Ω be an open subset of $\mathbb{R}^n$, and let $1 < p < \infty$. We say that the *p-Hardy inequality* holds in Ω if there is a uniform constant c_H such that

$$\int_\Omega \left(\frac{|u(x)|}{\delta(x)} \right)^p dx \leqslant c_H \int_\Omega |\nabla u(x)|^p \, dx \tag{1.1}$$

for all $u \in W_0^{1,p}(\Omega)$, where $\delta(x) = \operatorname{dist}(x, \partial\Omega)$.

By density arguments, it suffices to consider (1.1) for compactly supported smooth functions $u \in C_0^\infty(\Omega)$.

Sufficient Lipschitz and Hölder type boundary conditions under which the Hardy inequality holds were obtained by Nečas [41], Kufner [26], Kufner and Opic [42]. In [37, Chapter 2], Maz'ya gave capacitary characterizations of

Juha Kinnunen
Institute of Mathematics, P.O. Box 1100, FI-02015 Helsinki University of Technology, Finland
e-mail: juha.kinnunen@tkk.fi

Riikka Korte
Department of Mathematics and Statistics, P.O. Box 68, Gustaf Hällströmin katu 2 b, FI-00014 University of Helsinki, Finland
e-mail: riikka.korte@helsinki.fi

A. Laptev (ed.), *Around the Research of Vladimir Maz'ya I: Function Spaces*, International Mathematical Series 11, DOI 10.1007/978-1-4419-1341-8_10,

the Hardy inequality (cf. Section 2). Ancona [2] in the case $p = 2, n \geqslant 2$, Lewis [31], and Wannebo [49] in the case $p \geqslant 1, n \geqslant 2$ proved that the Hardy inequality holds if the complement of Ω satisfies the uniform capacity density condition

$$\operatorname{cap}_p\big((\mathbb{R}^n \setminus \Omega) \cap \overline{B}(x,r), B(x,2r)\big) \geqslant c_T \operatorname{cap}_p\big(\overline{B}(x,r), B(x,2r)\big) \qquad (1.2)$$

for all $x \in \mathbb{R}^n \setminus \Omega$ and $r > 0$. The definition and properties of variational capacity can be found in [37, Chapter 2] or [16, Chapter 2]. If (1.2) holds, we say that $\mathbb{R}^n \setminus \Omega$ is *uniformly p-thick* or *p-fat.*

The class of open sets whose complements satisfy the uniform capacity density condition is relatively large. Every nonempty $\mathbb{R}^n \setminus \Omega$ is uniformly p-thick for $p > n$, and hence the condition is nontrivial only if $p \leqslant n$. In particular, in the case $p > n$, the Hardy inequality holds for every proper open subset of $\mathbb{R}^n$. Hence it suffices to consider the case $1 < p \leqslant n$.

The capacity density condition has several applications in the theory of partial differential equations. It is stronger than the Wiener criterion

$$\int_0^1 \left(\frac{\operatorname{cap}_p\big((\mathbb{R}^n \setminus \Omega) \cap \overline{B}(x,r), B(x,2r)\big)}{\operatorname{cap}_p\big(\overline{B}(x,r), B(x,2r)\big)}\right)^{1/(p-1)} \frac{dr}{r} = \infty$$

characterizing regular boundary points for the Dirichlet problem for the p-Laplace equation. Sufficiency was proved by Maz'ya [36] and necessity was established by Lindqvist and Martio [32] in the case $p > n - 1$ and by Kilpeläinen and Malý [19] in the case $1 < p \leqslant n$.

Hajłasz [13] showed that the capacity density condition is sufficient for the validity of a pointwise version of the Hardy inequality, in terms of the Hardy–Littlewood maximal function. A similar result was also obtained in [21]. Recently Lehrbäck [27] showed that the pointwise Hardy inequality is equivalent to the uniform thickness of the complement (cf. also [22]).

In this paper, we also consider a characterization through a boundary Poincaré inequality.

In the bordeline case $p = n$, there are several characterizations of the Hardy inequality. In this case, rather surprisingly, certain analytic, metric, and geometric conditions turn out to be equivalent. Ancona [2] proved that uniform p-thickness is also necessary for the validity of the Hardy inequality when $p = n = 2$, and Lewis [31] generalized this result for $p = n \geqslant 2$. Sugawa [46] proved that, in the case $p = n = 2$, the Hardy inequality is equivalent to the uniform perfectness of the complement. This result was recently generalized [23] for other values of n.

We outline the main points of the argument in this work and discuss other characterizations of Hardy inequalities in the borderline case.

The following variational problem is naturally related to the p-Hardy inequality. Consider the Rayleigh quotient

$$\lambda_p = \lambda_p(\Omega) = \inf\left\{\left[\int_\Omega |\nabla u(x)|^p\,dx\right] \Big/ \left[\int_\Omega \left(\frac{|u(x)|}{\delta(x)}\right)^p dx\right]\right\}, \tag{1.3}$$

where the infimum is taken over all $u \in W_0^{1,p}(\Omega)$. Note that $\lambda_p(\Omega) > 0$ if and only if the p-Hardy inequality holds in Ω. This approach have attracted a lot of attention (cf., for example, [3, 4, 5, 7, 8, 33, 34, 12, 35, 43, 45, 44, 47, 48]).

Hardy [14, 15] observed that, in the one-dimensional case,

$$\lambda_p = (1 - 1/p)^p$$

and the minimum is not attained in (1.3) In higher dimensions, the constant λ_p generally depends on p and Ω.

Note that $u \in W_0^{1,p}(\Omega)$ is a minimizer of (1.3) if and only if it is a weak solution to the nonlinear eigenvalue problem

$$\operatorname{div}(|\nabla u(x)|^{p-2}\nabla u(x)) + \lambda_p \frac{|u(x)|^{p-2}u(x)}{\delta(x)^p} = 0. \tag{1.4}$$

Ancona [2] characterized the Hardy inequality for $p = 2$ via supersolutions, called *strong barriers* of (1.4). In this paper, we generalize this characterization for other values of p.

We also consider the selfimproving phenomena related to the Hardy inequalities. It is easy to see that if $\mathbb{R}^n \setminus \Omega$ is uniformly p-thick, then it is uniformly q-thick for every $q > p$ as well. Lewis [31] showed that p-thickness has a deep selfimproving property: p-thickness implies the same condition for some smaller value of p. For another proof we refer to [40].

The Hardy inequality is selfimproving as well. Indeed, as Koskela and Zhong [25] showed, if the Hardy inequality holds for some value of p, then it also holds for other sufficiently close values of p. In contrast with the capacity density condition, the Hardy inequality can fail for some values of p. Indeed, in a punctured ball, the p-Hardy inequality holds for $p \neq n$ and does not hold for $p = n$. More generally, as was shown in [25], the Hardy inequality cannot hold if the boundary contains $(n-p)$-dimensional parts. Roughly speaking, the Hardy inequality can hold if the complement of the domain is either large or small in a neighborhood of each boundary point.

Many arguments related to the Hardy inequality are based on general principles and some of them apply on metric measure spaces (cf. [6, 18, 23, 22]). The Hardy inequalities are studied in Carnot–Carathéodory spaces [11] and in Orlicz–Sobolev spaces [9, 10].

2 Maz'ya Type Characterization

In this section, we describe a characterization of the Hardy inequality in terms of inequalities connecting measures and capacities. Recall the definition of variational capacity. Let Ω be an open subset of $\mathbb{R}^n$, and let K be a compact subset of Ω. The *variational p-capacity* of K with respect to Ω is defined as follows:

$$\operatorname{cap}_p(K,\Omega) = \inf \int_\Omega |\nabla u(x)|^p \, dx,$$

where the infimum is taken over all $u \in C_0^\infty(\Omega)$ such that $u(x) \geqslant 1$ for every $x \in K$. The same quantity is obtained if the infimum is taken over compactly supported continuous functions in $W_0^{1,p}(\Omega)$ instead of smooth functions.

The proof the following result is based on an elegant truncation argument [37, p.110] (cf. [37, Chapter 2] and [38] for more information about such characterizations and [39, 20] for generalizations).

Theorem 2.1. *An open set Ω satisfies the p-Hardy inequality if and only if there is a constant c_M such that*

$$\int_K \delta(x)^{-p} \, dx \leqslant c_M \operatorname{cap}_p(K,\Omega) \tag{2.1}$$

for every compact subset K of Ω.

Proof. Assume that the p-Hardy inequality holds in Ω. Let $u \in C_0^\infty(\Omega)$ be such that $u(x) \geqslant 1$ for every $x \in K$. By (1.1),

$$\int_K \delta(x)^{-p} \, dx \leqslant \int_\Omega \left(\frac{|u(x)|}{\delta(x)} \right)^p dx \leqslant c_H \int_\Omega |\nabla u(x)|^p \, dx.$$

Taking the infimum over all such functions u, we obtain (2.1) with $c_M = c_H$.

Then assume that (2.1) holds. By a density argument it is enough to prove (1.1) for compactly supported smooth functions in Ω. Let $u \in C_0^\infty(\Omega)$. For $k \in \mathbb{Z}$ denote

$$E_k = \{x \in \Omega : |u(x)| > 2^k\}.$$

By (2.1), we have

$$\begin{aligned}\int_\Omega \left(\frac{|u(x)|}{\delta(x)} \right)^p dx &\leqslant \sum_{k=-\infty}^{\infty} 2^{(k+1)p} \int_{E_k \setminus E_{k+1}} \delta(x)^{-p} \, dx \\ &\leqslant c_M \sum_{k=-\infty}^{\infty} 2^{(k+1)p} \operatorname{cap}_p(\overline{E}_k, \Omega) \leqslant c_M 2^p \sum_{k=-\infty}^{\infty} 2^{(k+1)p} \operatorname{cap}_p(\overline{E}_{k+1}, E_k).\end{aligned}$$

Introduce $u_k : \Omega \to [0,1]$ as

$$u_k(x) = \begin{cases} 1 & \text{if } |u(x)| \geqslant 2^{k+1}, \\ \dfrac{|u(x)|}{2^k} - 1 & \text{if } 2^k < |u(x)| < 2^{k+1}, \\ 0 & \text{if } |u(x)| \leqslant 2^k. \end{cases}$$

Then $u_k \in W_0^{1,p}(\Omega)$ is a continuous function, $u_k = 1$ in $\overline{E}_{k+1}$, and $u_k = 0$ in $\mathbb{R}^n \setminus E_k$. Therefore, we can apply take it for a test function for the capacity:

$$\operatorname{cap}_p(\overline{E}_{k+1}, E_k) \leqslant \int_{E_k \setminus E_{k+1}} |\nabla u_k(x)|^p \, dx \leqslant 2^{-pk} \int_{E_k \setminus E_{k+1}} |\nabla u(x)|^p \, dx.$$

Consequently,

$$\begin{aligned} \sum_{k=-\infty}^{\infty} 2^{(k+1)p} \operatorname{cap}_p(\overline{E}_{k+1}, E_k) &\leqslant 2^p \sum_{k=-\infty}^{\infty} \int_{E_k \setminus E_{k+1}} |\nabla u(x)|^p \, dx \\ &= 2^p \int_\Omega |\nabla u(x)|^p \, dx, \end{aligned}$$

and the claim follows with $c_H = 2^{2p} c_M$. □

Remark 2.1. A result of Koskela and Zhong [25] shows that the Hardy inequality is an open ended condition in the following sense: If the Hardy inequality holds in Ω for some $1 < p < \infty$, then there exists $\varepsilon > 0$ such that the Hardy inequality holds in Ω for every q with $p - \varepsilon < q < p + \varepsilon$ (relative the weighted case cf. [24]). This implies that the Maz'ya type condition (2.1) is an open ended condition as well. Indeed, if (2.1) holds, then there are $c > 0$ and $\varepsilon > 0$ such that

$$\int_K \delta(x)^{-q} \, dx \leqslant c \operatorname{cap}_q(K, \Omega)$$

for all q with $p - \varepsilon < q < p + \varepsilon$.

3 The Capacity Density Condition

In this section, we consider a sufficient condition for the Hardy inequality in terms of the uniform thickness of complements (cf. (1.2)). The capacity density condition has a deep selfimproving property, essential in many aspects. The following result is due to Lewis [31, Theorem 1] (cf. also [2] and [40, Section 8]).

Theorem 3.1. *If $\mathbb{R}^n \setminus \Omega$ is uniformly p-thick, then there is $q < p$ such that $\mathbb{R}^n \setminus \Omega$ is uniformly q-thick.*

Assume that $\mathbb{R}^n \setminus \Omega$ is uniformly p-thick. Let $u \in C_0^\infty(\Omega)$. Denote

$$A = \{y \in B(x, r) : u(y) = 0\}.$$

By a capacitary version of a Poincaré type inequality, there is $c = c(n, p)$ such that

$$\Big(\frac{1}{|B(x,r)|}\int_{B(x,r)}|u(y)|^p\,dy\Big)^{1/p}$$
$$\leqslant\left(\frac{c}{\operatorname{cap}_p\big(A\cap\overline{B}(x,r),B(x,2r)\big)}\int_{B(x,r)}|\nabla u(y)|^p\,dy\right)^{1/p}$$
$$\leqslant c\,r\left(\frac{1}{|B(x,r)|}\int_{B(x,r)}|\nabla u(y)|^p\,dy\right)^{1/p}. \tag{3.1}$$

Such inequalities were studied in [37, Chapter 10] (cf. also [1, Chapter 8]). We also need the equality [37, Section 2.2.4]

$$\operatorname{cap}_p\big(\overline{B}(x,r),B(x,2r)\big)=c\,r^{n-p},$$

where $c=c(n,p)$

The inequality (3.1) implies the pointwise estimate

$$|u(x)|\leqslant c\,\delta(x)\big(M_\Omega(|\nabla u|^p)(x)\big)^{1/p} \tag{3.2}$$

for every $x\in\Omega$ with $c=c(n,p)$. The restricted Hardy–Littlewood maximal function is defined as

$$M_\Omega f(x)=\sup\frac{1}{|B(x,r)|}\int_{B(x,r)}|f(y)|\,dy,$$

where the supremum is taken over $r>0$ such that $r\leqslant 2\delta(x)$. Such pointwise Hardy inequalities were considered in [13, 21] (cf. also [11, 24]). By Theorem 3.1, the pointwise Hardy inequality (3.2) also holds for some $q<p$. Integrating this inequality over Ω and using the maximal function theorem, we have

$$\int_\Omega\left(\frac{|u(x)|}{\delta(x)}\right)^p dx\leqslant c\int_\Omega\big(M_\Omega(|\nabla u|^q)(x)\big)^{p/q}\,dx\leqslant c\int_\Omega|\nabla u(x)|^p\,dx$$

for every $u\in C_0^\infty(\Omega)$ with $c=c(n,p,q)$.

This proof relies heavily on a rather deep Theorem 3.1. Wannebo [49] gave a more direct proof in the case where the complement of the domain is uniformly thick (cf. also [50, 51, 52]). Following to Wannebo, one should first use a Poincaré type inequality (3.4) and the a Whitney type covering argument to show that

$$\int_{\{x\in\Omega:2^{-k-1}<\delta(x)<2^{-k}\}}|u(x)|^p\,dx\leqslant c2^{-kp}\int_{\{x\in\Omega:\delta(x)\leqslant 2^{-k+1}\}}|\nabla u(x)|^p\,dx$$

for every $k\in\mathbb{Z}$. Multiplying both sides by $\delta(x)^{-p-\beta}$, with $\beta>0$, and summing up over k, we obtain the weighted Hardy inequality

$$\int_{\Omega} \frac{|u(x)|^p}{\delta(x)^{p+\beta}}\,dx \leqslant \frac{c}{\beta} \int_{\Omega} \frac{|\nabla u(x)|^p}{\delta(x)^{\beta}}\,dx. \tag{3.3}$$

Applying this inequality to

$$u(x)\delta(x)^{\beta/p}$$

with sufficiently small $\beta > 0$, we obtain the unweighted p-Hardy inequality. Weighted Hardy inequalities were also studied in [30, 28, 29].

The pointwise Hardy inequality is not equivalent to the Hardy inequality. For example, the punctured ball satisfies the pointwise Hardy inequality only if $p > n$, whereas the usual Hardy inequality holds for $1 < p < n$. A recent result of Lehrbäck [27] shows that uniform thickness is not only sufficient, but also necessary for the pointwise Hardy inequality. Before formulating the result, we recall a definition from [27]. An open set Ω in $\mathbb{R}^n$ satisfies an *inner boundary density condition* with exponent α if there exists a constant $c > 0$ such that

$$\mathcal{H}^{\alpha}_{\infty}\big(B(x, 2\delta(x)) \cap \partial\Omega\big) \geqslant c\delta(x)^{\alpha}$$

for every $x \in \Omega$. Here,

$$\mathcal{H}^{\alpha}_{\infty}(E) \left\{ \sum_{i=1}^{\infty} r_i^{\alpha} : E \subset \bigcup_{i=1}^{\infty} B(x_i, r_i) \right\}$$

is the spherical Hausdorff content of a set E. We formulate the main result of [27].

Theorem 3.2. *The following conditions are equivalent:*

(1) *The set $\mathbb{R}^n \setminus \Omega$ is uniformly p-thick.*

(2) *The set Ω satisfies the pointwise Hardy inequality with some $q < p$.*

(3) *There exists α with $n-p < \alpha \leqslant n$ so that Ω satisfies the inner boundary density condition with exponent α.*

Note that if conditions (1)–(3) hold for some parameter, then they also hold for all larger parameters. However, the Hardy inequality does not share this property with them.

Remark 3.1. A recent result of [22] shows that condition (2) in Theorem 3.2 can be replaced with the pointwise p-Hardy inequality. By Theorem 3.1, the pointwise p-Hardy inequality implies the pointwise q-Hardy inequality for some $q < p$. This means that the pointwise Hardy inequality is a selfimproving property. It would be interesting to obtain a direct proof, without the use of Theorem 3.1, for this selfimproving result.

We present yet another characterization of uniform thickness through a boundary Poincaré inequality of type (3.1).

Theorem 3.3. *The set $\mathbb{R}^n \setminus \Omega$ is uniformly p-thick if and only if*

$$\int_{B(x,r)} |u(y)|^p \, dy \leqslant c\, r^p \int_{B(x,r)} |\nabla u(y)|^p \, dy \tag{3.4}$$

for every $x \in \mathbb{R}^n \setminus \Omega$ and $u \in C_0^\infty(\Omega)$.

Proof. The uniform p-thickness implies (3.4) by the capacitary version of the Poincaré inequality (3.1).

To prove the reverse implication, let $u \in C_0^\infty(B(x,2r))$ be such that $u(x) = 1$ for every $x \in (\mathbb{R}^n \setminus \Omega) \cap \overline{B}(x,r)$. If

$$\frac{1}{|B(x,r/2)|} \int_{B(x,r/2)} |u(y)|^p \, dy \geqslant \frac{1}{2^p},$$

then, by the standard Poincaré inequality, we have

$$\frac{1}{2^p} |B(x,r/2)| \leqslant c \int_{B(x,2r)} |u(y)|^p \, dy lec\, r^p \int_{B(x,2r)} |\nabla u(y)|^p \, dy,$$

which implies

$$\int_{B(x,2r)} |\nabla u(y)|^p \, dy \geqslant c\, r^{n-p}.$$

Assume that

$$\frac{1}{|B(x,r/2)|} \int_{B(x,r/2)} |u(y)|^p \, dy < \frac{1}{2^p}.$$

It is clear that

$$|B(x,r/2)| \leqslant 2^{p-1} \left(\int_{B(x,r/2)} |u(y)|^p \, dy + \int_{B(x,r/2)} |1 - u(y)|^p \, dy \right)$$

and, consequently,

$$\int_{B(x,r/2)} |1 - u(y)|^p \, dy \geqslant 2^{1-p} |B(x,r/2)| - \int_{B(x,r/2)} |u(y)|^p \, dy \geqslant c r^n.$$

Let $v = (1-u)\varphi$, where $\varphi \in C_0^\infty(B(x,r))$ is a cutoff function such that $\varphi = 1$ in $B(x,r/2)$. Then $v \in C_0^\infty(\Omega)$ and, by (3.4), we have

$$\begin{aligned}\int_{B(x,r/2)} |1 - u(y)|^p \, dy &= \int_{B(x,r/2)} |v(y)|^p \, dy \leqslant c\, r^p \int_{B(x,r/2)} |\nabla v(y)|^p \, dy \\ &\leqslant c\, r^p \int_{B(x,2r)} |\nabla u(y)|^p \, dy.\end{aligned}$$

It follows that

$$\int_{B(x,2r)} |\nabla u(y)|^p \, dy \geqslant c\, r^{n-p}.$$

Taking the infimum over all such functions u, we conclude that

$$Ca_p\big((\mathbb{R}^n \setminus \Omega) \cap \overline{B}(x,r), B(x,2r)\big) \geqslant c\, r^{n-p}.$$

Hence Ω is uniformly p-thick. □

Remark 3.2. Again, by Theorem 3.1, we conclude that the boundary Poincaré inequality 3.4 implies the same inequality for some $q < p$. Hence the boundary Poincaré inequality is a selfimproving property.

4 Characterizations in the Borderline Case

In the borderline case $p = n$, there are several characterizations for the Hardy inequality. We begin with the following metric definition. A set $\mathbb{R}^n \setminus \Omega$ is *uniformly perfect* if it contains more than one point and there is a constant $c_P \geqslant 1$ such that for any $x \in \mathbb{R}^n \setminus \Omega$ and $r > 0$

$$\mathbb{R}^n \setminus \Omega) \cap \big(B(x, c_P r) \setminus B(x,r)\big) \neq \varnothing,$$

whenever $(\mathbb{R}^n \setminus \Omega) \setminus B(x, c_P r) \neq \varnothing$ (cf. [17, 46] for details).

We give four variants of characterization for the Hardy inequality in the borderline case. Needless to say that by Theorem 2.1, Theorem 3.2, and Theorem 3.3 we have four more characterizations.

Theorem 4.1. *The following conditions are quantitatively equivalent:*

(1) *Ω satisfies the n-Hardy inequality,*

(2) *$\mathbb{R}^n \setminus \Omega$ is uniformly perfect,*

(3) *$\mathbb{R}^n \setminus \Omega$ is uniformly n-thick,*

(4) *$\mathbb{R}^n \setminus \Omega$ is uniformly $(n - \varepsilon)$-thick for some $\varepsilon > 0$.*

Proof. The scheme of the proof is as follows. Conditions (3) and (4) are equivalent in view of Theorem 3.1. The fact that (3) and (4) imply (1) can be shown as above. The equivalence of (1) and (3) was proved by Ancona [2] for $n = 2$ and by Lewis [31] for $n \geqslant 2$. Sugawa [46] proved that conditions (1)–(4) are equivalent for $n = 2$. This result was recently generalized for $n \geqslant 2$ in [23].

The first step of the proof is to show that the n-Hardy inequality implies the uniform perfectness (and unboundedness) of the complement. The method is indirect. First, assume that $\mathbb{R}^n \setminus \Omega$ is not uniformly perfect with some large constant $M > 1$. This means that there exists $x_0 \in \mathbb{R}^n \setminus \Omega$ and $r_0 > 0$ such that $B(x_0, Mr_0) \setminus B(x_0, r_0)$ is contained in Ω. The test function

$$u(x) = \begin{cases} \left(\dfrac{|x-x_0|}{r_0} - 1\right)_+ & \text{if } |x-x_0| \leqslant 2r_0, \\ 1 & \text{if } 2r_0 < |x-x_0| < \dfrac{Mr_0}{2}, \\ \left(2 - 2\dfrac{|x-x_0|}{Mr_0}\right)_+ & \text{if } |x-x_0| \geqslant \dfrac{Mr_0}{2} \end{cases}$$

shows that if Ω satisfies the n-Hardy inequality with some constant c_H, then $c_H \geqslant c \log M$.

The following step is to show that the uniform perfectness of the complement further implies a boundary density condition similar to condition (3) in Theorem 3.2. More precisely, there exists $\alpha > 0$ such that

$$\mathcal{H}^\alpha_\infty(\overline{B}(x_0, r_0) \setminus \Omega) \geqslant c r_0^\alpha \tag{4.1}$$

for any $x_0 \in \mathbb{R}^n \setminus \Omega$ and $r_0 > 0$. For the argument to work it is essential that $\overline{B}(x_0, r_0) \setminus \Omega$ is compact. Indeed, there are uniformly perfect countable sets with zero Hausdorff-dimension. To estimate the Hausdorff content of the set, we take a cover $\mathcal{F}$ for $\overline{B}(x_0, r_0) \setminus \Omega$ with balls $B(x, r)$. By compactness, we can choose a finite cover $\mathcal{F}$. We can also assume that the balls in $\mathcal{F}$ are centered in $\overline{B}(x_0, r_0) \setminus \Omega$, which can increase (4.1) at most by factor 2^α. Next, we reduce the number of balls in $\mathcal{F}$ in such a way that the sum

$$\sum_{B(x,r)\in\mathcal{F}} r^\alpha \tag{4.2}$$

does not increase: suppose that $\mathbb{R}^n \setminus \Omega$ is uniformly perfect with constant M. If $\alpha > 0$ is sufficiently small, the elementary inequality

$$r^\alpha + s^\alpha \geqslant (r + s + 2M \min\{r, s\})^\alpha$$

holds for all $r, s > 0$. If there exists balls $B(x, r)$ and $B(y, s)$ in $\mathcal{F}$ such that $r \leqslant 2s$ and $B(x, Mr) \cap B(y, s) \neq \varnothing$, then

$$B(x, r) \cup B(y, s) \subset B(z, r + s + 2M \min\{r, s\})$$

for some $z \in \{x, y\}$. Thus, we can replace the original balls $B(x, r)$ and $B(y, s)$ by a single ball of larger radius so that the sum (4.2) does not increase. We continue this replacement procedure until there are no balls satisfying the condition left. Since $\mathcal{F}$ is finite, the process terminates in a finite number of steps.

Let a ball $B(x_1, r_1) \in \mathcal{F}$ contain x_0. By the uniform perfectness of $\mathbb{R}^n \setminus \Omega$, the set

$$A_1 \setminus \Omega = (B(x_1, Mr_1) \setminus B(x_1, r_1)) \setminus \Omega$$

is not empty. There are two possibilities: A_1 intersects either the complement of $\overline{B}(x_0, r_0)$ or some ball $B(x_2, r_2) \in \mathcal{F}$. In the first case, $r_1 \geqslant r_0/(M+1)$.

In the second case, we know that $r_2 \leqslant r_1/2$; otherwise, the balls $B(x_1, r_1)$ and $B(x_2, r_2)$ could been replaced by a single ball in the above iteration. We continue in the same way: For a ball $B(x_k, r_k)$

$$A_k = B(x_k, Mr_k) \setminus B(x_k, r_k)$$

intersects either the complement of $B(x_0, r_0)$ or some ball $B(x_{k+1}, r_{k+1}) \in \mathcal{F}$ with radius $r_{k+1} \leqslant r_k/2$. This procedure terminates when the first alternative occurs. This happens after a finite number of steps since $\mathcal{F}$ is finite. Let K be an index at which the iteration stops. Since $x_0 \in B(x_1, r_1)$ and $B(x_K, Mr_K)$ intersects the complement of $B(x_0, r_0)$, we have

$$r_0 \leqslant \sum_{i=1}^{K} (M+1) r_i \leqslant (M+1) \sum_{i=1}^{K} 2^{1-i} r_1 \leqslant 2(M+1) r_1.$$

Thus, $r_1 \geqslant r_0/(2(M+1))$ and

$$\sum_{B(x,r) \in \mathcal{F}} r^\alpha \geqslant r_1^\alpha \geqslant \frac{r_0^\alpha}{2(M+1)}.$$

Since this holds for all covers of $\overline{B}(x_0, r_0) \setminus \Omega$, we obtain a lower bound for its Hausdorff α-content depending only on the uniform perfectness constant M. The uniform estimate for the Hausdorff α-content further implies the uniform p-thickness for every $p > n - \alpha$ (cf., for example, [16, Lemma 2.31]). □

Remark 4.1. Theorem 4.1 gives a relatively elementary proof of Theorem 3.1 with $p = n$. It is of interest to obtain an elementary proof for other values of p as well.

5 Eigenvalue Problem

This section gives a characterization of the p-Hardy inequality in terms of weak supersolutions to the nonlinear eigenvalue problem (1.4). This generalizes Proposition 1 of [2], where the result for $p = 2$ was established. We recall that $v \in W^{1,p}_{\mathrm{loc}}(\Omega)$ is a *weak solution* to the problem (1.4) if

$$\int_\Omega \left(|\nabla v(x)|^{p-2} \nabla v(x) \cdot \nabla \varphi(x) - \lambda_p \frac{|v(x)|^{p-2} v(x)}{\delta(x)^p} \varphi(x) \right) dx = 0 \qquad (5.1)$$

for all $\varphi \in C_0^\infty(\Omega)$ and a *weak supersolution* if the integral in (5.1) is nonnegative for $\varphi \geqslant 0$.

Theorem 5.1. *Let $1 < p < \infty$. The inequality* (1.1) *holds with a finite c_H if and only if there is a positive eigenvalue $\lambda_p = \lambda_p(\Omega) > 0$ and a positive weak supersolution $v \in W_0^{1,p}(\Omega)$ to the problem* (1.4) *in Ω.*

Proof. First, suppose that there exists a positive weak supersolution v to the problem (1.4). Take $u \in C_0^\infty(\Omega)$. We can assume that $u \geqslant 0$ in Ω. Let $\varepsilon > 0$. We take

$$\varphi(x) = \frac{u(x)^p}{(v(x)+\varepsilon)^{p-1}}$$

for a test function. It follows that

$$\begin{aligned}
\lambda_p \int_\Omega \frac{u(x)^p v(x)^{p-1}}{\delta(x)^p (v(x)+\varepsilon)^{p-1}}\,dx &\leqslant \int_\Omega |\nabla v(x)|^{p-2}\nabla v(x)\cdot\nabla\varphi(x)\,dx \\
&= (1-p)\int_\Omega |\nabla v(x)|^p (v(x)+\varepsilon)^{-p} u(x)^p\,dx \\
&\quad + p\int_\Omega u(x)^{p-1}(v(x)+\varepsilon)^{1-p}|\nabla v(x)|^{p-2}\nabla v(x)\cdot\nabla u(x)\,dx \\
&\leqslant (1-p)\int_\Omega \left|\frac{u(x)\nabla v(x)}{v(x)+\varepsilon}\right|^p dx + p\int_\Omega \left|\frac{u(x)\nabla v(x)}{v(x)+\varepsilon}\right|^{p-1} |\nabla u(x)|\,dx.
\end{aligned}$$

Using the Young inequality, we conclude that

$$\begin{aligned}
p\int_\Omega &\left|\frac{u(x)\nabla v(x)}{v(x)+\varepsilon}\right|^{p-1} |\nabla u(x)|\,dx \\
&\leqslant (p-1)\int_\Omega \left|\frac{u(x)\nabla v(x)}{v(x)+\varepsilon}\right|^p dx + \int_\Omega |\nabla u(x)|^p\,dx.
\end{aligned}$$

Combining these estimates, passing to the limit as $\varepsilon \to 0$, and taking into account the Lebesgue dominated convergence theorem, we find

$$\int_\Omega \left(\frac{u(x)}{\delta(x)}\right)^p dx \leqslant \frac{1}{\lambda_p}\int_\Omega |\nabla u(x)|^p\,dx.$$

Thus, Ω satisfies the p-Hardy inequality with constant $c_H = 1/\lambda_p$.

To prove the opposite implication, we assume that Ω satisfies the p-Hardy inequality with some finite constant c_H. Let $\lambda_p < 1/c_H$. Introduce the function space

$$X = \{f \in L^p_{\text{loc}}(\Omega) : \nabla f \in L^p(\Omega),\ f\delta^{-1} \in L^p(\Omega)\}$$

equipped with the norm

$$\|f\|_X = \left\|\frac{f}{\delta}\right\|_{L^p(\Omega)} + \|\nabla f\|_{L^p(\Omega)}.$$

Since Ω satisfies the p-Hardy inequality, $X = W_0^{1,p}(\Omega)$ and the norms are equivalent. Therefore, X is a reflexive Banach space. We define the operator $T : X \to X^*$ by

$$(Tf, g) = \int_\Omega |\nabla f(x)|^{p-2} \nabla f(x) \cdot \nabla g(x)\, dx - \lambda_p \int_\Omega \frac{|f(x)|^{p-2} f(x)}{\delta(x)^p} g(x)\, dx,$$

where $f, g \in X$. We will apply the following result from functional analysis. Let $T : X \to X^*$ satisfy the following boundedness, demicontinuity, and coercivity properties:

(i) T is bounded,

(ii) if $f_j \to f$ in X, then $(Tf_j, g) \to (Tf, g)$ for all $g \in X$,

(iii) if (f_j) is a sequence in X with $\|f_j\|_X \to \infty$ as $j \to \infty$, then

$$\frac{(Tf_j, f_j)}{\|f_j\|_X} \to \infty.$$

Then for every $f \in X^*$ there is $v \in X$ such that $Tv = f$.

Now, we check these conditions. Condition (i) holds because

$$\begin{aligned}
\|Tf\|_{X^*} &= \sup_{\|g\|_X \leqslant 1} (Tf, g) \\
&= \sup_{\|g\|_X \leqslant 1} \left| \int_\Omega |\nabla f(x)|^{p-2} \nabla f(x) \cdot \nabla g(x)\, dx - \lambda_p \int_\Omega \frac{|f(x)|^{p-2} f(x)}{\delta(x)^p} g(x)\, dx \right| \\
&\leqslant \sup_{\|g\|_X \leqslant 1} \left[\|\nabla f\|_{L^p(\Omega)}^{p-1} \|\nabla g\|_{L^p(\Omega)} + \lambda_p \left\| \frac{f}{\delta} \right\|_{L^p(\Omega)}^{p-1} \left\| \frac{g}{\delta} \right\|_{L^p(\Omega)} \right] \\
&\leqslant (1 + \lambda_p) \|f\|_X^{p-1} \|g\|_X.
\end{aligned}$$

Assume that $f_j \to f$ in X and $g \in X$. By the Hölder inequality,

$$\begin{aligned}
|(Tf_j, g) - (Tf, g)| &= \left| \int_\Omega (|\nabla f_j(x)|^{p-2} \nabla f_j(x) - \nabla f(x)|^{p-2} \nabla f(x)) \cdot \nabla g(x)\, dx \right. \\
&\quad \left. - \lambda_p \int_\Omega \frac{|f_j(x)|^{p-2} f_j(x) - |f(x)|^{p-2} f(x)}{\delta(x)^p} g(x)\, dx \right| \\
&\leqslant (p-1) \|f_j - f\|_X \max\{\|f_j\|_X^{p-2}, \|f\|_X^{p-2}\} \|g\|_X \to 0
\end{aligned}$$

as $f_j \to f$ in X. Thus, the second condition is satisfied. Finally, let $(f_j)_j$ be a sequence in X such that $\|f_j\|_X \to \infty$ as $j \to \infty$. By the Hardy inequality,

$$
\begin{aligned}
(Tf_j, f_j) &= \int_\Omega \left(|\nabla f_j(x)|^p - \lambda_p \frac{|f_j(x)|^p}{\delta(x)^p} \right) dx \quad \geqslant (1 - \lambda_p c_H) \int_\Omega |\nabla f_j(x)|^p \, dx \\
&\geqslant \frac{1 - \lambda_p c_H}{1 + c_H} \|f_j\|_X^p.
\end{aligned}
$$

Hence $\dfrac{(Tf_j, f_j)}{\|f_j\|_X} \to \infty$, as $j \to \infty$. Thus, condition (iii) is also satisfied.

We fix a function $w \in X^*$ such that $w \geqslant 0$ and $w \not\equiv 0$. Then there exists $v \in X = W_0^{1,p}(\Omega)$ such that $Tv = w$. This means that v is a weak supersolution to the problem (1.4). Moreover, v is positive since

$$
\begin{aligned}
0 \leqslant (Tv, v_-) &= \int_\Omega |\nabla v(x)|^{p-2} \nabla v(x) \cdot \nabla v_-(x) \, dx \\
&\quad - \lambda_p \int_\Omega |v(x)|^{p-2} v(x) v_-(x) \delta(x)^{-p} \, dx \\
&= - \int_\Omega |\nabla v_-(x)|^p \, dx + \lambda_p \int_\Omega v_-(x)^p \delta(x)^{-p} \, dx \\
&\leqslant (\lambda_p - 1/c_H) \int_\Omega v_-(x)^p \delta(x)^{-p} \, dx \leqslant 0.
\end{aligned}
$$

Here, $v_-(x) = -\min(v(x), 0)$ is the negative part of v. Now, the strict positivity of v follows from the weak Harnack inequality. □

Remark 5.1. The eigenvalue problem (1.4) has the following stability property. If there is a positive eigenvalue $\lambda_p = \lambda_p(\Omega) > 0$ and a positive weak supersolution to the problem (1.4) in Ω, then there is ε such that for every q with $p - \varepsilon < q < p + \varepsilon$ there is an eigenvalue $\lambda_q = \lambda_q(\Omega) > 0$ and a positive weak supersolution to the problem (1.4) with p replaced by q in Ω. This fact directly follows from the selfimproving result for the Hardy inequality (cf. Remark 2.1).

References

1. Adams, D.R., Hedberg, L.I.: Function Spaces and Potential Theory. Springer, Berlin (1995)
2. Ancona, A.: On strong barriers and an inequality of Hardy for domains in $\mathbf{R}^n$. J. London Math. Soc. (2) **34**, 274–290 (1986)
3. Barbatis, G., Filippas, S., Tertikas, A.: Series expansion for L^p Hardy inequalities. Indiana Univ. Math. J. **52**, no. 1, 171–190 (2003)
4. Barbatis, G., Filippas, S., Tertikas, A.: Refined geometric L^p Hardy inequalities. Commun. Contemp. Math. **5**, 869–883 (2003)
5. Barbatis, G., Filippas, S., Tertikas, A.: A unified approach to improved L^p Hardy inequalities with best constants. Trans. Am. Math. Soc. **356**, 2169–2196 (2004)

6. Björn, J., MacManus, P., Shanmugalingam, N.: Fat sets and pointwise boundary estimates for p-harmonic functions in metric spaces. J. Anal. Math. **85**, 339–369 (2001)
7. Brezis, H., Marcus, M.: Hardy's inequalities revisited. Ann. Sc. Norm. Super. Pisa, Cl. Sci. (4) **25**, 217–237 (1997)
8. Brezis, H., Marcus, M., Shafrir, I.: Extremal functions for Hardy's inequality with weights. J. Funct. Anal. **171**, 177-191 (2000)
9. Buckley, S.M., Koskela, P.: Orlicz–Hardy inequalities, Illinois J. Math. **48**, 787–802 (2004)
10. Cianchi, A.: Hardy inequalities in Orlicz spaces. Trans. Am. Math. Soc. **351**, 2459-2478 (1999)
11. Danielli, D., Garofalo, N., Phuc, N.C.: Inequalities of Hardy–Sobolev type in Carnot–Carathéodory paces. In: Maz'ya, V. (ed.), Sobolev Spaces in Mathematics. I: Sobolev Type Inequalities. Springer, New York; Tamara Rozhkovskaya Publisher, Novosibirsk. International Mathematical Series **8**, 117-151 (2009)
12. Davies, E.B.: The Hardy constant. Quart. J. Math. Oxford (2) **46**, 417–431 (1995)
13. Hajłasz, P.: Pointwise Hardy inequalities. Proc. Am. Math. Soc. **127**, 417–423 (1999)
14. Hardy, G.H.: Note on a theorem of Hilbert. Math. Z. **6**, 314–317 (1920)
15. Hardy, G.H.: An inequality between integrals. Mess. Math. **54**, 150–156 (1925)
16. Heinonen, J., Kilpeläinen, T., Martio, O.: Nonlinear Potential Theory of Degenerate Elliptic Equations. Oxford Univ. Press, Oxford (1993)
17. Järvi, P., Vuorinen, M.: Uniformly perfect sets and quasiregular mappings. J. London Math. Soc. (2) **54**, 515–529 (1996)
18. Kilpeläinen, T., Kinnunen, J., Martio, O: Sobolev spaces with zero boundary values on metric spaces. Potential Anal. **12**, no. 3, 233–247 (2000)
19. Kilpeläinen, T., Malý, J.: The Wiener test and potential estimates for quasilinear elliptic equations. Acta Math. **172**, 137–161 (1994)
20. Kinnunen, J., Korte, R.: Characterizations of Sobolev inequalities on metric spaces. J. Math. Anal. Appl. **344**, 1093–110 (2008)
21. Kinnunen, J., Martio, O.: Hardy's inequalities for Sobolev functions. Math. Research Lett. **4**, 489–500 (1997)
22. Korte, R., Lehrbäck, J., Tuominen, H.: The equivalence between pointwise Hardy inequalities and uniform fatness. [In preparation]
23. Korte, R., Shanmugalingam, N.: Equivalence and self-improvement of p-fatness and Hardy's inequality, and association with uniform perfectness. Math. Z. [To appear]
24. Koskela, P., Lehrbäck, J.: Weighted pointwise Hardy inequalities. J. London Math. Soc. [To appear]
25. Koskela, P., Zhong, X.: Hardy's inequality and the boundary size. Proc. Am. Math. Soc. **131**, no. 4, 1151–1158 (2003)
26. Kufner, A.: Weighted Sobolev Spaces. John Wiley and Sons, Inc. New York (1985)
27. Lehrbäck, J.: Pointwise Hardy inequalities and uniformly fat sets. Proc. Am. Math. Soc. **136**, 2193–2200 (2008)
28. Lehrbäck, J.: Self-improving properties of weighted Hardy inequalities. Adv. Calc. Var. **1**, no. 2, 193–203 (2008)
29. Lehrbäck, J.: Weighted Hardy inequalities and the size of the boundary. Manuscripta Math. **127**, no. 2, 249-27 (2008)
30. Lehrbäck, J.: Necessary conditions for weighted pointwise Hardy inequalities. Ann. Acad. Sci. Fenn. Math. [To appear]

31. Lewis, J.L.: Uniformly fat sets. Trans. Am. Math. Soc. **308**, 177–196 (1988)
32. Lindqvist, P., Martio, O.: Two theorems of N. Wiener for solutions of quasilinear elliptic equations. Acta Math. **155**, 153–171 (1985)
33. Marcus, M., Mizel, V.J., Pinchover, Y.: On the best constant for Hardy's inequality in $\mathbb{R}^n$. Trans. Am. Math. Soc. **350**, 3237–3255 (1998)
34. Marcus, M., Shafrir, I.: An eigenvalue problem related to Hardy's L^p inequality. Ann. Sc. Norm. Super. Pisa, Cl. Sci. (4) **29**, 581–604 (2000)
35. Matskewich, T., Sobolevskii, P.E.: The best possible constant in a generalized Hardy's inequality for convex domains in $\mathbb{R}^n$. Nonlinear Anal. **28**, 1601-1610 (1997)
36. Maz'ya, V.G.: The continuity at a boundary point of solutions of quasilinear equations (Russian). Vestnik Leningrad. Univ. **25**, no. 13, 42–55 (1970). Correction, ibid. **27:1**, 160 (1972); English transl.: Vestnik Leningrad Univ. Math. **3**, 225–242 (1976)
37. Maz'ya, V.G.: Sobolev Spaces. Springer, Berlin etc. (1985)
38. Maz'ya, V.G.: Lectures on isoperimetric and isocapacitary inequalities in the theory of Sobolev spaces. In: Heat Kernels and Analysis on Manifolds, Graphs, and Metric Spaces, pp. 307–340. Am. Math. Soc., Providence, RI (2003)
39. Maz'ya, V.G.: Conductor and capacitary inequalities for functions on topological spaces and their applications to Sobolev type imbeddings. J. Funct. Anal. **224**, 408–430 (2005)
40. Mikkonen, P.: On the Wolff potential and quasilinear elliptic equations involving measures. Ann. Acad. Sci. Fenn. Ser. A I Math. Disser. 104 (1996)
41. Nečas, J.: Sur une méthode pour résoudre les équations aux dérivées partielles du type elliptique, voisine de la variationelle. Ann. Scuola Norm. Sup. Pisa **16**, 305–326 (1962)
42. Opic, B., Kufner, A.: Hardy-Type Inequalities. Longman Scientific & Technical, Harlow (1990)
43. Pinchover, Y., Tintarev, K.: Ground state alternative for p-Laplacian with potential term. Calc. Var. Partial Differ. Equ. **28**, 179–201 (2007)
44. Pinchover, Y., Tintarev, K.: On positive solutions of minimal growth for singular p-Laplacian with potential term. Adv. Nonlin. Studies **8**, 213–234 (2008)
45. Pinchover, Y., Tintarev, K.: On the Hardy–Sobolev–Maz'ya inequality and its generalizations. In: Maz'ya, V. (ed.), Sobolev Spaces in Mathematics. I: Sobolev Type Inequalities. Springer, New York; Tamara Rozhkovskaya Publisher, Novosibirsk. International Mathematical Series **8**, 281–297 (2009)
46. Sugawa, T.: Uniformly perfect sets: analytic and geometric aspects. Sugaku Expos. **16**, 225–242 (2003)
47. Tidblom, J.: A geometrical version of Hardy's inequality for $W_0^{1,p}(\Omega)$. Proc. Am. Math. Soc. **132**, 2265-2271 (2004) (electronic)
48. Tidblom, J.: A Hardy inequality in the half-space. J. Funct. Anal. **221**, 482-495 (2005)
49. Wannebo, A.: Hardy inequalities. Proc. Am. Math. Soc. **109**, no. 1, 85–95 (1990)
50. Wannebo, A.: Hardy inequalities and imbeddings in domains generalizing $C^{0,\lambda}$ domains. Proc. Am. Math. Soc. **122**, no. 4, 1181–1190 (1994)
51. Wannebo, A.: Hardy and Hardy PDO type inequalities in domains. Part 1. math.AP/0401253, arxiv (2004)
52. Wannebo, A.: A remark on the history of Hardy inequalities in domains. math.AP/0401255, arxiv (2004)

Geometric Properties of Planar BV-Extension Domains

Pekka Koskela, Michele Miranda Jr., and
Nageswari Shanmugalingam

Dedicated to Professor Vladimir G. Maz'ya

Abstract We investigate geometric properties of those planar domains that are extension for functions with bounded variation. We start from a characterization of such domains given by Burago–Maz'ya and prove that a bounded, simply connected domain is a BV-extension domain if and only if its complement is quasiconvex. We further prove that the extension property is a bi-Lipschitz invariant and give applications to Sobolev extension domains.

1 Introduction

Let $\Omega \subset \mathbb{R}^2$ be a domain and $1 \leqslant p \leqslant \infty$. Recall that

$$BV(\Omega) = \{u \in L^1(\Omega) : \ |Du|(\Omega) < \infty\},$$

where

$$|Du|(\Omega) = \sup\Big\{\int_\Omega u \operatorname{div} \ v \ dx : \ v = (v_1, v_2) \in C_0^\infty(\Omega; \mathbb{R}^2), |v| \leqslant 1\Big\}$$

Pekka Koskela
Department of Mathematics and Statistics, P.O. Box 35 (MaD), FIN–40014, University of Jyväskylä, Finland
e-mail: pkoskela@maths.jyu.fi

Michele Miranda Jr.
Department of Mathematics, University of Ferrara, via Machiavelli 35, 44100, Ferrara, Italy
e-mail: michele.miranda@unife.it

Nageswari Shanmugalingam
Department of Mathematical Sciences, P.O.Box 210025, University of Cincinnati, Cincinnati, OH 45221–0025, USA
e-mail: nages@math.uc.edu

A. Laptev (ed.), *Around the Research of Vladimir Maz'ya I: Function Spaces*,
International Mathematical Series 11, DOI 10.1007/978-1-4419-1341-8_11,

and
$$W^{1,p}(\Omega) = \{u \in L^p(\Omega) : \nabla u \in L^p(\Omega, \mathbb{R}^2)\}.$$
Here, ∇u is the distributional gradient of u. We employ these spaces with the norms
$$\|u\|_{BV(\Omega)} = \|u\|_{L^1(\Omega)} + |Du|(\Omega)$$
and
$$\|u\|_{W^{1,p}(\Omega)} = \|u\|_{L^p(\Omega)} + \|\nabla u\|_{L^p(\Omega)}.$$
From the discussion in [3] and [11]
$$|Du|(\Omega) = \inf\Big\{\liminf_{k\to\infty} \int_\Omega |\nabla u_k| dx \,:\, u_k \in W^{1,1}_{loc}(\Omega), u_k \to u \text{ in } L^1(\Omega)\Big\}, \tag{1.1}$$
where we also may replace $W^{1,1}_{\text{loc}}(\Omega)$ with $C^\infty(\Omega)$.

In this paper, we study geometric properties of those bounded, simply connected planar domains Ω that are extension domains for BV or for $W^{1,1}$. We say that a domain $\Omega \subset \mathbb{R}^2$ is a *BV-extension domain* if there exists a constant c and an extension operator $T : BV(\Omega) \to BV(\mathbb{R}^2)$, not necessarily linear, so that $Tu|_\Omega = u$ and $\|Tu\|_{BV(\mathbb{R}^2)} \leqslant c\|u\|_{BV(\Omega)}$ for each $u \in BV(\Omega)$. Replacing BV by $W^{1,p}$ above gives the definition of a $W^{1,p}$-extension domain. In the case $p > 1$, $W^{1,p}$-extension domains admit a linear extension operator, but it appears to be unknown if this holds for $p = 1$ or for BV-extension domains. For other possible definitions of extension domains see Section 2 below.

The geometry of bounded, simply connected $W^{1,p}$-extension domains for $p = 2$ is well understood. Indeed, this class of domains coincides with the thoroughly investigated class of quasidisks (cf. [14, 4, 5, 8]) that allows us for a number of geometric characterizations. For $p > 2$, one also has rather good geometric criteria for the extension property [9]. In the remaining range $1 \leqslant p < 2$ for bounded, simply connected domains, it is known that Ω has to be a so-called John domain (cf. [4, 12]), but no geometric characterization is available. Finally, Burago and Maz'ya [2] have given a characterization for an extension property related to BV in terms of extendability of sets of finite perimeter in the domain. In fact, this seminal result by Burago and Maz'ya was the first characterization for Sobolev type extensions and should be viewed as the predecessor of all the results mentioned above.

Our first result that partly relies on the work of Burago and Maz'ya [2] (cf. also [11, Section 6.3.5]) gives a concrete characterization for bounded, simply connected BV-extension domains.

Theorem 1.1. *Let $\Omega \subset \mathbb{R}^2$ be a bounded, simply connected domain. Then Ω is a BV-extension domain if and only if there exists a constant $C > 0$ such that for all $x, y \in \mathbb{R}^2 \setminus \Omega$ there is a rectifiable curve $\gamma \subset \mathbb{R}^2 \setminus \Omega$ connecting x and y with length $\ell(\gamma) \leqslant C\,|x - y|$. That is, Ω is a BV-extension domain if and only if the complement of Ω is quasiconvex.*

As a corollary of this theorem and Lemma 2.4 we obtain a new necessary condition for a bounded, simply connected domain to be a $W^{1,1}$-extension domain.

Corollary 1.2. *Let $\Omega \subset \mathbb{R}^2$ be a bounded, simply connected domain that is a $W^{1,1}$-extension domain. Then the complement of Ω is quasiconvex.*

Simple examples such as a slit disk show that the quasiconvexity of the complement does not characterize $W^{1,1}$-exendability. However, it is easy to check that the quasiconvexity of the complement of Ω is a stronger requirement than Ω being a John domain or the complement of Ω being of bounded turning [4]. Note also that the complement of a quasidisk is quasiconvex. Consequently, the claim of Corollary 1.2 holds also in the $W^{1,2}$-extension setting. We conjecture that it, in fact, holds for all $1 \leqslant p \leqslant 2$.

Our second corollary deals with the invariance of the extension property under bi-Lipschitz mappings of Ω onto Ω'. This may seem trivial as bi-Lipschitz mappings preserve the spaces in question. The novelty here is that our bi-Lipschitz mapping is *a priori* only defined in the domain in question and extendability requires information in the entire plane.

Corollary 1.3. *Let $\Omega \subset \mathbb{R}^2$ be a bounded, simply connected domain that is a BV-extension domain (or a $W^{1,1}$-extension domain), and let $f : \Omega \to \Omega' \subset \mathbb{R}^n$ be a bi-Lipschitz mapping. Then Ω' is also a BV-extension domain (or a $W^{1,1}$-extension domain).*

Corollary 1.3 leaves open the case $1 < p \leqslant \infty$, but the analog holds also in this case by a recent result from [6]. We conjecture that the assumption that Ω be simply connected in Corollary 1.3 is superfluous.

This paper is organized as follows. In Section 2, we give necessary preliminaries and discuss an alternative definition for an extension domain. Section 3 contains proofs of the main results stated above. Finally, in Section 4, we discuss the meaning of Theorem 1.1 in a special case and briefly comment on possible generalizations of our result.

2 Preliminaries

The notation used in this paper is as follows. Given $x \in \mathbb{R}^2$ and $r > 0$, the (open) disk centered at x with radius r is denoted by $B_r(x)$, and $S(x,r)$ denotes its boundary $\partial B_r(x)$. The 2-dimensional Lebesgue measure of a measurable set $A \subset \mathbb{R}^2$ is denoted by $|A|$.

Burago and Maz'ya [2] consider extension operators for

$$BV_l(\Omega) = \{u \in L^1_{\text{loc}}(\Omega) : |Du|(\Omega) < +\infty\}.$$

They provide a necessary and sufficient condition for the existence of an extension operator $T_l : BV_l(\Omega) \to BV_l(\mathbb{R}^2)$ such that for all $u \in BV_l(\Omega)$

$$|DT_l(u)|(\mathbb{R}^2) \leqslant c|Du|(\Omega). \tag{2.1}$$

We call such a domain a *BV_l-extension domain.* If $E \subset \mathbb{R}^2$ is a measurable set whose characteristic function χ_E lies in $BV_l(\Omega)$, then we say that E has *finite perimeter* in Ω and denote $P(E,\Omega) := |D\chi_E|(\Omega)$. From the Burago–Maz'ya characterization and the subadditivity property of the perimeter measure of a given set it follows that it is necessary and sufficient to know that for every set $E \subset \Omega$ of finite perimeter $P(E,\Omega)$ in Ω there is a set $F \subset \mathbb{R}^2$ of finite perimeter such that $F \cap \Omega = E$ and $P(F,\mathbb{R}^2) \leqslant C\,P(E,\Omega)$.

Let us begin by pointing out that a bounded domain is a BV-extension domain in our sense if and only if it is a BV_l-extension domain. This can be seen, for example, via a modification of an argument of Herron and Koskela [7].

Lemma 2.1. *A bounded domain $\Omega \subset \mathbb{R}^2$ is a BV-extension domain if and only if it is a BV_l-extension domain.*

Towards the proof, we record a Poincaré type inequality resulting from the compactness of a suitable embedding. It can be obtained by combining some results in [11, Sections 6.1.7, 3.2.3, 3.5.2] For the convenience of the reader we give a simple proof below. Recall that a normed space X is said to be *embed compactly* into another normed space Y if there is a bounded embedding map $\iota : X \to Y$ such that whenever $(a_k)_k$ is a norm-bounded sequence in X, the limit $\lim_j \iota(a_{k_j})$ exists in Y for some subsequence $(a_{k_j})_j$. We call this embedding *natural* if ι can be taken to be the identity map.

We continue with a simple observation.

Lemma 2.2. *Suppose that $\Omega \subset \mathbb{R}^2$ is a domain such that $BV(\Omega)$ embeds naturally compactly in $L^1(\Omega)$. Then $|\Omega| < \infty$.*

Proof. We define a function $m_\Omega : [0,\infty) \to \mathbb{R}$ by setting

$$m_\Omega(r) = |\Omega \cap B_r(0)|. \tag{2.2}$$

Then $m_\Omega \in Lip_{\mathrm{loc}}([0,+\infty))$ (with $m_\Omega(r) \leqslant \pi r^2$). Therefore, m_Ω is differentiable almost everywhere and, by the coarea formula applied to the function $u(x) = (|x| - r)/h$ in the annular region $\Omega \cap B_{r+h}(0) \setminus B_r(0)$, at almost all points r of differentiability of m_Ω we have

$$m'_\Omega(r) = P(B_r(0),\Omega). \tag{2.3}$$

Let I be the set of all $r > 0$ that are points of differentiability of m_Ω and for which (2.3) holds. We claim that

$$\liminf_{I \ni r \to +\infty} \frac{m'_\Omega(r)}{m_\Omega(r)} = 0.$$

In fact, if there are positive numbers M and r_M so that $m'_\Omega(r)/m_\Omega(r) \geqslant M$ for all $r \geqslant r_M$, then $m_\Omega(r) \geqslant m_\Omega(r_M)e^{M(r-r_M)}$, which contradicts (2.2).

From the above discussion it follows that there exists $C > 0$ and a sequence $(r_n)_n$ from I with $r_n \to \infty$ such that $P(B_{r_n}(0), \Omega) = m'_\Omega(r_n) \leqslant C m_\Omega(r_n)$. We define a sequence of functions by setting

$$u_n = \frac{1}{m_\Omega(r_n)} \chi_{\Omega \cap B_{r_n}(0)}.$$

Then $\|u_n\|_{L^1(\Omega)} = 1$ and

$$|Du_n|(\Omega) = \frac{1}{m_\Omega(r_n)} P(B_{r_n}, \Omega) \leqslant C.$$

If the area of Ω were infinite, the sequence $(u_n)_n$ would converge uniformly to the zero function, and so this would be the only potential L^1-limit of a subsequence of $(u_n)_n$. Since $\|u_n\|_{L^1(\Omega)} = 1$, we would conclude that there is no subsequence that converges in $L^1(\Omega)$, which contradicts our assumption. □

In the next result and in what follows, for sets A with $0 < |A| < +\infty$ we write

$$u_A = \fint_A u\,dx = \frac{1}{|A|} \int_A u\,dx,$$

whenever $u \in L^1(A)$.

Lemma 2.3. *If $\Omega \subset \mathbb{R}^2$ is a domain and $BV(\Omega)$ embeds naturally compactly into $L^1(\Omega)$, then there is a constant $C > 0$ such that whenever $u \in BV(\Omega)$,*

$$\int_\Omega |u - u_\Omega|\,dx \leqslant C\,|Du|(\Omega). \tag{2.4}$$

Proof. By Lemma 2.2 and the hypothesis of this lemma, the measure of Ω must necessarily be finite. Suppose that for every positive integer n there is a function $u_n \in BV(\Omega)$ such that

$$\int_\Omega |u_n - (u_n)_\Omega|\,dx \geqslant n\,|Du_n|(\Omega).$$

Replacing u_n with

$$\left(\int_\Omega |u_n - (u_n)_\Omega|\,dx\right)^{-1} (u_n - (u_n)_\Omega),$$

we may also assume that

$$\|u_n\|_{L^1(\Omega)} = 1 \quad \text{and} \quad (u_n)_\Omega = \fint_\Omega u_n\,dx = 0.$$

Then, by the above assumption, $|Du_n|(\Omega) \leqslant n^{-1}$, and so the sequence (u_n) is bounded in $BV(\Omega)$, and hence there exists $w \in L^1(\Omega)$ such that $u_{n_j} \to w$ in $L^1(\Omega)$ for some subsequence $(u_{n_j})_j$. Because

$$\lim_{n\to\infty} |Du_n|(\Omega) = 0,$$

we have $w \in BV(\Omega)$ with $|Dw|(\Omega) = 0$. As Ω is connected, it follows (using the Poincaré inequality for BV cf. [3, 11]) that w is constant on Ω. On the other hand,

$$\int_\Omega w\,dx = \lim_j \int_\Omega u_{n_j}\,dx = 0,$$

but

$$\int_\Omega |w|\,dx = \lim_j \int_\Omega |u_{n_j}|\,dx = 1,$$

which is impossible if w is a constant function. This leads to a contradiction. □

Proof of Lemma 2.1. First suppose that Ω is a bounded BV_l-extension domain, and let $T_l : BV_l(\Omega) \to BV_l(\mathbb{R}^2)$ be the bounded extension operator. Since $BV(\Omega) \subset BV_l(\Omega)$, for every $f \in BV(\Omega)$ the function $T_l f$ belongs to $BV_l(\mathbb{R}^2)$, with $|DT_l f|(\mathbb{R}^2) \leqslant C|Df|(\Omega)$. Let B be a ball in $\mathbb{R}^2$ such that Ω is a relatively compact subdomain of B. Let $c_0 = (T_l f)_B$. By the Poincaré inequality,

$$\int_B |T_l f - c_0|\,dx \leqslant C\,\mathrm{diam}(B)|DT_l f|(B) \leqslant C\,\mathrm{diam}(B)|Df|(\Omega).$$

Thus,

$$\begin{aligned}
|c_0| &\leqslant \fint_\Omega |f - c_0|\,dx + \fint_\Omega |f|\,dx \\
&\leqslant \frac{1}{|\Omega|}\int_B |T_l f - c_0|\,dx + \fint_\Omega |f|\,dx \\
&\leqslant C|\Omega|^{-1}\,\mathrm{diam}(B)\left(|Df|(\Omega) + \int_\Omega |f|\,dx\right).
\end{aligned}$$

Fix a Lipschitz function $\eta : \mathbb{R}^2 \to [0,1]$ with compact support in B such that $\eta = 1$ on Ω. We define our extension operator $E : BV(\Omega) \to BV(\mathbb{R}^2)$ by setting $Ef = \eta T_l f$. Now,

$$\begin{aligned}\int_{\mathbb{R}^2} |Ef|\,dx &\leqslant \int_B |T_l f|\,dx \\ &\leqslant \int_B |T_l f - c_0|\,dx + |B|\,|c_0| \\ &\leqslant C\,\operatorname{diam}(B)\Big(|Df|(\Omega) + |B||\Omega|^{-1}|Df|(\Omega) + |B| \fint_\Omega |f|\,dx\Big) \\ &\leqslant C_0\,\|f\|_{BV(\Omega)}.\end{aligned}$$

Furthermore,

$$\begin{aligned}|DEf|(\mathbb{R}^2) &\leqslant |DT_l f|(B) + \int_{\mathbb{R}^2} |T_l f||\nabla\eta|\,dx \\ &\leqslant C\,|Df|(\Omega) + C\int_B |T_l f - c_0|\,dx + C|B|\,|c_0| \\ &\leqslant C_1\,\|f\|_{BV(\Omega)}.\end{aligned}$$

This proves that E is bounded, and hence Ω is a BV-extension domain.

Now suppose that Ω is a bounded BV-extension domain. Let $T : BV(\Omega) \to BV(\mathbb{R}^2)$ be an extension operator. Fix a ball B so that $\Omega \subset B$. By the Rellich theorem for BV (cf. [3, 11]), $T(BV(\Omega))\big|_B$ embeds naturally compactly into $L^1(B)$. Especially, $BV(\Omega)$ embeds naturally compactly into $L^1(\Omega)$. Hence, by Lemma 2.3, we have a constant $C > 0$ for which the inequality (2.4) is satisfied by every $u \in BV(\Omega)$. For $u \in BV_l(\Omega)$ and every positive integer n we set $u_n(x) = \max\{-n, \min\{n, u(x)\}\}$. Then $u_n \in BV(\Omega)$ with $|Du_n|(\Omega) \leqslant |Du|(\Omega)$ and $u_n \to u$ pointwise. Let

$$c_n = \fint_\Omega u_n\,dx.$$

Then $u_n - c_n \in BV(\Omega)$, and, by the inequality (2.4),

$$\|u_n - c_n\|_{BV(\Omega)} \leqslant C\,|Du_n|(\Omega) \leqslant C\,|Du|(\Omega).$$

Hence, by the compactness of the embedding $BV(\Omega)$ into $L^1(\Omega)$, there is a subsequence $(u_{n_k} - c_{n_k})_k$ converging in $L^1(\Omega)$ to a function $w \in L^1(\Omega)$. Passing to a further subsequence if necessary, we may also assume that $u_{n_k} - c_{n_k} \to w$ pointwise almost everywhere in Ω as well. Since $u_{n_k} \to u$ pointwise in Ω, it follows that the sequence $(c_{n_k})_k$ of real numbers converges to some $c_0 \in \mathbb{R}$. Therefore, $w = u - c_0$, $u \in L^1(\Omega)$ and hence $u \in BV(\Omega)$, and $u_{n_k} - c_{n_k} \to u - c_0$ in $L^1(\Omega)$. Furthermore,

$$c_0 = \fint_\Omega u\,dx.$$

Because $u \in BV(\Omega)$, we have $Tu \in BV(\mathbb{R}^2)$, but it is not clear if we can control the BV_l norm of Tu purely in terms of the BV_l norm of u. To fix this, we modify our extension operator by setting by $E(u) = T(u - c_0) + c_0$, where

$$c_0 = \fint_\Omega u\,dx.$$

Then $E : BV_l(\Omega) \to BV_l(\mathbb{R}^2)$. Moreover,

$$|DE(u)|(\mathbb{R}^2) = |DT(u - c_0)|(\mathbb{R}^2) \leqslant C\,\|u - c_0\|_{BV(\Omega)} \leqslant C\,|Du|(\Omega),$$

where we again used the inequality (2.4) to obtain the last inequality. This completes the proof. □

For the sake of completeness, we include a simple proof for the following connection between Sobolev- and BV-extension domains.

Lemma 2.4. *A $W^{1,1}$-extension domain is necessarily a BV-extension domain.*

Proof. Let Ω be a $W^{1,1}$-extension domain, with a bounded extension operator $T : W^{1,1}(\Omega) \to W^{1,1}(\mathbb{R}^2)$, and let $u \in BV(\Omega)$. Then there is a sequence $(u_k)_k \subset W^{1,1}(\Omega)$ such that $u_k \to u$ in $L^1(\Omega)$,

$$\int_\Omega |\nabla u_k| dx \leqslant 2|Du|(\Omega),$$
$$\int_\Omega |u_k| dx \leqslant 2\int_\Omega |u| dx,$$
$$\lim_k \int_\Omega |\nabla u_k| dx = |Du|(\Omega).$$

Let $v_k = Tu_k \in W^{1,1}(\mathbb{R}^2)$.

Since $\|u_k\|_{W^{1,1}(\Omega)} \leqslant 2\|u\|_{BV(\Omega)}$, we see that $\|v_k\|_{W^{1,1}(\mathbb{R}^2)} \leqslant C\,\|u\|_{BV(\Omega)}$. Again, fix a ball $B_j(0)$ so that Ω is a relatively compact subdomain of $B_j(0)$. By the Rellich theorem, there is a subsequence $(v_k^{(j)})_k$ that converges in $L^1(B_j(0))$ and almost everywhere in $B_j(0)$ to some function $w_j \in L^1(B_j(0))$. We repeat the argument for this subsequence and $B_{j+1}(0)$, and continue by induction. Then the diagonal sequence $(v_k^{(k)})_k$ converges almost everywhere to a function w with $w = w_l$ on $B_l(0)$, $l > j$, and the convergence holds also with respect to $L^1(B_l(0))$. It follows that $\|v_k\|_{L^1(B_l(0))} \leqslant C\,\|u\|_{BV(\Omega)}$ for all $l \geqslant j$ and, consequently, $w \in L^1(\mathbb{R}^2)$ with the same bound. Secondly,

$$\int_{\mathbb{R}^2} |\nabla v_k^k| \leqslant 2\|u\|_{BV(\Omega)},$$

and it thus easily follows that $w \in BV(\mathbb{R}^2)$ with the desired norm bound. The claim follows when we set $E(u) = w$. □

In [2], Burago and Maz'ya gave a characterization of BV_l-extension domains. A general Euclidean spaces version of the following result can be found in [2] or [11, p. 314]. A more general metric space version of this statement was recently given in [1].

Theorem 2.5 (Burago–Maz'ya). *A domain $\Omega \subset \mathbb{R}^2$ is a BV_l-extension domain if and only if there is a constant $C > 0$ such that whenever $E \subset \Omega$ is a Borel set of finite perimeter in Ω,*

$$\tau_\Omega(E) \leqslant C\, P(E, \Omega), \tag{2.5}$$

where $\tau_\Omega(E) = \inf\{P(F, \mathbb{R}^2 \setminus \Omega) : F \cap \Omega = E\}$.

Note that

$$P(F, \mathbb{R}^2 \setminus \Omega) = \inf\{P(F, U) : U \text{ is open and } \mathbb{R}^2 \setminus \Omega \subset U\}.$$

The following lemma of Burago–Maz'ya [2] gives an analogous characterization for a variant of bounded BV-extension domains (cf. also [11, Section 6.3.5]). For a self-contained proof of this lemma in a more general setting, also see [1].

Lemma 2.6. *If $\Omega \subset \mathbb{R}^2$ is a bounded domain, then there is a bounded extension map $T : BV(\Omega) \to BV_l(\mathbb{R}^2)$ if and only if there exist constants $C, \delta > 0$ such that for all Borel sets $E \subset \Omega$ of finite perimeter in Ω with* $\operatorname{diam}(E) \leqslant \delta$

$$\tau_\Omega(E) \leqslant C\, P(E, \Omega).$$

The next lemma allows us to approximate sets of finite perimeter by smooth sets of finite perimeter. The statement and proof of this theorem for domains in $\mathbb{R}^n$ can be found in [11, Section 6.1.3]. Recall that for sets F and G their symmetric difference is denoted by $F \Delta G$.

Lemma 2.7. *If $F \subset \mathbb{R}^2$ is a set of finite perimeter, then there exist sets $F_k \subset \mathbb{R}^2$ such that ∂F_k is smooth, $\chi_{F_k} \to \chi_E$ in $L^1_{loc}(\mathbb{R}^2)$, and $\lim_k P(F_k, \mathbb{R}^2) = P(F, \mathbb{R}^2)$. Furthermore, this sequence can be chosen so that*

$$F_k \Delta F \subset \bigcup_{x \in \partial F} B_{1/k}(x). \tag{2.6}$$

Recall that by the isoperimetric inequality in $\mathbb{R}^2$, if F is a set of finite perimeter, then either $|F|$ or $|\mathbb{R}^2 \setminus F|$ is finite. If $|F|$ is finite, the expression (2.6) follows from the construction in [11] of F_k as certain level sets of smooth convolution approximations of χ_F. If $|\mathbb{R}^2 \setminus F|$ is finite, then (2.6) follows from setting F_k to be the complement of the construction in [11] that approximates $\mathbb{R}^2 \setminus F$.

Suppose that ∂F_k is smooth in $\mathbb{R}^2$. If F_k is bounded (or its complement is bounded), then the number of connected components of ∂F_k is finite. If both F_k and its complement are unbounded, then there can be infinitely (but countably) many components, but only finitely many ones can intersect any given disc. If ∂F_k is only assumed to be smooth in a domain Ω, then the corresponding analog is that the connected components cannot accumulate in any compact part of Ω, though they could accumulate toward $\partial\Omega$.

From now on, we use the abbreviation $\partial F \cap \Omega \in \mathcal{C}^\infty$ for the statement that $\partial F \cap \Omega$ is smooth.

Lemma 2.8. *Let $\Omega \subset \mathbb{R}^2$ be a bounded domain. Suppose that there is a constant $C > 0$ such that for every closed set $F \subset \mathbb{R}^2$ with $\partial F \cap \Omega \in \mathcal{C}^\infty$ there exists a set $\widehat{F} \subset \mathbb{R}^2$ with $\widehat{F} \cap \Omega = F \cap \Omega$ and*

$$|D\chi_{\widehat{F}}|(\mathbb{R}^2) \leqslant C|D\chi_F|(\Omega).$$

Then Ω is a BV-extension domain.

Proof. By Lemma 2.1, it suffices to show that Ω is a BV_l-extension domain, i.e., Ω satisfies the Burago–Maz'ya condition of Lemma 2.5.

Let E be any set such that $\chi_\Omega \neq \chi_E \in BV(\Omega)$. Then, by [11, Section 6.1.3], there exists a sequence $(F_k)_k$ of sets in Ω so that $\partial F_k \cap \Omega \in \mathcal{C}^\infty$ and

$$\chi_{F_k} \to \chi_E \text{ in } L^1(\Omega), \quad |D\chi_{F_k}|(\Omega) \to |D\chi_E|(\Omega). \tag{2.7}$$

By the regularity of F_k, we may assume that F_k is closed. Now, by hypothesis, there exist sets $\widehat{F}_k$ so that $\widehat{F}_k \cap \Omega = F_k \cap \Omega$ and

$$|D\chi_{\widehat{F}_k}|(\mathbb{R}^2) \leqslant C|D\chi_{F_k}|(\Omega). \tag{2.8}$$

By (2.7) and (2.8), we get

$$\limsup_k |D\chi_{\widehat{F}_k}|(\mathbb{R}^2) \leqslant C|D\chi_E|(\Omega).$$

By the Rellich theorem applied to balls containing Ω and an application of a diagonalization argument, we may assume that there is F_∞ such that $\chi_{\widehat{F}_k} \to \chi_{F_\infty}$ in $L^1_{loc}(\mathbb{R}^2)$. For this set, by the lower semicontinuity of the BV_l norm, we have

$$|D\chi_{F_\infty}|(\mathbb{R}^2) \leqslant \limsup_k |D\chi_{\widehat{F}_k}|(\mathbb{R}^2) \leqslant C|D\chi_E|(\Omega).$$

Since for every k we have $\widehat{F}_k \cap \Omega = F_k \cap \Omega$, we conclude that $\chi_{F_\infty \cap \Omega} = \chi_E$ almost everywhere. Thus, such an extension χ_{F_∞} of χ_E proves that Ω satisfies the Burago–Maz'ya condition (2.5). □

We will need a lower bound for the perimeters of certain sets. The following lemma provides a suitable one.

Lemma 2.9. *Let $E \subset \mathbb{R}^2$ be an open set with finite perimeter. Suppose that there exist two curves $\gamma_1, \gamma_2 : [0,1] \to \mathbb{R}^2$ with $\gamma_1([0,1]) \subset E$ and $\gamma_2([0,1]) \subset \mathbb{R}^2 \setminus \overline{E}$ with $\min\{|\gamma_1(1) - \gamma_1(0)|, |\gamma_2(1) - \gamma_2(0)|\} \geqslant \tau$. Then $P(E, \mathbb{R}^2) \geqslant 2\tau$.*

Proof. Let us first assume that E is an open, bounded, connected smooth subset of $\mathbb{R}^2$, and let $x, y \in E$. Then $P(E) \geqslant 2|x-y|$. In fact, if we consider

$$t_1 = \inf\{t \in \mathbb{R} : x + t(y-x) \in E\},$$
$$t_2 = \sup\{t \in \mathbb{R} : x + t(y-x) \in E\}$$

with our hypothesis on E, the points $x_i = x + t_i(y-x)$, $i = 1, 2$, belong to the same connected component β of ∂E and they divide it into two curves β_1 and β_2 each with length $l(\beta_i) \geqslant |x_1 - x_2|$, and then

$$P(E) \geqslant l(\beta_1) + l(\beta_2) \geqslant 2|x_1 - x_2| \geqslant 2|x-y|.$$

If now E is any open set with finite perimeter, then either E or $\mathbb{R}^2 \setminus E$ has finite area. Let us assume that $|\mathbb{R}^2 \setminus E| < +\infty$. We then consider $F = \mathbb{R}^2 \setminus E$ and the curve γ_2 (in the case $|E| < +\infty$, we have to consider γ_1). By assumption, $\delta = \operatorname{dist}(\gamma_2, \overline{E}) > 0$. Let F_k be an approximation of F obtained as in Lemma 2.7, with $k > 2/\delta$. With this choice, the curve γ_2 is eventually contained in one of the connected components $\widetilde{F}_\varepsilon$ of F_ε. Now, by the discussion in the previous paragraph,

$$P(E) = P(F) = \lim_{k\to\infty} P(F_k) \geqslant \limsup_{k\to\infty} P(\widetilde{F}_k) \geqslant 2\tau. \qquad \square$$

Lemma 2.10. *Let $\Omega \subset \mathbb{R}^2$ be a BV_l-extension domain. Then there exist constants $c, c_1, c_2 \in (0,1)$ and $r_0 > 0$ such that for any $x \in \partial\Omega$ and $0 < r < r_0$*

$$|\Omega \cap B_r(x)| \geqslant c|B_r(x)|. \tag{2.9}$$

Moreover, for each connected component E of $\Omega \cap B_r(x)$ that intersects $B_{r/5}(x)$

$$|E| \geqslant c_1|B_r(x)| \quad \text{and} \quad \mathcal{H}^1(\Omega \cap \partial E) \geqslant c_2\, r.$$

Proof. We choose $r_0 > 0$ such that whenever $x \in \partial\Omega$ and $0 < r < r_0$, $\Omega \setminus \overline{B}_r(x)$ contains a connected subset of diameter at least r_0.

Suppose that there exists a sequence $(x_k)_k \subset \partial\Omega$, $0 < r_k < r_0$, and a sequence $\varepsilon_k \to 0$ such that there is a connected component E_k of $\Omega \cap B_{r_k}(x_k)$ intersecting $B_{r_k/5}(x_k)$ with

$$|E_k| = \varepsilon_k |B_{r_k}(x_k)| = \pi\varepsilon_k r_k^2.$$

Since

$$|E_k| = \int_0^{r_k} \mathcal{H}^1(E_k \cap \partial B_t(x_k))dt,$$

there exists $\overline{t} \in [r_k/2, r_k]$ such that

$$P(B_{\overline{t}}(x_k) \cap E_k, \Omega) = \mathcal{H}^1(E_k \cap \partial B_{\overline{t}}(x_k)) \leqslant 2\pi\varepsilon_k r_k. \tag{2.10}$$

Observe that as E_k contains a curve connecting a point in $S(x_k, r_k)$ to some point in $S(x_k, r_k/5)$, it is clear that $E_k \cap B_{\overline{t}}(x_k)$ contains a curve connecting some point in $S(x_k, \overline{t})$ to a point in $S(x_k, r_k/5)$. Hence the extension $\widehat{E}_k$ of $E_k \cap B_{\overline{t}}(x_k)$ has a connected component of diameter at least $3r_k/10$. Furthermore, as $r_k < r_0$ and $\widehat{E}_k \cap \Omega = E_k \cap B_{\overline{t}}(x_k) \subset B_r(x_k)$, it follows that $\mathbb{R}^2 \setminus \overline{\widehat{E}_k}$ also contains a connected set of diameter at least $3r_k/10$. It therefore follows by Lemma 2.9 that $P(\widehat{E}_k, \mathbb{R}^2) \geqslant 3r_k/10$. This means that

$$\frac{P(\widehat{E}_k, \mathbb{R}^2)}{P(E_k \cap B_{\overline{t}}(x_k), \Omega)} = \frac{P(\widehat{E}_k, \mathbb{R}^2)}{\mathcal{H}^1(E_k \cap \partial B_{\overline{t}}(x_k))} \geqslant \frac{3r_k}{20\pi\varepsilon_k r_k} = \frac{3}{20\pi\varepsilon_k}.$$

Letting $k \to \infty$ and recalling that $\varepsilon_k \to 0$, we obtain a contradiction with the extension property.

Now, fix a connected component E of $B_r(x) \cap \Omega$ that intersects $B_{r/5}(x)$. Then, by the above argument and the BV-extension property with an extension $\widehat{E}$ of E given by the BV-extension property,

$$C \geqslant \frac{P(\widehat{E}, \mathbb{R}^2)}{\mathcal{H}^1(\Omega \cap \partial E)} = \frac{P(\widehat{E}, \mathbb{R}^2)}{P(E, \Omega)} \geqslant \frac{3r}{10P(E, \Omega)},$$

which completes the proof. □

We complete this section by pointing out that Lemmas 2.1, 2.3–2.8, as well as their proofs given here, hold in higher dimensional Euclidean spaces as well.

3 Proofs of the Results

Proof of Theorem 1.1. We first prove the quasiconvexity of a bounded, simply connected BV_l-extension domain. The same for BV-extension domains then follows from Lemma 2.1.

Suppose that Ω is a bounded, simply connected BV_l-extension domain. It suffices to prove the quasiconvexity estimate for all $x, y \in \partial\Omega$ such that $d(x, y) \leqslant r_0$ for some fixed $r_0 > 0$ (recall that we assume the domain to be bounded). Let $\delta_0 > 0$ be the constant from Lemma 2.6, and let $r_0 = \min\{\delta_0, \operatorname{diam}(\Omega)\}/(2C)$, where C is the maximum of all the constants from the previous section. We denote by L_{xy} the line segment joining x and y. If

$L_{xy} \cap \Omega$ is empty, then we can set $\gamma = L_{xy}$. Hence we may assume that L_{xy} intersects Ω.

Since Ω is an open set, $L_{xy} \cap \Omega$ is the disjoint union of countably many line segments $L_{x_i y_i}$, $i \in I \subset \mathbb{N}$, with end points $x_i, y_i \in \partial\Omega$. Let $L_{x_i y_i}$ be one of them. Because Ω is simply connected, $\Omega \setminus L_{x_i y_i}$ has exactly two components, say E_1 and E_2. Assume that $|E_1| \leqslant |E_2|$. Since $\Omega \cap \partial E_1 = L_{x_i y_i}$ and hence $P(E_1, \Omega) = \mathcal{H}^1(L_{x_i y_i}) = |x_i - y_i|$, by Theorem 2.5 and subadditivity of the perimeter measure, there is a set $F \subset \mathbb{R}^2$ of finite perimeter such that $F \cap \Omega = E_1$ and

$$P(F, \mathbb{R}^2) \leqslant C\,|x_i - y_i|. \tag{3.1}$$

By Lemma 2.7, there is a sequence of smooth sets F_k with $\chi_{F_k} \to \chi_F$ both in $L^1_{loc}(\mathbb{R}^2)$ and pointwise almost everywhere, $P(F_k, \mathbb{R}^2) \to P(F, \mathbb{R}^2)$, $F_k \Delta F \subset \bigcup_{x \in \partial F} B(x, 1/k)$, and as vector-valued signed Radon measures, $D\chi_{F_k}$ converge weakly to $D\chi_F$.

Since F_k is smooth, ∂F_k consists of countably many smooth simple loops $\beta_{k,1}, \ldots$ (these curves are loops because they are of finite length). Recall from the discussion following the statement of Lemma 2.7 that the sets F_k are certain level sets of convolution approximations to χ_F. Hence for sufficiently large k (by passing to a subsequence if necessary) we may assume that $\partial F_k \subset \bigcup_{x \in \partial F} B(x, 1/k)$ and one of the loops $\beta_{k,1}, \ldots$, say $\beta_{k,1}$, has the property that all of the line segment L_{x_i, y_i} except perhaps a $1/k$-neighborhood of x_i and y_i lies in a $1/k$-neighborhood of $\beta_{k,1}$, i.e.,

$$\beta_{k,1} \subset \bigcup_{x \in \partial F} B(x, 1/k) \tag{3.2}$$

and

$$L_{x_i, y_i} \setminus (B(x_i, 1/k) \cup B(y_i, 1/k)) \subset \bigcup_{x \in \beta_{k,1}} B(x, 1/k). \tag{3.3}$$

Furthermore,

$$\ell(\beta_{k,1}) \leqslant P(F_k, \mathbb{R}^2) \leqslant 2\,P(F, \mathbb{R}^2),$$

and so we can use the Arzela–Ascoli theorem (and pass to a further subsequence if necessary) to obtain a loop β such that $\beta_{k,1} \to \beta$ uniformly and $\ell(\beta) \leqslant 2P(F, \mathbb{R}^2)$. By (3.2) and (3.3), it follows that $L_{x_i, y_i} \subset \beta$ and $\beta \subset \partial F$. Hence $\beta \cap \Omega = L_{x_i, y_i}$. Furthermore, by the inequality (3.1),

$$\ell(\beta) \leqslant 2P(F, \mathbb{R}^2) \leqslant 2C\,P(E, \Omega) = 2C\,|x_i - y_i|.$$

Since β is a loop containing E_1 and not containing E_2, by [10, Theorem 5. p. 513]) there is a simple subloop β_0 containing L_{x_i, y_i}. The curve $\gamma_i := \beta_0 \setminus L_{x_i, y_i} \subset \mathbb{R}^2 \setminus \Omega$ with $\ell(\gamma_i) \leqslant (2C - 1)|x_i - y_i|$ is a curve in $\mathbb{R}^2 \setminus \Omega$ connecting x_i to y_i.

The concatenated curve $\gamma = (L_{xy} \setminus \Omega) *_{i \in I} \beta_i$ is a curve in $\mathbb{R}^2 \setminus \Omega$ connecting x and y, with

$$\ell(\gamma) \leqslant \ell(L_{xy}) + \sum_{i\in I} \ell(\beta_i) \leqslant |x-y| + \sum_{i\in I} C\,|x_i - y_i| \leqslant (1+C)\,|x-y|.$$

Next suppose that $\mathbb{R}^2 \setminus \Omega$ is quasiconvex. By Lemma 2.8, we only need to verify the extension property for the characteristic functions of sets $E \subset \mathbb{R}^2$ such that $\partial E \cap \Omega$ is smooth. Therefore, $P(E, \Omega) = P(\overline{E}, \Omega) = P(\mathrm{int}(\overline{E}), \Omega)$, and so without loss of generality we may assume that $\mathrm{int}(\overline{E}) \cap \Omega = E \cap \Omega$ is open. Again, without loss of generality we may assume that $E \subset \Omega$ and that E is connected; recall that only a finite number of the components of $\partial E \cap \Omega$ can intersect a given relatively compact open set $U \subset \Omega$ and $P(E, \Omega)$ can be computed as the supremum of the perimeters $P(E, U)$ over all such U. From the smoothness of E it follows that $\Omega \cap \partial E$ consists of a collection of closed curves in Ω and a collection of at most a countable union of smooth curves γ_i, $i \in I \subset \mathbb{N}$, with end points $x_i, y_i \in \partial\Omega$ (indeed, if $\partial E \cap \partial\Omega$ is empty, i.e., no such points x_i, y_i exist, then E or $\mathbb{R}^2 \setminus E$ is the extension of E or $\Omega \setminus E$ respectively, and we need not do anything). Again, without loss of generality, we may assume that $|E| \leqslant |\Omega \setminus E|$ since otherwise we replace E with $\Omega \setminus E$. By assumption, there is a curve $\beta_i \subset \mathbb{R}^2 \setminus \Omega$ connecting x_i and y_i with $\ell(\beta_i) \leqslant C|x_i - y_i|$. The concatenated curve $\gamma_i * \beta_i$ is a simple loop (Jordan curve) in $\mathbb{R}^2$. Let F_i be the bounded subset of $\mathbb{R}^2$ enclosed by this loop. Since E is connected, if $E \cap F_i \neq \varnothing$, then $E \subset F_i$. Let J be the collection of all indices $i \in I$ for which this holds. If J is not empty, then we define

$$F := \Big(\bigcap_{i\in J} F_i\Big) \setminus \Big(\bigcup_{i\in I\setminus J} F_i\Big) \setminus (\text{all regions bounded by loops lying in } \Omega).$$

If J is empty, then we set

$$F := \mathbb{R}^2 \setminus \Big(\bigcup_{i\in I} F_i\Big) \setminus (\text{all regions bounded by loops lying in } \Omega).$$

With the above selection of F, we see that $F \cap \Omega = E$, and, by the construction of the curves β_i, we have

$$\begin{aligned} P(F, \mathbb{R}^2) &\leqslant \sum_{i\in I} \ell(\gamma_i * \beta_i) = \sum_{i\in I} \ell(\gamma_i) + \sum_{i\in I} \ell(\beta_i) \\ &\leqslant \sum_{i\in I} \ell(\gamma_i) + C\sum_{i\in I} |x_i - y_i| \leqslant \sum_{i\in I} \ell(\gamma_i) + C\sum_{i\in I} \ell(\gamma_i) = (1+C)P(E, \Omega), \end{aligned}$$

which completes the proof. □

Proof of Corollary 1.2. The claim follows from Lemma 2.4 and Theorem 1.1. □

We record the following recent result by Väisälä [13, Section 2.8].

Lemma 3.1 (Väisälä, 2008). *If Ω is a bounded, simply connected planar domain whose complement is quasiconvex, and if Ω' is a planar domain with $f : \Omega \to \Omega'$ a bi-Lipschitz mapping, then there are open sets $U \supset \overline{\Omega}$, $V \supset \overline{\Omega'}$, and a bi-Lipschitz mapping $F : U \to V$ such that $F = f$ on Ω.*

Proof of Corollary 1.3. By Theorem 1.1 and Corollary 1.2, the complement of Ω is quasiconvex. Hence, by the above lemma, the bi-Lipschitz map f on Ω can be extended to a bi-Lipschitz map F on a neighborhood U of the compact set $\overline{\Omega}$. Hence if Ω is a BV-extension domain (or $W^{1,1}$-extension domain) and u is a function in $BV(\Omega')$ (or $W^{1,1}(\Omega')$ respectively), then $u \circ f$ is in the class $BV(\Omega)$ (or $W^{1,1}(\Omega)$ respectively), and hence can be extended to a function $T(u \circ f)$ that lies in the class $BV(\mathbb{R}^2)$ (or $W^{1,1}(\mathbb{R}^2)$ respectively), with norm controlled by the norm of u. Thus, $T(u \circ f) \circ F^{-1}$ lies in the class $BV(V)$ (or $W^{1,1}(V)$ respectively), where $V = F(U)$ is a neighborhood of the compact set $\overline{\Omega'}$, with norm controlled by the norm of $T(u \circ f)$, and hence by the norm of u.

Let $\eta : \mathbb{R}^2 \to [0,1]$ be an L-Lipschitz function with compact support in V such that $\eta = 1$ on Ω'. Let $E(u) := \eta T(u \circ f) \circ F^{-1}$. Then $E(u) \in BV(\mathbb{R}^2)$ (or in $W^{1,1}(\mathbb{R}^2)$ respectively). Note that

$$\|E(u)\|_{L^1(\mathbb{R}^2)} \leqslant \|T(u \circ f) \circ F^{-1}\|_{L^1(V)} \leqslant C\|u\|_X,$$

where $X = BV(\Omega')$ (or $X = W^{1,1}(\Omega')$ respectively). Furthermore,

$$\begin{aligned} |DE(u)|(V) &\leqslant \mathrm{Lip}(\eta)\|T(u \circ f) \circ F^{-1}\|_{L^1(V)} + |DT(u \circ f) \circ F^{-1}|(V) \\ &\leqslant C\|u\|_X, \end{aligned}$$

where $\mathrm{Lip}\,\eta = \sup |\eta(x) - \eta(y)|/|x - y|$, the supremum taken over all distinct pairs of points $x, y \in \mathbb{R}^2$. This completes the proof. □

4 Examples

The characterization given by Theorem 1.1 is easy to verify for planar Jordan domains. We now explore some specific examples of bounded, simply connected planar BV-extension domains by answering the following question. Suppose that $\Omega \subset \mathbb{R}^2$ is a bounded BV_l-extension domain. Let $\gamma \subset \Omega$ be a curve such that $\Omega \setminus \gamma$ is also a domain. When is $\Omega \setminus \gamma$ also a BV-extension domain? It follows from Theorem 1.1 that γ has to be a rectifiable curve. However, the rectifiability of γ by itself does not guarantee the BV-extension property of $\Omega \setminus \gamma$, as the following example demonstrates.

Example 4.1. Let $\Omega = (-a, a) \times (-2, 2)$ be a rectangular region centered at the origin, where $a = \sum_{j=1}^{\infty} \frac{1}{j^3}$. Further, let $\gamma : [0, 1) \to \Omega$ with $\gamma(0) = (1, 0)$ be defined as follows: for each $n \in \mathbb{N}$ with $n \geqslant 2$

$$\gamma(1-1/n) = \Big(\sum_{j=1}^{n} 1/j^3, 0\Big),$$

and the open interval $(1-1/n, 1-1/(n+1)) \subset [0,1]$ is mapped to the curve obtained by joining two segments, the line segment joining $(\sum_{j=1}^{n} 1/j^3, 0)$ and $(1/(2n+2)^3 + \sum_{j=1}^{n} 1/j^3, 1/n^2)$, and the line segment joining $(1/(2n+2)^3 + \sum_{j=1}^{n} 1/j^3, 1/n^2)$ and $(\sum_{j=1}^{n+1} 1/j^3, 0)$. This γ is a saw-tooth curve for which the height of the nth tooth, $1/n^2$, is substantially larger than the width $1/n^3$ of the tooth. It can be seen that γ is rectifiable and $\Omega \setminus \gamma$ is *not* a BV-extension domain.

Example 4.2. Let $\Omega = (0,2) \times (-2,2)$, and let γ be the curve given by $\gamma : (0,1] \to \Omega$, $\gamma(t) = (t^2, t)$. Again it can be seen, via the use of sets

$$E_t = \{(x,y) \in \Omega : 0 < y < t, 0 < x < y^2\},$$

that $\Omega \setminus \gamma$ is not a BV-extension domain, even though γ is rectifiable.

The following answer to the above question is a corollary to Theorem 1.1. Here, $\delta_\Omega(x) = \mathrm{dist}(x, \partial\Omega)$ for $x \in \Omega$.

Corollary 4.3. *Suppose that $\Omega \subset \mathbb{R}^2$ is a bounded, simply connected BV-extension domain and γ is a curve in Ω so that $\Omega \setminus \gamma$ is also a simply connected domain. Then $\Omega \setminus \gamma$ is a BV-extension domain if and only if the following two conditions hold for γ.*

(i) *There is a constant $C > 0$ such that for all $x, y \in \gamma$ and for all subcurves γ_{xy} of γ with end points x and y, we have $\ell(\gamma_{xy}) \leqslant C|x-y|$ (i.e., γ is quasiconvex).*

(ii) *There is a constant $C > 0$ such that for all $x, y \in \gamma$ and a subcurve γ_{xy} of γ with end points x and y, we have $|x-y| \leqslant C \max\{\delta_\Omega(x), \delta_\Omega(y)\}$ (i.e., γ satisfies a double cone condition in Ω).*

If γ is not rectifiable, then $\Omega \setminus \gamma$ is not a BV-extension domain as $\mathbb{R}^2 \setminus (\Omega \setminus \gamma)$, and hence γ, has to be quasiconvex. This is in contrast to the fact that $\partial\Omega$ need not be rectifiable even if Ω is a BV-extension domain; as shown by the von Koch snowflake domain, which is a uniform domain and hence (cf. [8]) is a $W^{1,1}$- and further a BV-extension domain. It should be noted that the assumption that $\Omega \setminus \gamma$ is a domain ensures that γ does not have loops. The first condition above ensures that γ is quasiconvex. Observe that the curve in Example 4.1 fails to satisfy this condition, though it does satisfy the second condition. Note also that the curve in Example 4.2 fails to satisfy the second condition of the above corollary, but does satisfy the first condition. Hence both conditions above are essential in the above result, though if γ does not intersect the boundary of Ω, the second condition will follow from the first condition.

The conclusion of the corollary remains valid if we replace the condition that $\Omega \setminus \gamma$ is a simply connected domain with the condition that $\Omega \setminus \gamma$ is a domain; however, in this case, the result is not a direct consequence of Theorem 1.1.

Remark 4.4. Lemma 2.1 fails for some unbounded domains. For example, the domain

$$\Omega = \{(x, y) \in \mathbb{R}^2 : |y| > x \text{ if } x \geqslant 0 \text{ and } |y| + 1 > -x \text{ if } x \leqslant -1\}$$

is a BV-extension domain because it has uniformly Lipschitz boundary, but is not a BV_l-extension domain as the set $E = \{(x, y) \in \Omega : y > 0\}$ has no extension satisfying the Burago–Mazya characterization. Therefore, Theorem 1.1 might fail for unbounded, simply connected domains. However, the actual proof of this theorem demonstrates that the complement of a planar simply connected domain is quasiconvex if and only if the domain itself is a BV_l-extension domain. The above example also shows that if Ω is an unbounded, simply connected planar domain, the conclusion of Corollary 1.2 may fail. The domain in the above counterexample has the property that the complement of the domain in $\mathbb{R}^2$ is not connected; however, the example $\Omega = (0, \infty) \times (0, 1) \subset \mathbb{R}^2$ also is a BV-extension domain, but is not a BV_l-extension domain, even though $\mathbb{R}^2 \setminus \Omega$ is indeed connected. If $\Omega \subset \mathbb{R}^2$ is such that $\mathbb{R}^2 \setminus \Omega$ is connected, then the proof of Theorem 1.1 also shows that $\mathbb{R}^2 \setminus \Omega$ is quasiconvex if and only if Ω is a BV_l-extension domain. We point out here that, in this case, there is no reason to assume that Ω needs to be simply connected.

Acknowledgments. P. K. was supported by the Academy of Finland (grant 120972). N.S. was partially supported by NSF (grant DMS-0355027). Part of the research was conducted during the second author's visit to the University of Cincinnati and the third author's visit to the University of Jyväskylä and to the Scuola Normale Superiore, Pisa. They wish to thank these institutions for their generous hospitality.

References

1. Baldi, A., Montefalcone, F.: A note on the extension of BV functions in metric measure spaces. J. Math. Anal. Appl. **340**, 197–208 (2008)
2. Burago, Yu.D., Maz'ya, V.G.: Certain questions of potential theory and function theory for irregular regions. Zap. Nauchn. Semin. LOMI **3** (1967); English transl.: Consultants Bureau, New York (1969)
3. Evans, L.C., Gariepy, R.F.: Measure Theory and Fine Properties of Functions. CRC Press, Boca Raton, FL (1992)
4. Gol'dshtein, V.M., Reshetnyak, Yu.G.: Quasiconformal Mappings and Sobolev Spaces. Kluwer Academic Publishers Group, Dordrecht (1990)

5. Gol'dstein, V.; Vodop'janov, S.: Prolongement de fonctions différentiables hors de domaines plans (French). C. R. Acad. Sci. Paris I Math. **293**, 581–584 (1981)
6. Hajłasz, P., Koskela, P., Tuominen, H.: Sobolev embeddings, extensions and measure density condition. J. Funct. Anal. **254**, 197–205 (2008)
7. Herron, D., Koskela, P.: Uniform, Sobolev extension and quasiconformal circle domains. J. Anal. Math. **57**, 172–202 (1991)
8. Jones, P.W.: Quasiconformal mappings and extendability of Sobolev functions. Acta Math. **47**, 71–88 (1981)
9. Koskela, P.: Extensions and imbeddings. J. Funct. Anal. **159**, 1–15 (1998)
10. Kuratowski, K.: Topology. Vol. II. Acad. Press, New York (1968)
11. Maz'ya, V.G.: Sobolev Spaces. Springer, Berlin (1985)
12. Näkki, R., Väisälä, J.: John disks. Exposition. Math. **9**, 3–43 (1991)
13. Väisälä, J.: Holes and maps of Euclidean spaces. Conform. Geom. Dyn. **12**, 58–66 (2008)
14. Vodop'yanov, S.K., Gol'dshtein, V.M., Latfullin, T.G.: Criteria for extension of functions of the class L_2^1 from unbounded plane domains. Sib. Math. J. **20**, 298–301 (1979)

On a New Characterization of Besov Spaces with Negative Exponents

Moshe Marcus and Laurent Véron

Dedicated to Vladimir G. Maz'ya with high esteem

Abstract It is proved that for all $1 < q < \infty$ any negative Besov spaces $B^{-s,q}(\Sigma)$ can be described by an integrability condition on the Poisson potential of its elements.

1 Introduction

Let B denote the unit N-ball, and let $\Sigma = \partial B$. If μ is a distribution on Σ, we denote by $\mathbb{P}(\mu)$ its Poisson potential in B, i.e.,

$$\mathbb{P}(\mu)(x) = \langle \mu, P(x,.)\rangle_\Sigma \quad \forall\, x \in B, \tag{1.1}$$

where $\langle\ ,\ \rangle_\Sigma$ denotes the pairing between distributions on Σ and functions in $C^\infty(\Sigma)$. In the particular case where μ is a measure, this can be written as follows

$$\mathbb{P}(\mu)(x) = \int_\Sigma P(x,y)d\mu(y) \quad \forall\, x \in B. \tag{1.2}$$

In [3], it is proved that for $q > 1$ the Besov space $W^{-2/q,q}(\Sigma)$ is characterized by an integrability condition on $\mathbb{P}(\mu)$ with respect to a weight function

Moshe Marcus
Department of Mathematics, Israel Institute of Technology-Technion, 33000 Haifa, Israel
e-mail: marcusm@math.technion.ac.il

Laurent Véron
Laboratoire de Mathématiques et Physique Théorique, CNRS UMR 6083, Fédération Denis Poisson and Faculté des Sciences, Université de Tours, Parc de Grandmont, 37200 Tours, France e-mail: veronl@univ-tours.fr

A. Laptev (ed.), *Around the Research of Vladimir Maz'ya I: Function Spaces*, International Mathematical Series 11, DOI 10.1007/978-1-4419-1341-8_12,

involving the distance to the boundary, and more precisely that there exists a positive constant $C = C(N, q)$ such that for any distribution μ on Σ

$$C^{-1}\|\mu\|_{W^{-2/q,q}(\Sigma)} \leqslant \left(\int_B |\mathbb{P}(\mu)|^q (1-|x|)dx\right)^{1/q} \leqslant C\|\mu\|_{W^{-2/q,q}(\Sigma)}. \quad (1.3)$$

The aim of this article is to prove that for all $1 < q < \infty$ any negative Besov spaces $B^{-s,q}(\Sigma)$ can be described by an integrability condition on the Poisson potential of its elements. More precisely, we prove

Theorem 1.1. *Let $s > 0$, $q > 1$, and let μ be a distribution on Σ. Then*

$$\mu \in B^{-s,q}(\Sigma) \Longleftrightarrow \mathbb{P}(\mu) \in L^q(B; (1-|x|)^{sq-1}dx).$$

Moreover, there exists a constant $C > 0$ such that for any $\mu \in B^{-s,q}(\Sigma)$,

$$C^{-1}\|\mu\|_{B^{-s,q}(\Sigma)} \leqslant \left(\int_B |\mathbb{P}(\mu)|^q (1-|x|)^{sq-1}dx\right)^{1/q} \leqslant C\|\mu\|_{B^{-s,q}(\Sigma)}. \quad (1.4)$$

The key idea for proving such a result is to use a lifting operator which reduces the estimate question to an estimate between Besov spaces with positive exponents. In one direction, the main technique relies on interpolation theory between domain of powers of analytic semigroups. In the other direction, we use a new representation formula for harmonic functions in a ball. Similar type of results, with more complicated proofs, have alreadty been obtained in the case of harmonic functions in a half space (cf. [3, Theorem 1.14.4]).

2 The Left-Hand Side Inequality (1.4)

We recall that for $1 \leqslant p < \infty$, $r \notin \mathbb{N}$, $r = k + \eta$ with $k \in \mathbb{N}$ and $0 < \eta < 1$,

$$B^{r,p}(\mathbb{R}^d) = \left\{\varphi \in W^{k,p}(\mathbb{R}^d) : \int_{\mathbb{R}^d}\int_{\mathbb{R}^d} \frac{|D^\alpha\varphi(x) - D^\alpha\varphi(y)|^p}{|x-y|^{d+\eta p}}dxdy < \infty \right.$$
$$\left. \forall\, \alpha \in \mathbb{N}^d, |\alpha| = k \right\}$$

with the norm

$$\|\varphi\|^p_{B^{r,p}} = \|\varphi\|^p_{W^{k,p}} + \sum_{|\alpha|=k}\int_{\mathbb{R}^d}\int_{\mathbb{R}^d} \frac{|D^\alpha\varphi(x+y) - D^\alpha\varphi(x)|^p}{|y|^{d+\eta p}}dxdy.$$

When $r \in \mathbb{N}$,

$$B^{r,p}(\mathbb{R}^d) = \Big\{\varphi \in W^{r-1,p}(\mathbb{R}^d) :$$

$$\int_{\mathbb{R}^d}\int_{\mathbb{R}^d} \frac{|D^\alpha\varphi(x+2y) + D^\alpha\varphi(x) - 2D^\alpha\varphi(x+y)|^p}{|y|^{p+d}} dxdy < \infty$$

$$\forall\, \alpha \in \mathbb{N}^d, |\alpha| = r-1\Big\}$$

with the norm

$$\|\varphi\|^p_{B^{r,p}} = \|\varphi\|^q_{W^{k,p}}$$

$$+ \sum_{|\alpha|=r-1} \int_{\mathbb{R}^d}\int_{\mathbb{R}^d} \frac{|D^\alpha\varphi(x+2y) + D^\alpha\varphi(x) - 2D^\alpha\varphi(x+y)|^p}{|y|^{p+d}} dxdy.$$

The relation of the Besov spaces with integer order of differentiation and the classical Sobolev spaces is the following [2], [1]

$$\begin{aligned} &B^{r,p}(\mathbb{R}^d) \subset W^{r,p}(\mathbb{R}^d) \quad \text{if } 1 \leqslant p \leqslant 2,\\ &W^{r,2}(\mathbb{R}^d) = B^{r,2}(\mathbb{R}^d),\\ &W^{r,p}(\mathbb{R}^d) \subset B^{r,p}(\mathbb{R}^d) \quad \text{if } p \geqslant 2. \end{aligned} \tag{2.1}$$

Since for $r\mathbb{N}_*$ and $1 \leqslant p < \infty$ the space $B^{-r,p}(\mathbb{R}^d)$ is the space of derivatives of $L^p(\mathbb{R}^d)$-functions, up to the total order k, for noninteger r, $r = k + \eta$ with $k \in \mathbb{N}$ and $0 < \eta < 1$ $B^{-r,p}(\mathbb{R}^d)$ can be defined by using the real interpolation method [2] by

$$\big[W^{-k,p}(\mathbb{R}^d), W^{-k-1,p}(\mathbb{R}^d)\big]_{\eta,p} = B^{-r,p}(\mathbb{R}^d).$$

The spaces $B^{-r,p}(\mathbb{R}^d)$, or $1 < p < \infty$ and $r > 0$ can also be defined by duality with $B^{-r,p'}(\mathbb{R}^d)$. The Sobolev and Besov spaces $W^{k,p}(\Sigma)$ and $B^{r,p}(\Sigma)$ are defined by using local charts from the same spaces in $\mathbb{R}^{N-1}$.

Now, we present the proof of the left-hand side inequality in the case $N \geqslant 3$. However, with minor modifications, the proof applies also to the case $N = 2$ (cf. Remark 3.1 below). Let $(r, \sigma) \in [0, \infty) \times S^{N-1}$ (with $S^{N-1} \approx \Sigma$) be spherical coordinates in B. We put $t = -\ln r$. Suppose that $\mu \in B^{-s,q}(S^{N-1})$. Let $u = \mathbb{P}(\mu)$. Denote by $\widetilde{u}$ the function u expressed in terms of the coordinates (t, σ). Then

$$u_{rr} + \frac{N-1}{r}u_r + \frac{1}{r^2}\Delta_\sigma u = 0 \quad \text{in } (0,1) \times S^{N-1} \tag{2.2}$$

and

$$\widetilde{u}_{tt} - (N-2)\widetilde{u}_t + \Delta_\sigma \widetilde{u} = 0 \quad \text{in } (0,\infty) \times S^{N-1}. \tag{2.3}$$

Then the right inequality in (1.4) takes the form

$$\int_0^\infty \int_{S^{N-1}} |\widetilde{u}|^q (1-e^{-t})^{sq-1} e^{-Nt} d\sigma\, dt \leqslant C \|\mu\|^q_{B^{-s,q}(S^{N-1})}. \tag{2.4}$$

Clearly, it suffices to establish this inequality in the case $\mu \in \mathfrak{M}(S^{N-1})$ (or even $\mu \in C^\infty(S^{N-1})$), which is assumed in the sequel. We define $k \in \mathbb{N}^*$ by

$$2(k-1) \leqslant s < 2k, \tag{2.5}$$

with the restriction $s > 0$ if $k = 1$. We denote by $\mathbb{B}$ the elliptic operator of order 2k

$$\mathbb{B} = \left(\frac{(N-2)^2}{4} - \Delta_\sigma \right)^k$$

and call f the unique solution of

$$\mu = \mathbb{B} f \quad \text{in } S^{N-1}.$$

Then $f \in W^{2k-s,q}(S^{N-1})$ since $\mathbb{B}$ is an isomorphism between the spaces $B^{2k-s,q}(S^{N-1})$ and $B^{-s,q}(S^{N-1})$. Put $v = \mathbb{P}(f)$ in B. Then v satisfies the same equation as u in $(0,1) \times S^{N-1}$. Let $\widetilde{v}$ denote this function in terms of the coordinates (t, σ). Then

$$\begin{aligned} \widetilde{L}\widetilde{v} &:= \widetilde{v}_{tt} - (N-2)\widetilde{v}_t + \Delta_\sigma \widetilde{v} = 0 \quad \text{in } \mathbb{R}_+ \times S^{N-1}, \\ \widetilde{v}\big|_{t=0} &= f \quad \text{in } S^{N-1}. \end{aligned} \tag{2.6}$$

Since the operator $\mathbb{B}$ commutes with Δ_σ and $\partial/\partial t$, and this problem has a unique solution which is bounded near $t = \infty$, it follows that

$$\mathbb{P}(\mathbb{B} f) = \mathbb{B}\widetilde{v}. \tag{2.7}$$

Hence

$$\widetilde{u} = \mathbb{P}(\mu) = \mathbb{P}(\mathbb{B} f) = \mathbb{B}\widetilde{v}. \tag{2.8}$$

If $v^* := e^{-t(N-2)/2}\widetilde{v}$, then

$$\begin{aligned} & v^*_{tt} - \frac{(N-2)^2}{4} v^* + \Delta_\sigma v^* = 0 \quad \text{in } \mathbb{R}_+ \times S^{N-1}, \\ & v^*(0,\cdot) = f \quad \text{in } S^{N-1}. \end{aligned} \tag{2.9}$$

Note that

$$v^* = e^{tA}(f), \quad \text{where} \quad A = -\left(\frac{(N-2)^2}{4} I - \Delta_\sigma \right)^{1/2} \Longleftrightarrow A^{2k} = \mathbb{B},$$

where e^{tA} is the semigroup generated by A in $L^q(S^{N-1})$. By the Lions–Peetre real interpolation method [2],

$$[W^{2k,q}(S^{N-1}), L^q(S^{N-1})]_{1-s/2k,q} = B^{2k-s,q}(S^{N-1}).$$

Since $D(A^2) = W^{2,q}(S^{N-1})$,

$$D(A^{2k}) = W^{2k,q}(S^{N-1}).$$

The semigroup generated by A is analytic as any semigroup generated by the square root of a closed operator. Therefore, by [4, p. 96],

$$\begin{aligned}
\|f\|^q_{W^{2k-s,q}} &\sim \|f\|^q_{L^q(S^{N-1})} + \int_0^\infty \left(t^{(2kqs/2kq)} \left\|A^{2k}v^*\right\|_{L^q(S^{N-1})}\right)^q \frac{dt}{t} \\
&\sim \|f\|^q_{L^q(S^{N-1})} + \int_0^1 \left(t^s \left\|A^{2k}v^*\right\|_{L^q(S^{N-1})}\right)^q \frac{dt}{t} \\
&= \|f\|^q_{L^q(S^{N-1})} + \int_0^1 \left(t^s e^{-t(N-2)/2} \|\mathbb{B}\widetilde{v}\|_{L^q(S^{N-1})}\right)^q \frac{dt}{t} \qquad (2.10)
\end{aligned}$$

where the symbol $\sim$ denotes equivalence of norms. Therefore, by (2.10),

$$\begin{aligned}
\|f\|^q_{W^{2k-s,q}(S^{n-1})} &\geqslant C\|f\|^q_{L^q(S^{N-1})} \\
&\quad + C\int_0^1 \left(t^s e^{-t(N-2)/2} \|\widetilde{u}\|_{L^q(S^{N-1})}\right)^q \frac{dt}{t} \\
&\geqslant C\|f\|^q_{L^q(S^{N-1})} + C\int_0^1 \|\widetilde{u}\|^q_{L^q(S^{N-1})} e^{-Nt} t^{sq-1} dt. \qquad (2.11)
\end{aligned}$$

Furthermore,

$$\begin{aligned}
&\int_0^\infty \|\widetilde{u}\|^q_{L^q(S^{N-1})} (1-e^t)^{sq-1} e^{-Nt} dt \\
&\qquad \leqslant C\int_0^1 \|\widetilde{u}\|^q_{L^q(S^{N-1})} (1-e^{-t})^{sq-1} e^{-Nt} dt \\
&\qquad \leqslant C\int_0^1 \|\widetilde{u}\|^q_{L^q(S^{N-1})} e^{-Nt} t^{sq-1} dt. \qquad (2.12)
\end{aligned}$$

This is a consequence of the inequality

$$\int_{\partial B_r} |u|^q dS \leqslant (r/\rho)^{N-1} \int_{\partial B_\rho} |u|^q dS,$$

which holds for $0 < r < \rho$, for every harmonic function u in B. By a straightforward computation, this inequality implies that

$$\int_{|x|<1} |u|^q(1-r)\,dx \leqslant c(\gamma)\int_{\gamma<|x|<1} |u|^q(1-r)\,dx,$$

for every $\gamma \in (0,1)$.

In view of the definition of f,

$$\|\mu\|^q_{B^{-s,q}(S^{n-1})} \sim \|f\|^q_{W^{2k-s,q}(S^{n-1})}. \tag{2.13}$$

Therefore, the right-hand side inequality (2.4) follows from (2.11), (2.12), and (2.13).

3 The Right-Hand Side Inequality (1.4)

Suppose that μ is a distribution on S^{N-1} such that $\mathbb{P}(\mu) \in L^q(B;(1-|x|^{sq-1})$. Then we claim that $\mu \in B^{-s,q}(S^{N-1})$ and

$$C^{-1}\|\mu\|_{B^{-s,q}(\Sigma)} \leqslant \left(\int_B |\mathbb{P}(\mu)|^q(1-|x|)^{sq-1}dx\right)^{1/q}. \tag{3.1}$$

Because of the estimate (2.10) it suffices to prove that

$$\|f\|_{L^q(S^{N-1})} \leqslant C\,\|u\|_{L^q(B,(1-r)^{sq-1}\,dx)}. \tag{3.2}$$

With $u = \mathbb{B}v$ this relation becomes

$$\begin{aligned}\|f\|_{L^q(S^{N-1})} &\leqslant C\,\|\mathbb{B}v\|_{L^q(B;(1-r)^{sq-1}\,dx)}\\ &\leqslant C\left(\int_0^1 \|v\|^q_{W^{2k,q}(S^{N-1})}(1-r)^{sq-1}r^{N-1}dr\right)^{1/q}.\end{aligned} \tag{3.3}$$

In order to simplify the exposition, we first present the case $0 < s < 2$.

3.1 The case $0 < s < 2$

We take $k = 1$. Since the imbedding of $B^{2-s,q}(S^{N-1})$ into $L^q(S^{N-1})$ is compact, for any $\varepsilon > 0$ there is $C_\varepsilon > 0$ such that

$$\|\varphi\|_{L^q(S^{N-1})} \leqslant \varepsilon\|\varphi\|_{B^{2-s,q}(S^{N-1})} + C_\varepsilon\|\varphi\|_{L^1(S^{N-1})} \quad \forall\varphi \in B^{2-s,q}(S^{N-1}).$$

Therefore, the following norm for $B^{2-s,q}(S^{N-1})$ is equivalent to the one given in (2.4)

$$\|f\|^q_{B^{2-s,q}} = \|f\|^q_{L^1(S^{N-1})} + \int_0^1 \left(t^s \left\|A^2 v^*\right\|_{L^q(S^{N-1})}\right)^q \frac{dt}{t}, \tag{3.4}$$

and the estimate (3.3) is a consequence of

$$\|f\|^q_{L^1(S^{N-1})} \leqslant \int_0^1 \left(t^s \left\|A^2 v^*\right\|_{L^q(S^{N-1})}\right)^q \frac{dt}{t}. \tag{3.5}$$

Integrating (2.9) and using the fact that

$$\lim_{t\to\infty} \|v^*\|_{L^\infty(S^{N-1})} = \lim_{t\to\infty} \|v^*_t\|_{L^\infty(S^{N-1})} = 0, \tag{3.6}$$

we find

$$v^*_t(t,\sigma) = -\int_t^\infty A^2 v^*(s,\sigma)ds \quad \forall\,(t,\sigma) \in (0,\infty)\times S^{N-1},$$

and

$$\begin{aligned} v^*(t,\sigma) &= \int_t^\infty \int_s^\infty A^2 v^*(\tau,\sigma)d\tau ds \\ &= \int_t^\infty A^2 v^*(\tau,\sigma)(\tau - t)d\tau \quad \forall\,(t,\sigma) \in (0,\infty)\times S^{N-1}. \end{aligned} \tag{3.7}$$

Letting $t \to 0$ and integrating over S^{N-1}, one obtains

$$\begin{aligned} \int_{S^{N-1}} |f|\, d\sigma &\leqslant \int_0^\infty \int_{S^{N-1}} \left|A^2 v^*\right| \tau d\sigma d\tau \\ &\leqslant C(N,s,q,\delta) \left(\int_0^\infty \int_{S^{N-1}} \left|A^2 v^*\right|^q e^{\delta\tau} \tau^{sq-1} d\sigma d\tau\right)^{1/q} \end{aligned} \tag{3.8}$$

for any $\delta > 0$ (δ will be taken smaller than $(N-2)q/2$) in the sequel), where

$$C(N,s,q,\delta) = \left(\left|S^{N-1}\right| \int_0^\infty \tau^{(q+1-sq)/(q-1)} e^{-\delta\tau/(q-1)} d\tau\right)^{1/q'}.$$

Note that the integral is convergent since $(q+1-sq)/(q-1) > -1$ if and only if $s < 2$. Going back to $\widetilde{v}$, we have

$$\int_0^\infty \int_{S^{N-1}} \left|A^2 v^*\right|^q e^{\delta\tau} \tau^{sq-1} d\sigma d\tau = \int_0^\infty \int_{S^{N-1}} \left|A^2 \widetilde{v}\right|^q e^{(\delta-(N-2)q/2)\tau} \tau^{sq-1} d\sigma d\tau.$$

Since u is harmonic,

$$\int_{S^{N-1}} |\widetilde{u}(\tau_1,\cdot)|^q d\sigma \leqslant \int_{S^{N-1}} |\widetilde{u}(\tau_2,\cdot)|^q d\sigma \quad \forall\, 0 < \tau_2 \leqslant \tau_1$$

or, equivalently,

$$\int_{S^{N-1}} \left|A^2\widetilde{v}(\tau_1,\cdot)\right|^q d\sigma \leqslant \int_{S^{N-1}} \left|A^2\widetilde{v}(\tau_2,\cdot)\right|^q d\sigma \quad \forall\, 0<\tau_2\leqslant\tau_1. \tag{3.9}$$

Applying (3.9) between τ and $1/\tau$ for $\tau \geqslant 1$, we get

$$\int_1^\infty \int_{S^{N-1}} \left|A^2\widetilde{v}\right|^q e^{(\delta-(N-2)q/2)\tau} \tau^{sq-1} d\sigma d\tau$$
$$\leqslant \int_0^1 \int_{S^{N-1}} \left|A^2\widetilde{v}\right|^q e^{(\delta-(N-2)q/2)\tau^{-1}} \tau^{-sq-1} d\sigma d\tau. \tag{3.10}$$

Moreover, there exists $C = C(N,q,\delta) > 0$ such that

$$e^{(\delta-(N-2)q/2)t^{-1}} t^{-sq-1} \leqslant Ce^{(\delta-(N-2)q/2)t} t^{sq-1} \quad \forall\, 0<t\leqslant 1.$$

Plugging this inequality into (3.9) and using (3.8), one derives

$$\int_{S^{N-1}} |f|\, d\sigma \leqslant C\left(\int_0^1 \int_{S^{N-1}} |A^2 v^*|^q e^{\delta\tau} \tau^{sq-1} d\sigma d\tau\right)^{1/q} \tag{3.11}$$

for some positive constant C, from which (3.5) follows.

3.2 The general case

We assume that $k \geqslant 1$. Since the imbedding of $B^{2k-s,q}(S^{N-1})$ into $L^q(S^{N-1})$ is compact, for any $\varepsilon > 0$ there is $C_\varepsilon > 0$ such that

$$\|\varphi\|_{L^q(S^{N-1})} \leqslant \varepsilon\|\varphi\|_{B^{2k-s,q}(S^{N-1})} + C_\varepsilon\|\varphi\|_{L^1(S^{N-1})} \quad \forall \varphi \in B^{2k-s,q}(S^{N-1}).$$

Thus, the following norm for $B^{2k-s,q}(S^{N-1})$ is equivalent to the one given in (2.4)

$$\|f\|^q_{B^{2k-s,q}} = \|f\|^q_{L^1(S^{N-1})} + \int_0^1 \left(t^s \left\|A^{2k}v^*\right\|_{L^q(S^{N-1})}\right)^q \frac{dt}{t}, \tag{3.12}$$

and the estimate (3.3) follows from

$$\|f\|^q_{L^1(S^{N-1})} \leqslant \int_0^1 \left(t^s \left\|A^{2k}v^*\right\|_{L^q(S^{N-1})}\right)^q \frac{dt}{t}. \tag{3.13}$$

From (3.7)

$$v^*(t,\sigma) = \int_t^\infty A^2 v^*(\tau,\sigma)(\tau - t)d\tau \quad \forall\ (t,\sigma) \in (0,\infty) \times S^{N-1}. \tag{3.14}$$

Since the operator A^2 is closed,

$$A^2 v^*(t,\sigma) = \int_t^\infty A^4 v^*(\tau,\sigma)(\tau - t)d\tau$$

and

$$\begin{aligned} v^*(t,\sigma) &= \int_t^\infty (t_1 - t) \int_{t_1}^\infty A^4 v^*(t_2,\sigma)(t_2 - t_1)dt_2 dt_1 \\ &= \int_t^\infty \int_{t_1}^\infty (t_1 - t)(t_2 - t_1) A^4 v^*(t_2,\sigma) dt_2 dt_1 \\ &\qquad \forall\ (t,\sigma) \in (0,\infty) \times S^{N-1}. \end{aligned} \tag{3.15}$$

Iterating this process, for every $(t,\sigma) \in (0,\infty) \times S^{N-1}$ one gets

$$v^*(t,\sigma) = \int_t^\infty \int_{t_1}^\infty \cdots \int_{t_{k-1}}^\infty \prod_{j=1}^k (t_j - t_{j-1}) A^{2k} v^*(t_k,\sigma) dt_k dt_{k-1} \ldots dt_1. \tag{3.16}$$

where we set $t = t_0$ in the product symbol. The following representation formula is valid for any $k \in \mathbb{N}_*$.

Lemma 3.1. *For any* $(t,\sigma) \in (0,\infty) \times S^{N-1}$,

$$v^*(t,\sigma) = \int_t^\infty \frac{(s-t)^{2k-1}}{(2k-1)!} A^{2k} v^*(s,\sigma) ds. \tag{3.17}$$

Proof. We proceed by induction. By the Fubini theorem,

$$\begin{aligned} &\int_t^\infty \int_{t_1}^\infty (t_1 - t)(t_2 - t_1) A^4 v^*(t_2,\sigma) dt_2 dt_1 \\ &\quad = \int_t^\infty A^4 v^*(t_2,\sigma) \int_t^{t_2} (t_1 - t)(t_2 - t_1) dt_1 dt_2 \\ &\quad = \int_t^\infty \frac{(t_2 - t)^3}{6} A^4 v^*(t_2,\sigma) dt_2. \end{aligned}$$

Suppose that for $t > 0$, $\ell < k$ and any smooth function φ defined on $(0,\infty)$

$$\int_t^\infty \int_{t_1}^\infty \cdots \int_{t_{\ell-1}}^\infty \prod_{j=1}^\ell (t_j - t_{j-1}) \varphi(t_\ell) dt_\ell dt_{\ell-1} \ldots dt_1 = \int_t^\infty \frac{(t_\ell - t)^{2\ell-1}}{(2\ell-1)!} \varphi(t_\ell) dt_\ell. \tag{3.18}$$

Then

$$\int_t^\infty\int_{t_1}^\infty\dots\int_{t_\ell}^\infty\prod_{j=1}^{\ell+1}(t_j-t_{j-1})\varphi(t_{\ell+1})dt_{\ell+1}dt_\ell\dots dt_1$$

$$=\int_t^\infty\int_{t_1}^\infty\dots\int_{t_{\ell-1}}^\infty\prod_{j=1}^{\ell}(t_j-t_{j-1})\Phi(t_\ell)dt_\ell dt_{\ell-1}\dots dt_1$$

$$=\int_t^\infty\frac{(t_\ell-t)^{2\ell-1}}{(2\ell-1)!}\Phi(t_\ell)dt_\ell$$

with

$$\Phi(t_\ell)=\int_{t_\ell}^\infty(t_{\ell+1}-t_\ell)\varphi(t_{\ell+1})dt_{\ell+1}.$$

But

$$\int_t^\infty\frac{(t_\ell-t)^{2\ell-1}}{(2\ell-1)!}\int_{t_\ell}^\infty(t_{\ell+1}-t_\ell)\varphi(t_{\ell+1})dt_{\ell+1}dt_\ell$$

$$=\int_t^\infty\varphi(t_{\ell+1})\int_t^{t_{ell+1}}\frac{(t_\ell-t)^{2\ell-1}}{(2\ell-1)!}(t_{\ell+1}-t_\ell)dt_\ell dt_{\ell+1}$$

$$=\int_t^\infty\varphi(t_{\ell+1})\int_0^{t_{ell+1}-t}\frac{\tau^{2\ell-1}}{(2\ell-1)!}(t_{\ell+1}-t-\tau)d\tau dt_{\ell+1}$$

$$=\int_t^\infty\varphi(t_{\ell+1})\frac{(t_{2\ell+1}-\tau)^{2\ell+1}}{(2\ell+1)!}dt_{\ell+1}$$

as

$$\frac{1}{(2\ell-1)!}\left(\frac{1}{2\ell}-\frac{1}{2\ell+1}\right)=\frac{1}{(2\ell+1)!}.$$

Taking $\varphi(t_{\ell+1})=A^{2\ell}v^*(t_{\ell+1},\sigma)$, we obtain (3.17).

End of the proof. From (3.16) and Lemma 3.1 with $t=0$ we get

$$\int_{S^{N-1}}|f|\,d\sigma\leqslant\int_0^\infty\int_{S^{N-1}}\left|A^{2k}v^*\right|\frac{\tau^{2k-1}}{(2k-1)!}d\sigma d\tau$$

$$\leqslant C(N,s,k,q,\delta)\left(\int_0^\infty\int_{S^{N-1}}\left|A^{2k}v^*\right|^q e^{\delta\tau}\tau^{sq-1}d\sigma d\tau\right)^{1/q}\qquad(3.19)$$

for any $\delta>0$ (δ will be taken smaller than $(N-2)q/2$) in the sequel), where

$$C(N,s,k,q,\delta)=\left(\left|S^{N-1}\right|\int_0^\infty\tau^{(2k-s-1/q')q'}e^{-\delta\tau/(q-1)}d\tau\right)^{1/q'}.$$

Note that the integral is convergent since $(2k - s - 1/q')q' > -1$ if and only if $s < 2k$. As in the case $s < 2$, we return to $\widetilde{v}$ and $\widetilde{u} = A^{2k}\widetilde{u}$, use the harmonicity of u in order to derive

$$\int_1^\infty \int_{S^{N-1}} |A^{2k}\widetilde{v}|^q e^{(\delta-(N-2)q/2)\tau} \tau^{sq-1} d\sigma d\tau$$

$$\leqslant \int_0^1 \int_{S^{N-1}} |A^{2k}\widetilde{v}|^q e^{(\delta-(N-2)q/2)\tau^{-1}} \tau^{-sq-1} d\sigma d\tau \tag{3.20}$$

as in (3.10) and, finally,

$$\begin{aligned}\int_{S^{N-1}} |f|\, d\sigma &\leqslant C\left(\int_0^1 \int_{S^{N-1}} |A|^{2k}\, {v^*}^q \tau^{sq-1} d\sigma d\tau\right)^{1/q} \\ &\leqslant C'\left(\int_0^1 \int_{S^{N-1}} |\widetilde{u}|^q \tau^{sq-1} d\sigma d\tau\right)^{1/q},\end{aligned} \tag{3.21}$$

which completes the proof of Theorem 1.1. □

Remark 3.1. If $N = 2$, the lifting operator is

$$\mathbb{B} = \left(1 - \frac{d^2}{d\sigma^2}\right)^k,$$

and the proof is similar. Moreover, since $\mathbb{B}$ is an isomorphism between $B^{2k-s,1}(S^1)$ and $B^{-s,1}(S^1)$, the result of Theorem 1.1 also holds in the case $q = 1$.

4 A Regularity Result for the Green Operator

Put $(1-|x|) = \delta(x)$. By duality between $L^q(B; \delta^{sq-1}dx)$ and $L^{q'}(B; \delta^{sq-1}dx)$, we write

$$\int_B \mathbb{P}(\mu)\psi\delta^{sq-1}dx = -\int_B \mathbb{P}(\mu)\Delta\zeta dx = -\int_\Sigma \frac{\partial\zeta}{\partial\nu} d\mu, \tag{4.1}$$

where ζ is the solution of the problem

$$\begin{aligned} -\Delta\zeta &= \delta^{sq-1}\psi \quad \text{in } B, \\ \zeta &= 0 \quad \text{on } \partial B. \end{aligned} \tag{4.2}$$

In (4.1), the boundary term should be written as $\langle \mu, \partial\zeta/\partial\nu\rangle_\Sigma$ if μ is a distribution on Σ. Then the adjoint operator $\mathbb{P}^*$ is defined by

$$\mathbb{P}^*(\psi) = -\frac{\partial}{\partial\nu}\mathbb{G}(\delta^{sq-1}\psi), \tag{4.3}$$

where $\mathbb{G}(\delta^{sq-1}\psi)$ is the Green potential of $\delta^{sq-1}\psi$. Consequently, Theorem 1.1 implies that there exists a constant $C > 0$ such that

$$C^{-1} \|\psi\|_{L^{q'}(B;\delta^{sq-1}dx)} \leqslant \left\| \frac{\partial}{\partial\nu}\mathbb{G}(\delta^{sq-1}\psi) \right\|_{B^{s,q'}(\Sigma)} \leqslant C \|\psi\|_{L^{q'}(B;\delta^{sq-1}dx)}. \tag{4.4}$$

But

$$\psi \in L^{q'}(B;\delta^{sq-1}dx) \Longleftrightarrow \delta^{sq-1}\psi \in L^{q'}(B;\delta^{(sq-1)(1-q')}dx).$$

Putting $\varphi = \delta^{sq-1}\psi$ and replacing q' by p, we obtain the following result.

Theorem 4.1. *Let $s > 0$ and $1 < p < \infty$. Then*

$$\varphi \in L^p(B;\delta^{p(1-s)-1})dx) \Longleftrightarrow \frac{\partial}{\partial\nu}\mathbb{G}(\varphi) \in B^{s,p}(\Sigma).$$

Moreover, there exists a constant $C > 0$ such that for any function $\varphi \in L^p(B;\delta^{p(1-s)-1})dx)$

$$C^{-1}\|\varphi\|_{L^p(B;\delta^{p(1-s)-1})dx)} \leqslant \left\| \frac{\partial}{\partial\nu}\mathbb{G}(\varphi) \right\|_{B^{s,p}(\Sigma)} \leqslant C\|\varphi\|_{L^p(B;\delta^{p(1-s)-1})dx)}. \tag{4.5}$$

Acknowledgment. The research of the first author was supported by the Israel Science Foundation (grant 174/97).

References

1. Grisvard P.: Commutativité de deux foncteurs d'interpolation et applications. J. Math. Pures Appl. **45**, 143-290 (1966)
2. Lions, J.F., Peetre, J.: Sur une classe d'espaces d'interpolation. Publ. Math. IHES **19**, 5-68 (1964)
3. Marcus, M., Véron, L.: Removable singularities and boundary trace. J. Math. Pures Appl. **80**, 879-900 (2001)
4. Triebel, H.: Interpolation Theory, Function Spaces, Differential Operators. North–Holland Publ. Co. (1978)

Isoperimetric Hardy Type and Poincaré Inequalities on Metric Spaces

Joaquim Martín and Mario Milman

Abstract We give a general construction of manifolds for which Hardy type operators characterize Poincaré inequalities. We also show a class of spaces where this property fails. As an application, we extend recent results of E. Milman to our setting.

1 Introduction

While working on sharp Sobolev–Poincaré inequalities in the classical Euclidean setting (cf. [17]) as well as the Gaussian setting (cf. [14]), we observed that the symmetrization methods we were developing could be readily extended to the more general setting of metric spaces (cf. [6, 14, 15, 16]). However, in the metric setting we found that we could not always decide if the results we had obtained were "sharp" or best possible.

Indeed, generally speaking, the methods that we use to show sharpness require the construction of special rearrangements and thus our spaces need to exhibit sufficient symmetries. In fact, in all the examples where we know how to prove sharpness, the "winning" rearrangements are those that are somehow connected with the solution of the underlying isoperimetric problems (for example, the symmetric decreasing rearrangements in the Euclidean case, which are associated with balls (cf. [17]), while in the Gaussian case one uses special rearrangements associated with half spaces (cf. [8, 14]) and, likewise,

---

Joaquim Martín
Department of Mathematics, Universitat Autònoma de Barcelona, Bellaterra, 08193 Barcelona, Spain
e-mail: jmartin@mat.uab.cat

Mario Milman
Florida Atlantic University, Boca Raton, Fl. 33431 USA
e-mail: extrapol@bellsouth.net

A. Laptev (ed.), *Around the Research of Vladimir Maz'ya I: Function Spaces*, International Mathematical Series 11, DOI 10.1007/978-1-4419-1341-8_13,

in the more general model cases of log concave measures (cf. [5, 7] and the more recent [1, 15, 16]). In particular, these special rearrangements allow us to show that there exist "special symmetrizations that do not increase the norm of the gradient," i.e., that a suitable version of the Pólya–Szegö principle holds.

In preparation for a systematic study we observed that, in all the model cases we could treat, a key role was played by the boundedness of certain Hardy operators, which we termed "isoperimetric Hardy operators." This led us to isolate the concept of "isoperimetric Hardy type spaces." This property can be formulated in very general metric spaces and can be applied if we have estimates on the isoperimetric profiles. By formulating the problem in this fashion, while we may lose information about best constants, we gain the possibility of obtaining positive results that would be hard to obtain by other methods.

In this note, we continue this program and we address the question: which metric spaces are of "Hardy isoperimetric type"? On the positive side we show, using ideas of Ros [26], how to construct a class of metric spaces of "Hardy isoperimetric type" that contains all the model cases mentioned above. Therefore this construction provides us with a large class of spaces where our inequalities are sharp.

As another application we continue the discussion of the connection between our results and the recent work of E. Milman [23, 22, 24]), who has shown the equivalence, under convexity assumptions, of certain estimates for isoperimetric profiles. In [16], we extended and simplified E. Milman's results to the setting of metric spaces of isoperimetric Hardy type. The construction presented in this paper thus gives a general concrete class of model spaces where Milman's equivalences hold.

Finally on the negative side we also construct spaces that do not satisfy the "isoperimetric Hardy type condition."

In the spirit of this book, we now comment briefly on the influence of Maz'ya's work in our development. Underlying the equivalences of Theorem 2.1 below are two deep insights due to Maz'ya: Maz'ya's fundamental result showing the equivalence between the Gagliardo–Nirenberg inequality and the isoperimetric inequality (cf. [18] and also [9]), and Maz'ya's technique of showing self improvement of Sobolev's inequality via smooth cut-offs (cf. [20]). Indeed, one of the themes of Theorem 2.1 is to develop the explicit connection of these two ideas using pointwise symmetrization inequalities ("symmetrization by truncation" cf. [17, 14]). Another theme of our method is that we formulate our inequalities incorporating directly geometric information, an idea that one can also already find in Maz'ya's fundamental work characterizing Sobolev inequalities in rough domains [20] as well as in Maz'ya's method characterizing Sobolev–Poincaré inequalities via isocapacitory inequalities[1] (cf. [21]).

---

[1] A detailed discussion of the connection between Maz'ya's isocapacitory inequalities and symmetrization inequalities is given in [16].

As this outline shows, and we hope the rest of the paper proves, our methods owe a great deal to the pioneering work of Professor Maz'ya and we are grateful and honored by the opportunity to contribute to this book.

2 Background

Let (Ω, d, μ) be a metric probability space equipped with a separable Borel probability measure μ. Let $A \subset \Omega$ be a Borel set. Then the boundary measure or *Minkowski content* of A is by definition

$$\mu^+(A) = \lim\inf_{h\to 0} \frac{\mu(A_h) - \mu(A)}{h},$$

where $A_h = \{x \in \Omega : d(x, y) < h\}$ denotes the h-neighborhood of A.

The *isoperimetric profile* $I_{(\Omega,d,\mu)}$ is defined as the pointwise maximal function $I_{(\Omega,d,\mu)} : [0,1] \to [0,\infty)$ such that

$$\mu^+(A) \geqslant I_{(\Omega,d,\mu)}(\mu(A))$$

holds for all Borel sets A.

Condition. We assume throughout that our metric spaces have isoperimetric profile functions $I_{(\Omega,d,\mu)}$ which are: continuous, concave, increasing on $(0, 1/2)$, symmetric about the point $1/2$, and vanish at zero[2].

A continuous function $I : [0,1] \to [0,\infty)$, with $I(0) = 0$, concave, increasing on $(0, 1/2)$ and symmetric about the point $1/2$, and such that

$$I \geqslant I_{(\Omega,d,\mu)},$$

is called an *isoperimetric estimator* on (Ω, d, μ).

For measurable functions $u : \Omega \to \mathbb{R}$ the distribution function of u is given by

$$\lambda_u(t) = \mu\{x \in \Omega : |u(x)| > t\} \quad (t > 0).$$

The *decreasing rearrangement* u^* of u is defined, as usual, by

$$u^*_\mu(s) = \inf\{t \geqslant 0 : \lambda_u(t) \leqslant s\} \quad (t \in (0, \mu(\Omega)]),$$

and we set

$$u^{**}_\mu(t) = \frac{1}{t}\int_0^t u^*_\mu(s)ds.$$

Given a locally Lipschitz real function, f defined on (Ω, d) (we write in what follows $f \in \mathrm{Lip}(\Omega)$), the *modulus of the gradient* of f is defined by

[2] For a large class of examples, where these assumptions are satisfied we refer to [6, 23] and the references therein.

$$|\nabla f(x)| = \limsup_{d(x,y)\to 0} \frac{|f(x)-f(y)|}{d(x,y)}$$

and zero at isolated points[3].

A Banach function space $X = X(\Omega)$ on (Ω, d, μ) is called a *rearrangement-invariant* (r.i.) space if $g \in X$ implies that all μ-measurable functions f with the same rearrangement function with respect to the measure μ, i.e., such that $f_\mu^* = g_\mu^*$, also belong to X; moreover, $\|f\|_X = \|g\|_X$. An r.i. space $X(\Omega)$ can be represented by an r.i. space $\overline{X} = \overline{X}(0,1)$ on the interval $(0,1)$, with Lebesgue measure, such that

$$\|f\|_X = \|f_\mu^*\|_{\overline{X}},$$

for every $f \in X$. Typical examples of r.i. spaces are the L^p-spaces, Lorentz spaces, and Orlicz spaces. For more information we refer to [4].

In our recent work on symmetrization of Sobolev inequalities, we showed the following general theorem (cf. [15, 16] and the references therein)

Theorem 2.1. *Let $I : [0,1] \to [0,\infty)$ be an isoperimetric estimator on (Ω, d, μ). The following statements hold and are in fact equivalent.*

1. *Isoperimetric inequality*

$$\forall A \subset \Omega, \text{ Borel set}, \ \mu^+(A) \geqslant I(\mu(A)).$$

2. *Ledoux's inequality*

$$\forall f \in \mathrm{Lip}(\Omega), \ \int_0^\infty I(\lambda_f(s))ds \leqslant \int_\Omega |\nabla f(x)|\, d\mu(x).$$

3. *Maz'ya's inequality*[4]

$$\forall f \in \mathrm{Lip}(\Omega), \ (-f_\mu^*)'(s)I(s) \leqslant \frac{d}{ds}\int_{\{|f|>f_\mu^*(s)\}} |\nabla f(x)|\, d\mu(x).$$

4. *Pólya–Szegö's inequality*

$$\forall f \in \mathrm{Lip}(\Omega), \quad \int_0^t ((-f_\mu^*)'(.)I(.))^*(s)ds \leqslant \int_0^t |\nabla f|_\mu^*(s)ds.$$

(The second rearrangement on the left hand side is with respect to the Lebesgue measure).

---

[3] In fact, it is enough in order to define $|\nabla f|$ that f will be Lipschitz on every ball in (Ω, d) cf. [6, pp. 184,189] for more details.

[4] See [19]; one can also find this inequality in [27, 28].

6. *Oscillation inequality*

$$\forall f \in \mathrm{Lip}(\Omega), \quad (f_\mu^{**}(t) - f_\mu^*(t)) \leqslant \frac{t}{I(t)} |\nabla f|_\mu^{**}(t). \tag{2.1}$$

Given any rearrangement invariant space $X(\Omega)$, it follows readily from (2.1) that for all $f \in \mathrm{Lip}(\Omega)$

$$\|f\|_{LS(X)} := \left\| (f_\mu^{**}(t) - f_\mu^*(t)) \frac{I(t)}{t} \right\|_{\overline{X}} \preceq \|\nabla f\|_X .$$

One salient characteristic of these spaces is that they explicitly incorporate in their definition the isoperimetric profiles associated with the geometry in question and thus they can automatically select the correct optimal spaces for different geometries (cf. [14, 15, 16] for more on this). While the $LS(X)$ spaces are not necessarily normed, often they are equivalent to normed spaces (cf. [25]), and, in the classical cases, lead to optimal Sobolev–Poincaré inequalities and embeddings (cf. [17, 13, 14] as well as [3, 2, 29] and the references therein).

3 Hardy Isoperimetric Type

Let Q_I be the operator defined on measurable functions on $(0, 1)$ by

$$Q_I f(t) = \int_t^1 f(s) \frac{ds}{I(s)},$$

where I is an isoperimetric estimator. We consider the possibility of completely characterizing Poincaré inequalities in terms of the boundedness of Q_I as an operator from $\overline{X}$ to $\overline{Y}$.

In order to motivate what follows, we recall the following result obtained in [15, 16] for classical settings (cf. [14]).

Theorem 3.1. *Let X, Y be two r.i. spaces on Ω. Suppose that there exists a constant $c = c(X, Y)$ such that for every positive function $f \in \overline{X}$ with* $\mathrm{supp}\, f \subset (0, 1/2)$

$$\|Q_I f(t)\|_{\overline{Y}} \leqslant c \left\| f_\mu^* \right\|_{\overline{X}}.$$

Then for all $g \in \mathrm{Lip}(\Omega)$[5]

$$\left\| g - \int_\Omega g d\mu \right\|_Y \preceq \|\nabla g\|_X . \tag{3.1}$$

[5] We note for future use that Poincaré inequalities can be equivalently formulated replacing $\int_\Omega g d\mu$ by a median value m of g, i.e., $\mu\,(g \geqslant m) \geqslant 1/2$ and $\mu\,(g \leqslant m) \geqslant 1/2$.

Furthermore, if the space $\overline{X}$ *is such that* $\|f_\mu^*\|_{\overline{X}} \simeq \|f_\mu^{**}\|_{\overline{X}}$, *then*

$$\|f\|_Y \preceq \|f\|_{LS(X)} + \|f\|_{L^1}.$$

In fact, $LS(X)$ *is an optimal space in the sense that if* (3.1) *holds, then for all* $g \in \operatorname{Lip}(\Omega)$

$$\left\|g - \int_\Omega g d\mu\right\|_Y \preceq \left\|g - \int_\Omega g d\mu\right\|_{LS(X)} \preceq \|\nabla g\|_X.$$

We give a simple, but nontrivial example that illustrates how the preceding developments allow us to transplant Sobolev–Poincaré inequalities to the metric setting.

Example. Suppose that (Ω, μ) has an isoperimetric estimator

$$I(s) \simeq s^{1-1/n} \quad (0 < s < 1/2).$$

It follows that, on functions supported on $(0, 1/2)$,

$$Q_I f(t) \simeq \int_t^{1/2} s^{1/n} f(t) \frac{ds}{s}.$$

Since the conditions for the boundedness Q_I on r.i. spaces are well understood, we can transplant the classical Sobolev inequalities to (Ω, μ). Furthermore, we note that, in the borderline case $q = n$, the corresponding result using the optimal $LS(L^n)$ spaces is sharper than the classical Sobolev theorems (cf. [2]).

As was mentioned, it is known that the converse to Theorem 3.1 is true in a number of important classical cases. In other words, the operator Q_I in those cases gives a complete characterization of the Poincaré inequalities (for the most recent results cf. [15, 16]).

This led us to introduce the following definition.

Definition. We say that a probability metric space (Ω, d, μ) is of *isoperimetric Hardy type* if for any given isoperimetric estimator I, the following assertions are equivalent for all r.i. spaces $X = X(\Omega)$, $Y = Y(\Omega)$.

1. There exists $c = c(X, Y)$ such that

$$\forall f \in \operatorname{Lip}(\Omega), \quad \left\|f - \int_\Omega f d\mu\right\|_Y \leqslant c \|\nabla f\|_X. \tag{3.2}$$

2. There exists $c = c(X, Y)$ such that

$$\|Q_I f\|_{\overline{Y}} \preceq \|f\|_{\overline{X}}, \quad f \in \overline{X} \text{ with } \operatorname{supp}(f) \subset (0, 1/2).$$

4 Model Riemannian Manifolds

In this section, we construct a class of spaces of isoperimetric Hardy type spaces that includes the n-sphere $\mathbb{S}^n$ $(\mathbb{R}^n, \gamma_n)$ ($\mathbb{R}^n$ with Gaussian measure) and symmetric log-concave probability measures on $\mathbb{R}$.

We follow the construction of Ros (cf. [26]). Let M_0 be a complete smooth oriented n_0-dimensional Riemannian manifold with distance d. An absolutely continuous probability measure μ_0 w.r.t. to dV in M_0 is called a *model measure* if there exists a continuous family (in the sense of the Hausdorff distance on compact subsets) $\mathcal{D} = \{D^t : 0 \leqslant t \leqslant 1\}$ of closed subsets of M_0 satisfying the following conditions.

1. $\mu_0(D^t) = t$ and $D^s \subset D^t$, for $0 \leqslant s < t \leqslant 1$.

2. D^t is a smooth isoperimetric domain of μ_0 and $I_{\mu_0}(t) = \mu_0^+(D^t)$ is positive and smooth for $0 < t < 1$, where I_{μ_0} denotes the isoperimetric profile of M_0.

3. The r-enlargement of D^t, defined by $(D^t)_r = \{x \in M_0 : d(x, D^t) \leqslant r\}$ verifies $(D^t)_r = D^s$ for some $s = s(t, r)$, $0 \leqslant t \leqslant 1$.

4. $D^1 = M_0$ and D^0 is either a point or the empty set.

Theorem 4.1. *Let (M_0, d) be an n_0-dimensional Riemannian manifold endowed with a model measure μ_0. Then (M_0, d) is of isoperimetric Hardy type.*

Proof. Consider the function defined by

$$\begin{aligned} p : M_0 &\to [0,1], \\ x \in \partial D^t &\to t. \end{aligned}$$

Let $x, y \in M_0$ be such that $0 < p(y) < p(x)$. Let $D \in \mathcal{D}$ be such that $y \notin D$. Consider the function $h(r) = \mu_0(D_r)$, which is continuous and smooth for $0 < h(r) < 1$ and, in this range (cf. [26]),

$$h'(r) = I_{\mu_0}(h(r)). \tag{4.1}$$

From the definition of p it follows that $p(x) = h(d(x, D))$ and $p(y) = h(d(y, D))$. Since $d(x, D) - d(y, D) \leqslant d(x, y)$, we see that

$$\frac{p(x) - p(y)}{d(x, y)} \leqslant \frac{h(d(x, D)) - h(d(y, D))}{d(x, D) - d(y, D)} \leqslant \sup_s h'(s),$$

i.e., $p \in \mathrm{Lip}(M_0)$ and

$$|\nabla p(x)| = \limsup_{y \to x} \left| \frac{p(x) - p(y)}{d(x, y)} \right|$$

is finite, it follows that $|\nabla p(x)|$ exists a.e. w.r. to dV (cf. [6, p. 2]) and hence a.e. w.r. μ_0. Let us now compute $|\nabla p|$. Given $x \in M_0$ such that $p(x) = t < 1$, let $D \in \mathcal{D}$ so that $x \notin D$ and, as above, consider the function $h(r) = \mu_0(D_r)$. Let $z(x) \in M_0$ be such that

$$d(x, D) = d(x, z(x)).$$

Select y_n on the geodesic joining $z(x)$ and x such that $y_n \to x$. Then

$$\begin{aligned}\lim_{n\to\infty}\left|\frac{p(x)-p(y_n)}{d(x,y_n)}\right| &= \lim_{n\to\infty}\left|\frac{h(d(x,D))-h(d(y_n,D))}{d(x,D)-d(y_n,D)}\right| = h'(d(x,D))\\ &= I_{\mu_0}(h(d(x,D))) \qquad \text{[by (4.1)]}\\ &= I_{\mu_0}(p(x)). \end{aligned} \tag{4.2}$$

Let $f \in \overline{X}$ be a positive function with $\operatorname{supp} f \subset (0, 1/2)$. Define

$$F(x) = \int_{p(x)}^{1} f(s)\frac{ds}{I_{\mu_0}(s)}.$$

It is obvious that $F \in \operatorname{Lip}(M_0)$ and, by (4.2),

$$|\nabla F(x)| = f(p(x))\frac{1}{I_{\mu_0}(p(x))}|\nabla p(x)| = f(p(x)) \ \text{ a.e.}$$

We claim that the map $p : (M_0, \mu_0) \to ([0,1], ds)$ is a measure-preserving transformation. To prove this claim, we need to see that for any measurable subset $R \subset [0,1]$

$$\mu_0\left(p^{-1}(R)\right) = \int_R ds. \tag{4.3}$$

It is enough to see (4.3) for a closed interval. Let $[a,b] \subset [0,1]$ $(0 \leqslant a < b \leqslant 1)$. Then

$$\mu_0\left(p^{-1}([a,b])\right) = \mu_0(D^b) - \mu_0(D^a) = b - a.$$

Using this claim (cf. [4, Proposition 7.2, p. 80]) then a.e. we have

$$|F|^*_{\mu_0}(s) = \int_t^1 f(s)\frac{ds}{I_{\mu_0}(s)}, \quad |\nabla F|^*_{\mu_0}(s) = f^*(s).$$

It is obvious that the condition (3.2) is equivalent to

$$\|u - m\|_Y \preceq \|\nabla u\|_X,$$

where m is a median[6] of f, now since $\mu_0(F = 0) \geqslant 1/2$, 0 is a median of F, and from

[6] i.e., $\mu_0(f \geqslant m) \geqslant 1/2$ and $\mu_0(f \leqslant m) \geqslant 1/2$.

$$\|F - 0\|_Y \preceq \|\nabla F\|_X$$

we obtain

$$\left\| \int_t^1 f(s) \frac{ds}{I_{\mu_0}(s)} \right\|_{\overline{Y}} \preceq \|f\|_{\overline{X}}$$

as we wished to show. □

5 E. Milman's Equivalence Theorems

In the next result, we have a list of progressively weaker statements that nevertheless have been shown by E. Milman (cf. [23, 22, 24]) to be equivalent under certain convexity assumptions. Likewise, E. Milman also has formulated similar results in the context of Orlicz spaces.

In [16], we have simplified and extended Milman's results to the context of metric spaces with Hardy isoperimetric type, as well as considering general r.i. spaces.

Theorem 5.1. *Let (Ω, d, μ) be a space of Hardy isoperimetric type. Then the following statements are equivalent:*

(E1) *Cheeger's inequality*

$$\exists C > 0 \ s.t. \quad I_{(\Omega,d,\mu)} \geqslant Ct, \quad t \in (0, 1/2].$$

(E2) *Poincaré inequality*

$$\exists P > 0 \ s.t. \quad \|f - m\|_{L^2(\Omega)} \leqslant P \|f\|_{L^2(\Omega)}.$$

(E3) *Exponential concentration: for all $f \in \mathrm{Lip}(\Omega)$ with $\|f\|_{\mathrm{Lip}(\Omega)} \leqslant 1$*

$$\exists c_1, c_2 > 0 \ s.t. \quad \mu\{|f - m| > t\} \leqslant c_1 e^{-c_2 t}, \quad t \in (0, 1).$$

(E4) *First moment inequality: for all $f \in \mathrm{Lip}(\Omega)$ with $\|f\|_{\mathrm{Lip}(\Omega)} \leqslant 1$*

$$\exists F > 0 \ s.t. \quad \|f - m\|_{L^1(\Omega)} \leqslant F.$$

Theorem 5.2. *Let (Ω, d, μ) be a space of isoperimetric Hardy type. Let $1 \leqslant q \leqslant \infty$,, and let N be a Young function such that $\frac{N(t)^{1/q}}{t}$ is nondecreasing and there exists $\alpha > \max\{\frac{1}{q} - \frac{1}{2}, 0\}$ such that $\frac{N(t^\alpha)}{t}$ is nonincreasing. Then the following statements are equivalent.*

(E5) (L_N, L^q) *Poincaré inequality holds*

$$\exists P > 0 \ s.t. \quad \|f - m\|_{L_N(\Omega)} \leqslant P \|f\|_{L^q(\Omega)}.$$

(E6) *Any isoperimetric profile estimator I satisfies: there exists a constant $c > 0$ such that*

$$I(t) \geqslant c \frac{t^{1-1/q}}{N^{-1}(1/t)}, \quad t \in (0, 1/2].$$

The construction of the previous section thus provides a class of spaces were the previous theorems apply.

6 Some Spaces That Are not of Isoperimetric Hardy Type

In this section, we show that, unfortunately, not all metric spaces are of isoperimetric Hardy type.

Let $I : [0,1] \to [0,\infty)$ be concave, continuous, increasing on $(0,1/2)$, symmetric about the point $1/2$, and such that $I(0) = 0$. Let $0 \leqslant \beta \leqslant 1$. We say that I has *β-asymptotic behavior* if $\lim_{s\to 0^+} \frac{I(s)}{s^{1-\beta}}$ exists and lies on $(0,\infty)$.

Theorem 6.1. *Suppose that I is of β-asymptotic behavior. Then the following assertions hold.*

(i) *Given $0 < \beta < 1/2$, there is a metric space (Ω_0, d, μ) with $I(s) \simeq I_{(\Omega_0,d_0,\mu_0)}(s)$, a pair of r.i. spaces X, Y on Ω_0, and a constant $c = c(X,Y)$ such that*

$$\left\| g - \int_{\Omega_0} g d\mu_0 \right\|_Y \leqslant c \left\| \nabla g \right\|_X, \quad g \in \operatorname{Lip}(\Omega_0),$$

but $Q_I : \overline{X} \to \overline{Y}$ is not bounded.

(ii) *Given $0 < \beta < 1$, there is a metric space (Ω_1, d_1, μ_1) such that*

$$I(s) \simeq I_{(\Omega_1,d_1,\mu_1)}(s)$$

and (Ω_1, d, μ) is of isoperimetric Hardy type.

Proof. (i) (cf. [13] for a more general result). Let $1 < \alpha < 2$,, and let Ω be an α-John domain on $\mathbb{R}^2$ ($|\Omega| = 1$). Then (cf. [11])

$$I_\Omega(s) \simeq s^{\alpha/2} = s^{1-(1-\alpha/2)}, \quad 0 \leqslant s \leqslant 1/2.$$

Let $t > 1$ be such that $\alpha > t-1$, and let $r = \frac{2t}{\alpha+(1-t)}$. Note that $1 < t < r$. Then (cf. [12])

$$\left\| g - \int_\Omega g \right\|_{L^r} \preceq \left\| \nabla g \right\|_{L^t}.$$

In this case, the operator Q_{I_Ω} is given by

$$Q_{I_\Omega} f(t) = \int_t^1 u^{-\alpha/2} f(u) du.$$

Note that Q_{I_Ω} is not bounded from L^t to L^r. Indeed, the boundedness of Q_{I_Ω} can be reformulated as a weighted norm inequality for the operator $g \to \int_x^1 g(u)du$; namely,

$$\left\| \int_x^1 g(u)du \right\|_{L^r} \leqslant c \left\| g(x)x^{\alpha/2} \right\|_{L^t}. \tag{6.1}$$

It is well known that (6.1) holds if and only if (cf. [20])

$$\sup_{a>0} \left(\int_0^a 1 \right)^{1/r} \left(\int_a^1 \left(u^{\alpha t/2} \right)^{\frac{-1}{t-1}} du \right)^{\frac{t-1}{t}} < \infty. \tag{6.2}$$

Now, since $\alpha < 2$, it follows that $\frac{-\alpha t}{2(t-1)} + 1 < 0$, and for a near zero we have

$$\begin{aligned} \left(\int_0^a 1 \right)^{1/r} \left(\int_a^1 \left(u^{\alpha t/2} \right)^{\frac{-1}{t-1}} du \right)^{\frac{t-1}{t}} &\simeq a^{1/r} \left(a^{\frac{-\alpha t + 2(t-1)}{2(t-1)}} - 1 \right)^{\frac{t-1}{t}} \\ &\simeq a^{\frac{(1-t)(\alpha-1)}{2t}}. \end{aligned}$$

Consequently, since $\frac{(1-t)(\alpha-1)}{2t} < 0$, (6.2) cannot hold.

(ii) We follow Gallot's method (cf. [10]) in order to construct (Ω_1, d_1, μ_1). Let

$$B(r) = \int_r^1 \frac{ds}{I(s)}, \quad 0 \leqslant r \leqslant 1.$$

Since I is of β-asymptotic behavior, we see that $L = B(0) < \infty$. Since B is decreasing, it has an inverse which we denote by A. Consider the revolution surface $M = (0, L) \times \mathbb{S}^1$ (compactified by adjoining the points $\{0\} \times \mathbb{S}^1$ and $\{L\} \times \mathbb{S}^1$)) provided with the Riemannian metric

$$g = dr^2 + I(A(r))^2 d\theta^2,$$

where $\theta \in \mathbb{S}^1$ and $d\theta^2$ is the canonical Riemannian metric on $\left(\mathbb{S}^1, \mathrm{can}\right)$. Notice that $I(A(0)) = I(A(L)) = 0$. We denote by Vol_M the volume of (M, g). Multiplying the metric g by a constant, we can and will assume without loss that $\mathrm{Vol}_M(M) = 1$. Denote by I_M the isoperimetric profile of (M, g, Vol_M). Then (cf. [10, Appendix A.1.]) we can find a constant c, depending only on I, such that

$$cI(s) \leqslant I_M(s) \leqslant I(s).$$

Let X, Y be two r.i. spaces on M such that

$$\left\| g - \int_M g d\mathrm{Vol}_M \right\|_Y \preceq \|\nabla g\|_X, \quad g \in \mathrm{Lip}(M).$$

Let f be a positive Lebesgue measurable function on $(0,1)$ with supp$f \subset (0,1/2)$. Define

$$u(r,\theta_1,\theta_2) = \int_{A(r)}^1 f(s) \frac{ds}{I(s)}, \quad (r,\theta_1,\theta_2) \in M.$$

It is plain that u is a Lipschitz function on M such that $\mathrm{Vol}_M \{u = 0\} \geqslant 1/2$. Hence 0 is a median of u.

On the other hand, recall that (cf. [10, Page 57]) for any domain of revolution $\Omega(\lambda) = (0,\lambda) \times \mathbb{S}^1 \subset M$ we have that

$$\mathrm{Vol}_M^+ (\partial\Omega(\lambda)) = I\left(\mathrm{Vol}_M (\Omega(\lambda))\right).$$

In other words,

$$A'(r) = I(A(r)).$$

Therefore,

$$|\nabla u(r,\theta_1,\theta_2)| = \left|\frac{\partial}{\partial r} u(r)\right| = \left|-f(A(r))\frac{A'(r)}{I(A(r))}\right| = f(A(r)).$$

Now, since

$$u^*_{\mathrm{Vol}_M}(t) = \int_t^1 f(s)\frac{ds}{I(s)}, \quad |\nabla u|^*_{\mathrm{Vol}_M}(t) = f^*(t),$$

from

$$\|u - 0\|_Y \preceq \|\nabla u\|_X$$

we deduce that

$$\left\| \int_t^1 f(s)\frac{ds}{I(s)} \right\|_Y \preceq \|f\|_X,$$

as we wished to show. □

By Theorem 6.1 the verification of a Sobolev–Poincaré inequality cannot be reduced, in general, to establish the boundedness of the associated isoperimetric Hardy operator. However, if the profiles are of β-asymptotic behavior, we have the following weaker positive result.

Theorem 6.2. *Let I be of β-asymptotic behavior $(0 < \beta < 1)$. Let $\mathcal{M}_I$ be the set of metric probability spaces (Ω, d, μ) such that*

$$I_{(\Omega,d,\mu)} \geqslant I.$$

Let $\overline{X}, \overline{Y}$ be two r.i. spaces on $[0,1]$. Then the following statements are equivalent.

1.

$$\inf_{(\Omega,d,\mu)\in\mathcal{M}_h} \inf_{g\in\mathrm{Lip}(\Omega)} \frac{\left\| |\nabla g|_\mu^* \right\|_{\overline{X}}}{\left\| \left(g - \int_\Omega g d\mu\right)_\mu^* \right\|_{\overline{Y}}} = c > 0.$$

2.

$$Q_I : \overline{X} \to \overline{Y} \quad \text{is bounded.} \tag{6.3}$$

Proof. 1 → 2. Given I of β-asymptotic behavior, consider the revolution surface M constructed in part (ii) of the previous theorem. Since $M \in \mathcal{M}_I$, by hypothesis:

$$\left\| \left(g - \int_\Omega g d\mu \right)_\mu^* \right\|_{\overline{Y}} \leqslant c \left\| (\nabla g)_\mu^* \right\|_{\overline{X}}, \quad g \in \mathrm{Lip}(M),$$

which is equivalent to (6.3) since M is of Hardy isoperimetric type.

2 → 1. The implication is a direct consequence of Theorem 3.1. □

Acknowledgement. The first author was supported in part by Grants MTM2007-60500, MTM2008-05561-C02-02 and by 2005SGR00556.

References

1. Barthe, F., Cattiaux, P., Roberto, C.: Interpolated inequalities between exponential and Gaussian, Orlicz hypercontractivity and isoperimetry. Rev. Mat. Iberoam. **22**, 993–1067 (2006)
2. Bastero. J., Milman, M., Ruiz, F.: A note on $L(\infty, q)$ spaces and Sobolev embeddings. Indiana Univ. Math. J. **52**, 1215–1230 (2003)
3. Bennett, C., DeVore, R., Sharpley, R.: Weak L^∞ and BMO. Ann. Math. **113**, 601–611 (1981)
4. Bennett, C., Sharpley, R.: Interpolation of Operators. Academic Press, Boston (1988)
5. Bobkov, S.G.: Extremal properties of half-spaces for log-concave distributions. Ann. Probab. **24**, 35–48 (1996)
6. Bobkov, S.G., Houdré, C.: Some connections between isoperimetric and Sobolev type inequalities. Mem. Am. Math. Soc. **129**, no. 616 (1997)
7. Borell, C.: Intrinsic bounds on some real-valued stationary random functions. In: Lect. Notes Math. **1153**, pp. 72–95. Springer (1985)
8. Ehrhard, A.: Symétrisation dans le space de Gauss. Math. Scand. **53**, 281–301 (1983)
9. Federer, H.: Geometric Measure Theory. Springer, Berlin (1969)

10. Gallot, S.: Inégalités isopérimétriques et analytiques sur les variétés riemanniennes. Astérisque. **163–164**, 31–91 (1988)
11. Hajlasz, P., Koskela, P.: Isoperimetric inequalities and Imbedding theorems in irregular domains. J. London Math. Soc. **58**, 425–450 (1998)
12. Kilpeläinen, T., Malý, J.: Sobolev inequalities on sets with irregular boundaries. Z. Anal. Anwen.**19**, 369–380 (2000)
13. Martín J., Milman, M.: Self improving Sobolev–Poincaré inequalities, truncation and symmetrization. Potential Anal. **29**, 391–408 (2008)
14. Martín J., Milman, M.: Isoperimetry and Symmetrization for Logarithmic Sobolev inequalities. J. Funct. Anal. **256**, 149–178 (2009)
15. Martín J., Milman, M.: Isoperimetry and symmetrization for Sobolev spaces on metric spaces. Compt. Rend. Math. **347**, 627–630 (2009)
16. Martín J., Milman, M.: Pointwise Symmetrization Inequalities for Sobolev Functions and Applications. Preprint (2009)
17. Martín, J., Milman, M., Pustylnik, E.: Sobolev inequalities: Symmetrization and self-improvement via truncation. J. Funct. Anal. **252**, 677–695 (2007)
18. Maz'ya, V.G.: Classes of domains and imbedding theorems for function spaces (Russian). Dokl. Akad. Nauk SSSR **3**, 527–530 (1960); English transl.: Sov. Math. Dokl. **1**, 882–885 (1961)
19. Maz'ya, V.G.: Weak solutions of the Dirichlet and Neumann problems (Russian). Tr. Mosk. Mat. O-va. **20**, 137–172 (1969)
20. Maz'ya, V.G.: Sobolev Spaces. Springer, Berlin etc. (1985)
21. Maz'ya, V.G.: Conductor and capacitary inequalities for functions on topological spaces and their applications to Sobolev type imbeddings. J. Funct. Anal. **224**, 408–430 (2005)
22. Milman, E.: Concentration and isoperimetry are equivalent assuming curvature lower bound. C. R. Math. Acad. Sci. Paris **347**, 73–76 (2009)
23. Milman, E.: On the role of Convexity in Isoperimetry, Spectral-Gap and Concentration. Invent. Math. **177**, 1–43 (2009)
24. Milman, E.: On the role of convexity in functional and isoperimetric inequalities. Proc. London Math. Soc. **99**, 32–66 (2009)
25. Pustylnik, E.: A rearrangement-invariant function set that appears in optimal Sobolev embeddings. J. Math. Anal. Appl. **344**, 788–798 (2008)
26. Ros, A.: The isoperimetric problem. In: Global Theory of Minimal Surfaces. Clay Math. Proc. Vol. 2, pp. 175–209. Am. Math. Soc., Providence, RI (2005)
27. Talenti, G.: Elliptic Equations and Rearrangements. Ann. Scuola Norm. Sup. Pisa Cl. Sci. **3**, 697–718 (1976)
28. Talenti, G.: Inequalities in rearrangement-invariant function spaces. In: Nonlinear Analysis, Function Spaces and Applications 5, pp. 177–230. Prometheus, Prague (1995)
29. Tartar, L.: Imbedding theorems of Sobolev spaces into Lorentz spaces. Boll. Unione Mat. Ital. Sez B Artic. Ric. Mat. **8**, 479–500 (1998)

Gauge Functions and Sobolev Inequalities on Fluctuating Domains

Eric Mbakop and Umberto Mosco

Dedicated to Professor Vladimir G. Maz'ya

Abstract We study Sobolev inequalities on domains with fluctuating geometry. Fluctuations are expressed in terms of suitable sub-linear *gauge functions*, typically of iterated logarithmic type. The same gauge function describes the metric oscillations of the domain, as well as the resulting oscillations observed in the Sobolev exponent. The theory applies to fractal domains displaying random self-similarity. The gauge function is then produced by the scale fluctuations of the fractal.

1 Introduction

The object of this study is to consider Sobolev and Poincaré inequalities in a setting including certain fluctuating fractal domains. There is a vast literature on Sobolev and Poincaré inequalities in metric spaces (cf., for example, [9, 8, 3] and [19]). The aim of these theories is to show that if the volume $\mu(R)$ of a ball B_R of radius R has a power growth $\mu(R) \sim R^\nu$ for some exponent $\nu > 0$, then for every $1 \leqslant p < \nu$ the Sobolev space $W^{1,p}$ on B_R is imbedded into the Lebesgue space L^{p^*} on a comparable smaller ball, say $B_{R/q}$ for some constant $q \geqslant 1$ independent of R. The Sobolev exponent p^* is given by $p^* = \frac{\nu-p}{p\nu}$. This

Eric Mbakop
Worcester Polytechnic Institute, 100 Institute Road, Worcester, MA 01609, USA
e-mail: steve055@WPI.EDU

Umberto Mosco
Worcester Polytechnic Institute, 100 Institute Road, Worcester, MA 01609, USA e-mail: mosco@WPI.EDU

A. Laptev (ed.), *Around the Research of Vladimir Maz'ya I: Function Spaces*,
International Mathematical Series 11, DOI 10.1007/978-1-4419-1341-8_14,

relation between p and p^* is the analogue of the one occurring in the familiar case of Euclidean space of (integer) dimension $\nu = D \in \mathbb{N}$.

In the more general situation dealt with in this paper, the volume $\mu(B_R)$ is subject to *fluctuations* in R. This affects the Sobolev imbedding by inducing fluctuations on the Sobolev exponent. The main goal of the paper is to describe these fluctuations in detail.

Following [22], we rely on a *two-scale* formalism. In addition to power-scaling, we introduce a finer metric scale, typically of logarithmic type, incorporated in a suitable *gauge function* g.

In our setting, a gauge function is defined to be a (Lipschitz) nondecreasing function on the real line such that $g(0) = 1$ and

$$g(t) \leqslant g_0 t^{1-\epsilon_0} \quad \forall\, t \geqslant 1 \tag{1.1}$$

for some constants $g_0 \geqslant 1$ and $0 < \epsilon_0 \leqslant 1$.

A significative example, from probability theory, is the function $g(n) = cn^a(\log\log n)^b$, with $a < 1$ and $b > 0$, which for large $n \in \mathbb{N}$ satisfies (1.1) with $0 < \epsilon_0 < 1 - a$. We will meet this function in Section 5 in connection with the fractal examples which motivate our theory.

Related scaling functions are the function

$$f(R) = \exp\left(cg\left(c\,|\log R|\right)\right) \tag{1.2}$$

for $R > 0$ and some constant $c > 0$ and, given an exponent $\nu > 0$, the function

$$\Theta(x) = x\left(S^{(1)}\left(Rx^{1/\nu}\right)\right)^{-\frac{\nu}{\nu-1}}$$

for $x > 0$, with

$$S^{(1)}(r) = \sum_{k\geqslant 0} 2^{-k} f(2^{-k}r)\,, \tag{1.3}$$

$0 < r \leqslant 1$.

The function f describes the fluctuations of the volume of balls $\mu(B_R)$. The function Θ describes the fluctuations of the Sobolev imbedding. Both functions f and Θ are build upon the *same* gauge function g. Therefore, the two functions together provide a quantitative, explicit description of how volume fluctuations carry over to summability fluctuations.

We consider functions defined on a space X. We assume X to be a *quasi-metric space*, i.e., is a space endowed with a *quasidistance*. While distances are Lipschitz continuous functions, quasidistances are Hölder continuous functions. This provides an additional parameter scale. Moreover, we allow for gradient of functions be only defined in a *measure-valued* sense, namely as *local energy forms*, called *Lagrangeans*. Both these features, quasidistances and Lagrangeans, together with the gauged functions mentioned before, play a crucial role in our applications to fluctuating fractals.

Our main results are stated in Theorems 2.3, 3.1, and 4.1. Theorem 2.3 establishes some summability properties of the gauge functions. Theorem 3.1, which refines and extends some results from [22], shows the equivalence of gauged Sobolev and Poincaré inequalities on the balls of X. Theorem 4.1, inspired by recent work of Mazy'a and others [17, 18, 6], deals with the equivalence of gauged Sobolev inequalities with suitable gauged capacitary inequalities.

All inequalities involve the *same* gauge function g. Taken together, they describe the interrelation between the structural fluctuations of the space X and the analytical fluctuations of the function spaces on X.

Section 2 is dedicated to the study of gauge functions. Section 3 deals with Sobolev and Poincaré inequalities, Section 4 with Sobolev and capacitary inequalities, and in Section 5 we give examples of fluctuating domains in which our results apply.

2 Gauge Functions

A *gauge function* is a nondecreasing continuous function g on the real line such that $g(0) = g_0 \geqslant 1$ and

$$g(t) \leqslant g_0 t^{1-\epsilon_0} \quad \forall\, t \geqslant 1, \tag{2.1}$$

$0 < \epsilon_0 \leqslant 1$. We will have as an additional assumption that g is Lipschitz on $[1, \infty)$, with Lipschitz constant M. We identify g with its constants g_0, ϵ_0, and write $g = (g_0, \epsilon_0)$. In some parts of our paper, we assume that g satisfies the following condition

Sub-additivity condition. There exists a constant $C_g > 0$ and $t \geqslant 1$ such that for all $x, y \geqslant t$

$$g(x + y) \leqslant C_g(g(x) + g(y)). \tag{2.2}$$

Given a gauge function g, we define an associated *scaling function* f as follows:

$$f(s) \equiv f(f_0, f_1; s) := \exp\left(f_0\; g(f_1 h(s))\right), \quad s > 0,$$

where

$$h(s) = \max\{1, |\log(s)|\} \tag{2.3}$$

with $f_0 > 0$ and $f_1 \geqslant 1$ two constants. For such a function f and for a given $R > 0$ we consider the function f_R defined by

$$f_R = f\Big(f_0, f_1\; m(R); \frac{s}{m(R)}\Big). \tag{2.4}$$

Here, the function $m(R)$ is defined by

$$m(R) = \max\{1, R\}. \tag{2.5}$$

We identify a scaling function f, associated to the gauge function g, with its constants f_0 and f_1 and write $f = (f_0, f_1)$.

Also associated to the gauge function g is the *fluctuation modulus* $S_R^{(\gamma)}$ defined for some positive constants R and γ as

$$S_R^{(\gamma)}(r) = \sum_{k \geqslant 0} 2^{-\gamma k} f_R(2^{-k} r). \tag{2.6}$$

It is important to note that f is nonincreasing on $(0, e)$ and f_R is nonincreasing on $(0, em(R))$.

Also, the scaling function f satisfies

$$f(qs) \leqslant f(f_0, f_1 m(q); s) \tag{2.7}$$

for all $q, s > 0$.

We proceed by giving some estimates on the growth of the fluctuation modulus S_R^{γ} near the origin.

Proposition 2.1. *Let $R_0 > 0$ and $\gamma > 0$ be given constants. Let f be a scaling function associated to a gauge function $g = (g_0, \epsilon_0)$ with constants $f_0 > 0$ and $f_1 \geqslant 1$. Then there exists two positive constants $C_0, C_1 > 0$ such that for all $0 < R \leqslant R_0$ we have*

$$S_R^{(\gamma)}(r) \leqslant C_0 \exp\Big(C_1 \Big(\log\Big(\frac{m(R)}{r}\Big)\Big)^{1-\epsilon_0}\Big) \qquad \text{for all} \quad 0 < r \leqslant \min\Big\{R, \frac{m(R)}{e}\Big\}, \tag{2.8}$$

where

$$C_1 = f_0 g_0 (f_1 m(R_0))^{1-\epsilon_0}, \quad C_0 = \sum_{k \geqslant 0} 2^{-\gamma k} \exp\left(C_1 (k \log 2)^{1-\epsilon_0}\right).$$

Proof. Let $0 < r \leqslant \min\{R, \frac{m(R)}{e}\}$, and $0 < R \leqslant R_0$. We have

$$\begin{aligned}
S_R^{(\gamma)}(r) &= \sum_{k \geqslant 0} 2^{-\gamma k} f_R(2^{-k} r) \\
&= \sum_{k \geqslant 0} 2^{-\gamma k} \exp\left(f_0 g\Big(f_1 m(R) \log \frac{2^k m(R)}{r}\Big)\right) \\
&= \sum_{k \geqslant 0} 2^{-\gamma k} \exp\left(f_0 g\Big(f_1 m(R)\Big(k \log 2 + \log \frac{m(R)}{r}\Big)\Big)\right) \\
&\leqslant \sum_{k \geqslant 0} 2^{-\gamma k} \exp\left(f_0 g\Big(f_1 m(R_0)\Big(k \log 2 + \log \frac{m(R)}{r}\Big)\Big)\right)
\end{aligned}$$

$$\leqslant \sum_{k\geqslant 0} 2^{-\gamma k} \exp\Big(f_0 g_0 \Big(f_1 m(R_0)\Big(k\log 2 + \log\frac{m(R)}{r}\Big)\Big)^{1-\epsilon_0}\Big)$$

$$\leqslant \sum_{k\geqslant 0} 2^{-\gamma k} \exp\Big(C_1 (k\log 2)^{1-\epsilon_0} + C_1\Big(\log\frac{m(R)}{r}\Big)^{1-\epsilon_0}\Big)$$

$$= C_0 \exp\Big(C_1\Big(\log\Big(\frac{m(R)}{r}\Big)\Big)^{1-\epsilon_0}\Big).$$

Note that we used the fact that g is increasing, $g(t) \leqslant g_0 t^{1-\epsilon_0}$ for all $t \geqslant 1$, and $0 < \epsilon_0 \leqslant 1$. Now, it only remains to show that C_0 is finite.

Since $0 < \epsilon_0 \leqslant 1$, there exists $K^* \in \mathbb{N}$ such that for all $k \geqslant K^*$ we have

$$\frac{1}{k^{\epsilon_0}} < \frac{\gamma(\log 2)^{\epsilon_0}}{2C_1}.$$

Hence

$$C_0 < \sum_{k<K^*} 2^{-\gamma k} \exp(C_1(k\log 2)^{1-\epsilon_0}) + \sum_{k\geqslant K^*} 2^{-\frac{k\gamma}{2}} < \infty,$$

where the second term is part of a geometric sum. □

Corollary 2.2. *For all $R > 0$ and for all $\beta > 0$ the fluctuation modulus satisfies*

$$S_R^\gamma(r) \leqslant C_0\Big(\frac{m(R)}{r}\Big)^{\beta}$$

$$\text{for all } \ 0 < r \leqslant \min\Big\{R, \frac{m(R)}{e}, m(R)\exp\Big(-\Big(\frac{C_1}{\beta}\Big)^{\frac{1}{\epsilon_0}}\Big)\Big\}, \tag{2.9}$$

where

$$C_1 = f_0 g_0 (f_1 m(R))^{1-\epsilon_0}, \quad C_0 = \sum_{k\geqslant 0} 2^{-\gamma k} \exp(C_1(k\log 2)^{1-\epsilon_0}).$$

Proof. The inequality follows from the fact that

$$C_1(\log\frac{m(R)}{r})^{1-\epsilon_0} \leqslant \beta \log\frac{m(R)}{r}$$

if $r \leqslant m(R)\exp(-(C_1/\beta)^{\frac{1}{\epsilon_0}})$. □

We define a function ϕ_R related to the gauge function g that will be used in the statement of our results. For $R > 0$ and $0 < \gamma < \nu$ let $\phi_R := \phi_R^{\nu,\gamma}$ be defined by

$$\phi_R(x) = x\big(S_R^{(\gamma)}(Rx^{\frac{1}{\nu}})\big)^{-\frac{\nu}{\nu-\gamma}} \tag{2.10}$$

We establish some summability properties of the function ϕ_R.

Theorem 2.3. *Let R_0 and R_1 be two given positive constants that satisfy $0 < R_0 \leqslant R_1$. Then for all $p > 1$ with $\frac{1}{p} + \frac{1}{p'} = 1$ and for all R such that $R_0 \leqslant R \leqslant R_1$ the functions $(\phi_R')^{-\frac{1}{p}}$ belong to $L^{p'}((0,1))$ with $L^{p'}$ – norm uniformly bounded by a constant $C_p = C_p(p, R_0, R_1)$, i.e.,*

$$\left(\int_0^1 (\phi_R'(x))^{1-p'}\,dx\right)^{\frac{1}{p'}} \leqslant \left(\int_0^1 \left(\frac{x}{\phi_R(x)}\right)^{p'-1} dx\right)^{\frac{1}{p'}} \leqslant C_p. \tag{2.11}$$

Here, ϕ_R' denotes the derivative of ϕ_R. If, in addition, g satisfies the sub-additivity condition, then for all $R_1 > 0$ and $p > 1$ there exists $\widetilde{C}_p = \widetilde{C}_p(p, R_1) > 0$ such that

$$\left(\int_0^1 (\phi_R'(x))^{1-p'}\,dx\right)^{\frac{1}{p'}} \leqslant \widetilde{C}_p \widetilde{f}(R) \quad \forall\, 0 < R \leqslant R_1. \tag{2.12}$$

Here, the new scaling function $\widetilde{f}$ is related to the original scaling function $f = (f_0, f_1)$ by

$$\widetilde{f} = (\frac{C_g \zeta}{p} f_0, f_1), \quad \zeta = \frac{\nu}{\nu - \gamma}.$$

Proof. We first prove that ϕ_R' exists almost everywhere (a.e) on the interval $(0,1)$. With that end in mind, we first show that for all $R > 0$, $\gamma > 0$, and $0 < r_0 < R$ the sequence of partial sums ${}^{(m)}S_R^{(\gamma)}$ defined by

$${}^{(m)}S_R^{(\gamma)}(r) = \sum_{k=0}^{k=m} 2^{-\gamma k} f_R(2^{-k} r)$$

converges uniformly to $S_R^{(\gamma)}$ on $[r_0, R]$.

In fact,

$$(S_R^{(\gamma)} - {}^{(m)}S_R^{(\gamma)})(r) = \frac{1}{2^{\gamma(m+1)}} S_R^{(\gamma)}(\frac{r}{2^{m+1}}).$$

Since $S_R^{(\gamma)}$ is nonincreasing, we get

$$\|S_R^{(\gamma)} - {}^{(m)}S_R^{(\gamma)}\|_{L^\infty([r_0,R])} \leqslant \frac{1}{2^{\gamma(m+1)}} S_R^{(\gamma)}\left(\frac{r_0}{2^{m+1}}\right) \to 0 \text{ as } m \to \infty,$$

where the limit is obtained as a consequence of Corollary 2.2. Hence the continuity of the partial sums ${}^{(m)}S_R^{(\gamma)}$ on $(0, R]$ implies that of $S_R^{(\gamma)}$ on $(0, R]$. Since g is Lipschitz, every partial sums ${}^{(m)}S_R^{(\gamma)}$ is differentiable almost everywhere on $(0, R)$. Let ${}^{(m)}S_R'^{(\gamma)}$ denote the first order derivative of ${}^{(m)}S_R^{(\gamma)}$. We have

$$^{(m)}S_R'^{(\gamma)}(r) = -\frac{C_R}{r}\sum_{k=0}^{k=m} g'(f_1 m(R)\log\Big(\frac{2^k m(R)}{r}\Big))2^{-\gamma k} f_R(2^{-k}r)$$

for $0 < r < \min\{R, \frac{m(R)}{e}\}$, where $C_R = f_0 f_1 m(R)$. Define ψ_R as follows:

$$\psi_R(r) = -\frac{C_R}{r}\sum_{k=0}^{\infty} g'(f_1 m(R)\log\Big(\frac{2^k m(R)}{r}\Big))2^{-\gamma k} f_R(2^{-k}r)$$

for $0 < r < \min\{R, \frac{m(R)}{e}\}$. Using an argument similar to the one above, we obtain

$$\|\psi_R -^{(m)} S_R'^{(\gamma)}\|_{L^\infty([r_0, \frac{\max\{1,R\}}{e}])} \leqslant \frac{\widetilde{C}_R}{r_0}\frac{1}{2^{\gamma(m+1)}} S_R^{(\gamma)}\Big(\frac{r_0}{2^{m+1}}\Big) \to 0 \text{ as } m \to \infty$$

for all $0 < r_0 < \min\{R, \frac{m(R)}{e}\}$. Here, $\widetilde{C}_R = MC_r$, where M is the Lipschitz constant of g. Therefore, $S_R^{(\gamma)}$ is differentiable almost everywhere on $(0, \min\{R, \frac{m(R)}{e}\})$ and its derivative, which we denote by $S_R'^{(\gamma)}$, is the function ψ_R. Note that if $\min\{R, \frac{m(R)}{e}\} < R$, then $S_R'^{(\gamma)}(r) \equiv 0$ for all $x \in [\ \min\{R, \frac{m(R)}{e}\}, R)$.

Now,

$$\phi_R(x) = \frac{x}{(S_R^{(\gamma)}(Rx^{\frac{1}{\nu}}))^\zeta}$$

with $\zeta = \frac{\nu}{\nu-\gamma}$. Differentiating with respect to x, we obtain

$$(\phi_R'(x))^{-1} = \frac{(S_R^{(\gamma)}(Rx^{\frac{1}{\nu}}))^{\zeta+1}}{S_R^{(\gamma)}(Rx^{\frac{1}{\nu}}) - \frac{\zeta}{\nu}Rx^{\frac{1}{\nu}}S_R'^{(\gamma)}(Rx^{\frac{1}{\nu}})}$$

a.e $x \in (0,1)$. Using the fact that $S_R'^{(\gamma)}(Rx^{\frac{1}{\nu}}) \leqslant 0$ and $S_R^{(\gamma)}(Rx^{\frac{1}{\nu}}) \geqslant 1$, a.e $x \in (0,1)$, we get

$$(\phi_R'(x))^{-1} \leqslant \Big(S_R^{(\gamma)}(Rx^{\frac{1}{\nu}})\Big)^\zeta = \frac{x}{\phi_R(x)}$$

a.e $x \in (0,1)$. Therefore,

$$\begin{aligned}\int_0^1 (\phi_R'(x))^{-\frac{p'}{p}}dx &\leqslant \int_0^1 (S_R^{(\gamma)}(Rx^{\frac{1}{\nu}}))^{\zeta(p'-1)}dx\\ &= \frac{\nu}{R^\nu}\int_0^R (S_R^{(\gamma)}(x))^{\zeta(p'-1)}x^{\nu-1}dx\\ &\leqslant \frac{\nu}{R_0^\nu}\int_0^{R_1} (S_{R_1}^{(\gamma)}(x))^{\zeta(p'-1)}x^{\nu-1}dx = C_p^{p'} < \infty\end{aligned}$$

for all $R_0 \leqslant R \leqslant R_1$. Here, the finiteness follows from Corollary 2.2 and the fact that $S_R^{(\gamma)}$ is continuous for all $R > 0$. Note that we have use the inequality $S_R^\gamma(x) \leqslant S_{R_1}^\gamma(x)$ which holds for all $x \in (0, e)$.

If we assume, in addition, that g satisfies the sub-additivity condition, then the growth of the $L^{p'}$-norm of $(\phi'_R(x))^{-\frac{1}{p}}$ as R tends to zero can be controlled by the gauge function f as follows: For $R \leqslant 1/e$

$$\Big(\int_0^1 (\phi'_R(x))^{-\frac{p'}{p}} dx\Big)^{\frac{1}{p'}} \leqslant \Big(\int_0^1 (S_1^{(\gamma)}(Rx^{\frac{1}{\nu}}))^{\zeta(p'-1)}\Big)^{\frac{1}{p'}}$$
$$\leqslant \widetilde{f}(R)\Big(\int_0^1 (S_1^{(\gamma)}(x^{\frac{1}{\nu}}))^{\zeta(p'-1)}\Big)^{\frac{1}{p'}} dx = \widetilde{f}(R)C,$$

where we used the inequality

$$f\Big(\frac{Rx^{\frac{1}{\nu}}}{2^k}\Big) \leqslant \Big(\widetilde{f}(R)\widetilde{f}\Big(\frac{x^{\frac{1}{\nu}}}{2^k}\Big)\Big)^{\frac{p}{\zeta}}.$$

Note that Corollary 2.2 gives that

$$\Big(\int_0^1 (S_1^{(\gamma)}(x^{\frac{1}{\nu}}))^{\zeta(p'-1)}\Big)^{\frac{1}{p'}} = C < \infty.$$

Hence, given $R_1 > 0$, (2.12) holds with $\widetilde{C}_P = \max\{C_p(p, \frac{1}{e}, R_1), C\}$. □

Remark. If g is unbounded,

$$\lim_{R\to 0} \int_0^1 (\phi'_R(x))^{-\frac{p'}{p}} dx = \infty.$$

Indeed, for all $R < 1$ we have

$$\int_0^1 (\phi'_R(x))^{-\frac{p'}{p}} dx \geqslant C\frac{\nu}{R^\nu}\int_0^R (S_1^{(\gamma)}(x))^{\zeta(p'-1)} x^{\nu-1} dx,$$

where $C = (1 + f_0 f_1 M\zeta\nu^{-1})^{1-p'}$. Now,

$$\lim_{R\to 0}\frac{\nu}{R^\nu}\int_0^R (S_1^{(\gamma)}(x))^{\zeta(p'-1)} x^{\nu-1} dx = \lim_{R\to 0}(S_1^{(\gamma)}(R))^{\zeta(p'-1)}$$
$$\geqslant \lim_{R\to 0} C_\gamma\,(f_1(R))^{\zeta(p'-1)} = \infty,$$

where $C_\gamma = \big(\frac{2^\gamma}{2^\gamma-1}\big)^{\zeta(p'-1)}$. Therefore, (2.12) is an estimate of the divergence of the integral as $R \to 0$.

In the subsequent sections, we will make repeated use of the following lemma.

Lemma 2.4. *For all $\delta > 0$ define $C^{(\delta)}$ as*

$$C^{(\delta)} = \frac{1}{1 + |\log(\delta)|}. \tag{2.13}$$

Let $f^{(\delta)}$ be related to the scaling function $f = (f_0, f_1)$ by $f^{(\delta)} = (f_0, C^{(\delta)} f_1)$. Let $\phi_R^{(\delta)}$ be obtained from ϕ_R according to (2.3) by replacing f with $f^{(\delta)}$. Then for every $a, b \in [0,1]$ and $\delta > 0$ the following inequalities hold:

$$\phi_R(\delta a) \leqslant \delta \phi_R^{(\delta)}(a), \tag{2.14}$$

$$\phi_R(a+b) \leqslant 2(\phi_R^{(2)}(a) + \phi_R^{(2)}(b)). \tag{2.15}$$

Proof. Let h be defined as in (2.3). Then for every positive δ the following inequality holds:

$$h(\delta x) \geqslant C^{(\delta)} h(x) \quad \forall\, x > 0 \tag{2.16}$$

with $C^{(\delta)}$ as defined above. The proof is straightforward and we omit the details.

We first prove (2.14). From (2.16) it follows that

$$f_R(\delta x) = f\Big(f_0, f_1 m(R); \frac{\delta x}{m(R)}\Big) \geqslant f\Big(f_0, f_1 C^{(\delta)} m(R); \frac{\delta x}{m(R)}\Big) = f_R^{(\delta)}(x).$$

Therefore,

$$S_R^{(\gamma)}(\delta x) \geqslant^{(\delta)} S_R^{(\gamma)}(r),$$

where ${}^{(\delta)}S_R^{(\gamma)}$ is obtained from $S_R^{(\gamma)}$ in (2.6) by replacing the scaling function f with $f^{(\delta)}$. The inequality (2.14) easily follows from the definition of ϕ_R.

We now prove that (2.14) implies (2.15). If $a \leqslant b$, as ϕ_R is nondecreasing, from (2.14) we get

$$\phi_R(a+b) \leqslant \phi_R(2b) \leqslant 2\phi_R^{(2)}(b).$$

Therefore, $\phi_R(a+b) \leqslant 2\max\{\phi_R^{(2)}(a), \phi_R^{(2)}(b)\} \leqslant 2(\phi_R^{(2)}(a) + \phi_R^{(2)}(b))$ □

3 Gauged Poincaré Inequalities

In recent years, Sobolev and Poincaré inequalities have been extensively used and investigated by several authors in a variety of abstract setting. Some of the ideas used in these extensions date back to the work of Maz'ya [16], Hedberg [11], more recently, Long-Nie [13]. A general presentation of Sobolev inequalities and their applications can be found, for example, in [27]. For recent developments in regard to Sobolev inequalities in metric spaces, we refer to the survey of Hajlasz–Koskela [9] and to Heinonen [12] (cf. also the recent collection [19] edited by V. Maz'ya, devoted to Sobolev type inequalities).

In this section, we begin by describing the general setting of gauged homogeneous spaces which was first introduced in [22]. We then go on to establish the equivalence between gauged Sobolev and gauged Poincaré inequalities in that setting. The results we obtain are an extension to those obtained in [14], where the gauge function g is constant. The general setting of a *gauged homogeneous space* X is defined by the following properties:

H1 X is a locally compact space endowed with a quasidistance d, whose balls form a basis of open neighborhoods in X. Here, a quasidistance d is a function on the set X that has all the usual properties of a metric with the exception that we only have the following weaker form of the triangle inequality:

$$d(x,y) \leqslant c_T(d(x,z) + d(z,y)) \tag{3.1}$$

for all $x, y, z \in X$, with $c_T \geqslant 1$.

H2 $\mu \in \mathcal{M}^+$ satisfies the following *gauged eccentric volume doubling condition* on all balls $B = B(z,R) \subset\subset X$: for all $x \in B(z, \sigma R)$ such that $B(x,r) \subset B(z,R) \subset\subset X$

$$\frac{\mu(B(z,R))}{\mu(B(x,r))} \leqslant C_E\Big(\frac{R}{r}\Big)^{\nu} f(r)f(R), \tag{3.2}$$

where C_E and ν are fixed positive constants, $0 < \sigma < 1$, $\mathcal{M}^+$ is the space of nonnegative radon measures on X, and $f = (f_0, f_1)$ is a scaling function associated with a given gauge function g as defined in (2.1).

We will equip our space X with a p-Lagrangean $\mathcal{L}^{(p)}$, $p \geqslant 1$, having the following properties:

L1 $\mathcal{L}^{(p)} : \mathcal{C} \longrightarrow \mathcal{M}^+$ is a map which associates to each function u from a given subalgebra $\mathcal{C}$ of $C(X)$ a measure $\mathcal{L}^{(p)}[u] \in \mathcal{M}^+$.

L2 For all $u \in \mathcal{C}$ and $g \in C^1(\mathbb{R})$, with g' bounded, $g(u) \in \mathcal{C}$ and $\mathcal{L}^{(p)}[g(u)] \leqslant \max |g'(u)|^p \mathcal{X}_{\{g' \neq 0\}} \mathcal{L}^{(p)}[u]$, where $\mathcal{X}_A$ represents the characteristic function of the set A.

Remark. 1. In the general definition of homogeneous spaces [4, 24], only a similar concentric volume doubling condition is assumed, but for simplicity we will assume the eccentric condition stated above since it is the main tool used in our proofs. If the gauged eccentric volume doubling condition stated above does not hold for all the precompact balls in our space X, our results will be restricted to the "nice" balls where it holds.

2. Property L2 is a form of *weak chain rule* and expresses the gradient like dependence of the measure $\mathcal{L}^{(p)}[u]$ on the potential u.

We now introduce a class of measures on X that we will refer to as measure weights. This class of measure weights generalizes the Eucledian so-called A_∞ *Muckenhoupt weights.*

Definition. A *measure weight* w in a ball B of the gauged quasimetric space X is a measure $w \in \mathcal{M}^+(X)$ satisfying for some constants $c_w \geqslant 1$ and $\beta \geqslant 0$, the following gauged volume growth condition:

$$\frac{w(B_r)}{w(B)} \leqslant c_w \frac{\mu(B_r)}{\mu(B)} \left(\frac{R}{r}\right)^\beta f(R) f(r) \tag{W}$$

for every ball $B_r = B(y, r) \subset B = B(z, R)$, $0 < r \leqslant R$, with the same scaling function f used in the gauged eccentric volume growth condition.

Note that the measure weights w are not required to have an L^1 density with respect to the underlying measure μ.

We now define a function Θ_R related to the gauge function g and the measure weights w as

$$\Theta_R = \Theta_R^{\nu,\alpha,\beta} = \phi_R^{\nu,\alpha-\beta}, \tag{3.3}$$

where ϕ_R is as defined in (2.10), and β is the exponent of a measure weight w.

In what follows, $Rd(B)$ denotes the radius of a ball B, i.e.,

$$Rd(B) = a \Rightarrow \quad \exists\, x \in X \text{ such that } B = B(x, a)$$

and

$$|B| := \mu(B), \quad u_B := |B|^{-1} \int_B u \, d\mu, \quad \tau B := B(x, \tau a) \ \forall\, \tau > 0.$$

Our main result in this section is the equivalence between gauged Sobolev inequalities and gauged first order Poincaré inequalities, as well as the equivalence between first order and higher order gauged Poincaré inequalities. These results are summarized in the following theorem.

Theorem 3.1. *Let* $1 \leqslant p < \nu/\alpha$, *let* $R_0 > 0$ *be given constants, and let* g *be a given gauge function that satisfies the subaddivity condition. Let* $(\mathcal{X}, \mu, \mathcal{L})$ *satisfy the assumptions* H1, H2, L1 *and* L2. *Let* p^* *and* p^*_β *(Sobolev exponent associated to the measure weight* w*) be the Sobolev exponents defined by*

$$p^* = \frac{p\nu}{\nu - \alpha p}, \quad p^*_\beta = p^* \frac{\nu - \alpha}{\nu - \alpha + \beta},$$

and let f *be scaling functions associated with the given* g *and suitable constants* f_0 *and* f_1. *Then the following conditions are equivalent.*

(1) *The following inequalities hold:*

$$\fint_B |u - u_B| d\mu \leqslant C_p R^\alpha f(R) \left(\frac{1}{|\tau B|} \int_{\tau B} d\mathcal{L}^{(p)}[u] \right)^{\frac{1}{p}} \tag{3.4}$$

for every $u \in \mathcal{C}$ and for every ball B with $0 < R = Rd(B) \leqslant R_0$, for suitable constants $\tau \geqslant 1$ and $C_p > 0$, independent of u and B.

(2) *The following inequalities hold:*

$$\Big(\int_0^\infty \Theta_R \Big(\frac{w(x \in \sigma B | \, |u - \eta| > t)}{w(B)} \Big) d(t^{p^*_\beta}) \Big)^{\frac{1}{p^*_\beta}} \leqslant C_S R^\alpha f(R) \Big(\frac{1}{|B|} \int_B d\mathcal{L}^{(p)}[u] \Big)^{\frac{1}{p}} \tag{3.5}$$

for every $u \in \mathcal{C}$ and for every ball B with $0 < R = Rd(B) \leqslant R_0$, where $0 < c_T \sigma \ \tau < 1$ and $C_S > 0$ are suitable constants independent of u and B. Here, w is any measure weight that satisfies (W) *with $0 \leqslant \beta < \alpha$ and $\eta = \eta(B, u)$ is a constant that depends both on the ball B and the function u.*

(3) *The following inequalities hold:*

$$\left(\fint_B |u - u_B|^q d\mu \right)^{\frac{1}{q}} \leqslant C_{p,q} R^\alpha f(R) \left(\frac{1}{|\tau B|} \int_{\tau B} d\mathcal{L}^{(p)}[u] \right)^{\frac{1}{p}} \tag{3.6}$$

for every $1 \leqslant q < p^*$, for every $u \in \mathcal{C}$, and for every ball B with $0 < R = Rd(B) \leqslant R_0$, where $\tau \geqslant 1$ and $C_{p,q} > 0$ are suitable constants independent of u and B.

This theorem is a consequence of Theorem 3.2 ((1) implies (2)) and Theorem 3.3 ((2) implies (3)) stated below. These two theorems provide also additional information about the mutual relation between the constants and the gauged functions occurring in the inequalities of Theorem 3.1.

Theorem 3.2. *Let w be a measure weight that satisfies (W) with exponent β, $0 \leqslant \beta < \alpha$, and let $R_0 > 0$ be a fixed constant. Suppose there exists $C_p > 0$, $\tau \geqslant 1$ and $1 \leqslant p < \frac{\nu}{\alpha}$, such that*

$$\fint_B |u - u_B| d\mu \leqslant C_p R^\alpha f(R) \left(\frac{1}{|\tau B|} \int_{\tau B} d\mathcal{L}^{(p)}[u] \right)^{\frac{1}{p}} \tag{3.7}$$

for all $u \in \mathcal{C}$ and for every ball B with $Rd(B) \leqslant R_0$. Then there exists $C_s > 0$ such that

$$\Big(\int_0^\infty \Theta_R \Big(\frac{w(x \in \sigma B | \, |u - \eta| > t)}{w(B)} \Big) d(t^{p^*_\beta}) \Big)^{\frac{1}{p^*_\beta}}$$

$$\leqslant C_s R^{\alpha} f(R)\Big(\frac{1}{|B|}\int_B d\mathcal{L}^{(p)}[u]\Big)^{\frac{1}{p}} \tag{3.8}$$

for all $u \in \mathcal{C}$ and for every ball B with $Rd(B) \leqslant R_0$. Here, σ is a fixed constant satisfying $0 < c_T\sigma\ \tau < 1$ and $\eta = \eta(B,u)$ is a constant that depends both on the ball B and the function u.

Proof. Let $Y = (Y,\mu)$ be a measure space. Let $1 \leqslant q < \infty$, $1 \leqslant a \leqslant b < \infty$ and Θ be a nondecreasing continuous function on $[0,\infty)$. We define the Lorentz type spaces $L^{q,a}_{\Theta} = L^{q,a}_{\Theta}(Y,\mu)$ as the set of all measurable functions ν on Y such that

$$\int_0^{\infty} t^q(\Theta(\mu(\{|\nu| > t\})))^{\frac{a}{q}}\frac{dt}{t} < \infty.$$

As in the classical case, we have the continuous inclusion

$$L^{q,a}_{\Theta} \subset L^{q,b}_{\Theta}, \quad 1 \leqslant a \leqslant b < \infty. \tag{3.9}$$

Also, we define $L^q_{\Theta} = L^{q,q}_{\Theta}$. When $\Theta(s) = s$, the preceding spaces coincide with the classic Lorentz spaces $L^{q,a}$, $L^q = L^{q,q}$. It is shown in [22, Theorem 3] that, in our setting and under the assumptions of Theorem 3.2, the following inequality holds:

$$\int_{-\infty}^{\infty} |t|^{p-1}\Big(\Theta_R\Big(\frac{w(x\in\sigma B|\ \frac{u-\eta}{t} > 1)}{w(B)}\Big)\Big)^{\frac{p}{p^*_\beta}}dt \leqslant CR^{\alpha p}f(R)\frac{1}{|B|}\int_B d\mathcal{L}^{(p)}[u] \tag{3.10}$$

for a fixed positive constant C, where $\eta = \eta(B,u)$ is a constant that depends both on the ball B and the function u. Note that Lemma 2.4 gives

$$\widetilde{\Theta}_R\Big(\frac{w(x\in\sigma B|\ |u-\eta| > t)}{w(B)}\Big) \leqslant 2\Big(\Theta_R\Big(\frac{w(x\in\sigma B|\ u-\eta > t)}{w(B)}\Big) + \Theta_R\Big(\frac{w(x\in\sigma B|\ u-\eta < -t)}{w(B)}\Big)\Big),$$

where $\widetilde{\Theta}_R$ is defined as in (3.3) with scaling function $\widetilde{f} = (f_0, \frac{f_1}{C(2)})$. Hence

$$\int_0^{\infty} t^{p-1}\left(\widetilde{\Theta}_R\left(\frac{w(x\in\sigma B|\ |u-\eta| > t)}{w(B)}\right)\right)^{\frac{p}{p^*_\beta}} dt$$

$$\leqslant 2\int_{-\infty}^{\infty} |t|^{p-1}\left(\Theta_R\left(\frac{w(x\in\sigma B|\ \frac{u-\eta}{t} > 1)}{w(B)}\right)\right)^{\frac{p}{p^*_\beta}} dt.$$

Combining this inequality with the inclusion (3.9) (where $q = p^*_\beta$, $a = p$, and $b = p^*_\beta$) and the inequality (3.10) gives

$$\int_0^\infty \widetilde{\Theta}_R\Big(\frac{w(x\in\sigma B|\ |u-\eta|>t)}{w(B)}\Big)d(t^{p^*_\beta})$$

$$\leqslant CR^{\alpha p^*_\beta}\Big(\frac{f(R)}{|B|}\int_B d\mathcal{L}^{(p)}[u]\Big)^{\frac{p^*_\beta}{p}} \tag{3.11}$$

for some new constant C. □

Theorem 3.3. *Suppose there exists $C_s > 0$ and $p \geqslant 1$, $1 \leqslant p < \nu/\alpha$ such that*

$$\Big(\int_0^\infty \phi_R\Big(\frac{\mu(\{x\in\sigma B|\ |u-\eta|>t\})}{\mu(B)}\Big)d(t^{p^*})\Big)^{\frac{1}{p^*}}$$

$$\leqslant C_s R^\alpha f(R)\Big(\frac{1}{|B|}\int_B d\mathcal{L}^{(p)}[u]\Big)^{\frac{1}{p}} \tag{3.12}$$

for all $u \in \mathcal{C}$ and for every ball B. Here, σ is a fixed constant satisfying $0 < c_T\sigma < 1$ and $\eta = \eta(B,u)$ is a constant that depends both on the ball B and the function u.

Then for every pair of positive constants R_0 and R_1 satisfying $0 < R_0 < R_1$, and for all $1 \leqslant q < p^*$, there exists a constant $C_q > 0$, $C_q = C_q(q,p^*,R_0,R_1)$, such that

$$\Big(\fint_B |u-u_B|^q d\mu\Big)^{\frac{1}{q}} \leqslant C_q R^\alpha f(R)\Big(\frac{1}{|\tau B|}\int_{\tau B} d\mathcal{L}^{(p)}[u]\Big)^{\frac{1}{p}} \tag{3.13}$$

for all $u \in \mathcal{C}$ and for every ball B such that $R_0 \leqslant Rd(B) \leqslant R_1$. Here, τ is any constant satisfying $\tau \geqslant \frac{1}{\sigma}$.

If, in addition, g satisfies the sub-additivity condition, then for all $u \in \mathcal{C}$ and for all balls B such that $0 < Rd(B) \leqslant R_1$, the inequality (3.13) holds with a new constant C_q depending only on q, p^*, and R_1, and with a new scaling function f' that is a power of the original scaling function f, $f'(r) = (f(r))^{C_2}$ with

$$C_2 = 1 + \frac{1}{q} + \frac{C_g}{q} + \frac{C_g\nu}{pq(\nu-\alpha)}.$$

Proof. Fix $u \in \mathcal{C}$ and a ball B that verifies $R_0 \leqslant R = Rd(B) \leqslant R_1$. Let $\widetilde{\sigma} = \frac{1}{\sigma}$ and define the function u^* as:

$$u^*(s) = \sup\Big\{t \geqslant 0: \ \frac{\mu(\{x\in B:\ |u-\eta|\geqslant t\})}{\mu(\widetilde{\sigma}B)} \geqslant s\Big\}.$$

Note that u^* is a nonincreasing function on $(0,1)$ that has the same distribution with respect to the Lebesgue measure as u with respect to the measure $\frac{\mu(\cdot)}{\mu(\widetilde{\sigma}B)}$. Set $b = \frac{p^*}{q}$ and define b' to be the conjugate exponent of b, i.e., $1/b + 1/b' = 1$. We have

$$\fint_B |u - u_B|^q d\mu \leqslant 2^q \fint_B |u - u_{\widetilde{\sigma}B}|^q d\mu = 2^q \frac{\mu(\widetilde{\sigma}B)}{\mu(B)} \fint_{\widetilde{\sigma}B} |u - u_{\widetilde{\sigma}B}|^q d\mu$$

$$\leqslant 2^{2q} \frac{\mu(\widetilde{\sigma}B)}{\mu(B)} \fint_{\widetilde{\sigma}B} |u - \eta|^q d\mu = 2^{2q} \frac{\mu(\widetilde{\sigma}B)}{\mu(B)} \int_0^1 u^*(s)^q ds$$

$$= 2^{2q} \frac{\mu(\widetilde{\sigma}B)}{\mu(B)} \int_0^1 u^*(s)^q (\phi_R'(s))^{\frac{1}{b}} (\phi_R'(s))^{-\frac{1}{b}} ds$$

$$\leqslant 2^{2q} \frac{\mu(\widetilde{\sigma}B)}{\mu(B)} \Big(\int_0^1 (\phi_R'(s))^{-\frac{b'}{b}} ds \Big)^{\frac{1}{b'}} \Big(\int_0^1 u^*(s)^{p^*} \phi_R'(s) ds \Big)^{\frac{q}{p^*}}. \tag{3.14}$$

Note that we used the Hölder inequality ($b > 1$). We then obtain

$$\fint_B |u - u_B|^q d\mu$$

$$\leqslant 4^q \frac{\mu(\widetilde{\sigma}B)}{\mu(B)} \Big(\int_0^1 (\phi_R'(s))^{-\frac{b'}{b}} ds \Big)^{\frac{1}{b'}} \Big(\int_0^\infty \phi_R \Big(\frac{\mu(x \in B|\, |u - \eta| > t)}{\mu(\widetilde{\sigma}B)} \Big) d(t^{p^*}) \Big)^{\frac{q}{p^*}}$$

$$\leqslant C_1 \Big(\int_0^\infty \phi_R \Big(\frac{\mu(x \in B|\, |u - \eta| > t)}{\mu(\widetilde{\sigma}B)} \Big) d(t^{p^*}) \Big)^{\frac{q}{p^*}},$$

where $C_1 = 4^q C_E \widetilde{\sigma}^\nu C_0 C_b$ with $C_b = C_b(b, R_0, R_1)$ defined as in (2.11) and $C_0 = \max_{R_0 \leqslant R \leqslant R_1} f(\widetilde{\sigma}R) f(R)$. Combining this inequality with (3.12) gives

$$\Big(\fint_B |u - u_B|^q d\mu \Big)^{\frac{1}{q}} \leqslant C_q R^\alpha f(R) \Big(\frac{1}{|\tau B|} \int_{\tau B} d\mathcal{L}^{(p)}[u] \Big)^{\frac{1}{p}}$$

with $C_q = C_s C_1^{\frac{1}{q}}$ and $\tau = \widetilde{\sigma}$.

If, in addition, g satisfies the sub-additivity condition, then Theorem 2.3 gives

$$\frac{\mu(\widetilde{\sigma}B)}{\mu(B)} \Big(\int_0^1 (\phi_R'(s))^{-\frac{b'}{b}} ds \Big)^{\frac{1}{b'}} \leqslant C_E \widetilde{\sigma}^\nu f(\widetilde{\sigma}R) f(R) \widetilde{C}_b \widetilde{f}(R)$$

$$\leqslant C_E \widetilde{\sigma}^\nu \widetilde{f}(\widetilde{\sigma})^{\frac{p}{\zeta}} \widetilde{f}(R)^{\frac{p}{\zeta}} f(R) \widetilde{C}_b \widetilde{f}(R). \tag{3.15}$$

Therefore,

$$\Big(\fint_B |u - u_B|^q d\mu \Big)^{\frac{1}{q}} \leqslant C_q' R^\alpha (\widetilde{f}(R)^{\frac{p}{\zeta}} f(R) \widetilde{f}(R))^{\frac{1}{q}} f(R) \Big(\frac{1}{|\tau B|} \int_{\tau B} d\mathcal{L}^{(p)}[u] \Big)^{\frac{1}{p}}$$

$$= C_q' R^\alpha f'(R)\Big(\frac{1}{|\tau B|}\int_{\tau B} d\mathcal{L}^{(p)}[u]\Big)^{\frac{1}{p}},$$

where $C_q' = 4C_s(C_E\widetilde{\sigma}^\nu C_b)^{\frac{1}{q}}$ and $f = (C_2 f_0, f_1)$, $C_2 = 1 + \dfrac{1}{q} + \dfrac{C_g}{q} + \dfrac{C_g\zeta}{pq}$, with C_b and ζ as defined in Theorem 2.3. □

4 Gauged Capacitary Inequalities

In this section, we establish the equivalence between gauged Sobolev inequalities and certain *gauged capacitary inequalities*. These results are inspired by the seminal work of Vladimir Maz'ya in this field, in particular to the papers [17, 18]. The results are also influenced by the work of Fukushima and Uemura [6]. While the inequalities in the previous papers are global, the inequalities we obtain in this section are scaled inequalities on possible fluctuating quasimetric ball as in the setting of the previous section. The setting in this section is the same as in Section 3. Moreover we assume that, in addition to conditions L1 and L2, our p-Lagrangean satisfies the following condition:

L3 For every compact set K and every open set A such that $K \subset A$ there exists $\alpha \in \mathcal{C}_A$ such that $\alpha|_K \equiv 1$. Here, $\mathcal{C}_A = \{u \in \mathcal{C}$ such that $u|_{A^c} \equiv 0\}$ for any open set A in X, where A^c denotes the complement of A in X.

We consider the capacity function associated to the p-Lagrangean $\mathcal{L}^{(p)}$ which is defined in the following way:

Definition. For arbitrary compact sets K and open sets A such that $K \subset A$ we define the *capacity* of K relative to A as follows:

$$\mathrm{cap}_p(K, A) = \inf_{u \in \mathcal{A}_{K,A}} \int_X d\mathcal{L}^{(p)}[u],$$

where $\mathcal{A}_{K,A} = \{$nonnegative $u \in \mathcal{C}$ such that $u \geqslant 1$ on K and $u \equiv 0$ on $A^c\}$.

We now proceed to establish an equivalence between Sobolev and capacitary inequalities. As already mentioned, simplified gloval versions of these inequalities were obtained by Maz'ya [17, 18] and by Fukushima and Uemura [6] in different settings. Our proof is an adaptation of theirs to our setting.

Theorem 4.1. *Let w be a measure weight with exponent β satisfying $0 \leqslant \beta < 1$, and let $p^* = \dfrac{p\nu}{\nu - p}\dfrac{\nu - 1}{\nu - 1 + \beta}$ be the associated Sobolev exponent. The following conditions are equivalent on $(X, \mathcal{L})$.*

(1) *There exists $C_s > 0$ such that*

$$\Big(\int_0^\infty \Theta_R\Big(\frac{w(x\in\sigma B|\ |u|>t)}{w(B)}\Big)d(t^{p^*})\Big)^{\frac{1}{p^*}}$$

$$\leqslant C_s R f(R)\Big(\frac{1}{|B|}\int_B d\mathcal{L}^{(p)}[u]\Big)^{\frac{1}{p}} \tag{4.1}$$

for all $u\in\mathcal{C}_B$ *and for every ball* B *such that* $0<R=Rd(B)\leqslant R_0$. *Here,* σ *is a fixed constant satisfying* $0<c_T\sigma\ \tau<1$.

(2) *There exists a constant* $C_c>0$ *such that*

$$\Theta_R\left(\frac{w(K)}{w(B)}\right)^{\frac{1}{p^*}}\leqslant C_c R f(R)\left(\frac{\operatorname{cap}_p(K,\sigma B)}{\mu(B)}\right)^{\frac{1}{p}} \tag{4.2}$$

for every ball B *and for all* $K\subset\sigma B$ *such that* K *is compact.*

Proof. (1) $\Rightarrow$ (2) For $u\in\mathcal{A}_{K,\sigma B}$, with $K\subset\sigma B$ we have

$$\Theta_R\left(\frac{w(K)}{w(B)}\right)^{\frac{1}{p^*}}\leqslant\left(\int_0^1\Theta_R\left(\frac{w(x\in\sigma B|\ |u|>t)}{w(B)}\right)d(t^{p^*})\right)^{\frac{1}{p^*}}$$

$$\leqslant C_s R f(R)\left(\frac{1}{|B|}\int_B d\mathcal{L}^{(p)}[u]\right)^{\frac{1}{p}}.$$

We obtain the implication by minimizing the right-hand side over all admissible u.

(2) $\Rightarrow$ (1) Let $u\in\mathcal{C}_{\sigma B}$ and for each $j\in\mathbb{Z}$, let $M_j^+:=\{x\in B|u(x)\geqslant 2^j\}$, $M_j^-:=\{x\in B|-u(x)\geqslant 2^j\}$. Define accordingly the functions u_j^+, u_j^- as follows:

$$u_j^+=\phi\left(\frac{u}{2^j}\right),\quad u_j^-=\phi\left(\frac{-u}{2^j}\right).$$

Here, ϕ is a C^1 function on $\mathbb{R}$ such that $\phi=0$ on $[-\infty,1/2]$, $\phi=1$ on $[1,\infty)$, and the inequalities $0<\phi<1$, $0\leqslant\phi'\leqslant 4$ hold on $(1/2,1)$. It can be easily shown that $u_j^+\in\mathcal{A}_{M_j^+,\sigma B}$ and $u_j^-\in\mathcal{A}_{M_j^-,\sigma B}$. In addition, $u_j^+=0$ on $\left(M_{j-1}^+\right)^c$ and $u_j^-=0$ on $\left(M_{j-1}^-\right)^c$. Combining all these observations, we obtain

$$\Big(\int_0^\infty\Theta_R\left(\frac{w(x\in\sigma B|\ |u|>t)}{w(B)}\right)d(t^{p^*})\Big)^{\frac{1}{p^*}}$$

$$\leqslant\left((2^{p^*}-1)\sum_{j=-\infty}^{j=\infty}2^{p^*j}\Theta_R\left(\frac{w(x\in\sigma B|\ |u|>2^j)}{w(B)}\right)\right)^{\frac{1}{p^*}}$$

$$\leqslant\left((2^{p^*}-1)\sum_{j=-\infty}^{j=\infty}2^{p^*j}\Theta_R\left(\frac{w(M_j^+)}{w(B)}+\frac{w(M_j^-)}{w(B)}\right)\right)^{\frac{1}{p^*}}$$

$$\leqslant \Big(2(2^{p^*}-1)\sum_{j=-\infty}^{j=\infty} 2^{p^*j}\Big(\Theta_R^{(2)}\left(\frac{w(M_j^+)}{w(B)}\right)+\Theta_R^{(2)}\Big(\Big(\frac{w(M_j^-)}{w(B)}\Big)\Big)\Big)^{\frac{1}{p^*}}$$

$$\leqslant CRf^{'}(R)\mu(B)^{-\frac{1}{p}}\Big(\sum_{j=-\infty}^{j=\infty} 2^{p^*j}\left(\mathrm{cap}_{p,\sigma B}(M_j^+)^{\frac{p^*}{p}}+\mathrm{cap}_{p,\sigma B}(M_j^-)^{\frac{p^*}{p}}\right)\Big)^{\frac{1}{p^*}},$$

where $C = C_s(2(2^{p^*}-1))^{\frac{1}{p^*}}$ and $\mathrm{cap}_{p,B}(\cdot) = \mathrm{cap}_p(\cdot, B)$. Now, since $u_j^+ \in \mathcal{A}_{M_j^+,\sigma B}$ and $u_j^- \in \mathcal{A}_{M_j^-,\sigma B}$, we have

$$\Big(\int_0^\infty \Theta_R\left(\frac{w(x\in\sigma B|\ |u|>t)}{w(B)}\right)d(t^{p^*})\Big)^{\frac{1}{p^*}}$$
$$\leqslant CRf'(R)\mu(B)^{-\frac{1}{p}}\Big(\sum_{j=-\infty}^{j=\infty} 2^{p^*j}\,(\mathbf{I}+\mathbf{II})\Big)^{\frac{1}{p^*}},$$

where

$$\mathbf{I} = \left(\int_{M_{j-1}^+\setminus M_j^+} d\mathcal{L}^{(p)}[u_j^+]\right)^{\frac{p^*}{p}}$$

and

$$\mathbf{II} = \Big(\int_{M_{j-1}^-\setminus M_j^-} d\mathcal{L}[u_j^-]^p\Big)^{\frac{p^*}{p}}$$

$$\leqslant CRf'(R)\mu(B)^{-\frac{1}{p}}\Big(\sum_{j=-\infty}^{j=\infty}\Big(\Big(\int_{M_{j-1}^+\setminus M_j^+} d\mathcal{L}^{(p)}[u]\Big)^{\frac{p^*}{p}}$$
$$+\Big(\int_{M_{j-1}^-\setminus M_j^-} d\mathcal{L}^{(p)}[u]\Big)^{\frac{p^*}{p}}\Big)\Big)^{\frac{1}{p^*}}.$$

Using the fact that $p^*/p \geqslant 1$, we get

$$\Big(\int_0^\infty \Theta_R\Big(\frac{w(x\in\sigma B|\ |u|>t)}{w(B)}\Big)d(t^{p^*})\Big)^{\frac{1}{p^*}}$$

$$\leqslant CRf'(R)\mu(B)^{-\frac{1}{p}}\Big(\sum_{j=-\infty}^{j=\infty}\Big(\int_{M_{j-1}^+\setminus M_j^+} d\mathcal{L}^{(p)}[u]+\int_{M_{j-1}^-\setminus M_j^-} d\mathcal{L}^{(p)}[u]\Big)\Big)^{\frac{1}{p}}$$

$$= CRf'(R)\mu(B)^{-\frac{1}{p}}\Big(\int_B d\mathcal{L}^{(p)}[u]\Big)^{\frac{1}{p}}.$$

The proof is complete. □

5 Fluctuating Domains

To motivate our theory, we construct in this section a simple example of a gauged homogeneous space with a nonconstant gauge function of logarithmic type. The example is obtained by constructing in the plane a family of fluctuating curves of von Koch type. Our example is taken from [22]. A similar construction with curves of Sierpinski type is given in [21].

Let α be a fixed constant, $2 < \alpha \leqslant 4$. Consider the planar points $A = (0,0)$, $B = (\frac{1}{\alpha}, 0)$, $C = (\frac{1}{2}, \sqrt{\frac{1}{\alpha} - \frac{1}{4}})$, $D = (1 - \frac{1}{\alpha}, 0)$, $F = (1,0)$ and the segments $I = [A,F]$, $I_1 = [A,B]$, $I_2 = [B,C]$, $I_3 = [C,D]$ and $I_4 = [D,F]$. Here, for $x,y \in \mathbb{R}^2$, $[x,y] := \{z \in \mathbb{R}^2 | z = \lambda x + (1-\lambda)y,\ 0 \leqslant \lambda \leqslant 1\}$. Note that the segments I_1, I_2, I_3, and I_4 all have length $1/\alpha$. We refer to the constant α as the contraction factor. Let $F_i^{(\alpha)}$ be the similitude that carries the interval I into the segments I_i and define the set-to-set function $F^{(\alpha)}$ as follows:

$$F^{(\alpha)}(E) = \bigcup_{i=1}^{4} F_i^{(\alpha)}(E) \quad \text{for } E \in \mathbb{R}^2. \tag{5.1}$$

It can be shown that there is a unique compact set K that is $F^{(\alpha)}$ invariant, i.e., $F^{(\alpha)}(K) = K$. We now construct a family of Koch curves $K^{(\xi)}$ indexed by the binary sequence $\xi \in \{0,1\}^{\mathbb{N}}$. Let us first fix two values of the contraction factor α, namely $2 < \alpha_0 \leqslant \alpha_1 \leqslant 4$. For each value of α_j, $j \in \{0,1\}$ we denote by $F_i^{(j)} := F_i^{(\alpha_j)}$ the corresponding similitudes and by $F^{(j)}$ the corresponding set-to-set functions. For $\xi = (\xi_1, \xi_2, \xi_3, \ldots) \in \{0,1\}^{\mathbb{N}}$ we set $F_n^{(\xi)} := F^{(\xi_1)} \circ \ldots \circ F^{(\xi_n)}$ and V_n by $V_n^{(\xi)} := F_n^{(\xi)}(V_0)$ where $V_0 = \{(0,0),(1,0)\}$ is the "boundary" of the Koch curve. The Koch curve associated to the binary sequence ξ, $K^{(\xi)}$ is defined by

$$K^{(\xi)} = cl\Big(\bigcup_{n=0}^{\infty} V_n^{(\xi)}\Big), \tag{5.2}$$

where $cl(A)$ denotes the closure of a set A. We approximate the Koch curve $K^{(\xi)}$ by a sequence of graphs Γ_n with vertices V_n and edge relation $p \sim_n q$ defined by

$$p \sim_n q \quad \text{if } \{p,q\} = F_{i_1}^{(\xi_1)} \circ F_{i_2}^{(\xi_2)} \circ \ldots \circ F_{i_n}^{(\xi_n)} \tag{5.3}$$

for some finite sequence $(i_1, i_2, \ldots, i_n)$, where $i_j \in \{1,2,3,4\}$, $j = 1,2,\ldots,n$. Note that the sequence of sets $V_n^{(\xi)}$ is monotonically increasing. We now define a measure $\mu^{(\xi)}$ and Lagrangean $\mathcal{L}^{(\xi)}$ associated to the binary sequence ξ as follows: for every $\phi \in C(K^{(\xi)})$ we set

$$\int_{K^{(\xi)}} \phi \, d\mu^{(\xi)} = \lim_{n\to\infty} \frac{4^{-n}}{2} \sum_{p \in V_n^{(\xi)}} \phi(p) \tag{5.4}$$

and

$$\int_{K^{(\xi)}} \phi d\mathcal{L}^{(\xi)}(u,v) = \lim_{n\to\infty} \frac{4^n}{2} \sum_{p\in V_n^{(\xi)}} \phi(p) \sum_{p\sim_n q} (u(p)-u(q))(v(p)-v(q)), \quad (5.5)$$

where $u, v \in \mathcal{D}_{\mathcal{L}^{(\xi)}} := \left\{ u \in C(K^{(\xi)}) \middle| \int_{K^{(\xi)}} d\mathcal{L}^{(\xi)}(u,u) < +\infty \right\}$. Note that in the notation of this section, the index ξ in the Lagrangean denotes its dependence on the sequence ξ, not "summability" index p of the Lagrangean as in Sections 3 and 4. Indeed, all Lagrangeans $mathcalL^{(\xi)}$ of this section are 2-Lagrangeans of quadratic type, as defined in Section 3.

For each ξ we consider the frequency by which each contraction $F^{(j)}$ occurs in the finite sequence $\xi|n := (\xi_1, \xi_2, \dots, \xi_n)$, i.e.,

$$h_j^{(\xi)}(n) = \frac{1}{n} \sum_{k=1}^{n} \mathcal{I}_{\{\xi_k = j\}} \quad (5.6)$$

and set $p_j = \lim_{n\to\infty} h_j^{(\xi)}(n)$. Note that $0 \leqslant p_0, p_1 \leqslant 1$ and $p_0 + p_1 = 1$. The gauge function g as defined in Section 2 will determine the rate at which $h_j^{(\xi)}$ converges to p_j:

$$|h_j^{(\xi)} - p_j| \leqslant \frac{g(n)}{n} \quad \forall\, n \geqslant 1. \quad (5.7)$$

We now define a quasidistance $d^{(\xi)}$ on $K^{(\xi)}$. For $x, y \in K^{(\xi)}$ let

$$d^{(\xi)}(x,y) = \|x-y\|^{\delta^{(\xi)}}, \quad (5.8)$$

where $\| \,.\, \|$ denotes the Euclidean norm in the plane and

$$\delta^{(\xi)} = \frac{\log 4}{p_0 \log \alpha_0 + p_1 \log \alpha_1}.$$

It can be shown [21, Theorems 4 and 5] that there exists a constant $c \geqslant 1$, depending only on the contraction factors α_0 and α_1, such that for every ξ and for all $0 < r \leqslant R$ the following gauged volume growth condition and Poincaré inequality hold:

$$\frac{1}{c} \frac{r}{f_R(r)} \leqslant \mu^{(\xi)}\left(B_r^{(\xi)}(x)\right) \leqslant r f_R(r), \quad (5.9)$$

$$\int_{B_r^{(\xi)}(x)} |u - \overline{u}_{B_r^{(\xi)}(x)}|^2 d\mu(\xi) \leqslant r^2 f_R(r) \int_{B_R^{(\xi)}(x)} d\mathcal{L}^{(\xi)}(u,u). \quad (5.10)$$

We can extend our results from *arbitrary* fixed ξ to random $\xi = (\xi_1, \xi_2, \dots)$, where ξ_i are i.i.d. random variables on a probability space $(\Omega, \mathcal{F}, \mathcal{P})$ with distribution $p_0 > 0$ amd $p_1 > 0$. In this latter case, by the Kintchine law of iterated logarithm (cf., for example, [1]), the convergence in (5.7) is given by $g(n) = c(\xi)(n \log\log n)^{\frac{1}{2}}$ with $\mathcal{P}(\{c(\xi) < \infty\}) = 1$. Similar results hold for the random Sierpinski geometries as shown in [21, 20]. For every ξ, $K^{(\xi)} = (K^{(\xi)}, \mu^{(\xi)}, \mathcal{L})^{(\xi)}$ is a gauged homogeneous space equipped with a 2-Lagrangean. Note that in this context $\nu = 1$, $\alpha = 1$ and $p = 2$. By relying on the method of product Lagrangean first introduced by Mosco and Vivaldi [23] and then developed by Strichartz [25] and Tintarev [26], we can construct examples of gauged spaces with dimension $\nu \geqslant 2$, which produces a setting in which our theorems can be applied.

Random fractal constructions of the type described in this section are well known in fractal theory. They were first considered by Hambly [10] and Barlow–Hambly [2]. Further examples of similar constructions can be found in [5, 7, 15].

Acknowledgment. The article is based upon work supported by the National Science Foundation under Grant 0807840.

References

1. Bauer, H.: Measure and Integration Theory. Walter de Gruyter, Berlin etc. (2001)
2. Barlow, M.T., Hambly, B.M.: Transition density estimates for Brownian motion on scale irregular Sierpinski gaskets. Ann. l'Institut Henri Poincaré **33**, 531-557 (1997)
3. Biroli, M., Mosco, U.: Sobolev inequalities on homogeneous spaces. Potential Anal. **4**, no. 4, 311-324 (1995)
4. Coifman, R.R., Weiss, G.: Analyse Harmonique Non Commutative Sur Certains Espaces Homogenes. Lect. Notes Math. **242**, Springer, Berlin etc. (1971)
5. Falconer, K.J.: Random graphs. Math. Proc. Cambridge Phil. Soc. **100**, 559-582 (1986)
6. Fukushima, M., Uemura, T.: On Sobolev and capacitary inequalities for contractive Besov spaces over d-sets. Potential Anal. **18**, no. 1, 59-77 (2003)
7. Graf, S., Mauldin, R.D., Williams, S.C., Hambly, B.M.: The exact Hausdorff dimension in random recursive constructions. Mem. Am. Math. Soc. **381**, 126 (1988)
8. Hajlasz, P.: Sobolev spaces on an arbitrary metric space. Potential Anal. **5**, 403-415 (1996)
9. Hajlasz, P. Koskela, P.: Sobolev met Poincaré. Mem. Am. Math. Soc. **145**, 688 (2000)
10. Hambly, B.M.: Brownian motion on a homogeneous random fractal. Probab. Theory Related Fields **94**, 1–38 (1992)
11. Hedberg, L.: On certain convolution inequalities. Proc. Am. Math. Soc. **36**, 505-510 (1972)
12. Heinonen, J.: Lecture on Analysis on Metric Spaces. Springer, Berlin etc. (2000)

13. Long, R., Nie, F.: Weighted Sobolev inequality and eigenvalue estimates of Schrödinger operators. Lect. Notes Math. **1494**, pp. 131-141. Springer, Berlin etc. (1991)
14. Malý, J., Mosco, U.: Remarks on measure-valued Lagrangeans on homogeneous spaces. Res. Math. Napoli **48** (suppl.), 217–231 (1999)
15. Mauldin, R.D. Williams, S.C.: Hausdorff dimension in graph directed constructions. Trans. Am. Math. Soc. **309**, 811-829 (1988)
16. Maz'ya, V.G.: On the theory of the higher-dimensional Schrödinger operator (Russian). Izv. Akad. Nauk SSSR. Ser. Mat. **28**, 1145–1172 (1964)
17. Maz'ya, V.G.: Conductor and capacitary inequalities for functions on topological spaces and their applications to Sobolev type imbeddings. J. Funct. Anal. **224**, 408–430 (2005)
18. Maz'ya, V.: Conductor inequalities and criteria for Sobolev type two-weight imbeddings. J. Comput. Appl. Math. **114**, no. 1, 94–114 (2006)
19. Maz'ya, V. (ed.), Sobolev Spaces in Mathematics. I: Sobolev Type Inequalities. Springer, New York; Tamara Rozhkovskaya Publisher, Novosibirsk. International Mathematical Series **8** (2009)
20. Mosco, U.: Irregular Similarity and Quasi-Metric scaling. In: A. De la Pradelle and D. Feyel (eds) Conf. "Stochastic and Potential Theory," Saint-Priest-de-Gimel, France, 1-6. September. Lecture Notes of the Conference.
21. Mosco, U.: Harnack inequalities on scale irregular Sierpinski gaskets. In: Birman, M. Sh. et al (eds.), Nonlinear Problems in Mathematical Physics and Related Topics. II: in Honor of Professor O.A. Ladyzenskaya. Springer, New York. International Mathematical Series **2**, 305-328 (2003)
22. Mosco, U.: Gauged Sobolev inequalities. Appl. Anal. **86**, no. 3, 367–402 (2007)
23. Mosco, U., Vivaldi, M.A.: Variational problems with fractal layers. Mem. Mat. Appl. Rend. Acc. XL, 1210, XXVII Fasc 1, 237–251 (2003)
24. Stein, E.M.: Harmonic Analysis. Princeton Univ. Press, Princeton, NJ (1995)
25. Strichartz, R.S.: Analysis on products of fractals. Trans. Am. Math. Soc. **357**, 571–615 (2005)
26. Tintarev, K.: Permanence of metric fractals. Electron. J. Differ. Equ. **16**, 185–192 (2006)
27. Ziemer, W.: Weakly Differentiable Functions: Sobolev Spaces and Functions of Bounded Variations. Springer, Berlin etc. (1989)

A Converse to the Maz'ya Inequality for Capacities under Curvature Lower Bound

Emanuel Milman

Dedicated to V.G. Maz'ya
with great respect and admiration

Abstract We survey some classical inequalities due to Maz'ya relating isocapacitary inequalities with their functional and isoperimetric counterparts in a measure-metric space setting, and extend Maz'ya's lower bound for the q-capacity ($q > 1$) in terms of the 1-capacity (or isoperimetric) profile. We then proceed to describe results by Buser, Bakry, Ledoux and most recently by the author, which show that under suitable convexity assumptions on the measure-metric space, the Maz'ya inequality for capacities may be reversed, up to dimension independent numerical constants: a matching lower bound on 1-capacity may be derived in terms of the q-capacity profile. We extend these results to handle arbitrary $q > 1$ and weak semiconvexity assumptions, by obtaining some new delicate semigroup estimates.

1 Introduction

The notion of capacity, first systematically introduced by Frechet, has played a fundamental role in the theory developed by V.G. Maz'ya in the 1960's for the study of functional inequalities and embedding theorems, and has continued to play an important role in the development of the theory ever since (cf. [27] for an extended overview).

Before recalling the definition, let us first describe our setup.

Emanuel Milman
School of Mathematics, Institute for Advanced Study, Einstein Drive, Simonyi Hall, Princeton, NJ 08540, USA
e-mail: emilman@math.ias.edu

A. Laptev (ed.), *Around the Research of Vladimir Maz'ya I: Function Spaces*, International Mathematical Series 11, DOI 10.1007/978-1-4419-1341-8_15,

We denote by (Ω, d) a separable metric space and by μ a Borel *probability* measure on (Ω, d) which is not a unit mass at a point. Let $\mathcal{F} = \mathcal{F}(\Omega, d)$ denote the space of functions which are Lipschitz on every ball in (Ω, d); we call such functions "Lipschitz-on-balls." Given $f \in \mathcal{F}$, we denote by $|\nabla f|$ the following Borel function:

$$|\nabla f|(x) := \limsup_{d(y,x)\to 0+} \frac{|f(y) - f(x)|}{d(x,y)}$$

(and we define it as 0 if x is an isolated point; cf. [9, pp. 184,189] for more details). Although it is not essential for the ensuing discussion, it is more convenient to think of Ω as a complete smooth oriented n-dimensional Riemannian manifold (M, g) and of d as the induced geodesic distance, in which case $|\nabla f|$ coincides with the usual Riemannian length of the gradient.

Definition. Given two Borel sets $A \subset B \subset (\Omega, d)$ and $1 \leqslant q < \infty$, the *q-capacity* of A relative to B is defined as

$$\mathrm{Cap}_q(A, B) := \inf \left\{ \| |\nabla \Phi| \|_{L_q(\mu)} ; \Phi|_A \equiv 1 \, , \, \Phi|_{\Omega \setminus B} \equiv 0 \right\},$$

where the infimum is on all $\Phi : \Omega \to [0,1]$ which are Lipschitz-on-balls.

We remark that it is possible to give an even more general definition than the one above (cf. the monograph by Maz'ya [27]). Note that, in the case of a compact manifold (M, g) and the Riemannian (normalized) volume μ, $\mathrm{Cap}_2(A, B)^2$ coincides (up to constants) with the usual Newtonian capacity of a compact set A relative to the outer open set B. Following [3], we will only be interested in this work in the *q-capacity profile*:

Definition. Given a metric probability space (Ω, d, μ), $1 \leqslant q < \infty$, its *q-capacity profile* is defined for any $0 < a \leqslant b < 1$ as

$$\begin{aligned} \mathrm{Cap}_q(a, b) &:= \inf \left\{ \mathrm{Cap}_q(A, B) \, ; \, A \subset B \, , \, \mu(A) \geqslant a \, , \, \mu(B) \leqslant b \right\} \\ &= \inf \left\{ \| |\nabla \Phi| \|_{L_q(\mu)} \, ; \, \mu\{\Phi = 1\} \geqslant a \, , \, \mu\{\Phi = 0\} \geqslant 1 - b \right\}, \end{aligned}$$

where the latter infimum is on all $\Phi : \Omega \to [0,1]$ which are Lipschitz-on-balls.

The intimate relation between 1-capacity and the isoperimetric properties of a space was noticed by Fleming [18] in the Euclidean setting, using the co-area formula of Federer [16] (cf. also [17]) and generalized by Maz'ya [23]. An analogous relation between isoperimetric inequalities and functional inequalities involving the term $\| |\nabla f| \|_{L_1(\mu)}$ was discovered by Maz'ya [23] and independently by Federer and Fleming [17], leading, in particular, to the determination of the optimal constant in the Gagliardo inequality in Euclidean space $(\mathbb{R}^n, |\cdot|)$. Maz'ya continued to study these relations when 1 above is replaced by a general $q > 1$ [23, 24, 25, 27]: he showed how to pass from any lower bound on 1-capacity to an optimal lower bound on q-capacity, and demonstrated the equivalence between q-capacitary inequalities and functional inequalities involving the term $\||\nabla f|\|_{L_q(\mu)}$. Combining all these ingredients, Maz'ya discovered a way to pass from isoperimetric information to optimal (up to constants) functional inequalities. Especially useful is the case $q = 2$ since this corresponds to spectral information on the Laplacian and the Schrödinger operators, leading to many classical characterizations [25, 26, 27].

We define all of the above notions in Section 2 and sketch the proofs of their various relations in our metric probability space setting in Section 3. Moreover, we extend the transition from 1-capacity to q-capacity ($q > 1$) to handle arbitrary transition between p and q, when $p < q$, following [28].

It is easy to check that in general, one cannot deduce back information on p-capacity from q-capacity, when $p < q$. In other words, the above transition is only one-directional and cannot, in general, be reversed (cf. Subsection 2.3). We therefore need to add some additional assumptions in order to have any chance of obtaining a reverse implication. As we will see below, some type of convexity assumptions are a natural candidate. We start with two important examples when $(M, g) = (\mathbb{R}^n, |\cdot|)$ and $|\cdot|$ is some fixed Euclidean norm:

- Ω is an *arbitrary* bounded convex domain in $\mathbb{R}^n$ ($n \geqslant 2$), and μ is the uniform probability measure on Ω.
- $\Omega = \mathbb{R}^n$ ($n \geqslant 1$) and μ is an *arbitrary* absolutely continuous log-concave probability measure, meaning that $d\mu = \exp(-\psi)dx$, where $\psi : \mathbb{R}^n \to \mathbb{R} \cup \{+\infty\}$ is convex (we refer to [12] for more information).

In both cases, we say that "our convexity assumptions are fulfilled." More generally, we use the following definition from [30].

Definition. We say that our *smooth $\varkappa$-semiconvexity assumptions* are fulfilled if

- (M, g) denotes an n-dimensional ($n \geqslant 2$) oriented smooth complete connected Riemannian manifold or $(M, g) = (\mathbb{R}, |\cdot|)$, and $\Omega = M$,
- d denotes the induced geodesic distance on (M, g),
- $d\mu = \exp(-\psi)dvol_M$, $\psi \in C^2(M)$, and as tensor fields on M

$$\mathrm{Ric}_g + \mathrm{Hess}_g \psi \geqslant -\varkappa g \; . \tag{1.1}$$

We say that our *$\varkappa$-semiconvexity assumptions* are fulfilled if μ can be approximated in total-variation by measures $\{\mu_m\}$ so that each (Ω, d, μ_m) satisfies our smooth $\varkappa$-semiconvexity assumptions.

When $\varkappa = 0$, we say that our (*smooth*) *convexity assumptions* are satisfied.

The condition (1.1) is the well-known curvature–dimension condition $CD(-\varkappa, \infty)$ introduced by Bakry and Émery in their celebrated paper [1] (in the more abstract framework of diffusion generators). Here, Ric_g denotes the Ricci curvature tensor and Hess_g denotes the second covariant derivative.

Our main result from [29], as extended in [28], is that under our convexity assumptions ($\varkappa = 0$ case), the above transition can in fact be reversed, up to *dimension independent* constants. This can be formulated in terms of passing from q-capacity to p-capacity ($1 \leqslant p < q$), or equivalently, as passing from functional inequalities involving the term $\||\nabla f|\|_{L_q(\mu)}$ to isoperimetric inequalities.

In this work, we extend our previous results to handle the more general $\varkappa$-semiconvexity assumptions. For the case $q = 2$, this was previously done by Buser [13] in the case of a uniform density on a manifold with Ricci curvature bounded from below, and extended to the general smooth $\varkappa$-semiconvexity assumptions by Bakry and Ledoux [2] and Ledoux [22] using a diffusion semigroup approach. To handle the general $q > 1$ case, we follow the semigroup argument, as in our previous work [28]. Surprisingly, the case $\varkappa > 0$ requires proving new delicate semigroup estimates, which may be of independent interest, and which were not needed for the previous arguments. We formulate and prove this converse to the Maz'ya inequality for capacities in Section 4.

2 Definitions and Preliminaries

2.1 Isoperimetric inequalities

In Euclidean space, an isoperimetric inequality relates between the (appropriate notion of) surface area of a Borel set and its volume. To define an appropriate generalization of surface area in our setting, we use the Minkowski (exterior) boundary measure of a Borel set $A \subset (\Omega, d)$, denoted here by $\mu^+(A)$, which is defined as

$$\mu^+(A) := \liminf_{\varepsilon \to 0} \frac{\mu(A^d_\varepsilon) - \mu(A)}{\varepsilon} \;,$$

where $A^d_\varepsilon := \{x \in \Omega; \exists y \in A \;\; d(x,y) < \varepsilon\}$ denotes the ε-neighborhood of A with respect to the metric d. An isoperimetric inequality measures the relation between $\mu^+(A)$ and $\mu(A)$ by means of the isoperimetric profile $\mathcal{I} = \mathcal{I}_{(\Omega,d,\mu)}$, defined as the pointwise maximal function $\mathcal{I} : [0,1] \to \mathbb{R}_+$, so that

$$\mu^+(A) \geqslant \mathcal{I}(\mu(A)) \;, \tag{2.1}$$

for all Borel sets $A \subset \Omega$. Since A and $\Omega \setminus A$ typically have the same boundary measure, it is convenient to also define $\widetilde{\mathcal{I}} : [0,1/2] \to \mathbb{R}_+$ as $\widetilde{\mathcal{I}}(v) := \min(\mathcal{I}(v), \mathcal{I}(1-v))$.

Let us keep some important examples of isoperimetric inequalities in mind. We say that our space satisfies *linear isoperimetric inequality* if there exists a constant $D > 0$ so that

$$\widetilde{\mathcal{I}}_{(\Omega,d,\mu)}(t) \geqslant Dt$$

for all $t \in [0, 1/2]$. We denote the best constant D by $D_{\mathrm{Lin}} = D_{\mathrm{Lin}}(\Omega, d, \mu)$. Another useful example pertains to the standard Gaussian measure γ on $(\mathbb{R}, |\cdot|)$, where $|\cdot|$ is the Euclidean metric. We say that our space satisfies a *Gaussian isoperimetric inequality* if there exists a constant $D > 0$ so that

$$\mathcal{I}_{(\Omega,d,\mu)}(t) \geqslant D\mathcal{I}_{(\mathbb{R},|\cdot|,\gamma)}(t)$$

for all $t \in [0, 1]$. We denote the best constant D by $D_{\mathrm{Gau}} = D_{\mathrm{Gau}}(\Omega, d, \mu)$. It is known that $\widetilde{\mathcal{I}}_{(\mathbb{R},|\cdot|,\gamma)}(t) \simeq t \log^{1/2}(1/t)$ uniformly on $t \in [0, 1/2]$, where we use the notation $A \simeq B$ to signify that there exist universal constants $C_1, C_2 > 0$ so that $C_1 B \leqslant A \leqslant C_2 B$. Unless otherwise stated, all of the constants throughout this work are universal, independent of any other parameter, and, in particular, the dimension n in the case of an underlying manifold. The Gaussian isoperimetric inequality can therefore be equivalently stated as asserting that there exists a constant $D > 0$ so that

$$\widetilde{\mathcal{I}}_{(\Omega,d,\mu)}(t) \geqslant Dt \log^{1/2}(1/t)$$

for all $t \in [0, 1/2]$.

2.2 Functional inequalities

Let $f \in \mathcal{F}$. We consider functional inequalities which compare between $\|f\|_{N_1(\mu)}$ and $\||\nabla f|\|_{N_2(\mu)}$, where N_1, N_2 are some norms associated with the measure μ, like the $L_p(\mu)$ norms, or some other more general Orlicz quasi-norms associated to the class $\mathcal{N}$ of increasing continuous functions mapping $\mathbb{R}_+$ onto $\mathbb{R}_+$.

A function $N : \mathbb{R}_+ \to \mathbb{R}_+$ is called a *Young function* if $N(0) = 0$ and N is convex increasing. Given a Young function N, the *Orlicz norm* $N(\mu)$ associated to N is defined as

$$\|f\|_{N(\mu)} := \inf\left\{v > 0; \int_\Omega N(|f|/v)d\mu \leqslant 1\right\}.$$

For a general increasing continuous function $N : \mathbb{R}_+ \to \mathbb{R}_+$ with $N(0) = 0$ and $\lim_{t\to\infty} N(t) = \infty$ (this class is denoted by $\mathcal{N}$), the above definition still makes sense, although $N(\mu)$ will no longer necessarily be a norm. We say in this case that it is a *quasinorm*.

There is clearly no point to test constant functions in our functional inequalities, so it is natural to require that either the expectation $E_\mu f$ or median $M_\mu f$ of f are 0. Here, $E_\mu f = \int f d\mu$ and $M_\mu f$ is a value so that $\mu(f \geqslant M_\mu f) \geqslant 1/2$ and $\mu(f \leqslant M_\mu f) \geqslant 1/2$.

Definition. We say that the space (Ω, d, μ) satisfies an (N, q) *Orlicz–Sobolev inequality* ($N \in \mathcal{N}, q \geqslant 1$) if

$$\exists D > 0 \text{ s.t. } \forall f \in \mathcal{F} \quad D \left\| f - M_\mu f \right\|_{N(\mu)} \leqslant \left\| |\nabla f| \right\|_{L_q(\mu)} . \tag{2.2}$$

A similar (yet different) definition was given by Roberto and Zegarlinski [34] in the case $q = 2$ following the book by Maz'ya [27, p. 112]. Our preference to use the median M_μ in our definition (in place of the more standard expectation E_μ) is immaterial whenever N is a convex function, due to the following elementary lemma from [29].

Lemma 2.1. *Let $N(\mu)$ denote an Orlicz norm associated to the Young function N. Then*

$$\frac{1}{2} \left\| f - E_\mu f \right\|_{N(\mu)} \leqslant \left\| f - M_\mu f \right\|_{N(\mu)} \leqslant 3 \left\| f - E_\mu f \right\|_{N(\mu)} .$$

When $N(t) = t^p$, then $N(\mu)$ is just the usual $L_p(\mu)$ norm. If, in addition, M_μ in (2.2) is replaced by E_μ, the case $p = q = 2$ is then just the classical *Poincaré inequality*, and we denote the best constant in this inequality by D_{Poin}. Similarly, the case $q = 1, p = \frac{n}{n-1}$ corresponds to the *Gagliardo inequality*, and the case $1 < q < n$, $p = \frac{qn}{n-q}$ to the *Sobolev inequalities.* A limiting case where $q = 2$ and n tends to infinity is the so-called *log-Sobolev inequality.* More generally, we say that our space satisfies a *q-log-Sobolev inequality* ($q \in [1, 2]$) if there exists a constant $D > 0$ so that

$$\forall f \in \mathcal{F} \quad D \left(\int |f|^q \log |f|^q d\mu - \int |f|^q d\mu \log(\int |f|^q d\mu) \right)^{1/q} \leqslant \left\| |\nabla f| \right\|_{L_q(\mu)} . \tag{2.3}$$

The best possible constant D above is denoted by $D_{LS_q} = D_{LS_q}(\Omega, d, \mu)$. Although these inequalities do not precisely fit into our announced framework, it follows from the work of Bobkov and Zegarlinski [11, Proposition 3.1] (generalizing the case $q = 2$ due to Bobkov and Götze [8, Proposition 4.1]) that they are, in fact, equivalent to the following Orlicz–Sobolev inequalities:

$$\forall f \in \mathcal{F} \quad D_{\varphi_q} \left\| f - E_\mu f \right\|_{\varphi_q(\mu)} \leqslant \left\| |\nabla f| \right\|_{L_q(\mu)} , \tag{2.4}$$

where $\varphi_q(t) = t^q \log(1 + t^q)$ and $D_{LS_q} \simeq D_{\varphi_q}$ uniformly on $q \in [1, 2]$.

Various other functional inequalities admit an equivalent (up to universal constants) formulation using an appropriate Orlicz norm $N(\mu)$ on the left-hand side of (2.2). We refer the reader to the recent paper of Barthe and

Kolesnikov [4] and the references therein for an account of several other types of functional inequalities.

2.3 Known connections

It is well known that various isoperimetric inequalities imply their functional "counterparts." It was shown by Maz'ya [25, 26] and independently by Cheeger [14] that a linear isoperimetric inequality implies the Poincaré inequality $D_{\text{Poin}} \geqslant D_{\text{Lin}}/2$ (the Cheeger inequality). It was first observed by Ledoux [20] that a Gaussian isoperimetric inequality implies a 2-log-Sobolev inequality $D_{LS_2} \geqslant cD_{\text{Gau}}$ for some universal constant $c > 0$. This was later refined by Beckner (cf. [21]) using an equivalent functional form of the Gaussian isoperimetric inequality due to Bobkov [6, 7] (cf. also [5]): $D_{LS_2} \geqslant D_{\text{Gau}}/\sqrt{2}$. The constants 2 and $\sqrt{2}$ above are known to be optimal.

Another example is obtained by considering the isoperimetric inequality

$$\widetilde{\mathcal{I}}(t) \geqslant D_{\text{Exp}_q} t \log^{1/q} 1/t$$

for some $q \in [1, 2]$. This inequality is satisfied with some $D_{\text{Exp}_q} > 0$ by the probability measure μ_p with density $\exp(-|x|^p)/Z_p$ on $(\mathbb{R}, |\cdot|)$, for $p = q^* = q/(q-1)$ (where Z_p is a normalization factor). Bobkov and Zegarlinski [11] showed that $D_{LS_q} \geqslant cD_{\text{Exp}_q}$, for some universal constant $c > 0$, independent of q, in analogy to the inequality $D_{LS_2} \geqslant cD_{\text{Gau}}$ mentioned above. Another proof of this using capacities was given in our joint work with Sasha Sodin [31].

We will see in Section 3 how Maz'ya's general framework may be used to obtain all of these implications.

In general, however, it is known that these implications *cannot* be reversed. For instance, by using $([-1, 1], |\cdot|, \mu_\alpha)$, where $d\mu_\alpha = \frac{1+\alpha}{2}|x|^\alpha dx$ on $[-1, 1]$, it is clear that $\mu_\alpha^+([0, 1]) = 0$ so $D_{\text{Lin}} = D_{\text{Gau}} = 0$, whereas one can show that $D_{\text{Poin}}, D_{LS_2} > 0$ for $\alpha \in (0, 1)$ using criteria for the Poincaré and 2-log-Sobolev inequalities on $\mathbb{R}$ due to Artola, Talenti, and Tomaselli (cf. Muckenhoupt [32]) and Bobkov and Götze [8] respectively. These examples suggest that we must rule out the existence of narrow "necks" in our measure or space, for the converse implications to stand a chance of being valid. Adding some convexity type assumptions is therefore a natural path to take.

Indeed, under our $\varkappa$-semiconvexity assumptions, we see that a reverse implication *can* in fact be obtained. This extends some previously known results by several authors. Buser showed in [13] that $D_{\text{Lin}} \geqslant c \min(D_{\text{Poin}}, D_{\text{Poin}}^2/\sqrt{\varkappa})$ with $c > 0$ a universal constant, for the case of a compact Riemannian manifold (M, g) with uniform density whose Ricci curvature is bounded below by $-\varkappa g$. This was subsequently extended by Ledoux [22] to our more general smooth $\varkappa$-semiconvexity assumptions, following the semigroup approach he developed in [20] and refined by Bakry and Ledoux [2]. Similarly, the reverse inequality $D_{\mathrm{Gau}} \geqslant c\min(D_{LS_2}, D_{LS_2}^2/\sqrt{\varkappa})$ with $c > 0$ a universal constant, was obtained by Bakry and Ledoux (cf. also [22]) under our smooth $\varkappa$-semiconvexity assumptions. We will extend these results to handle general Orlicz–Sobolev inequalities in Section 4 and, in particular, show that $D_{\mathrm{Exp}_q} \geqslant c\min(D_{LS_q}, D_{LS_q}^2/\sqrt{\varkappa})$ for some universal constant $c > 0$, uniformly on $q \in [1,2]$. When $q > 2$, we will see that these formulas take on a different form.

3 Capacities

In this section, we formulate the various known connections mentioned in the Introduction between capacities and isoperimetric and functional inequalities, and provide for completeness most of the proofs, following [28].

Remark 3.1. A remark which will be useful for dealing with general metric probability spaces, is that in the definition of capacity, by approximating Φ appropriately, we may always assume that

$$\int_{\{\Phi=t\}} |\nabla \Phi|^q \, d\mu = 0$$

for any $t \in (0,1)$, even though we may have $\mu\{\Phi = t\} > 0$ (cf. [28, Remark 3.3] for more information).

3.1 1-*Capacity and isoperimetric profiles*

Our starting point is the following well-known co-area formula which in our setting becomes an inequality (cf. [9, 10]).

Lemma 3.2 (Bobkov–Houdré). *For any* $f \in \mathcal{F}$

$$\int_\Omega |\nabla f| \, d\mu \geqslant \int_{-\infty}^{\infty} \mu^+ \{f > t\} \, dt \, .$$

The following proposition [24, 17, 9] encapsulates the connection between the 1-capacity and isoperimetric profiles.

Proposition 3.3 (Federer–Fleming, Maz'ya, Bobkov–Houdré). *For all* $0 < a < b < 1$

$$\inf_{a\leqslant t\leqslant b} \mathcal{I}(t) \leqslant \mathrm{Cap}_1(a,b) \leqslant \inf_{a\leqslant t<b} \mathcal{I}(t) \, . \tag{3.1}$$

For completeness, we provide a proof following Sodin [35, Proposition A].

Proof. Given a function $\Phi : \Omega \to [0,1]$ which is Lipschitz-on-balls with $\mu\{\Phi = 1\} \geqslant a$ and $\mu\{\Phi = 0\} \geqslant 1 - b$, the co-area inequality implies

$$\int |\nabla \Phi| \, d\mu \geqslant \int_{-\infty}^{\infty} \mu^+ \{\Phi > t\} \, dt = \int_0^1 \mu^+ \{\Phi > t\} \, dt \geqslant \inf_{a \leqslant t \leqslant b} \mathcal{I}(t) \, .$$

Taking the infimum on all such functions Φ, we obtain the first inequality in (3.1). To obtain the second inequality, let A denote a Borel set with $a \leqslant \mu(A) < b$. We may exclude the case that $\mu^+(A) = \infty$ since it does not contribute to the definition of the isoperimetric profile $\mathcal{I}$. Now, denote for $r, s > 0$

$$\Phi_{r,s}(x) := \left(1 - s^{-1} d(x, A_r)\right) \vee 0 \, .$$

It is clear that $\mu\{\Phi_{r,s} = 1\} \geqslant \mu(A) \geqslant a$, and since $\mu^+(A) < \infty$, for $r + s$ small enough we have $\mu\{\Phi_{r,s} = 0\} \geqslant 1 - b$. Hence

$$\frac{\mu(A_{s+2r}) - \mu(A)}{s} \geqslant \frac{\mu\{r \leqslant d(x,A) \leqslant s + r\}}{s} \geqslant \int |\nabla \Phi_{r,s}| \, d\mu \geqslant \mathrm{Cap}_1(a,b) \, .$$

Taking the limit inferior as $r, s \to 0$ so that $r/s \to 0$, and taking the infimum on all sets A as above, we obtain the second inequality in (3.1). □

Since obviously $\mathrm{Cap}_1(a,b) = \mathrm{Cap}_1(1-b, 1-a)$, we have the following useful corollary.

Corollary 3.4. *For any nondecreasing continuous function* $J : [0,1/2] \to \mathbb{R}_+$

$$\widetilde{\mathcal{I}}(t) \geqslant J(t) \quad \forall t \in [0,1/2] \iff \mathrm{Cap}_1(t, 1/2) \geqslant J(t) \quad \forall t \in [0,1/2] \, .$$

Definition. A *q-capacitary inequality* is an inequality of the form

$$\mathrm{Cap}_q(t, 1/2) \geqslant J(t) \quad \forall t \in [0,1/2] \, ,$$

where $J : [0,1/2] \to \mathbb{R}_+$ is a nondecreasing continuous function.

3.2 q-Capacitary and weak Orlicz–Sobolev inequalities

Definition. Given $N \in \mathcal{N}$, denote by $N^\wedge : \mathbb{R}_+ \to \mathbb{R}_+$ the *adjoint* function

$$N^\wedge(t) := \frac{1}{N^{-1}(1/t)}.$$

Remark 3.5. Note that the operation $N \to N^\wedge$ is an involution on $\mathcal{N}$ and $N(\cdot^\alpha)^\wedge = (N^\wedge)^{1/\alpha}$ for $\alpha > 0$. It is also immediate to check that $N(t^\alpha)/t$ is nondecreasing if and only if $N^\wedge(t)^{1/\alpha}/t$ is nonincreasing ($\alpha > 0$).

We denote by $L_{s,\infty}(\mu)$ the weak L_s quasinorm defined as

$$\|f\|_{L_{s,\infty}(\mu)} := \sup_{t>0} \mu(|f| \geqslant t)^{1/s} t.$$

We now extend the definition of the weak L_s quasinorm to Orlicz quasinorms $N(\mu)$, using the adjoint function $N^\wedge$:

Definition. Given $N \in \mathcal{N}$, define the *weak* $N(\mu)$ *quasinorm* as

$$\|f\|_{N(\mu),\infty} := \sup_{t>0} N^\wedge(\mu\{|f| \geqslant t\}) t.$$

This definition is consistent with the one for $L_{s,\infty}$ and satisfies

$$\|f\|_{N(\mu),\infty} \leqslant \|f\|_{N(\mu)} , \tag{3.2}$$

as easily checked using the Markov–Chebyshev inequality. Also note that this is indeed a quasinorm by a simple union-bound:

$$\|f + g\|_{N(\mu),\infty} \leqslant 2 \left(\|f\|_{N(\mu),\infty} + \|g\|_{N(\mu),\infty} \right) .$$

Remark 3.6. The motivation for the definition of $N^\wedge$ stems from the immediate observation that for any Borel set A

$$\|\chi_A\|_{N(\mu)} = \|\chi_A\|_{N(\mu),\infty} = N^\wedge(\mu(A)) .$$

For this reason, the expression $1/N^{-1}(1/t)$ already appears in the works of Maz'ya [27, p. 112] and Roberto and Zegarlinski [34].

Definition. An inequality of the form:

$$\forall f \in \mathcal{F} \quad D \|f - M_\mu f\|_{N(\mu),\infty} \leqslant \||\nabla f|\|_{L_q(\mu)} \tag{3.3}$$

is called a *weak type Orlicz–Sobolev inequality.*

Lemma 3.7. *The weak type Orlicz–Sobolev inequality (3.3) implies*

$$\mathrm{Cap}_q(t, 1/2) \geqslant D N^\wedge(t) \quad \forall t \in [0, 1/2].$$

Proof. Apply (3.3) to $f = \Phi$, where $\Phi : \Omega \to [0,1]$ is any Lipschitz-on-balls function so that $\mu\{\Phi = 1\} \geqslant t$ and $\mu\{\Phi = 0\} \geqslant 1/2$. Since $M_\mu \Phi = 0$, it follows that

$$\||\nabla \Phi|\|_{L_q(\mu)} \geqslant D \|\Phi\|_{N(\mu),\infty} \geqslant D N^\wedge(\mu(\{\Phi = 1\})) \geqslant D N^\wedge(t).$$

Taking the infimum over all Φ as above, we obtain the required assertion. □

Proposition 3.8. *Let $1 \leqslant q < \infty$. Then the following statements are equivalent:*

(1)
$$\forall f \in \mathcal{F} \quad D_1 \|f - M_\mu f\|_{N(\mu),\infty} \leqslant \||\nabla f|\|_{L_q(\mu)} \quad , \tag{3.4}$$

(2)
$$Cap_q(t, 1/2) \geqslant D_2 N^\wedge(t) \quad \forall t \in [0, 1/2] \ ,$$

and the best constants D_1, D_2 above satisfy $D_1 \leqslant D_2 \leqslant 4D_1$.

Proof. We have $D_2 \geqslant D_1$ by Lemma 3.7. To see the other direction, note that as in Remark 3.1, we may assume that

$$\int_{\{f=t\}} |\nabla f|^q \, d\mu = 0$$

for all $t \in \mathbb{R}$, and by replacing f with $f - M_\mu f$, that $M_\mu f = 0$. Note that it suffices to show (3.4) with $D_1 = D_2$ for nonnegative functions for which $\mu\{f = 0\} \geqslant 1/2$ since for a general function as above we can apply (3.4) to $f_+ = f\chi_{f \geqslant 0}$ and to $f_- = -f\chi_{f \leqslant 0}$, which yields

$$\begin{aligned} \||\nabla f|\|_{L_q(\mu)} &= \left(\int |\nabla f_+|^q \, d\mu + \int |\nabla f_-|^q \, d\mu \right)^{1/q} \\ &\geqslant D_1 \left(\|f_+\|^q_{N(\mu),\infty} + \|f_-\|^q_{N(\mu),\infty} \right)^{1/q} \\ &\geqslant D_1 2^{1/q-1} \left(\|f_+\|_{N(\mu),\infty} + \|f_-\|_{N(\mu),\infty} \right) \geqslant \frac{D_1}{4} \|f\|_{N(\mu),\infty} \ . \end{aligned}$$

Given a nonnegative function f as above ($\mu\{f = 0\} \geqslant 1/2$ hence $M_\mu f = 0$) and $t > 0$, define $\Omega_t = \{f \leqslant t\}$ and $f_t := f/t \wedge 1$. Then

$$\begin{aligned} \left(\int_\Omega |\nabla f|^q \, d\mu \right)^{1/q} &\geqslant \left(\int_{\Omega_t} |\nabla f|^q \, d\mu \right)^{1/q} \geqslant t \left(\int_\Omega |\nabla f_t|^q \, d\mu \right)^{1/q} \\ &\geqslant t\mathrm{Cap}_q(\mu\{f_t \geqslant 1\}, 1/2) \geqslant D_2 t N^\wedge(\mu\{f \geqslant t\}) \ . \end{aligned}$$

Taking the supremum on $t > 0$, we obtain the required assertion. □

3.3 q-Capacitary and strong Orlicz–Sobolev inequalities

Proposition 3.9. *If $N(t)^{1/q}/t$ is nondecreasing on $\mathbb{R}_+$ with $1 \leqslant q < \infty$, then the following statements are equivalent:*

(1)
$$\forall f \in \mathcal{F} \quad D_1 \|f - M_\mu f\|_{N(\mu)} \leqslant \||\nabla f|\|_{L_q(\mu)} \quad , \tag{3.5}$$

(2)
$$\mathrm{Cap}_q(t,1/2) \geqslant D_2 N^{\wedge}(t) \quad \forall t \in [0,1/2] \ ,$$

and the best constants D_1, D_2 above satisfy $D_1 \leqslant D_2 \leqslant 4D_1$.

Remark 3.10. As already mentioned in Section 2, we call an inequality of the form (3.5) an Orlicz–Sobolev inequality (even though N may not be convex).

Remark 3.11. One may show (cf., for example, the proof of [34, Theorem 1]) that when $N(t^{1/q})$ is convex (so, in particular, $N(t)^{1/q}/t$ is nondecreasing), Proposition 3.9 is equivalent to a theorem of Maz'ya [27, p. 112] with a constant depending on q which is better than the constant 4 above. In particular, when $q = 1$, Maz'ya showed that the optimal constant is actually 1, so that $D_1 = D_2$. The latter conclusion was also independently derived by Federer and Fleming [17].

Proof. We have $D_2 \geqslant D_1$ by (3.2) and Lemma 3.7. To see the other direction, we assume again (as in Remark 3.1) that

$$\int_{\{f=t\}} |\nabla f|^q \, d\mu = 0$$

for all $t \in \mathbb{R}$, and by replacing f with $f - M_\mu f$, that $M_\mu f = 0$. Again, it suffices to show (3.5) for nonnegative functions for which $\mu\{f = 0\} \geqslant 1/2$, but now we do not lose in the constant. Indeed, for a general function as above, we can apply (3.5) to $f_+ = f\chi_{f\geqslant 0}$ and to $f_- = -f\chi_{f\leqslant 0}$, which yields

$$\begin{aligned} \||\nabla f|\|^q_{L_q(\mu)} &= \int |\nabla f_+|^q \, d\mu + \int |\nabla f_-|^q \, d\mu \\ &\geqslant D_1^q \left(\|f_+\|^q_{N(\mu)} + \|f_-\|^q_{N(\mu)} \right) \geqslant D_1^q \, \|f\|^q_{N(\mu)} \, . \end{aligned}$$

The last inequality follows from the fact that $N^{1/q}(t)/t$ is nondecreasing, so, denoting $v_\pm = \|f_\pm\|_{N(\mu)}$, we indeed verify that

$$\begin{aligned} &\int N\left(\frac{f_+ + f_-}{(v_+^q + v_-^q)^{1/q}}\right) d\mu \\ &= \int N\left(\frac{f_+}{v_+}\frac{v_+}{(v_+^q + v_-^q)^{1/q}}\right) d\mu + \int N\left(\frac{f_-}{v_-}\frac{v_-}{(v_+^q + v_-^q)^{1/q}}\right) d\mu \\ &\leqslant \frac{v_+^q}{v_+^q + v_-^q} \int N\left(\frac{f_+}{v_+}\right) d\mu + \frac{v_-^q}{v_+^q + v_-^q} \int N\left(\frac{f_-}{v_-}\right) d\mu \leqslant 1 \, . \end{aligned}$$

We first assume that f is bounded. Given a bounded nonnegative function f as above ($M_\mu f = 0$ and $\mu\{f = 0\} \geqslant 1/2$), we may assume by homogeneity that $\|f\|_{L_\infty} = 1$. For $i \geqslant 1$ denote $\Omega_i = \{1/2^i \leqslant f \leqslant 1/2^{i-1}\}$, $m_i = \mu(\Omega_i)$, $f_i = 2^i(f - 1/2^i) \vee 0 \wedge 1$ and set $m_0 = 0$. Also denote $J := N^{\wedge}$. Now,

$$\||\nabla f|\|_{L_q(\mu)}^q = \sum_{i=1}^{\infty} \int_{\Omega_i} |\nabla f|^q \, d\mu \geqslant \sum_{i=1}^{\infty} \frac{1}{2^{qi}} \int_{\Omega} |\nabla f_i|^q \, d\mu$$

$$\geqslant \sum_{i=1}^{\infty} \frac{1}{2^{qi}} \mathrm{Cap}_q^q(\mu\{f \geqslant 1/2^{i-1}\}, 1/2) \geqslant D_2^q \sum_{i=2}^{\infty} \frac{J^q(m_{i-1})}{2^{qi}} = \frac{D_2^q}{4^q} V^q,$$

where

$$V := \left(\sum_{i=1}^{\infty} \frac{J^q(m_i)}{2^{q(i-1)}} \right)^{1/q}.$$

It remains to show that $\|f\|_{N(\mu)} \leqslant V$. Indeed,

$$\int_{\Omega} N\left(\frac{f}{V}\right) d\mu \leqslant \sum_{i=1}^{\infty} m_i N\left(\frac{1}{2^{i-1}V}\right) = \sum_{i=1}^{\infty} \frac{J^{-1}(J(m_i))}{J^{-1}(2^{i-1}V)} \leqslant \sum_{i=1}^{\infty} \frac{J^q(m_i)}{2^{q(i-1)}V^q} = 1,$$

where in the last inequality we used the fact that $N(t)^{1/q}/t$ is nondecreasing, hence $(J^{-1})^{1/q}(t)/t$ is nondecreasing, and therefore

$$\frac{J^{-1}(x)}{J^{-1}(y)} \leqslant \left(\frac{x}{y}\right)^q,$$

whenever $x/y \leqslant 1$, which is indeed the case for us.

For an unbounded $f \in \mathcal{F}$ with $\mu\{f = 0\} \geqslant 1/2$, we may define $f_m = f \wedge b_m$ so that $\mu\{f > b_m\} \leqslant 1/m$ and (just for safety) $\mu\{f = b_m\} = 0$. It then follows by what was proved for bounded functions that

$$\||\nabla f|\|_{L_q(\mu)} \geqslant \lim_{m\to\infty} \||\nabla f_m|\|_{L_q(\mu)} \geqslant D_1 \lim_{m\to\infty} \|f_m\|_{N(\mu)} = D_1 Z,$$

where all the limits exist since they are nondecreasing.

To conclude, $Z \geqslant \|f\|_{N(\mu)}$ since N is continuous, so, by the monotone convergence theorem,

$$\int N(f/Z) d\mu = \int \lim_{m\to\infty} N(f_m/Z) d\mu = \lim_{m\to\infty} \int N(f_m/Z) d\mu \leqslant 1. \quad \square$$

3.4 Passing between q-capacitary inequalities

The case $q_0 = 1$ in the following proposition is due to Maz'ya [27, p. 105]. Following [28], we provide a proof which generalizes to the case of an arbitrary metric probability space and $q_0 > 1$. We denote the conjugate exponent to $q \in [1, \infty]$ by $q^* = q/(q-1)$.

Proposition 3.12. *Let* $1 \leqslant q_0 \leqslant q < \infty$. *We set* $p_0 = q_0^*, p = q^*$. *Then for all* $0 < a < b < 1$

$$\frac{1}{\mathrm{Cap}_q(a,b)} \leqslant \gamma_{p,p_0} \left(\int_a^b \frac{ds}{(s-a)^{p/p_0} \mathrm{Cap}_{q_0}^p(s,b)} \right)^{1/p} ,$$

where

$$\gamma_{p,p_0} := \frac{(p_0/p - 1)^{1/p_0}}{(1-p/p_0)^{1/p}} . \tag{3.6}$$

Proof. Let $0 < a < b < 1$ be given, and let $\Phi : \Omega \to [0,1]$ be a function in $\mathcal{F}$ such that $a' := \mu\{\Phi = 1\} \geqslant a$ and $1 - b' := \mu\{\Phi = 0\} \geqslant 1 - b$. As usual (cf. Remark 3.1), by approximating Φ, we may assume that

$$\int_{\{\Phi = t\}} |\nabla \Phi|^q \, d\mu = 0$$

for all $t \in (0,1)$. Let $C := \{t \in (0,1); \mu\{\Phi = t\} > 0\}$ denote the discrete set of atoms of Φ under μ. We set $\Gamma := \{f \in C\}$ and denote $\gamma = \mu(\Gamma)$.

We now choose $t_0 = 0 < t_1 < t_2 < \ldots < 1$, so that denoting for $i \geqslant 1$, $\Omega_i = \{t_{i-1} \leqslant \Phi \leqslant t_i\}$, and setting $m_i = \mu(\Omega_i \setminus \Gamma)$, we have $m_i = (b' - a' - \gamma)\alpha^{i-1}(1-\alpha)$, where $0 \leqslant q\alpha \leqslant 1$ will be chosen later. Denote, in addition, $\Phi_i = \left(\frac{\Phi - t_{i-1}}{t_i - t_{i-1}} \vee 0 \right) \wedge 1$ and $N_i = \sum_{j > i} m_j$. Applying the Hölder inequality twice, we estimate

$$\begin{aligned}
\left(\int_\Omega |\nabla \Phi|^q \, d\mu \right)^{1/q} &= \left(\sum_{i=1}^\infty \int_{\Omega_i \setminus \Gamma} |\nabla \Phi|^q \, d\mu \right)^{1/q} \\
&\geqslant \left(\sum_{i=1}^\infty m_i^{1 - \frac{q}{q_0}} \left(\int_{\Omega_i \setminus \Gamma} |\nabla \Phi|^{q_0} \, d\mu \right)^{q/q_0} \right)^{1/q} \\
&\geqslant \left(\sum_{i=1}^\infty m_i^{1 - \frac{q}{q_0}} (t_i - t_{i-1})^q \left(\int_\Omega |\nabla \Phi_i|^{q_0} \, d\mu \right)^{q/q_0} \right)^{1/q} \\
&\geqslant \left(\sum_{i=1}^\infty m_i^{1 - \frac{q}{q_0}} (t_i - t_{i-1})^q \mathrm{Cap}_{q_0}^q (\mu\{\Phi_i = 1\}, 1 - \mu\{\Phi_i = 0\}) \right)^{1/q} \\
&\geqslant \sum_{i=1}^\infty (t_i - t_{i-1}) \left(\sum_{i=1}^\infty \frac{m_i^{1-p/p_0}}{\mathrm{Cap}_{q_0}^p (\mu\{\Phi \geqslant t_i\}, b)} \right)^{-1/p} .
\end{aligned}$$

Since $\mu\{\Phi \geqslant t_i\} \geqslant a' + N_i$ and $\mathrm{Cap}_{q_0}(s,b)$ is nondecreasing in s, we continue to estimate as follows:

$$\left(\frac{1}{\int_\Omega |\nabla \Phi|^q \, d\mu} \right)^{p/q} \leqslant \sum_{i=1}^\infty \frac{m_i^{1-p/p_0}}{\mathrm{Cap}_{q_0}^p (\mu\{\Phi \geqslant t_i\}, b)}$$

$$\leqslant \sum_{i=1}^{\infty} \frac{m_i^{1-p/p_0}}{m_{i+1}} \int_{a'+N_{i+1}}^{a'+N_i} \frac{ds}{\mathrm{Cap}_{q_0}^p(s,b)} \leqslant \sum_{i=1}^{\infty} \frac{1}{\alpha m_i^{p/p_0}} \int_{a'+N_{i+1}}^{a'+N_i} \frac{ds}{\mathrm{Cap}_{q_0}^p(s,b)}$$

$$\leqslant \frac{1}{\alpha}\left(\frac{\alpha}{1-\alpha}\right)^{p/p_0} \sum_{i=1}^{\infty} \int_{a'+N_{i+1}}^{a'+N_i} \frac{ds}{(s-a')^{p/p_0}\mathrm{Cap}_{q_0}^p(s,b)}$$

$$\leqslant \frac{1}{\alpha}\left(\frac{\alpha}{1-\alpha}\right)^{p/p_0} \int_a^b \frac{ds}{(s-a)^{p/p_0}\mathrm{Cap}_{q_0}^p(s,b)},$$

where we used that $m_{i+1} = \alpha m_i$, $m_i = \frac{1-\alpha}{\alpha} N_i$ and, in the last inequality, the fact that $\mathrm{Cap}_{q_0}(s,b)$ is nondecreasing in s. The assertion now follows by taking the supremum on all Φ as above, and choosing the optimal $\alpha = 1 - p/p_0$. □

3.5 Combining everything

Combining all of the ingredients in this section, we see how to pass from isoperimetric inequalities to functional inequalities, simply by following the general diagram

Isoperimetric inequality	$\Leftrightarrow_{\text{Corollary 3.4}}$	1-capacitary inequality
		$\Downarrow$ Proposition 3.12
(N,q) Orlicz–Sobolev inequality with $N(t)^{1/q}/t$ nondecreasing	$\Leftrightarrow_{\text{Proposition 3.9}}$	q-capacitary inequality

In particular, it is an exercise to follow this diagram and obtain the previously mentioned inequalities of Subsection 2.3:

$$D_{\mathrm{Poin}} \geqslant c D_{\mathrm{Lin}}, \quad D_{LS_2} \geqslant c D_{\mathrm{Gau}}, \quad D_{LS_q} \geqslant c D_{\mathrm{Exp}_q},$$

$q \in [1,2]$, for some universal constant $c > 0$. In the first case, the optimal constant $c = 1/2$ may also be obtained by improving the constant in Proposition 3.9 as in Remark 3.11 (but, of course, easier ways are known to obtain this optimal constant; cf., for example, [29]). More generally, the following statement may easily be obtained (cf. [28] for more details and useful special cases).

Theorem 3.13. *Let $1 \leqslant q < \infty$, and let $p = q^*$. Let $N \in \mathcal{N}$, so that $N(t)^{1/q}/t$ is nondecreasing. Then*

$$\widetilde{\mathcal{I}}(t) \geqslant D t^{1-1/q} N^{\wedge}(t) \quad \forall t \in [0,1/2] \tag{3.7}$$

implies

$$\forall f \in \mathcal{F} \quad B_{N,q} D \left\|f - M_\mu f\right\|_{N(\mu)} \leqslant \left\|\left|\nabla f\right|\right\|_{L_q(\mu)}, \tag{3.8}$$

where

$$B_{N,q} := \frac{1}{4} \inf_{0<t<1/2} \frac{1}{\left(\int_t^{1/2} \frac{N^\wedge(t)^p ds}{s N^\wedge(s)^p} \right)^{1/p}} . \tag{3.9}$$

4 The Converse Statement

Our goal in this section is to prove the following converse to Theorem 3.13.

Theorem 4.1. *Let $1 < q \leqslant \infty$, and let $N \in \mathcal{N}$ denote a Young function so that $N(t)^{1/q}/t$ is nondecreasing. Then, under our $\varkappa$-semiconvexity assumptions, the statement*

$$\forall f \in \mathcal{F} \quad D \left\| f - M_\mu f \right\|_{N(\mu)} \leqslant \left\| |\nabla f| \right\|_{L_q(\mu)} \tag{4.1}$$

implies

$$\widetilde{\mathcal{I}}(t) \geqslant C_{N,q} \min \left(D, \frac{D^r}{\varkappa^{\frac{r-1}{2}}} \right) t^{1-1/q} N^\wedge(t) \ \ \forall t \in [0, 1/2] , \tag{4.2}$$

where $r = \max(q, 2)$ and $C_{N,q} > 0$ depends solely on N and q.

Remark 4.2. The assumption that N is a convex function is not essential for this result since it is possible to approximate N appropriately in the large using a convex function as in Subsection 4.2.2 below. We refer to [28, Theorem 4.5] for more details.

Remark 4.3. It is clear that, using the results of Section 3, we can reformulate Theorem 4.1 as a converse to the Maz'ya inequality relating Cap_q and Cap_1 (cf. Proposition 3.12) under our $\varkappa$-semiconvexity assumptions. For the case of our convexity assumptions ($\varkappa = 0$), this was explicitly written out in [28, Theorem 5.1].

The case $\varkappa = 0$ of Theorem 4.1 was proved in [28] by using the semigroup approach developed by Bakry and Ledoux [2] and Ledoux [20, 22]. Let us now recall this framework.

Given a smooth complete connected Riemannian manifold $\Omega = (M, g)$ equipped with a probability measure μ with density $d\mu = \exp(-\psi) dvol_M$, $\psi \in C^2(M, \mathbb{R})$, we define the associated Laplacian $\Delta_{(\Omega,\mu)}$ by

$$\Delta_{(\Omega,\mu)} := \Delta_\Omega - \nabla\psi \cdot \nabla, \tag{4.3}$$

where Δ_Ω is the usual Laplace–Beltrami operator on Ω; $\Delta_{(\Omega,\mu)}$ acts on $\mathcal{B}(\Omega)$, the space of bounded smooth real-valued functions on Ω. Let $(P_t)_{t \geqslant 0}$ denote the semigroup associated to the diffusion process with infinitesimal generator

$\Delta_{(\Omega,\mu)}$ (cf. [15, 21]), for which μ is its stationary measure. It is characterized by the following system of second order differential equations:

$$\frac{d}{dt}P_t(f) = \Delta_{(\Omega,\mu)}(P_t(f)) \quad P_0(f) = f \quad \forall f \in \mathcal{B}(\Omega) . \tag{4.4}$$

For each $t \geqslant 0$, $P_t : \mathcal{B}(\Omega) \to \mathcal{B}(\Omega)$ is a bounded linear operator in the L_∞ norm, and its action naturally extends to the entire $L_p(\mu)$ spaces ($p \geqslant 1$). We collect several elementary properties of these operators:

- $P_t 1 = 1$.
- $f \geqslant 0 \Rightarrow P_t f \geqslant 0$.
- $\int (P_t f) g d\mu = \int f(P_t g) d\mu$.
- $N(|P_t(f)|) \leqslant P_t(N(|f|))$ for any Young function N.
- $P_t \circ P_s = P_{t+s}$.

The following crucial dimension-free reverse Poincaré inequality was shown by Bakry and Ledoux [2, Lemma 4.2], extending Ledoux's approach [20] for proving the Buser theorem (cf. also [2, Lemma 2.4] and [22, Lemma 5.1]).

Lemma 4.4 (Bakry–Ledoux). *Assume that the following Bakry–Émery curvature–dimension condition holds on Ω :*

$$\mathrm{Ric}_g + \mathrm{Hess}_g \psi \geqslant -\varkappa g , \quad \varkappa \geqslant 0 . \tag{4.5}$$

Then for any $t \geqslant 0$ and $f \in \mathcal{B}(\Omega)$

$$K(\varkappa, t)\, |\nabla P_t f|^2 \leqslant P_t(f^2) - (P_t f)^2$$

pointwise, where

$$K(\varkappa, t) := \frac{1 - \exp(-2\varkappa t)}{\varkappa} \quad (= 2t \quad \textit{if } \varkappa = 0) .$$

Sketch of the proof following [22]. Note that

$$\Delta_{(\Omega,\mu)}(g^2) - 2g\Delta_{(\Omega,\mu)}(g) = 2|\nabla g|^2$$

for any $g \in \mathcal{B}(\Omega)$. Consequently, (4.4) implies

$$P_t(f^2) - (P_t f)^2 = \int_0^t \frac{d}{ds} P_s((P_{t-s} f)^2) ds = 2 \int_0^t P_s(|\nabla P_{t-s} f|^2) ds .$$

The main observation is that, under the Bakry–Émery condition (4.5), the function $\exp(2\varkappa s) P_s(|\nabla P_{t-s} f|^2)$ is nondecreasing, as verified by direct differentiation and use of the Bochner formula. Therefore,

$$P_t(f^2) - (P_t f)^2 \geqslant |\nabla P_t f|^2 \, 2 \int_0^t \exp(-2\varkappa s) ds ,$$

which concludes the proof. □

In fact, the proof of this lemma is very general and extends to the abstract framework of diffusion generators, as developed by Bakry and Émery in their celebrated paper [1]. In the Riemannian setting, it is known [33] (cf. also [19, 36]) that the gradient estimate of Lemma 4.4 remains valid when the support of μ is the closure of a locally convex domain (connected open set) $\Omega \subset (M, g)$ with C^2 boundary, and $d\mu|_\Omega = \exp(-\psi) dvol_M|_\Omega$, $\psi \in C^2(\overline{\Omega}, \mathbb{R})$. A domain Ω with C^2 boundary is called *locally convex* if the second fundamental form on $\partial\Omega$ is positive semidefinite (with respect to the normal field pointing inward). In this case, Δ_Ω in (4.3) denotes the Neumann Laplacian on $\overline{\Omega}$, $\mathcal{B}(\Omega)$ denotes the space of bounded smooth real-valued functions on $\overline{\Omega}$ satisfying the Neumann boundary conditions on $\partial\Omega$, and Lemma 4.4 remains valid.

Under these assumptions, Lemma 4.4 clearly implies that

$$\forall q \in [2, \infty] \quad \forall f \in \mathcal{B}(\Omega) \quad \| |\nabla P_t f| \|_{L_q(\mu)} \leqslant \frac{1}{\sqrt{K(\varkappa, t)}} \|f\|_{L_q(\mu)} , \tag{4.6}$$

and using $q = \infty$, Ledoux easily deduces the following dual statement (cf. [22, (5.5)]).

Lemma 4.5 (Ledoux). *Under the same assumptions as Lemma 4.4,*

$$\forall f \in \mathcal{B}(\Omega) \quad \|f - P_t f\|_{L_1(\mu)} \leqslant \int_0^t \frac{ds}{\sqrt{K(\varkappa, s)}} \| |\nabla f| \|_{L_1(\mu)} . \tag{4.7}$$

To use (4.6) and (4.7), it is convenient to note the following rough estimates:

$$t \in [0, 1/(2\varkappa)] \quad \Rightarrow \quad K(\varkappa, t) \geqslant t, \quad \int_0^t \frac{ds}{\sqrt{K(\varkappa, s)}} \leqslant 2\sqrt{t} . \tag{4.8}$$

It is also useful to introduce the following definition.

Definition. We denote by $N(\mu)^*$ the *dual norm* to $N(\mu)$, given by

$$\|f\|_{N(\mu)^*} := \sup \left\{ \int f g d\mu ; \|g\|_{N(\mu)} \leqslant 1 \right\} .$$

It is elementary to calculate the $N(\mu)^*$-norm of characteristic functions (cf. [27, p. 111]).

Lemma 4.6. *Let N denote a Young function. Then for any Borel set A with $\mu(A) > 0$*

$$\|\chi_A\|_{N(\mu)^*} = \mu(A)N^{-1}\left(\frac{1}{\mu(A)}\right) = \frac{\mu(A)}{N^\wedge(\mu(A))} \, .$$

4.1 Case $q \geqslant 2$

To handle the $\varkappa > 0$ case, we need the following new estimate, which may be of independent interest.

Proposition 4.7. *Assume that Bakry–Émery curvature-dimension condition (4.5) holds on Ω and the following (N, q) Orlicz-Sobolev inequality is satisfied for $N \in \mathcal{N}$ and $q \geqslant 2$:*

$$\forall f \in \mathcal{F} \quad D\,\|f - E_\mu f\|_{N(\mu)} \leqslant \||\nabla f|\|_{L_q(\mu)} \, . \tag{4.9}$$

Then for any $f \in L_\infty(\Omega)$ with

$$\int f d\mu = 0,$$

we have for all $t \geqslant 0$

$$\int |P_t f|^2 d\mu \leqslant \int f^2 d\mu \left(1 + (q-1)\frac{2D^q}{\|f\|^q_{N(\mu)^*}\,\|f\|^{q-2}_{L_\infty}}\left(\int f^2 d\mu\right)^{q-1}\right.$$
$$\left. \times \int_0^t K(\varkappa, s)^{\frac{q-2}{2}} ds\right)^{-\frac{1}{q-1}} .$$

Proof. Denote $u(t) := \int |P_t f|^2 d\mu = \int f P_{2t} f d\mu$ by self-adjointness and the semigroup property. By (4.4) and integration by parts, we have

$$u'(t) = 2\int P_t f \,\Delta_{(\Omega,\mu)}(P_t f) d\mu = -2\int |\nabla P_t f|^2 d\mu \, . \tag{4.10}$$

Note that $u(t)$ is decreasing. We now use the following estimate:

$$u(t)^q \leqslant u(t/2)^q = \left(\int f P_t f d\mu\right)^q \leqslant \|f\|^q_{N(\mu)^*}\,\|P_t f\|^q_{N(\mu)}$$
$$\leqslant \frac{\|f\|^q_{N(\mu)^*}}{D^q}\int |\nabla P_t f|^q d\mu \leqslant \frac{\|f\|^q_{N(\mu)^*}}{D^q}\int |\nabla P_t f|^2 d\mu\, \||\nabla P_t f|\|^{q-2}_{L_\infty} \, .$$

By (4.6) and (4.10), we obtain

$$u(t)^q \leqslant -\frac{\|f\|^q_{N(\mu)^*}}{2D^q}\frac{\|f\|^{q-2}_{L_\infty}}{K(\varkappa, t)^{\frac{q-2}{2}}}\, u'(t) \, .$$

Denoting $v(t) = u(t)^{-q+1}$, we see that this boils down to

$$v'(t) \geqslant (q-1)\frac{2D^q}{\|f\|_{N(\mu)^*}^q \|f\|_{L_\infty}^{q-2}} K(\varkappa, t)^{\frac{q-2}{2}} .$$

Integrating in t, the desired conclusion follows. □

Remark 4.8. When $N(x) = x^2$, or more generally, when $N(x^{1/2})$ is convex, it is possible to obtain a better dependence on q in Proposition 4.7, which, as $q \to 2$, would recover the exponential rate of convergence of $\int |P_t f|^2 d\mu$ to 0, as dictated by the spectral theorem. Unfortunately, it seems that this would not yield the correct dependence in N in the assertion of Theorem 4.1.

Proof of Theorem 4.1 for $q \geqslant 2$. We prove the theorem under the *smooth $\varkappa$-semiconvexity assumptions* of this section. The general case follows by an approximation argument which was derived in [28, Section 6] for the case $\varkappa = 0$, but holds equally true for any $\varkappa \geqslant 0$.

Since N is a Young function, we may invoke Lemma 2.1 and replace $M_\mu f$ in (4.1) by $E_\mu f$ as in (4.9), at the expense of an additional universal constant in the final conclusion.

Let A denote an arbitrary Borel set in Ω so that $\mu^+(A) < \infty$, and let $\chi_{A,\varepsilon,\delta}(x) := (1 - \frac{1}{\varepsilon} d(x, A^d_\delta)) \vee 0$ be a continuous approximation in Ω to the characteristic function χ_A of A (as usual d denotes the induced geodesic distance). Our assumptions imply that

$$\frac{\mu(A^d_{\varepsilon+2\delta}) - \mu(A)}{\varepsilon} \geqslant \int |\nabla \chi_{A,\varepsilon,\delta}| \, d\mu .$$

Applying Lemma 4.5 to functions in $\mathcal{B}(\Omega)$ which approximate $\chi_{A,\varepsilon,\delta}$ (in say $W^{1,1}(\Omega, \mu)$) and passing to the limit inferior as $\varepsilon, \delta \to 0$ so that $\delta/\varepsilon \to 0$, it follows that

$$\int_0^t \frac{ds}{\sqrt{K(\varkappa, s)}} \mu^+(A) \geqslant \int |\chi_A - P_t \chi_A| \, d\mu$$

(note that the assumption $\mu^+(A) < \infty$ guarantees that $\mu(\overline{A} \setminus A) = 0$, so $\chi_{A,\varepsilon,\delta}$ tends to χ_A in $L_1(\mu)$). We start by rewriting the right-hand side as

$$\int_A (1 - P_t\chi_A) d\mu + \int_{\Omega \setminus A} P_t \chi_A d\mu = 2\left(\mu(A) - \int_A P_t \chi_A d\mu\right)$$

$$= 2\left(\mu(A)(1 - \mu(A)) - \int_\Omega (P_t\chi_A - \mu(A))(\chi_A - \mu(A)) d\mu\right)$$

$$= 2\left(\int_\Omega |\chi_A - \mu(A)|^2 d\mu - \int_\Omega |P_{t/2}(\chi_A - \mu(A))|^2 d\mu\right) .$$

Denoting $f = \chi_A - \mu(A)$ and using Proposition 4.7 to estimate the right-most expression, we obtain after using the estimates in (4.8), that for $t \leqslant 1/(2\varkappa)$

$$2\sqrt{t}\mu^+(A) \geqslant 2\mu(A)(1-\mu(A))\left(1 - (1+(q-1)M_t)^{-\frac{1}{q-1}}\right) , \tag{4.11}$$

where

$$M_t := \frac{2D^q}{\|f\|^q_{N(\mu)^*} \|f\|^{q-2}_{L_\infty}} \left(\int f^2 d\mu\right)^{q-1} \frac{2}{q}\left(\frac{t}{2}\right)^{\frac{q}{2}} .$$

To estimate M_t, we employ Lemma 4.6:

$$\begin{aligned}
\|\chi_A - \mu(A)\|_{N(\mu)^*} &\leqslant (1-\mu(A)) \|\chi_A\|_{N(\mu)^*} + \mu(A) \left\|\chi_{\Omega\setminus A}\right\|_{N(\mu)^*} \\
&= \mu(A)(1-\mu(A)) \left(\frac{1}{N^\wedge(\mu(A))} + \frac{1}{N^\wedge(1-\mu(A))}\right) \\
&\leqslant 2\frac{\mu(A)(1-\mu(A))}{N^\wedge(\min(\mu(A), 1-\mu(A)))} ,
\end{aligned}$$

and using that

$$\int f^2 d\mu = \mu(A)(1-\mu(A)),$$

we conclude that

$$M_t \geqslant L_t := E\frac{2}{q}\left(\frac{t}{2}\right)^{\frac{q}{2}} , \quad E := \frac{D^q N^\wedge(\min(\mu(A), 1-\mu(A)))^q}{2^{q-1}\mu(A)(1-\mu(A))} . \tag{4.12}$$

It remains to optimize on t in (4.11). Denote

$$t_0 := 4\left(\frac{q}{2E}\right)^{2/q} = 4\left(\frac{q}{2}\right)^{2/q} \frac{2^{2/q^*}(\mu(A)(1-\mu(A)))^{2/q}}{D^2 N^\wedge(\min(\mu(A), 1-\mu(A)))^2} .$$

1. If $t_0 \leqslant 1/(2\varkappa)$, we see from (4.12) that $L_{t_0} = 2^{q/2}$, and we immediately obtain from (4.11)

$$\begin{aligned}
\mu^+(A) &\geqslant c_1 \frac{\mu(A)(1-\mu(A))}{\sqrt{t_0}} \\
&\geqslant c_2 D \min(\mu(A), 1-\mu(A))^{1-1/q} N^\wedge(\min(\mu(A), 1-\mu(A))) ,
\end{aligned}$$

where $c_1, c_2 > 0$ are some numeric constants.

2. If $t_0 > 1/(2\varkappa)$, we evaluate (4.12) and (4.11) at time $t_1 = 1/(2\varkappa)$, for which $L_{t_1} < 2^{q/2}$. Therefore, $(1+(q-1)L_{t_1})^{\frac{1}{q-1}} \geqslant 1 + c_3 2^{-q/2} L_{t_1}$ (recall that $q \geqslant 2$), and hence, by (4.11),

$$\mu^+(A) \geqslant c_4\sqrt{\varkappa}\mu(A)(1-\mu(A))2^{-q/2}L_{t_1} ,$$

where $c_3, c_4 > 0$ are numeric constants. Plugging in L_{t_1}, we obtain

$$\mu^+(A) \geqslant \frac{c_5}{q4^q} \frac{D^q}{\varkappa^{\frac{q-1}{2}}} N^\wedge(\min(\mu(A), 1 - \mu(A)))^q .$$

Using that $N(t)^{1/q}/t$ is nondecreasing, which is equivalent to $N^\wedge(t)/t^{1/q}$ being nonincreasing, we conclude that

$$\mu^+(A) \geqslant \frac{c_6 N^\wedge(1/2)^{q-1}}{q4^q} \frac{D^q}{\varkappa^{\frac{q-1}{2}}} \min(\mu(A), 1 - \mu(A))^{1-1/q} \times N^\wedge(\min(\mu(A), 1 - \mu(A))) \tag{4.13}$$

for some numeric constants $c_5, c_6 > 0$.

Combining both cases, we obtain the assertion in the case $q \geqslant 2$. □

We conclude the study of the case $q \geqslant 2$ by mentioning that, even though our estimates in (4.13) degrade as $q \to \infty$, it is also possible to study the limiting case $q = \infty$. In this case, the functional inequality (4.1) corresponds to an integrability property of *Lipschitz* functions f with $M_\mu(f) = 0$, or equivalently, to the *concentration* of the measure μ, in terms of the decay of $\mu(\Omega \setminus A_t^d)$ as a function of t for sets A with $\mu(A) \geqslant 1/2$. Using techniques from Riemannian geometry, it is still possible to deduce in this case an appropriate isoperimetric inequality under our $\varkappa$-semiconvexity assumptions (and an appropriate necessary assumption on the concentration); cf. [30].

4.2 Case $1 < q \leqslant 2$

We present two proofs of Theorem 4.1 for this case, each having its own advantages and drawbacks. The first runs along the same lines as in the previous subsection, and is new even in the $\varkappa = 0$ case. It has the advantage of working for arbitrary $N \in \mathcal{N}$ satisfying the assumptions of Theorem 4.1, but with this approach the estimates on $C_{N,q}$ degrade as $q \to 1$. This does not necessarily happen with the second proof, which is based on the idea in [28] of reducing the claim to the $q = 2$ case using Proposition 3.12. Nevertheless, some further conditions on N will need to be imposed for this approach to work.

4.2.1 Semigroup Approach

We need an analogue of Proposition 4.7, which again may be of independent interest.

Proposition 4.9. *Assume that the following (N, q) Orlicz–Sobolev inequality is satisfied for $N \in \mathcal{N}$ and $1 < q \leqslant 2$:*

$$\forall f \in \mathcal{F} \quad D \|f - E_\mu f\|_{N(\mu)} \leqslant \||\nabla f|\|_{L_q(\mu)} . \tag{4.14}$$

Then for any $f \in L_\infty(\Omega)$ with $\int f d\mu = 0$, we have for all $t \geqslant 0$

$$\int |P_t f|^q d\mu \leqslant \int f^q d\mu \left(1 + \frac{2D^2}{\|f\|_{L_1(\mu)}^{\frac{2(2-q)}{q-1}} \|f\|_{N(\mu)^*}^2} \left(\int f^q d\mu \right)^{\frac{2}{q(q-1)}} t \right)^{-\frac{q(q-1)}{2}} .$$

Note that the case $q = 2$ is identical to the one in Proposition 4.7.

Proof. Denote

$$u(t) = \int |P_t f|^q d\mu$$

and observe that

$$\begin{aligned} u'(t) &= q \int |P_t f|^{q-1} sign(P_t f) \Delta_{\Omega,\mu}(P_t f) d\mu \\ &= -q(q-1) \int |P_t f|^{q-2} |\nabla P_t f|^2 d\mu . \end{aligned}$$

Note that $u(t)$ is decreasing. Using the Hölder inequality (recall $q \leqslant 2$) twice, we estimate

$$\begin{aligned} u(t)^{\frac{q}{q-1}} &\leqslant \left(\int |P_{t/2} f|^q d\mu \right)^{\frac{q}{q-1}} \leqslant \left(\int |P_{t/2} f| d\mu \right)^{\frac{(2-q)q}{q-1}} \left(\int |P_{t/2} f|^2 d\mu \right)^q \\ &\leqslant \|f\|_{L_1(\mu)}^{\frac{(2-q)q}{q-1}} \left(\int f P_t f d\mu \right)^q \leqslant \|f\|_{L_1(\mu)}^{\frac{(2-q)q}{q-1}} \|f\|_{N(\mu)^*}^q \|P_t f\|_{N(\mu)}^q \\ &\leqslant \|f\|_{L_1(\mu)}^{\frac{(2-q)q}{q-1}} \|f\|_{N(\mu)^*}^q \frac{1}{D^q} \int |\nabla P_t f|^q d\mu \\ &\leqslant \frac{\|f\|_{L_1(\mu)}^{\frac{(2-q)q}{q-1}} \|f\|_{N(\mu)^*}^q}{D^q} \left(\int |P_t f|^q d\mu \right)^{\frac{2-q}{2}} \left(\int |P_t f|^{q-2} |\nabla P_t f|^2 d\mu \right)^{\frac{q}{2}} . \end{aligned}$$

Rearranging terms, we see that

$$u'(t) \leqslant -q(q-1) \frac{D^2}{\|f\|_{L_1(\mu)}^{\frac{2(2-q)}{q-1}} \|f\|_{N(\mu)^*}^2} u(t)^{1+\frac{2}{q(q-1)}} .$$

Setting $v(t) = u(t)^{-\frac{2}{q(q-1)}}$, we obtain

$$v'(t) \geqslant \frac{2D^2}{\|f\|_{L_1(\mu)}^{\frac{2(2-q)}{q-1}} \|f\|_{N(\mu)^*}^2} ,$$

and the assertion follows after integrating in t. □

Proof of Theorem 4.1 *in the case* $q \leqslant 2$. *Semigroup approach.* We begin as in the proof of the $q \geqslant 2$ case above, obtaining for a Borel set $A \subset \Omega$ and any time $0 \leqslant t \leqslant 1/(2\varkappa)$

$$2\sqrt{t}\mu^+(A) \geqslant \int |\chi_A - P_t \chi_A| \, d\mu.$$

Denote $f = \chi_A - \mu(A)$. Since $|x|^q - |y|^q \leqslant q|x-y|$ for all $|x|, |y| < 1$, we have

$$2\sqrt{t}\mu^+(A) \geqslant \int |f - P_t f| \, d\mu \geqslant \frac{1}{q} \left(\int |f|^q d\mu - \int |P_t f|^q d\mu \right) ,$$

and, using Proposition 4.9, we obtain for $0 \leqslant t \leqslant 1/(2\varkappa)$

$$2\sqrt{t}\mu^+(A) \geqslant \frac{1}{q} \int |f|^q d\mu \left(1 - (1 + 2Mt)^{-\frac{q(q-1)}{2}} \right) , \tag{4.15}$$

where

$$M := \frac{D^2 (\int |f|^q d\mu)^{\frac{2}{q(q-1)}}}{\|f\|_{L_1(\mu)}^{\frac{2(2-q)}{q-1}} \|f\|_{N(\mu)^*}^2} .$$

As in the proof of the $q \geqslant 2$ case, it is easy to verify that

$$M \geqslant \frac{D^2 \left(\mu(A)(1-\mu(A))\right)^{\frac{2}{q(q-1)}} N^\wedge(\min(\mu(A), 1-\mu(A)))^2}{(2\mu(A)(1-\mu(A)))^{\frac{2(2-q)}{q-1}+2}} ,$$

which simplifies to

$$M \geqslant E := \frac{D^2 N^\wedge(\min(\mu(A), 1-\mu(A)))^2}{2^{\frac{2}{q-1}} (\mu(A)(1-\mu(A)))^{\frac{2}{q}}} .$$

As usual, we need to optimize (4.15) in t. Set $t_0 := 1/E$.

1. If $t_0 \leqslant 1/(2\varkappa)$, we obtain

$$\begin{aligned} \mu^+(A) &\geqslant \frac{c_1(q-1)}{\sqrt{t_0}} \mu(A)(1-\mu(A)) \\ &\geqslant \frac{c_2(q-1)}{2^{\frac{1}{q-1}}} D \min(\mu(A), 1-\mu(A))^{1-1/q} N^\wedge(\min(\mu(A), 1-\mu(A))) , \end{aligned}$$

for some numeric constants $c_1, c_2 > 0$.

2. If $t_0 > 1/(2\varkappa)$, we evaluate (4.15) at $t_1 = 1/(2\varkappa)$. Since $Et_1 < 1$, we obtain

$$\begin{aligned}\mu^+(A) &\geqslant \frac{c_3(q-1)Et_1}{\sqrt{t_1}}\mu(A)(1-\mu(A)) \\ &\geqslant \frac{c_4(q-1)}{4^{\frac{1}{q-1}}}\frac{D^2}{\sqrt{\varkappa}}\min(\mu(A),1-\mu(A))^{1-2/q}N^\wedge(\min(\mu(A),1-\mu(A)))^2\ ,\end{aligned}$$

where $c_3, c_4 > 0$ are numeric constants. Using that $N^\wedge(t)/t^{1/q}$ is nonincreasing, we conclude that

$$\begin{aligned}\mu^+(A) \geqslant &\frac{c_5(q-1)N^\wedge(1/2)}{4^{\frac{1}{q-1}}}\frac{D^2}{\sqrt{\varkappa}}\min(\mu(A),1-\mu(A))^{1-1/q} \\ &\times N^\wedge(\min(\mu(A),1-\mu(A)))\ .\end{aligned}$$

Combining both cases, Theorem 4.1 follows for $1 < q \leqslant 2$. □

4.2.2 Capacity Approach

To complete a circle as we conclude this work, we present a second proof using capacities following [28, Theorem 4.5].

Proof of Theorem 4.1 for $q \leqslant 2$. Capacity approach. The assumption (4.1) implies by Proposition 3.8 that

$$\mathrm{Cap}_q(t,1/2) \geqslant DN^\wedge(t) \quad \forall t \in [0,1/2]\ ,$$

where $N^\wedge(t)/t^{1/q}$ is nonincreasing by our assumption on N. Using Proposition 3.12 (with $q_0 = q, q = 2$) to pass from Cap_q to Cap_2, we obtain

$$\mathrm{Cap}_2(t,1/2) \geqslant \frac{(1-\frac{2}{q^*})^{1/2}}{(\frac{q^*}{2}-1)^{1/q^*}}D\left(\int_t^{1/2}\frac{ds}{(s-t)^{2/q^*}N^\wedge(s)^2}\right)^{-1/2} \quad \forall t \in [0,1/2]\ .$$

Next, we modify $N^\wedge(t)$ when $t \geqslant 1/2$ as follows:

$$N_0^\wedge(t) = \begin{cases} N^\wedge(t) & t \in [0,1/2], \\ N^\wedge(1/2)2^{1/q}t^{1/q} & t \in [1/2,\infty), \end{cases}$$

so that $N_0^\wedge(t)/t^{1/q}$ is still nonincreasing. By [28, Lemma 4.2, Remark 4.3], it follows that there exists a numeric constant $c_1 > 0$ so that

$$\mathrm{Cap}_2(t,1/2) \geqslant c_1DN_2^\wedge(t) \quad \forall t \in [0,1/2]\ ,$$

where $N_2 \in \mathcal{N}$ is a function so that

$$N_2^\wedge(t) := \frac{1}{\left(\int_t^\infty \frac{ds}{s^{2/q^*} N_0^\wedge(s)^2}\right)^{1/2}} .$$

Moreover, by [28, Lemma 4.4], N_2 is, in fact, a convex function and $N_2(t)^{1/2}/t$ is nondecreasing. Proposition 3.9 then implies that

$$\forall f \in \mathcal{F} \quad \frac{c_1}{4} D \left\| f - M_\mu f \right\|_{N_2(\mu)} \leqslant \left\| |\nabla f| \right\|_{L_2(\mu)} .$$

We can now apply the case $q = 2$ of Theorem 4.1, and conclude that

$$\widetilde{\mathcal{I}}(t) \geqslant \min(c_2, N_2^\wedge(1/2)) \min\left(D, \frac{D^2}{\sqrt{\varkappa}}\right) t^{1/2} N_2^\wedge(t) \quad \forall t \in [0, 1/2]$$

with $c_2 > 0$ a numeric constant. Defining

$$C_{N,q} := \min(c_2, N_2^\wedge(1/2)) \inf_{0<t<1/2} \frac{t^{1/q-1/2}}{\left(\int_t^\infty \frac{N_0^\wedge(t)^2 ds}{s^{2/q^*} N_0^\wedge(s)^2}\right)^{1/2}} , \tag{4.16}$$

since $N_0^\wedge(t) = N^\wedge(t)$ for $t \in [0, 1/2]$, this implies

$$\widetilde{\mathcal{I}}(t) \geqslant C_{N,q} \min\left(D, \frac{D^2}{\sqrt{\varkappa}}\right) t^{1-1/q} N^\wedge(t) \quad \forall t \in [0, 1/2],$$

as required. This concludes the proof under the additional assumption that $C_{N,q} > 0$. □

Remark 4.10. Note the similarity between the definitions of $C_{N,q}$ in (4.16) and $B_{N,q}$ in (3.9).

This proof has the advantage that one may obtain estimates which do not degrade to 0 as $q \to 1$, if the constant $C_{N,q}$ in (4.16) may be controlled. Indeed, this is the case for the family of q-log-Sobolev inequalities (2.3) for $q \in [1, 2]$, discussed in Section 2. As already mentioned, it was shown by Bobkov and Zegarlinski that these inequalities may be put in an equivalent form, given by (2.4), corresponding to (φ_q, q) Orlicz–Sobolev inequalities, where $\varphi_q = t^q \log(1 + t^q)$. It is not hard to verify (cf. [28, Corollary 4.8]) that $C_{\varphi_q, q} \geqslant c > 0$ uniformly in $q \in [1, 2]$, and so we deduce the following assertion.

Corollary 4.11. *Under our $\varkappa$-semiconvexity assumptions, the q-log-Sobolev inequality*

$$\forall f \in \mathcal{F} \quad D\left(\int |f|^q \log |f|^q d\mu - \int |f|^q d\mu \log(\int |f|^q d\mu)\right)^{1/q} \leqslant \left\| |\nabla f| \right\|_{L_q(\mu)}$$

with $1 \leqslant q \leqslant 2$ implies the following isoperimetric inequality:

$$\widetilde{\mathcal{I}}(t) \geqslant c \min\left(D, \frac{D^2}{\sqrt{\varkappa}}\right) t \log^{1/q} 1/t \quad \forall t \in [0, 1/2] \ ,$$

where $c > 0$ *is a numeric constant (independent of* q*).*

Acknowledgments. I would like to thank Sasha Sodin for introducing me to Maz'ya's work on capacities.

The work was supported by NSF under agreement #DMS-0635607.

References

1. Bakry, D., Émery, M.: Diffusions hypercontractives. In: Séminaire de probabilités, XIX, 1983/84. Lect. Notes Math. **1123**, pp. 177–206. Springer (1985)
2. Bakry, D., Ledoux, M.: Lévy-Gromov's isoperimetric inequality for an infinite-dimensional diffusion generator. Invent. Math. **123**, no. 2, 259–281 (1996)
3. Barthe, F., Cattiaux, P., Roberto, C.: Interpolated inequalities between exponential and Gaussian, Orlicz hypercontractivity and isoperimetry. Rev. Mat. Iberoamericana **22**, no. 3, 993–1067 (2006)
4. Barthe, F., Kolesnikov, A.V.: Mass transport and variants of the logarithmic Sobolev inequality. J. Geom. Anal. **18**, no. 4, 921–979 (2008)
5. Barthe, F., Maurey, B.: Some remarks on isoperimetry of Gaussian type. Ann. Inst. H. Poincaré Probab. Statist. **36**, no. 4, 419–434 (2000)
6. Bobkov, S.G.: A functional form of the isoperimetric inequality for the Gaussian measure. J. Funct. Anal. **135**, no. 1, 39–49 (1996)
7. Bobkov, S.G.: An isoperimetric inequality on the discrete cube, and an elementary proof of the isoperimetric inequality in Gauss space. Ann. Probab. **25**, no. 1, 206–214 (1997)
8. Bobkov, S.G., Götze, F.: Exponential integrability and transportation cost related to logarithmic Sobolev inequalities. J. Funct. Anal. **163**, no. 1, 1–28 (1999)
9. Bobkov, S.G., Houdré, C.: Isoperimetric constants for product probability measures. Ann. Probab. **25**, no. 1, 184–205 (1997)
10. Bobkov, S.G., Houdré, C.: Some connections between isoperimetric and Sobolev type inequalities. Mem. Am. Math. Soc. **129**, no. 616 (1997)
11. Bobkov, S.G., Zegarlinski, B.: Entropy bounds and isoperimetry. Mem. Am. Math. Soc. **176**, no. 829 (2005)
12. Borell, Ch.: Convex measures on locally convex spaces. Ark. Mat. **12**, 239–252 (1974)
13. Buser, P.: A note on the isoperimetric constant. Ann. Sci. École Norm. Sup. (4) **15**, no. 2, 213–230 (1982)
14. Cheeger, J.: A lower bound for the smallest eigenvalue of the Laplacian. In: Gunning, R. (ed.), Problems in Analysis, pp. 195–199. Princeton Univ. Press, Princeton, NJ (1970)
15. Davies, E.B.: Heat Kernels and Spectral Theory. Cambridge Univ. Press, Cambridge (1989)
16. Federer, H.: Curvature measures. Trans. Am. Math. Soc. **93**, no. 418–491 (1959)

17. Federer, H., Fleming, W.H.: Normal and integral currents. Ann. Math. **72**, 458–520 (1960)
18. Fleming, W.H.: Functions whose partial derivatives are measures. Illinois J. Math. **4**, 52–478 (1960)
19. Hsu, E.P.: Multiplicative functional for the heat equation on manifolds with boundary. Michigan Math. J. **50**, no. 2, 351–367 (2002)
20. Ledoux, M.: A simple analytic proof of an inequality by P. Buser. Proc. Am. Math. Soc. **121**, no. 3, 951–959 (1994)
21. Ledoux, M.: The geometry of Markov diffusion generators. Ann. Fac. Sci. Toulouse Math. (6) **9**, no. 2, 305–366 (2000)
22. Ledoux, M.: Spectral gap, logarithmic Sobolev constant, and geometric bounds. In: Surveys in Differential Geometry. IX, pp. 219–240. Int. Press, Somerville, MA (2004)
23. Maz'ya, V.G.: Classes of domains and imbedding theorems for function spaces (Russian). Dokl. Akad. Nauk SSSR **3**, 527–530 (1960); English transl.: Sov. Math. Dokl. **1**, 882–885 (1961)
24. Maz'ya, V.G.: p-Conductivity and theorems on imbedding certain functional spaces into a C-space (Russian). Dokl. Akad. Nauk SSSR **140**, 299–302 (1961); English transl.: Sov. Math. Dokl. **2**, 1200–1203 (1961)
25. Maz'ya, V.G.: The negative spectrum of the higher-dimensional Schrödinger operator (Russian). Dokl. Akad. Nauk SSSR **144**, 721–722 (1962); English transl.: Sov. Math. Dokl. **3**, 808–810 (1962)
26. Maz'ya, V.G.: On the solvability of the Neumann problem (Russian). Dokl. Akad. Nauk SSSR **147**, 294–296 (1962); English transl.: Sov. Math. Dokl. **3**, 1595–1598 (1962)
27. Maz'ya, V.G.: Sobolev spaces. Springer, Berlin (1985)
28. Milman, E.: On the role of convexity in functional and isoperimetric inequalities. to appear in the Proc. London Math. Soc., arxiv.org/abs/0804.0453, 2008.
29. Milman, E.: On the role of convexity in isoperimetry, spectral gap and concentration. Invent. Math. **177**, no. 1, 1–43 (2009)
30. Milman, E.: Isoperimetric and concentration inequalities - equivalence under curvature lower bound. arxiv.org/abs/0902.1560 (2009)
31. Milman, E., Sodin, S.:. An isoperimetric inequality for uniformly log-concave measures and uniformly convex bodies. J. Funct. Anal. **254**, no. 5, 1235–1268 (2008) arxiv.org/abs/math/0703857
32. Muckenhoupt, B.: Hardy's inequality with weights. Studia Math. **44**, 31–38 (1972)
33. Qian, Z.: A gradient estimate on a manifold with convex boundary. Proc. Roy. Soc. Edinburgh Sect. A **127**, no. 1, 171–179 (1997)
34. Roberto, C., Zegarliński, B.: Orlicz–Sobolev inequalities for sub-Gaussian measures and ergodicity of Markov semigroups. J. Funct. Anal. **243**, no. 1, 28–66 (2007)
35. Sodin, S.: An isoperimetric inequality on the ℓ_p balls. Ann. Inst. H. Poincaré Probab. Statist. **44**, no. 2, 362–373 (2008)
36. Wang, F.-Y.: Gradient estimates and the first Neumann eigenvalue on manifolds with boundary. Stochastic Process. Appl. **115**, no. 9, 1475–1486 (2005)

Pseudo-Poincaré Inequalities and Applications to Sobolev Inequalities

Laurent Saloff-Coste

Abstract Most smoothing procedures are via averaging. Pseudo-Poincaré inequalities give a basic L^p-norm control of such smoothing procedures in terms of the gradient of the function involved. When available, pseudo-Poincaré inequalities are an efficient way to prove Sobolev type inequalities. We review this technique and its applications in various geometric setups.

1 Introduction

This paper is concerned with the question of proving the Sobolev inequality

$$\forall f \in \mathcal{C}_c^\infty(M), \quad \|f\|_q \leqslant S(M,p,q)\|\nabla f\|_p \tag{1.1}$$

when $M = (M,g)$ is a Riemannian manifold, perhaps with boundary ∂M, and $\mathcal{C}_c^\infty(M)$ is the space of smooth compactly supported functions on M (if M is a manifold with boundary ∂M, then points on ∂M are interior points in M and functions in $\mathcal{C}_c(M)$ do not have to vanish at such points). We say that (M,g) is *complete* when M equipped with the Riemannian distance is a complete metric space.

In (1.1), $p,q \in [1,\infty)$ and $q > p$. The norms $\|\cdot\|_p$ and $\|\cdot\|_q$ are computed with respect to some fixed reference measure, perhaps the Riemannian measure dv on M or, more generally, a measure $d\mu$ on M of the form $d\mu = \sigma dv$, where σ is a smooth positive function on M. We set $V(x,r) = \mu(B(x,r))$, where $B(x,r)$ is the geodesic ball of center $x \in M$ and radius $r \geqslant 0$. The gradient ∇f of $f \in \mathcal{C}^\infty(M)$ at x is the tangent vector at x defined by

Laurent Saloff-Coste
Department of Mathematics, Cornell University, Mallot Hall, Ithaca, NY 14853, USA
e-mail: lsc@math.cornell.edu

A. Laptev (ed.), *Around the Research of Vladimir Maz'ya I: Function Spaces*, International Mathematical Series 11, DOI 10.1007/978-1-4419-1341-8_16,

$$g_x(\nabla f(x), u) = df|_x (u)$$

for any tangent vector $u \in T_x$. Its length $|\nabla f|$ is given by $|\nabla f|^2 = g(\nabla f, \nabla f)$.

We will not be concerned here with the (interesting) problem of finding the best constant $S(M, p, q)$ but only with the validity of the Sobolev inequality (1.1), for some constant $S(M, p, q)$.

In $\mathbb{R}^n$, equipped with the Lebesgue measure dx, (1.1) holds for any $p \in [1, n)$ with $q = np/(n-p)$. The two simplest contexts where the question of the validity of (1.1) is meaningful is when $M = \Omega$ is a subset of $\mathbb{R}^n$, or when $\mathbb{R}^n$ is equipped with a measure $\mu(dx) = \sigma(x)dx$. In the former case, it is natural to relax our basic assumption and allow domains with nonsmooth boundary. It then becomes important to pay more attention to the exact domain of validity of (1.1) as approximation by functions that are smooth up to the boundary may not be available (cf., for example, [13, 14]).

The fundamental importance of the inequality (1.1) in analysis and geometry is well established. It is beautifully illustrated in the work of V. Maz'ya. One of the fundamental references on Sobolev inequalities is Maz'ya's treaty "Sobolev Spaces" [13] which discuss (1.1) and its many variants in $\mathbb{R}^n$ and in domains in $\mathbb{R}^n$ (cf. also [1, 3, 14] and the references therein). Maz'ya's treaty anticipates on many later works including [2]. More specialized works that discuss (1.1) in the context of Riemannian manifolds and Lie groups include [11, 19, 23] among many other possible references.

The aim of this article is to discuss a particular approach to (1.1) that is based on the notion of pseudo-Poincaré inequality. This technique is elementary in nature and quite versatile. It seems it has its origin in [4, 7, 17, 18] and was really emphasized first in [7, 18], and in [2]. To put things in some perspective, recall that the most obvious approach to (1.1) is via some "representation formula" that allows us to "recover" f from its gradient through an integral transform. One is them led to study the mapping properties of the integral transform in question.

However, this natural approach is not well suited to many interesting geometric setups because the needed properties of the relevant integral transforms might be difficult to establish or might even not hold true. For instance, its seems hard to use this approach to prove the following three (well-known) fundamental results.

Theorem 1.1. *Assume that (M, g) is a Riemannian manifold of dimension n equipped with its Riemannian measure and which is of one of the following three types:*

1. *A connected simply connected noncompact unimodular Lie group equipped with a left-invariant Riemannian structure.*
2. *A complete simply connected Riemannian manifold without boundary with nonpositive sectional curvature (i.e., a Cartan–Hadamard manifold).*
3. *A complete Riemannian manifold without boundary with nonnegative Ricci curvature and maximal volume growth.*

Then for any $p \in [1, n)$ the Sobolev inequality (1.1) *holds on M with $q = np/(n-p)$ for some constant $S(M, p, q) < \infty$.*

One remarkable thing about this theorem is the conflicting nature of the curvature assumptions made in the different cases. Connected Lie groups almost always have curvature that varies in sign, whereas the second and third cases we make opposite curvature assumptions. Not surprisingly, the original proofs of these different results have rather distinct flavors.

The result concerning unimodular Lie groups is due to Varopoulos and more is true in this case (cf. [22, 23]).

The result concerning Cartan–Hadamard manifolds is a consequence of a more general result due to Michael and Simon [15] and Hoffmann and Spruck [12]. A more direct prove was given by Croke [9] (cf. also [11, Section 8.1 and 8.2] for a discussion and further references).

The result concerning manifolds with nonnegative Ricci curvature and maximal volume growth (i.e., $V(x,r) \geqslant cr^n$ for some $c > 0$ and all $x \in M, r > 0$) was first obtained as a consequence of the Li-Yau heat kernel estimate using the line of reasoning in [22].

One of the aims of this paper is to describe proofs of these three results that are based on a common unifying idea, namely, the use of what we call pseudo-Poincaré inequalities. Our focus will be on how to prove the desired pseudo-Poincaré inequalities in the different contexts covered by this theorem. For relevant background on geodesic coordinates and Riemannian geometry see [5, 6, 10].

2 Sobolev Inequality and Volume Growth

There are many necessary conditions for (1.1) to hold and some are discussed in Maz'ya's treaty [13] in the context of Euclidean domains. For instance, if (1.1) holds for some fixed $p = p_0 \in [1, \infty)$ and $q = q_0 > p_0$ and we define m by $1/q_0 = 1/p_0 - 1/m$, then (1.1) also holds for all $p \in [p_0, m)$ with q given by $1/q = 1/p - 1/m$ (this easily follows by applying the p_0, q_0 inequality to $|f|^\alpha$ with a properly chosen $\alpha > 1$ and using the Hölder inequality). More importantly to us here is the following result (cf., for example, [2] or [19, Corollary 3.2.8]).

Theorem 2.1. *Let (M, g) be a complete Riemannian manifold equipped with a measure $d\mu = \sigma dv$, $0 < \sigma \in \mathcal{C}^\infty(M)$. Assume that* (1.1) *holds for some $1 \leqslant p < q < \infty$ and set $1/q = 1/p - 1/m$. Then for any $r \in (m, \infty)$ and any bounded open set $U \subset M$*

$$\forall f \in \mathcal{C}_c^\infty(U), \quad \|f\|_\infty \leqslant C_r \mu(U)^{1/m - 1/r} \|\nabla f\|_r. \tag{2.1}$$

Corollary 2.1. *If the complete Riemannian manifold* (M, g) *equipped with a measure* $d\mu = \sigma dv$ *satisfies* (1.1) *for some* $1 \leqslant p < q < \infty$, *then*

$$\inf\{s^{-m}V(x,s) : x \in M,\ s > 0\} > 0$$

with $1/q = 1/p - 1/m$.

Proof. Fix $r > m$ and apply (2.1) to the function

$$\phi_{x,s}(y) = y \mapsto (s - \rho(x,y))_+ = \max\{(s - \rho(x,y), 0\},$$

where ρ is the Riemannian distance on (M, g). Because (M, ρ) is complete, this function is compactly supported and can be approximated by smooth compactly supported functions in the norm $\|f\|_\infty + \|\nabla f\|_r$, justifying the use of (2.1). Moreover, $|\nabla \phi_{x,s}| \leqslant 1$ a.e. so that $\|\nabla \phi_{x,s}\|_r \leqslant V(x,r)^{1/r}$. This yields $s \leqslant C_r V(x,s)^{1/m-1/r} V(x,s)^{1/r} = C_r V(x,s)^{1/m}$ as desired. □

Remark 2.1. Let Ω be an unbounded Euclidean domain.

(a) If we assume that (1.1) holds but only for all traces $f|_\Omega$ of functions $f \in C_c^\infty(\mathbb{R}^n)$, then we can conclude that (2.1) holds for such functions. Applying (2.1) to $\psi_{x,s}(y) = (s - \|x - y\|)_+$, $x \in \Omega$, $s > 0$, yields

$$|\{z \in \Omega : \|x - z\| < s\}| \geqslant cs^m.$$

(b) If, instead, we consider the intrinsic geodesic distance $\rho = \rho_\Omega$ in Ω and assume that (1.1) holds for all ρ-Lipschitz functions vanishing outside some ρ-ball, then the same argument, properly adapted, yields $V(x,s) \geqslant cs^m$, where $V(x,s)$ is the Lebesgue measure of the ρ-ball of radius s around x in Ω.

For domains with rough boundary, the hypotheses made respectively in (a) and (b) may be very different.

3 The Pseudo-Poincaré Approach to Sobolev Inequalities

Our aim is to illustrate the following result which provide one of the most elementary and versatile ways to prove a Sobolev inequality in a variety of contexts (cf., for example, [2, Theorem 9.1]). The two main hypotheses in the following statement concern a family of linear operators A_r acting, say, on smooth compactly supported functions. The first hypothesis captures the idea that A_r is smoothing. The sup-norm of $A_r f$ is controlled in terms of the L^p-norm of f only and tends to 0 as r tends to infinity. The second hypothesis implies, in particular, that $A_r f$ is close to f if $|\nabla f|$ is in L^p and r is small.

Theorem 3.1. *Fix $m, p \geqslant 1$. Assume that for each $r > 0$ there is a linear map $A_r : \mathcal{C}_c^\infty(M) \to L^\infty(M)$ such that*

- $\forall f \in \mathcal{C}_c^\infty(M), r > 0, \ \|A_r f\|_\infty \leqslant C_1 r^{-m/p}\|f\|_p$.
- $\forall f \in \mathcal{C}_c^\infty(M), r > 0, \|f - A_r f\|_p \leqslant C_2 r \|\nabla f\|_p$.

Then, if $p \in [1, m)$ and $q = mp/(m-p)$, there exists a finite constant $S(M,p,q) = C(p,q)C_2 C_1^{1/m}$ such that the Sobolev inequality (1.1) holds on M.

Outline of the proof. The proof is entirely elementary and is given in [2]. For illustrative purpose and completeness, we explain the first step. Consider the distribution function of $|f|$, $F(s) = \mu(\{x : |f(x)| > s\})$. Then

$$F(s) \leqslant \mu(\{|f - A_r f| > s/2\}) + \mu(\{|A_r f| > s/2\}).$$

By hypothesis, if $s = 2C_1 r^{-m/p}\|f\|_p$, then $\mu(\{|A_r f| > s/2\}) = 0$ and

$$F(s) \leqslant \mu(\{|f - A_r f| > s/2\}) \leqslant 2^p C_2^p r^p s^{-p} \|\nabla f\|_p^p.$$

This gives

$$s^{p(1+1/m)} F(s) \leqslant 2^{p(1+1/m)} C_1^{p/m} C_2^p \|\nabla f\|_p^p \|f\|_p^{p/m}.$$

This is a weak form of the desired Sobolev inequality (1.1). But, as is already apparent in [13], such a weak form of (1.1) actually imply (1.1) (cf. also [2, 19]). □

Remark 3.1. When $p = 1$ and $\mu = v$ is the Riemannian volume, we get

$$s^{1+1/m} v(\{|f| > s\}) \leqslant 2^{1+1/m} C_1^{1/m} C_2 \|\nabla f\|_1 \|f\|_1^{1/m}.$$

For any bounded open set Ω with smooth boundary $\partial\Omega$ we can find a sequence of functions $f_n \in \mathcal{C}_c^\infty(M)$ such that $f_n \to \mathbf{1}_{\Omega_n}$ and $\|\nabla f_n\|_1 \to v_{n-1}(\partial\Omega)$. This yields the isoperimetric inequality

$$v(\Omega)^{1-1/m} \leqslant 2^{1+1/m} C_1^{1/m} C_2 v_{n-1}(\partial\Omega).$$

Of course, as was observed long ago by Maz'ya and others, the classical co-area formula and the above inequality imply

$$\forall f \in \mathcal{C}_c^\infty(M), \ \ \|f\|_{m/(m-1)} \leqslant 2^{1+1/m} C_1^{1/m} C_2 \|\nabla f\|_1.$$

There are many situations where one does not expect (1.1) to hold, but where one of the local versions

$$\forall f \in \mathcal{C}_c^\infty(M), \ \ \|f\|_q \leqslant S(M,p,q)(\|\nabla f\|_p + \|f\|_p), \tag{3.1}$$

or ($\Subset$ indicates an open relatively compact inclusion)

$$\forall \Omega \Subset M, \ \forall f \in \mathcal{C}_c^\infty(\Omega), \quad \|f\|_q \leqslant S(\Omega, p, q)(\|\nabla f\|_p + \|f\|_p) \tag{3.2}$$

may hold. This is handled by the following local version of Theorem 3.1 (cf. [2] and [19, Section 3.3.2]).

Theorem 3.2. *Fix an open subset $\Omega \subset M$. Assume that for each $r \in (0, R)$ there is a linear map $A_r : \mathcal{C}_c^\infty(\Omega) \to L^\infty(M)$ such that*

- $\forall f \in \mathcal{C}_c^\infty(\Omega), r \in (0, R), \ \|A_r f\|_\infty \leqslant C_1 r^{-m/p} \|f\|_p$.
- $\forall f \in \mathcal{C}_c^\infty(\Omega), r \in (0, R), \ \|f - A_r f\|_p \leqslant C_2 r \|\nabla f\|_p$.

Then, if $p \in [1, m)$ and $q = mp/(m-p)$, there exists a finite constant $S = S(p, q)$ such that

$$\forall f \in \mathcal{C}_c^\infty(\Omega), \quad \|f\|_q \leqslant S C_1^{1/m} (C_2 \|\nabla f\|_p + R^{-1} \|f\|_p). \tag{3.3}$$

Another useful version is as follows. For any open set Ω we let $W^{1,p}(\Omega)$ be the space of those functions in $L^p(\Omega)$ whose first order partial derivatives in the sense of distributions (in any local chart) can be represented by a locally integrable function and such that

$$\int_\Omega |\nabla f|^p dv < \infty.$$

We write $\|f\|_{\Omega,p}$ for the L^p-norm of f over Ω. Note that $\mathcal{C}^\infty(\Omega) \cap W^{1,p}(\Omega)$ is dense in $W^{1,p}(\Omega)$ for $1 \leqslant p < \infty$ (cf., for example, [1, 3, 13]).

Theorem 3.3. *Fix an open subset $\Omega \subset M$. Assume that for each $r \in (0, R)$ there is a linear map $A_r : \mathcal{C}^\infty(\Omega) \cap W^{1,p}(\Omega) \to L^\infty(M)$ such that*

- $\forall f \in \mathcal{C}^\infty(\Omega) \cap W^{1,p}(\Omega), r \in (0, R), \ \|A_r f\|_\infty \leqslant C_1 r^{-m/p} \|f\|_p$.
- $\forall f \in \mathcal{C}^\infty(\Omega) \cap W^{1,p}(\Omega), r \in (0, R), \ \|f - A_r f\|_p \leqslant C_2 r \|\nabla f\|_p$.

Then, if $p \in [1, m)$ and $q = mp/(m-p)$, there exists a finite constant $S = S(p, q)$ such that

$$\forall f \in W^{1,p}(\Omega), \quad \|f\|_q \leqslant S C_1^{1/m} (C_2 \|\nabla f\|_p + R^{-1} \|f\|_p). \tag{3.4}$$

4 Pseudo-Poincaré Inequalities

The term Poincaré inequality (say, with respect to a bounded domain $\Omega \subset M$) is used with at least two distinct meanings:

- The Neumann type L^p-Poincaré inequality for a bounded domain $\Omega \subset M$ is the inequality

$$\forall f \in W^{1,p}(\Omega), \ \inf_{\xi \in \mathbb{R}} \int_{\Omega} |f - \xi|^p dv \leqslant P_N(\Omega) \int_{\Omega} |\nabla f|^p dv.$$

- The Dirichlet type L^p-Poincaré inequality for a bounded domain $\Omega \subset M$ is the inequality

$$\forall f \in C_c^\infty(\Omega), \ \int_{\Omega} |f|^p dv \leqslant P_D(\Omega) \int_{\Omega} |\nabla f|^p dv.$$

When $p = 2$ and the boundary is smooth, the first (respectively, the second) inequality is equivalent to the statement that the lowest nonzero eigenvalue $\lambda_N(\Omega)$ (respectively, $\lambda_D(\Omega)$) of the Laplacian with the Neumann boundary condition (respectively, the Dirichlet boundary condition) is bounded below by $1/P_N(\Omega)$ (respectively, $1/P_D(\Omega)$). Note that if $M = \mathbb{S}^n$ is the unit sphere in $\mathbb{R}^{n+1}$ and $\Omega = B(o,r)$, $r < 2\pi$, is a geodesic ball, then $P_N(\Omega) \to 1/(n+1)$ and $P_D(\Omega) \to \infty$ as r tends to 2π.

Here, we will use the term Poincaré inequality for the collection of the Neumann type Poincaré inequalities on metric balls. More precisely, we say that the *L^p-Poincaré inequality* holds on the manifold M if there exists $P \in (0, \infty)$ such that

$$\forall B = B(x,r), \ \forall f \in W^{1,p}(B), \ \inf_{\xi \in \mathbb{R}} \int_B |f - \xi|^p dv \leqslant P r^p \int_B |\nabla f|^p dv. \tag{4.1}$$

The notion of pseudo-Poincaré inequality was introduced in [7, 18] to describe the inequality

$$\forall f \in C_c^\infty(M), \ \|f - f_r\|_p \leqslant C r \|\nabla f\|_p, \tag{4.2}$$

where

$$f_r(x) = V(x,r)^{-1} \int_{B(x,r)} f dv.$$

Although this looks like a version of the previous Poincaré inequality, it is quite different in several respects. The most important difference is the global nature of each of the members of the pseudo-Poincaré inequality family: in (4.2) all integrals are over the whole space.

We say the *doubling volume condition* holds on M if there exists $D \in (0, \infty)$ such that

$$\forall x \in M, \ r > 0, \ V(x, 2r) \leqslant D V(x,r). \tag{4.3}$$

The only known strong relation between (4.1) and (4.2) is the following result from [8, 18].

Theorem 4.1. *If a complete manifold M equipped with a measure $d\mu = \sigma dv$ satisfies the conjunction of* (4.3) *and* (4.1)*, then the pseudo-Poincaré inequality* (4.2) *holds on M.*

The most compelling reason for introducing the notion of pseudo-Poincaré inequality is that unimodular Lie groups always satisfy (4.2) with $C = 1$ (cf. [22] and the development in [7]). The proof is extremely simple and the result slightly stronger.

Theorem 4.2. *Let G be a connected unimodular Lie group equipped with a left-invariant Riemannian distance and Haar measure. For any group element y at distance $r(y)$ from the identity element e*

$$\forall f \in \mathcal{C}_c(G), \quad \|f - f_y\|_p \leqslant r(y)\|\nabla f\|_p,$$

where $f_y(x) = f(xy)$.

Proof. Indeed, let $\gamma_y : [0, r(y)] \to G$ be a (unit speed) geodesic joining e to y. Thus,

$$|f(x) - f(xy)|^p \leqslant r(y)^{p-1} \int_0^{r(y)} |\nabla f(x\gamma_y(s))|^p ds.$$

Integrating over $x \in G$ yields the desired result. □

With this simple observation and Theorem 3.1, we immediately find that any simply connected noncompact unimodular Lie group M of dimension n satisfies the Sobolev inequality

$$\|f\|_{np/(n-p)} \leqslant S(M,p)\|\nabla f\|_p.$$

This is because the volume growth function $V(x,r) = V(r)$ is always faster than cr^n (cf. [23] and the references therein). In fact, for $r \in (0,1)$, we obviously have $V(r) \simeq r^n$ and, for $r > 1$, either $V(r) \simeq r^N$ for some integer $N \geqslant n$ or $V(r)$ grows exponentially fast. This line of reasoning yields the following improved result (due to Varopoulos [22], with a different proof).

Theorem 4.3. *Let G be a connected unimodular Lie group equipped with a left-invariant Riemannian structure and Haar measure. If the volume $V(r)$ of the balls of radius r in G satisfies $V(r) \geqslant cr^m$ for some $m > 0$ and all $r > 0$, then* (1.1) *holds on G for all $p \in [1, m]$ and $q = mp/(m-p)$.*

In this article, we think of a pseudo-Poincaré inequality as an inequality of the more general form

$$\forall f \in \mathcal{C}_c^\infty(M), \quad \|f - A_r f\|_p \leqslant Cr\|\nabla f\|_p, \tag{4.4}$$

where $A_r : \mathcal{C}_c^\infty(M) \to L^\infty(M)$ is a linear operator. It is indeed very useful to replace the ball averages

$$f_r = V(x,r)^{-1} \int_{B(x,r)} f d\mu$$

by other types of averaging procedures. One interesting case is the following instance.

Theorem 4.4. *Let (M,g) be a Riemannian manifold, and let Δ be the Friedrichs extension of the Laplacian defined on smooth compactly supported functions on M. Let $H_t = e^{t\Delta}$ be the associated semigroup of selfadjoint operator on $L^2(M,dv)$ (the minimal heat semigroup on M). Then*

$$\forall f \in \mathcal{C}_c^\infty(M),\ \ \|f - H_t f\|_2 \leqslant \sqrt{t}\|\nabla f\|_2. \tag{4.5}$$

Consequently, if there are constants $C \in (0,\infty)$, $T \in (0,\infty]$ and $m > 2$ such that

$$\forall t \in (0,T),\ \ \|H_t f\|_\infty \leqslant C t^{-m/4}\|f\|_2, \tag{4.6}$$

then there exists a constant $S = S(C,m) \in (0,\infty)$ such that the Sobolev inequality

$$\forall f \in \mathcal{C}_c^\infty(M),\ \ \|f\|_{2m/(m-2)} \leqslant S(\|\nabla f\|_2 + T^{-1}\|f\|_2) \tag{4.7}$$

holds on M.

Proof. In order to apply Theorem 3.2 with $A_r = H_{r^2}$, it suffices to prove (4.5). But

$$H_t f - f = \int_0^t \partial_s H_s f ds$$

and

$$\langle \partial_s H_s f, H_\tau f\rangle = \langle \Delta H_s f, H_\tau f\rangle = -\|H_{(s+\tau)/2}(-\Delta)^{1/2} f\|_2^2 \geqslant -\|\nabla f\|_2^2.$$

Hence $\|H_t f - f\|_2^2 \leqslant t\|\nabla f\|_2^2$ as desired. □

Remark 4.1. One can show that (4.7) and (4.6) are, in fact, equivalent properties. This very important result was first proved by Varopoulos [21]. This equivalence holds in a much greater generality (cf. also [23]). When $m \in (0,2)$, one can replace (4.7) by the Nash inequality

$$\forall f \in \mathcal{C}_c^\infty(M),\ \ \|f\|_2^{2(1+2/m)} \leqslant N(\|\nabla f\|_2 + T^{-1}\|f\|_2)\|f\|_1^{4/m}$$

which is equivalent to (4.6) (for any fixed $m > 2$). See, for example, [2, 4, 19, 23] and the references therein.

5 Pseudo-Poincaré Inequalities and the Liouville Measure

Given a complete Riemannian manifold $M = (M,g)$ of dimension n (without boundary), we let $T_x M$ be the tangent space at x, $\mathbb{S}_x \subset T_x M$ the unit sphere, and SM the unit tangent bundle equipped with the Liouville measure defined by

$$\int_{SM} f d\mu = \int_M \int_{\mathbb{S}_x} f(x,u) d_{\mathbb{S}_x} u dv(x)$$

where we write $\xi = (x,u) \in SM$ and $d_{\mathbb{S}_x} u$ is the normalized measure on the unit sphere. We denote by Φ_t the geodesic flow on M (with phase space SM). For any t, $\Phi_t : SM \to SM$ is a diffeomorphism and the Liouville measure is invariant under Φ_t. By definition, for any $x \in M$, $u \in \mathbb{S}_x \subset T_x M$, we have $\Phi_t(x,u) = (\gamma_{x,u}(t), \dot\gamma_{x,u}(t))$, where $\gamma_{x,u} : [0,\infty) \to M$ is the (unit speed) geodesic starting at x with tangent unit vector u and $\dot\gamma_{x,u}(t)$ is the unit tangent vector to $\gamma_{x,u}$ at $\gamma_{x,u}(t)$ in the forward t direction.

If $f : SM \mapsto \mathbb{R}$ is a function on SM that depends on $\xi = (x,u) \in SM$ only through $x \in M$, we have (for any fixed $t > 0$, and with a slight abuse of notation, namely $f(\xi) = f(x)$)

$$\int_M f dv = \int_{SM} f d\mu = \int_{SM} f \circ \Phi_t d\mu = \int_{SM} f(\gamma_{x,u}(t)) d_{\mathbb{S}_x} u dv(x). \tag{5.1}$$

For any $(x,u) \in SM$, let $r(x,u)$ be the distance from x to the cutlocus in the direction of u. Namely,

$$r(x,u) = \inf\{t > 0 : d(x, \Phi_t(x,u)) < t\}.$$

The function r defined on SM is always upper semicontinuous and continuous when M is complete without boundary. Now, let $\psi : (x,u,s) \mapsto \psi(x,u,s) = \psi_{x,u}(s) \in L^{1,+}_{\mathrm{loc}}(SM \times [0,\infty))$ with $\psi_{x,u}(t) = 0$ if $t \geqslant r(x,u)$. Call such a function *admissible*. For $f \in \mathcal{C}^\infty_c(M)$ we set

$$A_r f(x) = w(x,r)^{-1} \int_0^r \int_{\mathbb{S}_x} f(\gamma_{x,u}(t)) \psi_{x,u}(t) dt d_{\mathbb{S}_x} u,$$

where

$$w(x,r) = \int_0^r \int_{\mathbb{S}_x} \psi_{x,u}(t) dt d_{\mathbb{S}_x} u.$$

In words, $A_r f$ is a weighted geodesic average of f over scales at most r. Note that, according to our definition, these averages never look past the cutlocus.

Example 5.1. Let $\psi_{x,u}(t) = J(x,u,t)$ be the density of the volume element dv in geodesic polar coordinate around x so that

$$dv(y) = J(x,u,t) dt d_{\mathbb{S}_x} u, \;\; y = \gamma_{x,u}(t) = \Phi_t(x,u), \; t < r(x,u).$$

By definition, we set $J(x,u,t) = 0$ for $t \geqslant r(x,u)$. Then $w(x,r) = V(x,r)$ and $A_r f(x) = f_r(x)$ is the mean of f in $B(x,r)$.

Theorem 5.1. *On any complete manifold without boundary and for any choice of admissible* $\psi \in L^{1,+}_{\mathrm{loc}}(SM \times [0,\infty))$, *we have*

$$\int_M |f - A_r f|^p dv \leqslant D(r) \int_M |\nabla f|^p dv$$

with

$$D_p(r) = \sup_{(x,u)\in SM} \left\{ r \int_0^r \frac{\psi_{x,u}(t)t^{p-1}}{w(x,r)} dt \right\}.$$

Proof. Write

$$\begin{aligned}
&\int_M |f - A_r f|^p dv \\
&\leqslant \int_M \frac{1}{w(x,r)} \int_0^r \int_{\mathbb{S}_x} |f(x) - f(\gamma_{x,u}(t))|^p \psi_{x,u}(t) dt d_{\mathbb{S}_x} u dv(x) \\
&\leqslant \int_M \frac{1}{w(x,r)} \int_0^r \int_{\mathbb{S}_x} \left(\int_0^t |\nabla f|(\gamma_{x,u}(s)) ds \right)^p \psi_{x,u}(t) dt d_{\mathbb{S}_x} u dv(x) \\
&\leqslant \int_M \frac{1}{w(x,r)} \int_0^r \int_{\mathbb{S}_x} \int_0^t |\nabla f|^p (\gamma_{x,u}(s)) ds \psi_{x,u}(t) t^{p-1} dt d_{\mathbb{S}_x} u dv(x) \\
&= \in t_0^r \int_M \int_{\mathbb{S}_x} |\nabla f|^p (\gamma_{x,u}(s)) \left(\int_s^r \frac{\psi_{x,u}(t)t^{p-1}}{w(x,r)} dt \right) d_{\mathbb{S}_x} u dv(x) ds \\
&\leqslant \left(\sup_{(x,u)\in SM} \left\{ \int_0^r \frac{\psi_{x,u}(t)t^{p-1}}{w(x,r)} dt \right\} \right) \int_0^r \int_M \int_{\mathbb{S}_x} |\nabla f|^p (\gamma_{x,u}(s)) d_{\mathbb{S}_x} u dv(x) ds \\
&\leqslant D(r) \int_M |\nabla f|^p dv,
\end{aligned}$$

where $D_p(r)$ is as defined in the theorem. Note the crucial use of (5.1) at the last step. □

Corollary 5.1. *Let* (M, g) *be an isotropic Riemannian manifold. Then, for any* $p \in [1, \infty]$, *the pseudo-Poincaré inequality*

$$\forall f \in \mathcal{C}_c^\infty(M), \;\; \|f - f_r\|_p \leqslant r \|\nabla f\|_p$$

is satisfied.

Proof. Use the previous theorem with $\psi_{x,u}(t) = J(x,u,t)$, in which case $A_r f = f_r$, $w(x,r) = V(x,r)$. Observe that $J(x,u,t)$ is independent of (x,u) because M is isotropic. It follows that

$$\int_0^r J(x,u,t)dt = \int_{\mathbb{S}_x} \int_0^r J(x,u,t) dt du = V(x,r).$$

Hence $D(r) \leqslant r^p$. □

Note that isotropic Riemannian manifolds are the same as two-point homogeneous Riemannian manifolds. They must be either $\mathbb{R}^n$ or a rank one symmetric space.

Corollary 5.2. *Assume that M is a simply connected complete n-dimensional manifold without boundary and with nonpositive sectional curvature (i.e., a Cartan–Hadamard manifold). Set*

$$A_r f(x) = nr^{-n} \int_0^r \int_{\mathbb{S}_x} f(\gamma_{x,u}(t)) t^{n-1} dt d_{\mathbb{S}_x} u.$$

Then, for any $p \in [1, \infty]$, the inequalities

$$\forall f \in \mathcal{C}_c^\infty(M), \ \ \|f - A_r\|_p \leqslant r \|\nabla f\|_p, \ \ \|A_r f\|_\infty \leqslant (\Omega_n r^n)^{-1/p} \|f\|_p.$$

are satisfied (Ω_n is the volume of the n-dimensional Euclidean unit ball).

Proof. Apply the theorem with $\psi_{x,u}(t) = t^{n-1}\mathbf{1}_{[0,r(x,u))}(t)$. This gives $D(r) \leqslant r^p$ on any manifold (i.e., we have not use nonpositive curvature yet). Now, since M has nonpositive sectional curvature and is simply connected, we have $r(x,u) = \infty$, and the classical comparison theorem gives $\omega_{n-1} t^{n-1} \leqslant J(x,u,t)$ (ω_{n-1} the volume of the unit sphere in $\mathbb{R}^n$). Hence

$$\begin{aligned} |A_r f(x)| &\leqslant \frac{1}{(\omega_{n-1}/n) r^n} \int_0^r \int_{\mathbb{S}_x} |f(\gamma_{x,u}(t)| J(x,u,t) dt d_{\mathbb{S}_x} u \\ &\leqslant \frac{1}{(\omega_{n-1}/n) r^n} \int_M |f| dv. \end{aligned}$$

The proof is complete. □

This and Theorem 3.1 yield the following classical result (case (2) of Theorem 1.1).

Corollary 5.3. *Assume that M is a simply connected complete n-dimensional manifold without boundary and with nonpositive sectional curvature (i.e., a Cartan–Hadamard manifold). Then the Sobolev inequality* (1.1) *holds on M with $q = np/(n-p)$ for any $p \in [1, n)$.*

This argument allow us to obtain a generalized version of this important result. Namely, for any $x \in M$, let

$$\mathcal{R}_x = \{u \in \mathbb{S}_x : r(x,u) = \infty\}.$$

In words, $\mathcal{R}_x$ is the set of unit tangent vectors $u \in T_x M$ associated with rays starting at x (a ray is a semiinfinite distance minimizing geodesic starting at x). Let

$$|\mathcal{R}_x| = \int_{\mathcal{R}_x} d_{\mathbb{S}_x} u$$

be the normalized volume of $\mathcal{R}_x$ as a subset of $\mathbb{S}_x$. Now, set

$$A_r^* f(x) = \frac{n}{|\mathcal{R}_x| r^n} \int_{\mathcal{R}_x} \int_0^r f(\gamma_{x,u}(t)) t^{n-1} dt d_{\mathbb{S}_x} u.$$

Obviously, Theorem 5.1 yields

$$\|f - A_r^* f\|_p \leqslant r^p \|\nabla f\|_p.$$

Further, if M has nonpositive sectional curvature along all rays,

$$|A_r^* f(x)| \leqslant \frac{n}{\omega_{n-1} |\mathcal{R}_x| r^n} \int_M |f| dv.$$

Hence we obtain the following statement.

Theorem 5.2. *Assume that M is a complete Riemannian n-manifold without boundary and with nonpositive curvature and such that $\rho = \min_x\{|\mathcal{R}_x|\} > 0$. Then M satisfies* (1.1) *with $q = np/(n-p)$.*

The simplest example of application of this result is to the surface of revolution known as the catenoid (it looks essentially like two planes connected through a compact cylinder) which is a celebrated example of minimal surface in $\mathbb{R}^3$. The theorem applies for $p \in [1, 2)$ and yields, for instance, the Sobolev inequality $\|f\|_2 \leqslant S \|\nabla f\|_1$.

6 Homogeneous Spaces

In this section, we revisit the pseudo-Poincaré inequality on unimodular Lie groups to extend it to a class of homogeneous spaces. The argument we will use contains similarities as well as serious differences with the argument based on the invariance of the Liouville measure that was described in Section 5. We present it in the context of sub-Riemannian geometry. For an introduction to sub-Riemannian geometry (cf. [16]).

Let G be a unimodular Lie group, and let K be a compact subgroup. Let $M = G/K$ be the associated homogeneous space equipped with its G invariant measure $d\mu$. Let $\tau_g : M \to M$ be the action of G on M, and let $\tau_g f(x) = f(\tau_g x)$, $f \in \mathcal{C}_c(M)$.

Assume that M is equipped with a (constant rank) sub-Riemannian structure, i.e., a vector subbundle $H \subset TM$ equipped with a fiber inner product $\langle \cdot, \cdot \rangle_H$ such that any local frame $(X_1, \ldots, X_k)$ for H is bracket generating (i.e., satisfies the Hörmander condition). For any function $f \in \mathcal{C}_c^\infty(M)$, let $\nabla_H f(x)$ be the vector in H_x such that $df|_x(u) = \langle \nabla_H f, u \rangle_{H_x}$ for any $u \in H_x$.

Assume further that $(H, \langle \cdot, \cdot \rangle_H)$ is G invariant, i.e., for all $g \in G$, $x \in M$, $X, Y \in H_x$ we have $d\tau_g(X) \in H_{\tau_g x}$ and

$$\langle d\tau_g(X), d\tau_g(Y)\rangle_{H_{\tau_g x}} = \langle X, Y\rangle_{H_x}.$$

The space H is called the *horizontal space*. Under the Hörmander condition, any two points can be joined by absolutely continuous curves in M that stay tangent to H almost surely. For any such $c : [0,T] \to M$ with $\dot{c}(t) \in H_{c(t)}$ we set

$$\ell_H(c) = \int_0^T \langle \dot{c}, \dot{c}\rangle_{H_{c(t)}}^{1/2} dt.$$

This is the horizontal length of c. By definition, for any two points $x, y \in M$, $d_H(x,y)$ is the infimum of the horizontal length of horizontal curves joining x to y. It is not hard to check that $d_H(x,y)$ is also equal to the infimum of all T such that there exists an absolutely continuous horizontal curve $c : [0,T] \to M$ with $\langle \dot{c}, \dot{c}\rangle_H \leqslant 1$ joining x to y. Since the action of G preserves horizontal length, it also preserves the distance d_H. We let $B_H(x,r) = \{y \in M : d_H(x,y) < r\}$ and $V_H(r) = \mu(B_H(x,r))$ which is, indeed, independent of x. Our aim is to prove the following result.

Theorem 6.1. *Let $M = G/K$ with G unimodular and K compact be an homogeneous manifold. Assume that M is equipped with a sub-Riemannian structure $(H, \langle\cdot,\cdot\rangle_H)$ satisfying the Hörmander condition and preserved by the action of G. Then, for any $1 \leqslant p \leqslant \infty$,*

$$\forall f \in \mathcal{C}_c(M), \quad \|f - f_r\|_p \leqslant r\|\nabla_H f\|_p.$$

Proof. We can choose the Haar measure on G and the G invariant measure μ on M so that for any $F \in \mathcal{C}_c(G)$

$$\int_G F(g)dg = \int_M \int_K F(g_x k)dk d\mu(x),$$

where dk is the normalized Haar measure on K. Here, g_x stands for any element of G such that $x = gK$. Note that

$$x \to \int_K F(g_x k)dk \in \mathcal{C}_c(M)$$

is indeed independent of the choice g_x, $x \in M$. We need to observe that there is such an integration formula for any choice of an origin in M. Namely, for any $z \in M$ there is a compact subgroup K_z in G that fixes z and we can write

$$\int_G F(g)dg = \int_M \int_{K_z} F(g_x k)d_{K_z}k d\mu(x) \tag{6.1}$$

for some choice of Haar measure on K_z. Because μ is invariant under the action of G and G is unimodular, the Haar measure on K_z must be taken to be the normalized Haar measure (cf., for example, [20]).

Now, for any $y \in g_y K \in M$ we pick an horizontal path $c : [0,T] \to M$ of horizontal length l with $\langle \dot{c}, \dot{c}\rangle_H \leqslant 1$ and joining $o = eK$ to y. For any $g \in G$

and $f \in \mathcal{C}_c(M)$

$$|f(\tau_g o) - f(\tau_g y)|^p \leqslant T^{p-1} \int_0^T |\nabla_H f(\tau_g c(t))|^p dt.$$

Hence

$$\int_G |f(\tau_g o) - f(\tau_g y)|^p dg \leqslant T^{p-1} \int_0^T \int_G |\nabla_H f(\tau_g c(t)|^p dg dt.$$

We now use (6.1) with $z = c(t)$ and $F(g) = |\nabla f(\tau_g c(t))|^p$ to compute

$$\int_G |\nabla_H f(gc(t))|^p dg = \int_M \int_{K_{c(t)}} |\nabla_H f(\tau_{g_x k} c(t))|^p d_{K_{c(t)}} k d\mu(x)$$

$$= \int_M |\nabla_H f(x)|^p d\mu(x).$$

Hence

$$\int_G |f(\tau_g o) - f(\tau_g y)|^p dg \leqslant T^p \|\nabla_H f\|_p^p.$$

Optimizing over the value of T yields

$$\int_G |f(\tau_g o) - f(\tau_g y)|^p dg \leqslant d(o,y) \|\nabla_H f\|_p^p.$$

Next, for any $g \in G$

$$V_H(r)^{-1} \int_{B_H(o,r)} f(\tau_g y) d\mu(y) = V_H(r)^{-1} \int_{B_H(\tau_g o, r)} f(y) d\mu(y).$$

Hence, if we set

$$f_r(x) = V_H(r)^{-1} \int_{B_H(x,r)} f d\mu,$$

we have

$$\|f - f_r\|_p^p = \int_M \left| f(x) - V_H(r)^{-1} \int_{B_H(x,r)} f d\mu \right|^p d\mu(x)$$

$$= \int_G \left| f(\tau_g o) - V_H(r)^{-1} \int_{B_H(\tau_g o, r)} f d\mu \right|^p dg$$

$$= \int_G \left| f(\tau_g o) - V_H(r)^{-1} \int_{B_H(o,r)} f(\tau_g y) d\mu(y) \right|^p dg$$

$$\leqslant V_H(r)^{-1} \int_{B_H(o,r)} \int_G |f(\tau_g o) - f(\tau_g y)|^p dg d\mu(y)$$

$$\leqslant r^p \|\nabla_H f\|_p^p.$$

This finishes the proof of Theorem 6.1. □

Corollary 6.1. *In the context of Theorem* 6.1, *if* $V_H(r) \geqslant cr^m$ *for all* $r > 0$, *then the Sobolev inequality*

$$\forall f \in \mathcal{C}_c(M), \quad \|f\|_q \leqslant S(M, H, p, q)\|\nabla_H f\|_p$$

holds on M *for all* $p \in [1, m)$ *with* $q = mp/(m - p)$ *and a finite constant* $S(M, H, c, p, q)$.

This covers the case of unimodular Lie groups ($M = G$, $K = \{e\}$) equipped with a family of left-invariant vector fields $\{X_1, \ldots, X_k\}$ that generates the Lie algebra (in this case, $|\nabla_H f|^2 = \sum_1^k |X_i f|^2$). It also covers the case of noncompact symmetric spaces $M = G/K$, $G = NAK$ semisimple, equipped with their canonical Riemannian structure. Note that when M is a noncompact symmetric space, the inequality $\|f\|_p \leqslant C(M, p)\|\nabla f\|_p$ holds as well for different reasons.

7 Ricci Curvature Bounded Below

This section offers variations on results from [19]. Let (M, g) be a Riemannian manifold of dimension n (without boundary) with Ricci curvature tensor Ric. Fix $K, R \geqslant 0$, an open set $\Omega \subset M$. We assume throughout that $\Omega_R = \{y \in M : d(x, y) \leqslant R\}$ is compact and that the Ricci tensor is bounded by $\mathrm{Ric} \geqslant -Kg$ over Ω_R. It follows that for any $x \in \Omega$ and $r \in (0, R)$ almost every point y in $B(x, r)$ can be joined to x by a unique minimizing geodesic $\gamma_{x,y} : [0, d(x, y)] \to \Omega_R$. Note that we use somewhat conflicting notation by letting $\gamma_{x,u}$ denote the unit speed geodesic starting at x in the direction $u \in \mathbb{S}_x$ and letting $\gamma_{x,y}$ denote the minimizing unit speed geodesic from x to y (when it exists). Note also that for any $x \in \Omega$ and $u \in \mathcal{S}_x$ and $r \in (0, R)$, the unit speed geodesic $\gamma_{x,u}$ (not necessarily minimizing) is defined at least on the interval $[0, r]$ because Ω_R is compact in M.

In this context, the Bishop–Gromov comparison theorem yields the following properties:

- For all $x \in \Omega$ and $0 < s < r < R$

$$V(x, r) \leqslant V(x, s)(r/s)^n e^{\sqrt{(n-1)K}\, r}.$$

- For any $x \in \Omega$, $u \in \mathbb{S}_x$, and $0 < s < r < R$ such that $\gamma_{x,u}$ is minimizing on $[0, r]$
$$J(x,u,r) \leqslant J(x,u,s)(r/s)^{n-1}e^{\sqrt{(n-1)K}\,r}.$$

We will use these properties to prove the following result.

Theorem 7.1. *Referring to the above setup concerning (M,g) and Ω, K, R, we have*

$$\forall r \in (0,R),\ \ \forall f \in \mathcal{C}_c^\infty(\Omega),\ \ \|f - f_r\|_p^p \leqslant 8^n e^{3\sqrt{(n-1)K}\,r} r^p \|\nabla f\|_p^p.$$

Proof. For simplicity, we write dx for the Riemannian measure $v(dx)$. It suffices to show that, for any $f \in \mathcal{C}_c^\infty(\Omega)$ and $r \in (0,R)$

$$\int_\Omega \int_\Omega |f(x) - f(y)|^p \frac{\mathbf{1}_{B(x,r)}(y)}{V(x,r)} dxdy \leqslant 8^n e^{3\sqrt{(n-1)K}\,r} r^p \|\nabla f\|_p^p.$$

By the Bishop–Gromov volume inequality, for $x, y \in \Omega$

$$\frac{\mathbf{1}_{B(x,r)}(y)}{V(x,r)} \leqslant 2^n e^{\sqrt{(n-1)K}\,r} \frac{\mathbf{1}_{B(x,r)}(y)}{\sqrt{V(x,r)V(y,r)}} \tag{7.1}$$

and it suffices to bound

$$I = \int_\Omega \int_\Omega |f(x) - f(y)|^p \frac{\mathbf{1}_{B(x,r)}(y)}{\sqrt{V(x,r)V(y,r)}} dxdy.$$

Let W be the (symmetric) subset of $\Omega \times \Omega$ of all (x,y) with $d(x,y) < r$ such that there exists a unique minimizing geodesic $\gamma_{x,y} : [0, d(x,y)] \to M$ joining x to y. As was noted above, for any $x \in \Omega$, almost all $y \in B(x,r)$ have this property and the image of $\gamma_{x,y}$ is contained in Ω_R. Hence

$$\begin{aligned} I &= \int_W |f(x) - f(y)|^p \frac{\mathbf{1}_{B(x,r)}(y)}{\sqrt{V(x,r)V(y,r)}} dxdy \\ &\leqslant \int_W \int_0^{d(x,y)} \frac{d(x,y)^{p-1} |\nabla f(\gamma_{x,y}(s))|^p \mathbf{1}_{B(x,r)}(y)}{\sqrt{V(x,r)V(y,r)}} dsdxdy. \end{aligned}$$

The following step is essential to the proof. By symmetry between x and y and since $\gamma_{x,y}(s) = \gamma_{y,x}(d(x,y) - s)$, we have

$$\int_W \int_0^{d(x,y)/2} \frac{d(x,y)^{p-1} |\nabla f(\gamma_{x,y}(s))|^p \mathbf{1}_{B(x,r)}(y)}{\sqrt{V(x,r)V(y,r)}} dsdxdy =$$

$$\int_W \int_{d(x,y)/2}^{d(x,y)} \frac{d(x,y)^{p-1} |\nabla f(\gamma_{x,y}(s))|^p \mathbf{1}_{B(x,r)}(y)}{\sqrt{V(x,r)V(y,r)}} dsdxdy.$$

Hence

$$I \leqslant 2r^{p-1} \int_W \int_{d(x,y)/2}^{d(x,y)} \frac{|\nabla f(\gamma_{x,y}(s))|^p \mathbf{1}_{B(x,r)}(y)}{\sqrt{V(x,r)V(y,r)}} dsdxdy.$$

By the Bishop–Gromov comparison theorem, we have

$$\forall x, y \in M,\ 0 < s < r,\ \ \frac{\mathbf{1}_{B(x,r)}(y)}{\sqrt{V(x,r)V(y,r)}} \leqslant 2^n e^{\sqrt{(n-1)K}\,r} \frac{\mathbf{1}_{B(x,r)}(\gamma_{x,y}(s))}{V(\gamma_{x,y}(s),r)}.$$

Moreover, again by the Bishop–Gromov comparison theorem, for all $(x,y) \in W$ and $d(x,y)/2 < s < d(x,y) \leqslant r$, the Jacobian $\mathrm{J}(\gamma_{x,y}(s))$ of the map $y \mapsto z = \phi(y) = \gamma_{x,y}(s)$ is bounded below by $2^{-n+1}e^{-\sqrt{(n-1)K}\,r}$. Note that we use here the fact that the image of the whole $\gamma_{x,y}$ lies in Ω_R, where the Ricci lower bound is satisfied.

For each $s \in [0,r]$ we set

$$W_s = \{(x,y) \in W : s \leqslant d(x,y)\}.$$

Using the two observations above in the previous upper bound for I and setting $C(r) = 4^n r^{p-1} e^{2\sqrt{(n-1)K}\,r}$, we obtain

$$\begin{aligned}
I &\leqslant C(r) \int_W \int_{d(x,y)/2}^{d(x,y)} \frac{|\nabla f(\gamma_{x,y}(s))|^p \mathrm{J}(\gamma_{x,y}(s)) \mathbf{1}_{B(x,r)}(\gamma_{x,y}(s))}{V(\gamma_{x,y}(s),r)} dsdxdy \\
&\leqslant C(r) \int_0^r \int_{W_s} \frac{|\nabla f(\gamma_{x,y}(s))|^p \mathrm{J}(\gamma_{x,y}(s)) \mathbf{1}_{B(\gamma_{x,y}(s),r)}(x)}{V(\gamma_{x,y}(s),r)} dxdyds \\
&\leqslant C(r) \int_0^r \int_{M\times M} \frac{|\nabla f(z)|^p \mathbf{1}_{B(z,r)}(x)}{V(z,r)} dxdzds \\
&= C(r) r \int_M |\nabla f(z)|^p dz.
\end{aligned}$$

Taking (7.1) into account, we obtain the desired result. □

As corollaries of Theorems 3.1 and 7.1, we obtain the following three well-known results.

Theorem 7.2. *For any relatively compact set Ω in a Riemannian manifold M (without boundary) of dimension n, the Sobolev inequality*

$$\forall f \in \mathcal{C}_c^\infty(\Omega),\ \ \|f\|_q \leqslant S(\Omega,p,q)\,(\|\nabla f\|_p + \|f\|_p)$$

holds for any $p \in [1,n)$ and $q = pn/(n-p)$.

For the proof of the next result, in addition to Theorems 3.1 and 7.1, one uses the Bishop–Gromov comparison theorem in the form of the volume lower bound

$$\forall x \in M,\ \forall s \in (0,r),\ \ V(x,s) \geqslant c(x,r)s^n$$

with $c(x,r) = e^{\sqrt{(n-1)K}\,r}V(x,r)$ which is valid as long as the closed ball $\overline{B(x,2r)}$ is compact and $K \geqslant 0$ is such that the Ricci curvature tensor is bounded below by $\mathrm{Ric} \geqslant -Kg$ on $B(x,2r)$.

Theorem 7.3. *On any Riemannian manifold M of dimension n (without boundary), the Sobolev inequality*

$$\forall\, f \in \mathcal{C}_c^\infty(B(x,r)),\ \ \|f\|_q \leqslant \frac{C(p,n,Kr^2)r}{V(x,r)^{1/n}}\left(\|\nabla f\|_p + r^{-1}\|f\|_p\right)$$

holds for any $p \in [1,n)$ and $q = pn/(n-p)$ as long as the closed ball $\overline{B(x,2r)}$ is compact and $K \geqslant 0$ is such that the Ricci curvature tensor is bounded below by $\mathrm{Ric} \geqslant -Kg$ *on $B(x,2r)$.*

If one follows the constants in the proof of Theorem 7.3, one finds that

$$C(p,n,Kr^2) \leqslant C_1(n,p)e^{C_2(n,p)\sqrt{Kr^2}}$$

for some finite constants $C_1(n,p)$ and $C_2(n,p)$.

The next result can be obtained from the previous theorem by letting r tend to infinity (which is possible when $K = 0$ since $Kr^2 = 0$ for all $r > 0$).

Theorem 7.4. *For any complete Riemannian manifold M of dimension n with nonnegative Ricci curvature and maximum volume growth (i.e., there exists $c > 0$ such that $V(x,r) \geqslant cr^n$ for all $x \in M$, $r > 0$) the Sobolev inequality*

$$\forall\, f \in \mathcal{C}_c^\infty(M),\ \ \|f\|_q \leqslant S(c,n,p)\|\nabla f\|_p$$

holds for any $p \in [1,n)$ and $q = pn/(n-p)$.

8 Domains with the Interior Cone Property

In this final section, we illustrate yet a slightly different use of the pseudo-Poincaré inequality. Let (M,g) be a complete Riemannian manifold without boundary.

Fix $\delta \in (0,1]$ and $r > 0$. A (δ,r)-cone at x is a set of the form

$$\mathfrak{C}(x,\omega_x,r) = \{y = \gamma_{x,u}(s) : u \in \omega_x, 0 \leqslant s < r\},$$

where ω_x is an open subset of $\mathbb{S}_x$ with the property that $r(x,u) > r$ for all $u \in \omega_x$ and $|\omega_x| \geqslant \delta$. Here, $|\omega_x|$ denotes the measure of ω_x with respect to the normalized measure on the sphere $\mathbb{S}_x$. We always assume further that for any continuous function f the function

$$x \mapsto \int_{\mathfrak{C}(x,\omega_x,r)} f(y)dv(y)$$

is measurable. In particular, $x \mapsto v(\mathfrak{C}(x,\omega_x,r))$ is measurable.

Note that the existence of a (δ,r) cone at x is a non trivial assumption. A domain Ω which contains an (δ,r) cone at x for any $x \in \Omega$ is said to satisfy the (δ,r) *interior cone condition.*

In the Euclidean space context, the interior cone condition is perhaps the most classical condition for the validity of various Sobolev embedding theorems (cf. [1, 13]). In the geometric context of complete Riemannian manifolds, we offer two results based on the interior cone conditions. The Sobolev inequalities stated in the following two theorems are of a different nature than those discussed earlier in this paper because the functions involved need not vanish at the boundary of Ω. To obtain these inequalities, we use Theorem 3.3.

Theorem 8.1. *Let Ω be a domain in an n-dimensional complete Riemannian manifold (M,g) (without boundary). Fix $K,R \geqslant 0$ and assume that*

- *The Ricci curvature is bounded by* $\mathrm{Ric} \geqslant -Kg$ *on Ω.*
- *There exists $\delta \in (0,1)$ such that for any $x \in \Omega$, there is a (δ,R)-cone $\mathfrak{C}(x,\omega_x,R)$ at x contained in Ω with the additional property that $v(\mathfrak{C}(x,\omega_x,r)) \geqslant cr^n$ for any $r \in (0,R)$.*

For any $f \in \mathcal{C}^\infty(\Omega) \cap W^{1,p}(\Omega)$ we set

$$\mathfrak{A}_r f(x) = \frac{1}{v(\mathfrak{C}(x,\omega_x,r))} \int_{\mathfrak{C}(x,\omega_x,r)} f dv.$$

Then for all $f \in \mathcal{C}^\infty(\Omega) \cap W^{1,p}(\Omega)$ the inequality

$$\forall\, r \in (0,R), \quad \|f - \mathfrak{A}_r f\|_{\Omega,p} \leqslant (\omega_{n-1}/cn) e^{2\sqrt{(n-1)K}\, r}\, r \|\nabla f\|_{\Omega,p}$$

holds. Further, for $p \in [1,n)$ and $q = np/(n-p)$ the Sobolev inequality

$$\forall\, f \in \mathcal{C}^\infty(\Omega) \cap W^{1,p}(\Omega), \quad \|f\|_{\Omega,q} \leqslant S(c,p,n,Kr^2)\left(\|\nabla f\|_{\Omega,p} + R^{-1}\|f\|_{\Omega,p}\right)$$

is satisfied.

Proof. Let $f \in \mathcal{C}^\infty(\Omega) \cap W^{1,p}(\Omega)$. For any $x \in \Omega$ we write

$$v(\mathfrak{C}(x,r))|f(x) - \mathfrak{A}_r f(x)| \leqslant \int_{\omega_x} \int_0^r \int_0^s |\nabla f(\gamma_{x,u}(\tau))d\tau| J(x,u,s) ds d_{\mathbb{S}_x} u.$$

By the Bishop–Gromov comparison theorem, for all $0 < \tau < s < r < r(x,u)$

$$J(x,u,\tau)(s/\tau)^{n-1} e^{\sqrt{(n-1)K}\, s} \geqslant J(x,u,s).$$

Hence

$$v(\mathfrak{C}(x,r))|f(x)-\mathfrak{A}_r f(x)|$$
$$\leqslant e^{\sqrt{(n-1)K}\,r}\int_{\omega_x}\int_0^r\int_0^s \tau^{1-n}|\nabla f(\gamma_{x,u}(\tau))|J(x,u,\tau)d\tau s^{n-1}dsd_{\mathbb{S}_x}u$$
$$\leqslant \frac{e^{\sqrt{(n-1)K}\,r}r^n}{n}\int_{\Omega\cap B(x,r)}\frac{|\nabla f(z)|}{d(x,z)^{n-1}}dz.$$

Using the hypothesis $v(\mathfrak{C}(x,r))\geqslant cr^n$, we obtain

$$|f(x)-\mathfrak{A}_r f(x)|^p$$
$$\leqslant \frac{e^{p\sqrt{(n-1)K}\,r}}{(cn)^p}\left(\int_{\Omega\cap B(x,r)}\frac{dz}{d(x,z)^{n-1}}\right)^{p-1}\int_{\Omega\cap B(x,r)}\frac{|\nabla f(z)|^p}{d(x,z)^{n-1}}dz.$$

The Bishop comparison theorem yields

$$\int_{\Omega\cap B(z,r)}\frac{dx}{d(z,x)^{n-1}}\leqslant \omega_{n-1}e^{\sqrt{(n-1)K}\,r}r.$$

The inequality

$$\|f-\mathfrak{A}_r f\|_{\Omega,p}\leqslant(\omega_{n-1}/cn)e^{2\sqrt{(n-1)K}\,r}r\|\nabla f\|_{\Omega,p}$$

follows, and Theorem 3.3 gives the desired Sobolev inequality. □

Theorem 8.2. *Fix $R>0$ and $\delta\in(0,1)$. Let Ω be a domain in an n-dimensional complete simply connected Riemannian manifold (M,g) without boundary and with nonpositive sectional curvature (i.e., a Cartan–Hadamard manifold). Assume that Ω as the (δ,R) interior cone property. Namely, for any $x\in\Omega$ there is a (δ,R)-cone $\{y=\gamma_{x,u}(s): u\in\omega_x, 0\leqslant s<R,\}$ at x contained in Ω. For any $f\in C^\infty(\Omega)$ we set*

$$A_r f(x)=\frac{n}{|\omega_x|r^n}\int_{\omega_x}\int_0^r f(\gamma_{x,u}(s)s^{n-1}dsd_{\mathbb{S}_x}u.$$

Then for all $f\in C^\infty(\Omega)\cap W^{1,p}(\Omega)$ the inequality

$$\forall\, r\in(0,R),\quad \|f-A_r f\|_{\Omega,p}\leqslant\delta^{-1/p}r\|\nabla f\|_{\Omega,p}$$

holds. Further, for $p\in[1,n)$ and $q=np/(n-p)$, the Sobolev inequality

$$\forall\, f\in C^\infty(\Omega)\cap W^{1,p}(\Omega),\quad \|f\|_{\Omega,q}\leqslant S(\delta,p,n)\left(\|\nabla f\|_{\Omega,p}+R^{-1}\|f\|_{\Omega,p}\right)$$

is satisfied.

Proof. Let $f \in \mathcal{C}^\infty(\Omega) \cap W^{1,p}(\Omega)$. For any $x \in \Omega$ we write

$$|f(x) - A_r f(x)| \leqslant \frac{n}{\delta r^n} \int_{\omega_x} \int_0^r \int_0^s |\nabla f(\gamma_{x,u}(\tau))| d\tau s^{n-1} ds d_{\mathbb{S}_x} u.$$

Hence, using (5.1) for the last step,

$$\begin{aligned}
\|f - A_r f\|_{\Omega,p}^p &\leqslant \frac{n}{\delta r^n} \int_0^r \int_\Omega \int_{\omega_x} \int_\tau^r |\nabla f(\gamma_{x,u}(\tau))|^p d_{\mathbb{S}_x} u s^{n+p-2} ds d\tau dx \\
&\leqslant \frac{n r^{p-1}}{(n+p-1)\delta} \int_0^r \int_\Omega \int_{\omega_x} |\nabla f(\gamma_{x,u}(\tau))|^p d_{\mathbb{S}_x} u dx d\tau \\
&= \frac{r^{p-1}}{\delta} \int_0^r \int_\Omega \int_{\omega_x} |\nabla f(\gamma_{x,u}(\tau))|^p \mathbf{1}_\Omega(\gamma_{x,u}(\tau)) d_{\mathbb{S}_x} u dx d\tau \\
&\leqslant \frac{r^{p-1}}{\delta} \int_0^r \int_M \int_{\mathbb{S}_x} |\nabla f(\gamma_{x,u}(\tau))|^p \mathbf{1}_\Omega(\gamma_{x,u}(\tau)) d_{\mathbb{S}_x} u dx d\tau \\
&= \frac{r^p}{\delta} \|\nabla f\|_{\Omega,p}^p.
\end{aligned}$$

This yields the desired pseudo-Poincaré inequality. To obtain the stated Sobolev inequality, we simply observe that

$$|A_r f(x)| \leqslant \frac{n}{\delta r^n} \int_{\omega_x} \int_0^r |f(\gamma_{x,u}(s)| s^{n-1} ds d_{\mathbb{S}_x} u \leqslant \frac{n}{\delta r^n} \int_\Omega |f| dv$$

because, on any Cartan–Hadamard manifold, $J(x,u,s) \geqslant s^{n-1}$ for all u. It then suffices to apply Theorem 3.3. □

Example 8.1. Let M be an n-dimensional Cartan–Hadamard manifold. Let C be a closed geodesically convex set, and let $\Omega = M \setminus C$. We claim that Ω has the $(1/2, \infty)$ interior cone property. Indeed, for any $x \in \Omega$, let ω_x be the subset of those unit tangent vectors v at x such that $C \cap \{y = \gamma_{x,v}(s) : s \geqslant 0\} = \varnothing$. Because C is geodesically convex and $x \in \Omega$, for any $u \in \mathbb{S}_x$, either u or $-u$ belongs to ω_x. This implies that $|\omega_x| \geqslant 1/2$. Applying Theorem 8.2 with $R = \infty$, we obtain the Sobolev inequality

$$\forall\, f \in \mathcal{C}^\infty(\Omega) \cap W^{1,p}(\Omega), \quad \|f\|_{\Omega,q} \leqslant S(p,n) \|\nabla f\|_{\Omega,p}, \;\; q = np/(n-p).$$

Note that the functions $f \in \mathcal{C}^\infty(\Omega) \cap W^{1,p}(\Omega)$ do not necessarily vanish along the boundary of Ω. Balls and sublevel sets of Busemann functions provide examples of geodesically convex sets in Cartan–Hadamard manifolds.

Example 8.2. Consider a geodesic ball of radius $\rho \geqslant 1$ in the hyperbolic plane. It is not hard to see that it has the $(\delta, 1)$ interior cone property. One may ask if, uniformly in $\rho \geqslant 1$, these balls have the $(\delta, a\rho)$ interior cone property for some $\delta, a \in (0,1)$. The answer is no. If one wants to fit cones of length $a\rho$ in a ball of radius ρ, then the aperture $\alpha(a, \rho)$ has to tend to 0 with ρ.

Acknowledgments. The research was partially supported by NSF (grant DMS 0603886).

References

1. Adams, R.A.: Sobolev spaces. Academic Press, New York etc. (1975)
2. Bakry, D., Coulhon, T., Ledoux, M., Saloff-Coste, L.: Sobolev inequalities in disguise. Indiana Univ. Math. J. **44**, no. 4, 1033–1074 (1995)
3. Burenkov, V.I.: Sobolev Spaces on Domains. B. G. Teubner Verlagsgesellschaft mbH, Stuttgart (1998)
4. Carlen, E.A., Kusuoka, S., Stroock, D.W.: Upper bounds for symmetric Markov transition functions. Ann. Inst. H. Poincaré Probab. Statist. **23**, no. 2, suppl., 245–287 (1987)
5. Chavel, I.: Eigenvalues in Riemannian geometry Academic Press Inc., Orlando, FL (1984)
6. Chavel, I.: Riemannian Geometry–a Modern Introduction. Cambridge Univ. Press, Cambridge (1993)
7. Coulhon, T., Saloff-Coste, L.: Isopérimétrie pour les groupes et les variétés. Rev. Mat. Iberoam. **9**, no. 2, 293–314 (1993)
8. Coulhon, T., Saloff-Coste, L.: Variétés riemanniennes isométriques à l'infini. Rev. Mat. Iberoam. **11**, no. 3, 687–726 (1995)
9. Croke, Ch.B.: A sharp four-dimensional isoperimetric inequality. Comment. Math. Helv. **59**, no. 2, 187–192 (1984)
10. Gallot, S., Hulin, D., Lafontaine, J.: Riemannian Geometry. Springer, Berlin (1990)
11. Hebey, E.: Nonlinear Analysis on Manifolds: Sobolev Spaces and Inequalities. Courant Lect. Notes **5**. Am. Math. Soc., Providence, RI (1999)
12. Hoffman, D., Spruck, J.: Sobolev and isoperimetric inequalities for Riemannian submanifolds. Commun. Pure Appl. Math. **27**, 715–727 (1974)
13. Maz'ya, V.G.: Sobolev Spaces. Springer, Berlin etc. (1985)
14. Maz'ya, V.G., Poborchi, S.V.: Differentiable Functions on Bad Domains. World Scientific, Singapure (1997)
15. Michael, J.H., Simon, L.M.: Sobolev and mean-value inequalities on generalized submanifolds of R^n. Commun. Pure Appl. Math. **26**, 361–379 (1973)
16. Montgomery, R.: A Tour of Subriemannian Geometries, Their Geodesics and Applications. Am. Math. Soc., Providence, RI (2002)
17. Robinson, D.W. Elliptic Operators and Lie Groups. Oxford Univ. Press, New York (1991)
18. Saloff-Coste, L.: A note on Poincaré, Sobolev, and Harnack inequalities. Int. Math. Res. Not. **2**, 27–38 (1992)

19. Saloff-Coste, L.: Aspects of Sobolev-Type Inequalities. Cambridge Univ. Press, Cambridge (2002)
20. Saloff-Coste, L., Woess, W.: Transition operators on co-compact G-spaces. Rev. Mat. Iberoam. **22**, no. 3, 747–799 (2006)
21. Varopoulos, N.Th.: Hardy-Littlewood theory for semigroups. J. Funct. Anal. **63**, no. 2, 240–260 (1985)
22. Varopoulos, N.Th.: Analysis on Lie groups. J. Funct. Anal. **76**, no. 2, 346–410 (1988)
23. Varopoulos, N.Th., Saloff-Coste, L., Coulhon, T.: Analysis and Geometry on Groups. Cambridge Univ. Press, Cambridge (1992)

The p-Faber-Krahn Inequality Noted

Jie Xiao

Abstract When revisiting the Faber-Krahn inequality for the principal p-Laplacian eigenvalue of a bounded open set in $\mathbb{R}^n$ with smooth boundary, we simply rename it as the p-Faber-Krahn inequality and interestingly find that this inequality may be improved but also characterized through Maz'ya's capacity method, the Euclidean volume, the Sobolev type inequality and Moser-Trudinger's inequality.

1 The p-Faber-Krahn Inequality Introduced

Throughout this article, we always assume that Ω is a bounded open set with smooth boundary $\partial\Omega$ in the $2 \leqslant n$-dimensional Euclidean space $\mathbb{R}^n$ equipped with the scalar product $\langle \cdot, \cdot \rangle$, but also dV and dA stand respectively for the n and $n-1$ dimensional Hausdorff measure elements on $\mathbb{R}^n$. For $1 \leqslant p < \infty$, the p-Laplacian of a function f on Ω is defined by

$$\Delta_p f = -\operatorname{div}(|\nabla f|^{p-2}\nabla f).$$

As usual, ∇ and $\operatorname{div}(|\nabla|^{p-2}\nabla)$ mean the gradient and p-harmonic operators respectively (cf. [8]). If $W_0^{1,p}(\Omega)$ denotes the p-Sobolev space on Ω – the closure of all smooth functions f with compact support in Ω (written as $f \in C_0^\infty(\Omega)$) under the norm

$$\Big(\int_\Omega |f|^p dV\Big)^{1/p} + \Big(\int_\Omega |\nabla f|^p dV\Big)^{1/p},$$

Jie Xiao
Department of Mathematics and Statistics, Memorial University of Newfoundland, St. John's, NL A1C 5S7, Canada
e-mail: jxiao@mun.ca

A. Laptev (ed.), *Around the Research of Vladimir Maz'ya I: Function Spaces*, International Mathematical Series 11, DOI 10.1007/978-1-4419-1341-8_17,

then the principal p-Laplacian eigenvalue of Ω is defined by

$$\lambda_p(\Omega) := \inf \left\{ \frac{\int_\Omega |\nabla f|^p dV}{\int_\Omega |f|^p dV} : \ 0 \neq f \in W_0^{1,p}(\Omega) \right\}.$$

This definition is justified by the well-known fact that $\lambda_2(\Omega)$ is the principal eigenvalue of the positive Laplace operator Δ_2 on Ω but also two kinds of observation that are made below. One is the normal setting: If $p \in (1, \infty)$, then according to [26] there exists a nonnegative function $u \in W_0^{1,p}(\Omega)$ such that the Euler-Lagrange equation

$$\Delta_p u - \lambda_p(\Omega)|u|^{p-2}u = 0 \ \text{ in } \Omega$$

holds in the weak sense of

$$\int_\Omega \langle |\nabla u|^{p-2}\nabla u, \nabla \phi\rangle dV = \lambda_p(\Omega) \int_\Omega |u|^{p-2}u\phi dV \quad \forall\, \phi \in C_0^\infty(\Omega).$$

The other is the endpoint setting: If $p = 1$, then since $\lambda_1(\Omega)$ may be also evaluated by

$$\inf \left\{ \frac{\int_\Omega |\nabla f| dV + \int_{\partial\Omega} |f| dA}{\int_\Omega |f| dV} : 0 \neq f \in BV(\Omega) \right\},$$

where $BV(\Omega)$, containing $W_0^{1,1}(\Omega)$, stands for the space of functions with bounded variation on Ω (cf. [9, Chapter 5]), according to [7, Theorem 4] (cf. [16]) there is a nonnegative function $u \in BV(\Omega)$ such that

$$\Delta_1 u - \lambda_1(\Omega)|u|^{-1}u = 0 \ \text{ in } \ \Omega$$

in the sense that there exists a vector-valued function $\sigma : \Omega \mapsto \mathbb{R}^n$ with

$$\|\sigma\|_{L^\infty(\Omega)} = \inf\{c : \ |\sigma| \leqslant c \ \text{ a.e. in } \Omega\} < \infty$$

and

$$\operatorname{div}(\sigma) = \lambda_1(\Omega), \quad \langle \sigma, \nabla u\rangle = |\nabla u| \ \text{ in } \ \Omega, \quad \langle \sigma, \mathbf{n}\rangle u = -|u| \ \text{ on } \ \partial\Omega,$$

where $\mathbf{n}$ represents the unit outer normal vector along $\partial\Omega$. Moreover, it is worth pointing out that

$$\lambda_1(\Omega) = \lim_{p\to\infty} \lambda_p(\Omega), \tag{1.1}$$

and so that $\Delta_1 u = \lambda_1(\Omega)|u|^{-1}u$ has no classical nonnegative solution in Ω: In fact, if not, referring to [18, Remark 7] we have that for $p > 1$ and $|\nabla u(x)| > 0$

$$\Delta_p u(x) = (1-p)|\nabla u(x)|^{p-4}\langle D^2 u(x)\nabla u(x), \nabla u(x)\rangle \\ + (n-1)H(x)|\nabla u(x)|^{p-1}, \tag{1.2}$$

where $D^2u(x)$ and $H(x)$ are the Hessian matrix of u and the mean curvature of the level surface of u respectively, whence getting by letting $p \to 1$ in (1.2) that $(n-1)H(x) = \lambda_1(\Omega)$ – namely all level surfaces of u have the same mean curvature $\lambda_1(\Omega)(n-1)^{-1}$ – but this is impossible since the level sets $\{x \in \Omega : u(x) \geqslant t\}$ are strictly nested downward with respect to $t > 0$.

Interestingly, Maz'ya's [23, Theorem 8.5] tells us that $\lambda_p(\Omega)$ has an equivalent description below:

$$\lambda_p(\Omega) \leqslant \gamma_p(\Omega) := \inf_{\Sigma \in AC(\Omega)} \mathrm{cap}_p(\overline{\Sigma}; \Omega)V(\Sigma)^{-1} \leqslant p^p(p-1)^{1-p}\lambda_p(\Omega). \tag{1.3}$$

Here and henceforth, for an open set $O \subseteq \mathbb{R}^n$, $AC(O)$ stands for the admissible class of all open sets Σ with smooth boundary $\partial\Sigma$ and compact closure $\overline{\Sigma} \subset \Omega$, and moreover

$$\mathrm{cap}_p(K; O) := \inf\left\{\int_O |\nabla f(x)|^p dx : \ f \in C_0^\infty(O) \ \ \& \ \ f \geqslant 1 \ \text{ in } K\right\}$$

represents the p-capacity of a compact set $K \subset O$ relative to O – this definition is extendable to any subset E of O via

$$\mathrm{cap}_p(E; O) := \sup\{\mathrm{cap}_p(K; O) : \text{compact } K \subseteq E\}$$

– of particular interest is that a combination of Maz'ya's [22, p. 107, Lemma] and the Hölder inequality yields

$$\mathrm{cap}_1(E; O) = \lim_{p \to 1} \mathrm{cap}_p(E; O). \tag{1.4}$$

The constant $\gamma_p(\Omega)$ is called the p-Maz'ya constant of Ω. Of course, if $p = 1$, then $(p-1)^{p-1}$ is taken as 1 and hence the equalities in (1.3) are valid – this situation actually has another description (cf. [25]):

$$\lambda_1(\Omega) = \gamma_1(\Omega) = h(\Omega) := \inf_{\Sigma \in AC(\Omega)} A(\partial\Sigma)V(\Sigma)^{-1}. \tag{1.5}$$

The right-hand-side constant in (1.5) is regarded as the Cheeger constant of Ω which has a root in [4]. As an extension of Cheeger's theorem in [4], Lefton and Wei [20] (cf. [18] and [14]) obtained the following inequality:

$$\lambda_p(\Omega) \geqslant p^{-p}h(\Omega)^p. \tag{1.6}$$

Generally speaking, the reversed inequality of (1.6) is not true at all for $p > 1$. In fact, referring to Maz'ya's first example in [25], we choose Q to be the open n-dimensional unit cube centered at the origin of $\mathbb{R}^n$. If K is a compact subset of Q with $A(K) = 0$ and $\mathrm{cap}_p(K; \mathbb{R}^n) > 0$, and if $\Omega = \mathbb{R}^n \setminus \cup_{z \in \mathbb{Z}^n}(K + z)$, i.e., the complement of the union of all integer shifts of K, then $h(\Omega)=\gamma_1(\Omega)=0$ and $\lambda_p(\Omega)>0$ thanks to Maz'ya's [22, p.425, Theorem], and hence there is no constant $c_1(p,n)>0$ only depending on $1<p<n$ such that $\lambda_p(\Omega)\leqslant c_1(p,n)h(\Omega)^p$. Moreover, Maz'ya's second example in [25] shows that if Ω is a subdomain of the unit open ball $B_1(o)$ of $\mathbb{R}^n$, star-shaped with respect to an open ball $B_\rho(o)\subset\mathbb{R}^n$ centered at the origin o with radius $\rho\in(0,1)$, then there is no constant $c_2(p,n)>0$ depending only on $1<p\leqslant n-1$ such that $\lambda_p(\Omega)\leqslant c_2(p,n)h(\Omega)^p$.

Determining the principal p-Laplacian eigenvalue of Ω is, in general, a really hard task that relies on the value of p and the geometry of Ω. However, the Faber-Krahn inequality for this eigenvalue of Ω, simply called the p-Faber-Krahn inequality, provides a good way to carry out the task. To be more precise, let us recall the content of the p-Faber-Krahn inequality: If Ω^* is the Euclidean ball with the same volume as Ω's, i.e., $V(\Omega^*)=V(\Omega)=r^n\omega_n$ (where ω_n is the volume of the unit ball in $\mathbb{R}^n$), then

$$\lambda_p(\Omega)\geqslant\lambda_p(\Omega^*) \tag{1.7}$$

for which equality holds if and only if Ω is a ball. A proof of (1.7) can be directly obtained by Schwarz's symmetrization – see for example [18, Theorem 1], but the equality treatment is not trivial – see [1] for an argument. Of course, the case $p=2$ of this result goes back to the well-known Faber-Krahn inequality (see also [3, Theorem III.3.1] for an account) with $\lambda_2(\Omega^*)$ being $(j_{(n-2)/2}/r)^2$, where $j_{(n-2)/2}$ is the first positive root of the Bessel function $J_{(n-2)/2}$ and r is the radius of Ω^*. Very recently, in [25] Maz'ya used his capacitary techniques to improve the foregoing special inequality. This paper of Maz'ya and his other two [23]–[24], together with some Sobolev type inequalities for $\lambda_2(\Omega)\geqslant\lambda_2(\Omega^*)$ described in [3, Chapter VI], motivate our consideration of not only a possible extension of Maz'ya's result – for details see Section 2 of this article, but also some interesting geometric-analytic properties of (1.7) – for details see Section 3 of this article.

2 The p-Faber-Krahn Inequality Improved

In order to establish a version stronger than (1.7), let us recall that if from now on $B_r(x)$ represents the Euclidean ball centered at $x\in\mathbb{R}^n$ of radius $r>0$, then (cf. [22, p. 106])

$$\operatorname{cap}_p(B_r(x);O)=\begin{cases} n\omega_n\left(\frac{n-p}{p-1}\right)^{p-1}r^{n-p} & \text{when } O=\mathbb{R}^n \ \&\ p\in[1,n),\\ 0 & \text{when } O=B_r(x)\ \&\ p=n,\\ n\omega_n\left(\frac{p-n}{p-1}\right)^{p-1}r^{n-p} & \text{when } O=B_r(x)\ \&\ p\in(n,\infty). \end{cases} \tag{2.1}$$

Proposition 2.1. *For $t \in (0,\infty)$ and $f \in C_0^\infty(\Omega)$, let $\Omega_t = \{x \in \Omega : |f(x)| \geqslant t\}$.*

(i) *If $p = 1$, then*

$$\lambda_1(\Omega^*) \leqslant \frac{(n\omega_n^{\frac{1}{n}})^{\frac{n}{n-1}} \int_\Omega |\nabla f| dV}{\int_0^\infty \min\{\mathrm{cap}_1(\Omega^*;\mathbb{R}^n)^{\frac{n}{n-1}}, \mathrm{cap}_1(\Omega_t;\Omega)^{\frac{n}{n-1}}\} dt}.$$

(ii) *If $p \in (1,n)$, then*

$$\lambda_p(\Omega^*) \leqslant \frac{(n^n\omega_n^p)^{\frac{1}{n-p}} \left(\frac{n-p}{p-1}\right)^{\frac{n(p-1)}{n-p}} \int_\Omega |\nabla f|^p dV}{\int_0^\infty \left(\mathrm{cap}_p(\Omega^*;\mathbb{R}^n)^{\frac{1}{1-p}} + \mathrm{cap}_p(\Omega_t;\Omega)^{\frac{1}{1-p}}\right)^{\frac{n(1-p)}{n-p}} dt^p}.$$

(iii) *If $p = n$, then*

$$\lambda_n(\Omega^*) \leqslant \frac{V(\Omega^*)^{-1} \int_\Omega |\nabla f|^n dV}{\int_0^\infty \exp\left(-n^{\frac{n}{n-1}} \omega_n^{\frac{1}{n-1}} \mathrm{cap}_n(\Omega_t;\Omega)^{\frac{1}{1-n}}\right) dt^n}.$$

(iv) *If $p \in (n,\infty)$, then*

$$\lambda_p(\Omega^*) \leqslant \frac{(n^n\omega_n^p)^{\frac{1}{n-p}} \left(\frac{p-n}{p-1}\right)^{\frac{n(p-1)}{n-p}} \int_\Omega |\nabla f|^p dV}{\int_0^\infty \left(\mathrm{cap}_p(\Omega^*;\Omega^*)^{\frac{1}{1-p}} - \mathrm{cap}_p(\Omega_t;\Omega)^{\frac{1}{1-p}}\right)^{\frac{n(1-p)}{p-n}} dt^p}.$$

(v) *The inequalities in* (i)–(ii)–(iii)–(iv) *imply the inequality* (1.7).

Proof. For simplicity, suppose that $r = (V(\Omega)\omega_n^{-1})^{\frac{1}{n}}$ is the radius of the Euclidean ball Ω^*, Ω_t^* is the Euclidean ball with $V(\Omega_t^*) = V(\Omega_t)$, and f^* equals $\int_0^\infty 1_{\Omega_t^*} dt$, where 1_E stands for the characteristic function of a set $E \subseteq \mathbb{R}^n$. Then

$$\int_\Omega |\nabla f^*|^p dV \leqslant \int_\Omega |\nabla f|^p dV \quad \& \quad \int_\Omega |f^*|^p dV = \int_\Omega |f|^p dV.$$

Consequently, from the definitions of $\lambda_p(\Omega^*)$ and f^* as well as [6, p.38, Exercise 1.4.1] it follows that

$$\lambda_p(\Omega^*) \int_0^r |a(t)|^p t^{n-1} dt \leqslant \int_0^r |a'(t)|^p t^{n-1} dt \tag{2.2}$$

holds for any absolutely continuous function a on $(0,r]$ with $a(r)=0$.

Case 1. Under $p\in(1,n)$, set

$$s=\frac{t^{\frac{p-n}{p-1}}-r^{\frac{p-n}{p-1}}}{\alpha},\quad \text{where}\ \ \alpha=(n\omega_n)^{\frac{1}{p-1}}\Big(\frac{n-p}{p-1}\Big).$$

This yields

$$t=\big(r^{\frac{p-n}{p-1}}+\alpha s\big)^{\frac{p-1}{p-n}}\quad\text{and}\quad \frac{dt}{ds}=\frac{\alpha(p-1)}{p-n}\Big(\alpha s+r^{\frac{p-n}{p-1}}\Big)^{\frac{n-1}{p-n}}.$$

If $b(s)=a(t)$, then

$$\int_0^r|a(t)|^pt^{n-1}dt=\Big(\frac{\alpha(p-1)}{n-p}\Big)\int_0^\infty|b(s)|^p\big(r^{\frac{p-n}{p-1}}+\alpha s\big)^{\frac{p(n-1)}{p-n}}ds$$

and

$$\int_0^r|a'(t)|^pt^{n-1}dt=\Big(\frac{\alpha(p-1)}{n-p}\Big)^{1-p}\int_0^\infty|b'(s)|^pds.$$

Consequently, (2.2) amounts to

$$\lambda_p(\Omega^*)\Big(\frac{\alpha(p-1)}{n-p}\Big)^p\int_0^\infty|b(s)|^p\big(r^{\frac{p-n}{p-1}}+\alpha s\big)^{\frac{p(n-1)}{p-n}}ds\leqslant\int_0^\infty|b'(s)|^pds. \tag{2.3}$$

Case 2. Under $p=n$, set

$$s=\frac{\ln\frac{r}{t}}{\beta},\quad\text{where}\ \ \beta=(n\omega_n)^{\frac{1}{n-1}}.$$

This gives

$$t=r\exp(-\beta s)\quad\text{and}\quad\frac{dt}{ds}=-\beta r\exp(-\beta s).$$

If $b(s)=a(t)$, then

$$\int_0^r|a(t)|^nt^{n-1}dt=\beta r^n\int_0^\infty|b(s)|^n\exp(-n\beta s)ds$$

and

$$\int_0^r|a'(t)|^nt^{n-1}dt=\beta^{1-n}\int_0^\infty|b'(s)|^nds.$$

As a result, (2.2) is equivalent to

$$\lambda_n(\Omega^*)\beta^n\int_0^\infty|b(s)|^n\exp(-n\beta s)ds\leqslant\int_0^\infty|b'(s)|^nds. \tag{2.4}$$

Case 3. Under $p\in(n,\infty)$, set

$$s = \frac{r^{\frac{p-n}{p-1}} - t^{\frac{p-n}{p-1}}}{\gamma}, \quad \text{where } \gamma = (n\omega_n)^{\frac{1}{p-1}}\Big(\frac{p-n}{p-1}\Big).$$

This produces

$$t = (r^{\frac{p-n}{p-1}} - \gamma s)^{\frac{p-1}{p-n}} \quad \text{and} \quad \frac{dt}{ds} = \Big(\frac{\gamma(p-1)}{n-p}\Big)(r^{\frac{p-n}{p-1}} - \gamma s)^{\frac{n-1}{p-n}}.$$

If $b(s) = a(t)$, then

$$\int_0^r |a(t)|^p t^{n-1} dt = \Big(\frac{\gamma(p-1)}{p-n}\Big)\int_0^{\frac{r^{\frac{p-n}{p-1}}}{\gamma}} |b(s)|^p \big(r^{\frac{p-n}{p-1}} - \gamma s\big)^{\frac{p(n-1)}{p-n}} ds$$

and

$$\int_0^r |a'(t)|^p t^{n-1} dt = \Big(\frac{\gamma(p-1)}{p-n}\Big)^{1-p}\int_0^{\frac{r^{\frac{p-n}{p-1}}}{\gamma}} |b'(s)|^p ds.$$

Thus, (2.2) can be reformulated as

$$\lambda_p(\Omega^*)\Big(\frac{\gamma(p-1)}{p-n}\Big)^p\int_0^{\frac{r^{\frac{p-n}{p-1}}}{\gamma}} |b(s)|^p \big(r^{\frac{p-n}{p-1}} - \gamma s\big)^{\frac{p(n-1)}{p-n}} ds \leqslant \int_0^{\frac{r^{\frac{p-n}{p-1}}}{\gamma}} |b'(s)|^p ds. \tag{2.5}$$

In the three inequalities (2.3)–(2.4)–(2.5), choosing

$$s = \int_0^T \Big(\int_{\{x\in\Omega: f(x)=t\}} |\nabla f|^{p-1} dA\Big)^{\frac{1}{1-p}} dt$$

and letting $\tau(s)$ be the inverse of the last function, we have two equalities:

$$\frac{ds}{d\tau} = \frac{1}{\tau'(s)} \quad \& \quad \int_0^\infty |s'(\tau)|^{-p} d\tau = \int_\Omega |\nabla f|^p dV \tag{2.6}$$

and Maz'ya's inequality for the p-capacity (cf. [22, p. 102]):

$$s \leqslant \operatorname{cap}_p\big(\Omega_{\tau(s)}; \Omega\big)^{\frac{1}{1-p}}. \tag{2.7}$$

The above estimates (2.1) and (2.3)–(2.4)–(2.5)–(2.6)–(2.7) give the inequalities in (ii)–(iii)–(iv).

Next, we verify (i). In fact, this assertion follows from formulas (1.1) and (1.4), taking the limit $p \to 1$ in the inequality established in (ii), and using the elementary limit evaluation

$$\lim_{p\to 1}(c_1^{\frac{1}{p-1}} + c_2^{\frac{1}{p-1}})^{p-1} = \max\{c_1, c_2\} \quad \text{for} \quad c_1, c_2 \geqslant 0.$$

Finally, we show (v). To do so, recall Maz'ya's lower bound inequality for $\mathrm{cap}_p(\cdot,\cdot)$ (cf. [22, p. 105]):

$$\mathrm{cap}_p(\Omega_t;\Omega) \geqslant \Big(\int_{V(\Omega_t)}^{V(\Omega)} \mu(v)^{\frac{p}{1-p}}\,dv\Big)^{1-p} \quad \text{for} \quad 0<t,p-1<\infty, \tag{2.8}$$

where $\mu(v)$ is defined as the infimum of $A(\partial\Sigma)$ over all open subsets $\Sigma \in AC(\Omega)$ with $V(\Sigma) \geqslant v$.

From the classical isoperimetric inequality with sharp constant

$$V(\Sigma)^{\frac{n-1}{n}} \leqslant (n\omega_n^{\frac{1}{n}})^{-1}A(\partial\Sigma) \quad \forall\, \Sigma \in AC(\mathbb{R}^n) \tag{2.9}$$

it follows that $\mu(v) \geqslant n\omega_n^{\frac{1}{n}} v^{\frac{n-1}{n}}$ and consequently

$$\int_{V(\Omega_t)}^{V(\Omega)} \mu(v)^{\frac{p}{1-p}}\,dv \leqslant \begin{cases} \dfrac{V(\Omega_t)^{\frac{p-n}{n(p-1)}} - V(\Omega)^{\frac{p-n}{n(p-1)}}}{\Big(\frac{n(p-1)}{(n-p)(n\omega^{1/n})^{p/(p-1)}}\Big)^{-1}} & \text{for } 1<p\neq n, \\ (n\omega_n^{1/n})^{n/(1-n)} \ln\Big(\dfrac{V(\Omega)}{V(\Omega_t)}\Big) & \text{for } p=n. \end{cases} \tag{2.10}$$

Using (2.10) and (ii)-(iii) we derive the following estimates.

Case 1. If $1<p<n$, then

$$\begin{aligned} I_{1<p<n} &:= \int_0^\infty \Big(\mathrm{cap}_p\,(\Omega^*;\mathbb{R}^n)^{\frac{1}{1-p}} + \mathrm{cap}_p\,(\Omega_t;\Omega)^{\frac{1}{1-p}}\Big)^{\frac{n(p-1)}{p-n}}\,dt^p \\ &\geqslant \int_0^\infty \left(\mathrm{cap}_p\,(\Omega^*;\mathbb{R}^n)^{\frac{1}{1-p}} + \frac{V(\Omega_t)^{\frac{p-n}{n(p-1)}} - V(\Omega)^{\frac{p-n}{n(p-1)}}}{\Big(\frac{n(p-1)}{(n-p)(n\omega^{1/n})^{p/(p-1)}}\Big)^{-1}}\right)^{\frac{n(p-1)}{p-n}} dt^p \\ &= \left(\frac{n(p-1)}{(n-p)(n\omega^{\frac{1}{n}})^{\frac{p}{p-1}}}\right)^{\frac{p-n}{n(p-1)}} \int_0^\infty V(\Omega_t)\,dt^p \\ &= \left(\frac{n(p-1)}{(n-p)(n\omega^{\frac{1}{n}})^{\frac{p}{p-1}}}\right)^{\frac{p-n}{n(p-1)}} \int_\Omega |f|^p dV. \end{aligned}$$

Case 2. If $p=n$, then

$$\begin{aligned} I_{p=n} &:= \int_0^\infty \exp\Big(-n^{\frac{n}{n-1}}\omega_n^{\frac{1}{n-1}}\,\mathrm{cap}_n\,(\Omega_t;\Omega)^{\frac{1}{1-n}}\Big)\,dt^n \\ &\geqslant V(\Omega)^{-1}\int_0^\infty V(\Omega_t)\,dt^n = V(\Omega)^{-1}\int_\Omega |f|^n dV. \end{aligned}$$

Case 3. If $n<p<\infty$, then

$$I_{n<p<\infty} := \int_0^\infty \left(\mathrm{cap}_p\,(\Omega^*;\Omega^*)^{\frac{1}{1-p}} - \mathrm{cap}_p\,(\Omega_t;\Omega)^{\frac{1}{1-p}}\right)^{\frac{n(p-1)}{p-n}} dt^p$$

$$\geqslant \int_0^\infty \left(\mathrm{cap}_p\,(\Omega^*;\Omega^*)^{\frac{1}{1-p}} - \frac{V(\Omega)^{\frac{p-n}{n(p-1)}} - V(\Omega_t)^{\frac{p-n}{n(p-1)}}}{\left(\frac{n(p-1)}{(p-n)(n\omega_n^{1/n})^{p/(p-1)}}\right)^{-1}}\right)^{\frac{(p-1)n}{p-n}} dt^p$$

$$= \left(\frac{n(p-1)}{(p-n)(n\omega_n^{\frac{1}{n}})^{\frac{p}{p-1}}}\right)^{\frac{(p-1)n}{p-n}} \int_0^\infty V(\Omega_t)dt^p$$

$$= \left(\frac{n(p-1)}{(p-n)(n\omega_n^{\frac{1}{n}})^{\frac{p}{p-1}}}\right)^{\frac{(p-1)n}{p-n}} \int_\Omega |f|^p dV.$$

Now the last three cases, along with (ii)-(iii)-(iv), yield (v) for $1 < p < \infty$. In order to handle the setting $p = 1$, letting $p \to 1$ in (2.8) we employ (1.4) and

$$\lim_{p\to 1}(1 - c^{\frac{1}{1-p}})^{1-p} = 1 \text{ for } c \geqslant 1$$

to achieve the following relative isocapacitary inequality with sharp constant:

$$\mathrm{cap}_1\,(\Omega_t;\Omega) \geqslant n\omega_n^{\frac{1}{n}} V(\Omega_t)^{\frac{n-1}{n}}. \tag{2.11}$$

As a consequence of (2.11), we find

$$I_{p=1} := \int_0^\infty \min\{\mathrm{cap}_1\,(\Omega^*;\mathbb{R}^n)^{\frac{n}{n-1}}, \mathrm{cap}_1\,(\Omega_t;\Omega)^{\frac{n}{n-1}}\}\,dt$$

$$\geqslant (n\omega_n^{\frac{1}{n}})^{\frac{n}{n-1}} \int_0^\infty \min\{V(\Omega), V(\Omega_t)\}\,dt$$

$$= (n\omega_n^{\frac{1}{n}})^{\frac{n}{n-1}} \int_0^\infty V(\Omega_t)\,dt$$

$$= (n\omega_n^{\frac{1}{n}})^{\frac{n}{n-1}} \int_\Omega |f|dV,$$

thereby getting the validity of (v) for $p = 1$ thanks to (i). □

Remark 2.1. Perhaps it is appropriate to mention that (ii)-(iii)-(iv) in Proposition 2.1 can be also obtained through choosing $q = p \in (1,\infty)$ and letting $\mathbf{M}(\theta)$-function in Maz'ya's [25, Theorem 2] be respectively

$$\frac{\lambda_p(\Omega^*)(n^n\omega_n^p)^{\frac{1}{p-n}}\left(\frac{n-p}{p-1}\right)^{\frac{n(p-1)}{p-n}}}{\left(\mathrm{cap}_p\,(\Omega^*;\mathbb{R}^n)^{\frac{1}{1-p}} + \theta\right)^{\frac{n(1-p)}{p-n}}} \quad \text{for } p \in (1,n),$$

$\lambda_n(\Omega^*)V(\Omega^*)\exp\big(-(n^n\omega_n)^{\frac{1}{n-1}}\theta\big)$ for $p=n$,

$$\frac{\lambda_p(\Omega^*)(n^n\omega_n^{\frac{p}{n}})^{\frac{1}{p-n}}\left(\frac{p-n}{p-1}\right)^{\frac{n(p-1)}{p-n}}}{\left(\mathrm{cap}_p\left(\Omega^*;\mathbb{R}^n\right)^{\frac{1}{1-p}}-\theta\right)^{\frac{n(1-p)}{p-n}}} \quad \text{for } \theta\leqslant \mathrm{cap}_p\left(\Omega^*;\mathbb{R}^n\right)^{\frac{1}{1-p}}\ \&\ p\in(n,\infty),$$

and 0 for $\theta > \mathrm{cap}_p\left(\Omega^*;\mathbb{R}^n\right)^{\frac{1}{1-p}}$ & $p\in(n,\infty)$.

3 The p-Faber-Krahn Inequality Characterized

When looking over the p-Faber-Krahn inequality (1.7), we get immediately its alternative (cf. [12, 13]) as follows:

$$\lambda_p(\Omega)V(\Omega)^{\frac{p}{n}} \geqslant \lambda_p\big(B_1(o)\big)\omega_n^{\frac{p}{n}}. \tag{3.1}$$

It is well known that (3.1) is sharp in the sense that if Ω is a Euclidean ball in $\mathbb{R}^n$, then equality of (3.1) is valid. Although the explicit value of $\lambda_p\big(B_1(o)\big)$ is so far unknown except

$$\lambda_1\big(B_1(o)\big)=n \quad \& \quad \lambda_2\big(B_1(o)\big)=j^2_{(n-2)/2}, \tag{3.2}$$

Bhattacharya's [1, Lemma 3.4] yields

$$\lambda_p\big(B_1(o)\big)\geqslant n^{2-p}p^{p-1}(p-1)^{1-p}, \tag{3.3}$$

whence giving $\lambda_1\big(B_1(o)\big)\geqslant n$. Meanwhile, from Proposition 2.1 we can get an explicit upper bound of $\lambda_p\big(B_1(o)\big)$ via selecting a typical test function in $W_0^{1,p}\big(B_1(o)\big)$, particularly finding $\lambda_1\big(B_1(o)\big)\leqslant n$ and hence the first formula in (3.2).

Although it is not clear whether Colesanti–Cuoghi–Salani's geometric Brunn–Minkowski type inequality of $\lambda_p(\Omega)$ for convex bodies Ω in [5] can produce (3.1), a geometrical-analytic look at (3.1) leads to the forthcoming investigation in accordance with four situations: $p=1$; $1<p<n$; $p=n$; $n<p<\infty$.

The case $p=1$ is so special that it produces sharp geometric and analytic isoperimetric inequalities indicated below.

Proposition 3.1. *The following statements are equivalent:*
(i) *The sharp* 1*-Faber-Krahn inequality*

$$\lambda_1(\Omega)V(\Omega)^{\frac{1}{n}}\geqslant n\omega_n^{\frac{1}{n}} \quad \forall\ \Omega\in AC(\mathbb{R}^n)$$

holds.
(ii) *The sharp* $(1,\frac{1-n}{n})$*-Maz'ya isocapacitary inequality*

$$\mathrm{cap}_1(\overline{\Omega};\mathbb{R}^n)V(\Omega)^{\frac{1-n}{n}} \geqslant n\omega_n^{\frac{1}{n}} \quad \forall\, \Omega \in AC(\mathbb{R}^n)$$

holds.
(iii) *The sharp* $(1, \frac{n}{n-1})$*-Sobolev inequality*

$$\Big(\int_{\mathbb{R}^n} |\nabla f| dV\Big)\Big(\int_{\mathbb{R}^n} |f|^{\frac{n}{n-1}} dV\Big)^{\frac{1-n}{n}} \geqslant n\omega_n^{\frac{1}{n}} \quad \forall\, f \in C_0^\infty(\mathbb{R}^n)$$

holds.

Proof. (i)⇒(ii) Noticing

$$V(\Omega)^{-1}A(\partial\Omega) \geqslant \lambda_1(\Omega) \quad \forall\, \Omega \in AC(\mathbb{R}^n),$$

we get (i)⇒(2.9). By Maz'ya's formula in [22, p. 107, Lemma] saying

$$\mathrm{cap}_1(\overline{\Omega};\mathbb{R}^n) = \inf_{\overline{\Omega}\subset\Sigma\in AC(\mathbb{R}^n)} A(\partial\Sigma) \quad \forall\, \Omega \in AC(\mathbb{R}^n),$$

we further find (2.9)⇒(ii).
(ii)⇒(iii) Under (ii), we use the end-point case of Maz'ya's inequality in [24, Proposition 1] (cf. [28, Theorems 1.1-1.2]) to obtain

$$\begin{aligned}
\int_{\mathbb{R}^n} |f|^{\frac{n}{n-1}}\, dV &= \int_0^\infty V\big(\{x \in \mathbb{R}^n :\ |f(x)| \geqslant t\}\big)\, dt^{\frac{n}{n-1}} \\
&\leqslant \int_0^\infty \Big((n\omega_n^{\frac{1}{n}})^{-1}\mathrm{cap}_1\big(\{x \in \mathbb{R}^n :\ |f(x)| \geqslant t\}\big)\Big)^{\frac{n}{n-1}}\, dt^{\frac{n}{n-1}} \\
&\leqslant (n\omega_n^{\frac{1}{n}})^{\frac{n}{1-n}}\Big(\int_{\mathbb{R}^n} |\nabla f|\, dV\Big)^{\frac{n}{n-1}},
\end{aligned}$$

whence getting (iii).
(iii)⇒(i) For $\Omega \in AC(\mathbb{R}^n)$ and $f \in W_0^{1,1}(\Omega)$, define a Sobolev function g on $\mathbb{R}^n$ via putting $g = f$ in Ω and $g = 0$ in $\mathbb{R}^n \setminus \Omega$. If (iii) holds, then the inequality in (iii) is valid for g. Using Hölder's inequality we have

$$\int_\Omega |f|\, dV \leqslant \Big(\int_\Omega |f|^{\frac{n}{n-1}}\, dV\Big)^{\frac{n-1}{n}} V(\Omega)^{\frac{1}{n}}$$

and consequently,

$$\frac{\int_\Omega |\nabla f|\, dV}{\int_\Omega |f|\, dV} \geqslant \frac{\int_{\mathbb{R}^n} |\nabla g|\, dV}{\big(\int_{\mathbb{R}^n} |g|^{\frac{n}{n-1}}\, dV\big)^{\frac{n-1}{n}} V(\Omega)^{\frac{1}{n}}} \geqslant \frac{n\omega_n^{\frac{1}{n}}}{V(\Omega)^{\frac{1}{n}}}.$$

This, along with the definition of $\lambda_1(\Omega)$, yields the inequality in (i). □

Remark 3.1. $n\omega_n^{\frac{1}{n}}$ is the best constant for (i)-(ii)-(iii) whose equalities occur when $\Omega = B_1(o)$ and $f \to 1_{B_1(o)}$. Moreover, the equivalence between the classical isoperimetric inequality (2.9) and the Sobolev inequality (iii) above is well known and due to Federer–Fleming [10] and Maz'ya [21].

Nevertheless, the setting $1 < p < n$ below does not yield optimal constants.

Proposition 3.2. *For $p \in (1, n)$, the statement (i) follows from the mutually equivalent ones (ii) and (iii) below:*
(i) *There is a constant $\varkappa_1(p,n) > 0$ depending only on p and n such that the p-Faber-Krahn inequality*

$$\lambda_p(\Omega)V(\Omega)^{\frac{p}{n}} \geqslant \varkappa_1(p,n) \quad \forall\ \Omega \in AC(\mathbb{R}^n)$$

holds.
(ii) *There is a constant $\varkappa_2(p,n) > 0$ depending only on p and n such that the $(p, \frac{p-n}{n})$-Maz'ya isocapacitary inequality*

$$\operatorname{cap}_p(\overline{\Omega}; \mathbb{R}^n)V(\Omega)^{\frac{p-n}{n}} \geqslant \varkappa_2(p,n) \quad \forall\ \Omega \in AC(\mathbb{R}^n)$$

holds.
(iii) *There is a constant $\varkappa_2(p,n) > 0$ depending only on p and n such that the $(p, \frac{pn}{n-p})$-Sobolev inequality*

$$\Big(\int_{\mathbb{R}^n} |\nabla f|^p dV\Big)\Big(\int_{\mathbb{R}^n} |f|^{\frac{pn}{n-p}} dV\Big)^{\frac{p-n}{n}} \geqslant \varkappa_3(p,n) \quad \forall\ f \in C_0^\infty(\mathbb{R}^n)$$

holds.

Proof. Note that (ii)⇔(iii) is a special case of Maz'ya's [23, Theorem 8.5]. So it suffices to consider the following implications.

(ii)⇒(i) This can be seen from [14]. In fact, for $\Sigma \in AC(\Omega)$ and $\Omega \in AC(\mathbb{R}^n)$ one has

$$\frac{\operatorname{cap}_p(\overline{\Sigma}; \Omega)}{V(\Sigma)} \geqslant \frac{\operatorname{cap}_p(\overline{\Sigma}; \mathbb{R}^n)}{V(\Sigma)} \geqslant \varkappa_2(p,n)V(\Sigma)^{-\frac{p}{n}} \geqslant \varkappa_2(p,n)V(\Omega)^{-\frac{p}{n}}$$

and thus by (1.3),

$$\lambda_p(\Omega)V(\Omega)^{\frac{p}{n}} \geqslant (p-1)^{p-1}p^{-p}\varkappa_2(p,n).$$

(iii)⇒(i) Suppose now that (iii) is true. Since there exists a nonzero minimizer $u \in W_0^{1,p}(\Omega)$ such that

$$\int_\Omega |\nabla u|^{p-2}\langle \nabla u, \nabla \phi\rangle\, dV = \lambda_p(\Omega)\int_\Omega |u|^{p-2}u\phi\, dV$$

holds for any $\phi \in C_0^\infty(\Omega)$. Letting ϕ approach u in the above equation, extending u from Ω to $\mathbb{R}^n$ via defining $u = 0$ on $\mathbb{R}^n \setminus \Omega$, and writing this extension as f, we employ (iii) and the Hölder inequality to get

$$\lambda_p(\Omega) = \frac{\int_\Omega |\nabla u|^p dV}{\int_\Omega |u|^p dV} = \frac{\int_{\mathbb{R}^n} |\nabla f|^p dV}{\int_{\mathbb{R}^n} |f|^p dV}$$

$$\geqslant \varkappa_3(p,n)\Big(\int_{\mathbb{R}^n} |f|^{\frac{pn}{n-p}} dV\Big)^{\frac{n-p}{n}}\Big(\int_{\mathbb{R}^n} |f|^p dV\Big)^{-1}$$

$$= \varkappa_3(p,n)\Big(\int_\Omega |u|^{\frac{pn}{n-p}} dV\Big)^{\frac{n-p}{n}}\Big(\int_\Omega |u|^p dV\Big)^{-1} \geqslant \varkappa_3(p,n)V(\Omega)^{-\frac{p}{n}},$$

whence reaching (i). □

Remark 3.2. It is worth remarking that the best values of $\varkappa_1(p,n)$, $\varkappa_2(p,n)$, and $\varkappa_3(p,n)$ are

$$\lambda_p(B_1(o))\omega_n^{\frac{p}{n}},\ n\omega_n^{\frac{p}{n}}\Big(\frac{n-p}{p-1}\Big)^{p-1},$$

and

$$n\omega_n^{\frac{p}{n}}\Big(\frac{n-p}{p-1}\Big)^{p-1}\Big(\frac{\Gamma(\frac{n}{p})\Gamma(n+1-\frac{n}{p})}{\Gamma(n)}\Big)^{\frac{p}{n}}$$

respectively. These constants tend to $n\omega_n^{\frac{1}{n}}$ as $p \to 1$. In addition, from Carron's paper [2] we see that (i) implies (ii) and (iii) under $p = 2$, and consequently conjecture that this implication is also valid for $p \in (1,n) \setminus \{2\}$.

Clearly, (ii) and (iii) in Proposition 3.2 cannot be naturally extended to $p = n$. However, they have the forthcoming replacements.

Proposition 3.3. *For $\Omega \in AC(\mathbb{R}^n)$, the statement (i) follows from the mutually equivalent ones (ii) and (iii) below:*
(i) *The n-Faber-Krahn type inequality*

$$\lambda_n(\Omega)V(\Omega) \geqslant \frac{n^n\omega_n}{(n-1)!}E_n(\Omega)^{-1}$$

holds where

$$E_n(\Omega) := \sup_{f\in C_0^\infty(\Omega),\ \int_\Omega |\nabla f|^n dV \leqslant 1} V(\Omega)^{-1}\int_\Omega \exp\Big(\frac{|f|^{\frac{n}{n-1}}}{(n^n\omega_n)^{\frac{1}{1-n}}}\Big)dV.$$

(ii) *The $(n,0)$-capacity-volume inequality*

$$V(\Sigma)V(\Omega)^{-1} \leqslant \exp\Big(-\Big(\frac{n^n\omega_n}{\mathrm{cap}_n(\overline{\Sigma};\Omega)}\Big)^{\frac{1}{n-1}}\Big) \quad \forall\ \Sigma \in AC(\Omega)$$

holds.
(iii) *The Moser-Trudinger inequality $E_n(\Omega) < \infty$ holds.*

Proof. (ii)⇒(iii) Suppose (ii) holds. For $f \in C_0^\infty(\Omega)$ and $t \geqslant 0$ with $\int_\Omega |\nabla f|^n\,dV \leqslant 1$ let $\Omega_t = \{x \in \Omega : |f(x)| \geqslant t\}$. Then the layer-cake formula gives

$$\int_\Omega \exp\Big(\frac{|f|^{\frac{n}{n-1}}}{(n^n\omega_n)^{\frac{1}{1-n}}}\Big)dV = \int_0^\infty V(\Omega_t)\,d\exp\big((n^n\omega_n)^{\frac{1}{n-1}}t^{\frac{n}{n-1}}\big)$$
$$\leqslant V(\Omega)\int_0^\infty \frac{d\exp\big((n^n\omega_n)^{\frac{1}{n-1}}t^{\frac{n}{n-1}}\big)}{\exp\big((n^n\omega_n)^{\frac{1}{n-1}}\operatorname{cap}_n(\Omega_t;\Omega)^{\frac{1}{1-n}}\big)}$$

where the last integral is finite by [24, Proposition 2]. As a result, we find

$$E_n(\Omega) \leqslant \int_0^\infty \frac{d\exp\big((n^n\omega_n)^{\frac{1}{n-1}}t^{\frac{n}{n-1}}\big)}{\exp\big((n^n\omega_n)^{\frac{1}{n-1}}\operatorname{cap}_n(\Omega_t;\Omega)^{\frac{1}{1-n}}\big)} < \infty,$$

thereby reaching (iii).

(iii)⇒(ii) If (iii) holds, then $f \in C_0^\infty(\Omega)$, $f \geqslant 1$ on $\overline{\Sigma}$ and $\Sigma \in AC(\Omega)$ imply

$$\int_\Omega \Big|\nabla\Big(\frac{f}{(\int_\Omega |\nabla f|^n\,dV)^{\frac{1}{n}}}\Big)\Big|^n dV = 1,$$

and hence

$$V(\Omega)E_n(\Omega) \geqslant \int_\Omega \exp\Big((n^n\omega_n)^{\frac{1}{n-1}}|f|^{\frac{n}{n-1}}\Big(\int_\Omega |\nabla f|^n dV\Big)^{\frac{1}{1-n}}\Big)dV$$
$$\geqslant V(\Sigma)\exp\Big((n^n\omega_n)^{\frac{1}{n-1}}\Big(\int_\Omega |\nabla f|^n dV\Big)^{\frac{1}{1-n}}\Big),$$

whence giving (ii) through the definition of $\operatorname{cap}_n(\overline{\Sigma};\Omega)$.

Of course, if either (ii) or (iii) is valid, then the elementary inequality

$$\exp t \geqslant \frac{t^{n-1}}{(n-1)!} \quad \forall t \geqslant 0$$

yields

$$V(\Omega)E_n(\Omega) \geqslant \frac{n^n\omega_n}{(n-1)!}\Big(\frac{\int_\Omega |f|^n\,dV}{\int_\Omega |\nabla f|^n\,dV}\Big) \quad \forall f \in C_0^\infty(\Omega),$$

whence giving (i) by the characterization of $\lambda_n(\Omega)$ in terms of $C_0^\infty(\Omega)$ – see also [19]. □

Remark 3.3. The equality of (ii) happens when Ω and Σ are concentric Euclidean balls – see also [11, p.15]. Moreover, the supremum defining $E_n(\Omega)$ becomes infinity when $(n^n\omega_n)^{\frac{1}{n-1}}$ is replaced by any larger constant – see also [11, p.97-98].

Next, let us handle the remaining case $p \in (n, \infty)$.

Proposition 3.4. *For $p \in (n, \infty)$ and $\Omega \in AC(\mathbb{R}^n)$, the statement (i) follows from the mutually equivalent ones (ii) and (iii) below:*
(i) *The p-Faber-Krahn inequality*

$$\lambda_p(\Omega)V(\Omega)^{\frac{p}{n}} \geqslant E_{n,p}(\Omega)$$

holds, where

$$E_{n,p}(\Omega) := \inf_{f \in C_0^\infty(\Omega), \|f\|_{L^\infty(\Omega)} \leqslant 1} V(\Omega)^{\frac{p-n}{n}} \int_\Omega |\nabla f|^p \, dV.$$

(ii) *The $(p, \frac{p-n}{n})$-capacity-volume inequality holds:*

$$\mathrm{cap}_p(\overline{\Sigma}; \Omega)V(\Omega)^{\frac{p-n}{n}} \geqslant E_{n,p}(\Omega) \quad \forall\ \Sigma \in AC(\Omega).$$

(iii) *The (p, ∞)-Sobolev inequality holds:*

$$\Big(\int_\Omega |\nabla f|^p \, dV\Big)\|f\|_{L^\infty(\Omega)}^{-p} V(\Omega)^{\frac{p-n}{n}} \geqslant E_{n,p}(\Omega) \quad \forall\ f \in C_0^\infty(\Omega).$$

Proof. (iii)$\Rightarrow$(ii) Suppose (iii) is valid. If $f \in C_0^\infty(\Omega)$, $\Sigma \in AC(\Omega)$ and $f \geqslant 1$ on $\overline{\Sigma}$, then $\|f\|_{L^\infty(\Omega)} \geqslant 1$ and hence

$$V(\Omega)^{\frac{p-n}{n}} \int_\Omega |\nabla f|^p \, dV \geqslant E_{n,p}(\Omega)\|f\|_{L^\infty(\Omega)}^p \geqslant E_{n,p}(\Omega).$$

This, plus the definition of $\mathrm{cap}_p(\overline{\Sigma}; \Omega)$, yields

$$V(\Omega)^{\frac{p-n}{n}} \mathrm{cap}_p(\overline{\Sigma}; \Omega) \geqslant E_{n,p}(\Omega).$$

Namely, (ii) holds.

(ii)$\Rightarrow$(iii) Suppose (ii) is valid. For $q > p$ and $\Sigma \in AC(\Omega)$ we have

$$\frac{\mathrm{cap}_p(\overline{\Sigma}; \Omega)}{V(\Sigma)^{\frac{q-p}{q}}} \geqslant E_{n,p}(\Omega)V(\Omega)^{\frac{p}{q}-\frac{p}{n}}.$$

For $f \in C_0^\infty(\Omega)$ and $t \geqslant 0$ let $\Omega_t = \{x \in \Omega : |f(x)| \geqslant t\}$. According to the layer-cake formula and [24, Proposition 1], we have

$$\begin{aligned}\int_\Omega |f|^{\frac{pq}{q-p}} \, dV &= \int_0^\infty V(\Omega_t) \, dt^{\frac{pq}{q-p}} \\ &\leqslant \Big(E_{n,p}(\Omega)V(\Omega)^{\frac{p}{q}-\frac{p}{n}}\Big)^{\frac{q}{p-q}} \int_0^\infty \mathrm{cap}_p(\Omega_t; \Omega)^{\frac{q}{q-p}} \, dt^{\frac{pq}{q-p}}\end{aligned}$$

$$\leqslant \left(E_{n,p}(\Omega)V(\Omega)^{\frac{p}{q}-\frac{p}{n}}\right)^{\frac{q}{p-q}}\left(\frac{\Gamma\left(\frac{pr}{r-p}\right)}{\Gamma\left(\frac{r}{r-p}\right)\Gamma\left(\frac{p(r-1)}{r-p}\right)}\right)^{\frac{r}{p}-1}\left(\int_\Omega |\nabla f|^p\,dV\right)^{\frac{r}{p}},$$

where $r=\frac{pq}{q-p}$ and $\Gamma(\cdot)$ is the classical gamma function. Simplifying the just-obtained estimates, we get

$$\left(\int_\Omega |f|^r\,dV\right)^{\frac{1}{r}}\leqslant\left(\frac{\Gamma\left(\frac{pr}{r-p}\right)}{\Gamma\left(\frac{r}{r-p}\right)\Gamma\left(\frac{p(r-1)}{r-p}\right)}\right)^{\frac{1}{p}-\frac{1}{r}}\left(\frac{\int_\Omega |\nabla f|^p\,dV}{E_{n,p}(\Omega)V(\Omega)^{\frac{p}{q}-\frac{p}{n}}}\right)^{\frac{1}{p}}.$$

Letting $q\to p$ in the last inequality, we find $r\to\infty$ and thus

$$\|f\|^p_{L^\infty(\Omega)}\leqslant\left(E_{n,p}(\Omega)V(\Omega)^{\frac{n-p}{n}}\right)^{-1}\int_\Omega |\nabla f|^p dV,$$

thereby establishing (iii).

(ii)/(iii)⇒(i) Due to (ii)⇔(iii), we may assume that (iii) is valid with $E_{n,p}(\Omega)>0$ (otherwise nothing is to prove). For $f\in C_0^\infty(\Omega)$ and $q>p$ we employ the Hölder inequality to get

$$\begin{aligned}\int_\Omega |f|^q\,dV &= \int_\Omega |f|^{q-p}|f|^p dV\\ &\leqslant \left(\left(\frac{\int_\Omega |\nabla f|^p\,dV}{E_{n,p}(\Omega)}\right)^{\frac{1}{p}}V(\Omega)^{\frac{1}{n}-\frac{1}{p}}\right)^{q-p}\int_\Omega |f|^p dV\\ &\leqslant \left(\frac{\int_\Omega |\nabla f|^p dV}{\int_\Omega |f|^p dV}\right)^{\frac{q-p}{p}}\left(E_{n,p}(\Omega)\right)^{\frac{p-q}{p}}V(\Omega)^{\frac{q-p}{n}}\int_\Omega |f|^q dV,\end{aligned}$$

thereby reaching

$$\frac{\int_\Omega |\nabla f|^p dV}{\int_\Omega |f|^p dV}\geqslant E_{n,p}(\Omega)V(\Omega)^{-\frac{p}{n}}.$$

Furthermore, the formulation of $\lambda_p(\Omega)$ in terms of $C_0^\infty(\Omega)$ (cf. [19]) is used to verify the validity of (i). □

Remark 3.4. The following sharp geometric limit inequality (cf. [18, Corollary 15] and [17]):

$$\lim_{p\to\infty}\lambda_p(\Omega)^{\frac{1}{p}}V(\Omega)^{\frac{1}{n}}\geqslant \omega_n^{\frac{1}{n}},$$

along with (1.3), induces a purely geometric quantity

$$\Lambda_\infty(\Omega) := \lim_{p\to\infty} \gamma_p(\Omega)^{\frac{1}{p}} = \lim_{p\to\infty} \lambda_p(\Omega)^{\frac{1}{p}} = \inf_{x\in\Omega} \operatorname{dist}(x, \partial\Omega)^{-1}.$$

Obviously, (1.7) is used to derive the ∞-Faber-Krahn inequality below:

$$\Lambda_\infty(\Omega) \geqslant \Lambda_\infty(\Omega^*). \tag{3.4}$$

Moreover, as the limit of $\Delta_p u = \lambda_p(\Omega)|u|^{p-2}u$ on Ω as $p \to \infty$, the Euler–Lagrange equation $\max\{\Lambda_\infty(\Omega) - |\nabla u|u^{-1}, \ \Delta_\infty u\} = 0$ holds in Ω in the viscosity sense (cf. [17]), where

$$\Delta_\infty u(x) := \sum_{j,k=1}^{n} \Big(\frac{\partial u(x)}{\partial x_j}\Big)\Big(\frac{\partial^2 u(x)}{\partial x_j \partial x_k}\Big)\Big(\frac{\partial u(x)}{\partial x_k}\Big) = \langle D^2u(x)\nabla u(x), \nabla u(x)\rangle$$

is the so-called ∞-*Laplacian.*

Last but not least, we would like to say that since the geometry of $\mathbb{R}^n$ – the isoperimetric inequality plays a key role in the previous treatment, the five propositions above may be generalized to a noncompact complete Riemannian manifold (substituted for $\mathbb{R}^n$) with nonnegative Ricci curvature and isoperimetric inequality of Euclidean type, using some methods and techniques from [3, 14, 15] and [27].

Acknowledgment. The work was partially supported by an NSERC (of Canada) discovery grant and a start-up fund of MUN's Faculty of Science as well as National Center for Theoretical Sciences (NCTS) located in National Tsing Hua University, Taiwan. The final version of this paper was completed during the author's visit to NCTS at the invitations of Der-Chen Chang (from both Georgetown University, USA and NCTS) and Jing Yu (from NCTS) as well as Chin-Cheng Lin (from National Central University, Taiwan).

References

1. Bhattacharia, T.: A proof of the Faber-Krahn inequality for the first eigenvalue of the p-Laplacian. Ann. Mat. Pura Appl. Ser. 4 **177**, 225–231 (1999)
2. Carron, G.: In'egakut'es isop'erim'etriques de Faber-Krahn et cons'equences. Publications de l'Institut Fourier. **220** (1992)
3. Chavel, I.: Isoperimetric Inequalities. Cambridge Univ. Press (2001)
4. Cheeger, J.: A lower bound for the smallest eigenvalue of the Laplacian. In: Gunning, R. (ed.), Problems in Analysis, pp. 195–199. Princeton Univ. Press, Princeton, NJ (1970)
5. Colesanti, A., Cuoghi, P., Salani, P.: Brunn–Minkowski inequalities for two functionals involving the p-Laplace operator of the Laplacian. Appl. Anal. **85**, 45–66 (2006)
6. Dacorogna, B.: Introduction to the Calculus of Variations. Imperical College Press (1992)

7. Demengel, F.: Functions locally almost 1-harmonic. Appl. Anal. **83**, 865–893 (2004)
8. D'Onofrio, L., Iwaniec, T.: Notes on p-harmonic analysis. Contemp. Math. **370**, 25–49 (2005)
9. Evans, L., Gariepy, R.: Measure Theory and Fine Properties of Functions. CRC Press LLC (1992)
10. Federer, H., Fleming, W. H.: Normal and integral currents. Ann. Math. **72**, 458–520 (1960)
11. Flucher, M.: Variational Problems with Concentration. Birkhäuser, Basel (1999)
12. Fusco, N., Maggi, F., Pratelli, A.: A note on Cheeger sets. Proc. Am. Math. Soc. electron.: January 26, 1–6 (2009)
13. Fusco, N., Maggi, F., Pratelli, A.: Stability estimates for certain Faber-Krahn, isocapacitary and Cheeger inequalities. Ann. Scuola Norm. Sup. Pisa Cl. Sci. (5) [To appear]
14. Grigor'yan, A.: Isoperimetric inequalities and capacities on Riemannian manifolds. In: The Maz'ya anniversary collection 1 (Rostock, 1998), pp. 139–153. Birkhäuser, Basel (1999)
15. Hebey, E.: Nonlinear Analysis on Manifolds: Sobolev Spaces and Inequalities. Courant Institute of Math. Sci. New York University. **5** (1999)
16. Hebey, E., Saintier, N.: Stability and perturbations of the domain for the first eigenvalue of the 1-Laplacian. Arch. Math. (Basel) (2007)
17. Juutinen, P., Lindqvist, P., Manfredi, J.: The ∞-eigenvalue problem. Arch. Ration. Mech. Anal. **148**, 89–105 (1999)
18. Kawohl, B., Fridman, V.: Isoperimetric estimates for the first eigenvalue of the p-Laplace operator and the Cheeger constant. Commun. Math. Univ. Carol. **44**, 659–667 (2003)
19. Kawohl, B., Lindqvist, P.: Positive eigenfunctions for the p-Laplace operator revisited. Analysis (Munich) **26**, 545–550 (2006)
20. Lefton, L., Wei, D.: Numerical approximation of the first eigenpair of the p-Laplacian using finite elements and the penalty method. Numer. Funct. Anal. Optim. **18**, 389–399 (1997)
21. Maz'ya, V.G.: Classes of domains and imbedding theorems for function spaces (Russian). Dokl. Akad. Nauk SSSR **3**, 527–530 (1960); English transl.: Sov. Math. Dokl. **1**, 882–885 (1961)
22. Maz'ya, V.: Sobolev Spaces. Springer, Berlin etc. (1985)
23. Maz'ya, V.: Lectures on isoperimetric and isocapacitary inequalities in the theory of Sobolev spaces. Contemp. Math. **338**, 307–340 (2003)
24. Maz'ya, V.: Conductor and capacitary inequalities for functions on topological spaces and their applications to Sobolev type imbeddings. J. Funct. Anal. **224**, 408–430 (2005)
25. Maz'ya, V.: Integral and isocapacitary inequalities. arXiv:0809.2511v1 [math.FA] 15 Sep 2008
26. Sakaguchi, S.: Concavity properties of solutions to some degenerate quasilinear elliptic Dirichlet problems. Ann. Scuola Norm. Sup. Pisa Cl. Sci. IV **14** (1987), 403–421 (1988)
27. Saloff-Coste, L.: Aspects of Sobolev-Type Inequalities. London Math. Soc. LMS. **289**, Cambridge Univ. Press, Cambridge (2002)
28. Xiao, J.: The sharp Sobolev and isoperimetric inequalities split twice. Adv. Math. **211**, 417–435 (2007)

Index

Barrier strong 241
Besov space 273
Bessel function 2
Bessel potential space 224
Brunn–Minkowski inequality 33

Carnot group 169
Cartan–Hadamard manifold 350
Cheeger inequality 27, 293, 327
capacitary inequality 16, 329
— gauged 314
capacity 28, 29, 222, 240, 322, 314
cone property
— interior 369
constant sharp 1, 161, 286

Dirichlet form 14, 221
Dirichlet space 221
divergence
— horizontal 174
— tangential 188
domain fluctuating 299

Estimator isoperimetric 287
extension operator 260

Gagliardo–Nirenberg inequality 286
Gauss map, horizontal 174
gauge function 300
Gibbs measure 14
Green formula 7
Green operator 283
Green potential 284
Gromov–Milman theorem 20

Hardy inequality 1, 107, 138, 139, 161, 239, 285
— fractional 162
— — sharp 162
— sharp 2
Hardy–Sobolev inequality 4, 138
Hardy–Sobolev–Maz'ya inequality 139, 140, 141, 163
inequality
— isoperimetric 24, 170, 263, 286, 324
— — Gaussian 325

Lagrangean 300
Laplace–Beltrami operator 336
— horizontal 192
layer horizontal 172
length horizontal 362
Lorentz norm 4
Lorentz space 217, 311
Lorentz–Zygmund space 84
Luxemburg norm 83

Markov kernel 14
Markov semigroup 15, 221
Markov–Chebyshev inequality 330
mean curvature
— horizontal 175
Mellin transform 111
Minkowski content 24, 287
Moser–Trudinger inequality 385

Muckenhoupt weight 309

Orlicz norm 20, 325
Orlicz space 36, 84
Orlicz–Lorentz space 85
Orlicz–Sobolev inequality 326
Orlicz–Sobolev space 81, 241

Poincaré inequality 295, 356
— gauged 337
Poincaré–Cartan
differential forms 171
Poincaré type inequality 13, 240, 258, 289
profile isoperimetric 287
pseudo-Poincaré inequality 349

Ricci tensor 171, 324, 364
Riemannian manifold 16, 291, 322, 349
rearrangement
— decreasing 82 , 285

Sobolev exponent 108, 138, 299
Sobolev homeomorphism 208
log-Sobolev inequality 326
Sobolev inequality 150, 349
Sobolev–Poincare inequality 89
Sobolev space 81, 106
— fractional 161

Trace inequality 100

Young function 20, 83, 293 325
(p, q)-composition property 207
H-perimeter 172
p-capacity 214, 242, 322
q-capacity profile 322
p-Laplacian 373

References to Maz'ya's Publications Made in Volume I

Monographs

- Maz'ya, V.G.: *Sobolev Spaces.* Springer, Berlin etc. (1985)
  - 16, 29, 30, 79   Bobkov–Zegarlinski [29]
  - 86, 92, 100, 103   Cianchi [35]
  - 138, 141, 153, 159   Filippas–Tertikas–Tidblom [22]
  - 163, 167   Frank–Seiringer [9]
  - 170, 206   Garofalo–Selby [25]
  - 209, 220   Gol'dshtein–Ukhlov [13]
  - 221, 237   Jacob–Schilling [18]
  - 239, 240, 242, 244, 254   Kinnunen–Korte [37]
  - 256, 258, 263, 264, 272   Koskela–Miranda Jr.–Shanmugalingam [11]
  - 286, 295, 298   Martín–M. Milman [20]
  - 321, 322, 323, 326, 330, 332, 333, 338, 348   E. Milman [27]
  - 350, 351, 354, 368, 371   Saloff-Coste [13]
  - 375, 376, 380, 383, 390   Xiao [22]
  - cf. also Vols. II and III
- Maz'ya, V.G., Shaposhnikova, T.O.: Theory of Multipliers in Spaces of Differentiable Functions. Pitman, Boston etc. (1985) Russian edition: Leningrad. Univ. Press, Leningrad (1986)
  - 207, 220   Gol'dshtein–Ukhlov [15]
  - cf. also Vol. II
- Kozlov, V.A., Maz'ya, V.G., Rossmann, J.: Elliptic Boundary Value Problems in Domains with Point Singularities. Am. Math. Soc., Providence, RI (1997)
  - 106, 107, 108, 109, 129, 136,   Costabel–Dauge–Nicaise [4]
  - cf. also Vol. II
- Maz'ya, V.G., Poborchi, S.V.: Differentiable Functions on Bad Domains. World Scientific, Singapore (1997)
  - 350, 371   Saloff-Coste [14]
  - cf. also Vol. II
- Kozlov, V.A., Maz'ya, V.G., Rossmann, J.: Spectral Problems Associated with Corner Singularities of Solutions to Elliptic Equations. Am. Math. Soc., Providence, RI (2001)
  - 108, 136   Costabel–Dauge–Nicaise [5]
- Maz'ya, V. (ed.), Sobolev Spaces in Mathematics. I: Sobolev Type Inequalities. Springer, New York; Tamara Rozhkovskaya Publisher, Novosibirsk. International Mathematical Series **8** (2009)
  - 299, 308, 320   Mbakop–Mosco [19]

Research papers

- Maz'ya, V.G.: Classes of domains and imbedding theorems for function spaces (Russian). Dokl. Akad. Nauk SSSR **3**, 527–530 (1960); English transl.: Sov. Math. Dokl. **1**, 882–885 (1961)
  - 286, 298   Martín–M. Milman [18]
  - 322, 348   E. Milman [23]
  - 384, 390   Xiao [21]
- Maz'ya, V.G.: p-Conductivity and theorems on imbedding certain functional spaces into a C-space (Russian). Dokl. Akad. Nauk SSSR **140**, 299–302 (1961); English transl.: Sov. Math. Dokl. **2**, 1200–1203 (1961)
  - 322, 328, 348   E. Milman [24]

- Maz'ya, V.G.: On the solvability of the Neumann problem (Russian). Dokl. Akad. Nauk SSSR **147**, 294–296 (1962); English transl.: Sov. Math. Dokl. **3**, 1595–1598 (1962)
  323, 327, 348 E. Milman [26]
- Maz'ya, V.G.: The negative spectrum of the higher-dimensional Schrödinger operator (Russian). Dokl. Akad. Nauk SSSR **144**, 721–722 (1962); English transl.: Sov. Math. Dokl. **3**, 808–810 (1962)
  322, 323, 327, 348 E. Milman [25]
  cf. also Vol. III
- Maz'ya, V.G.: On the theory of the higher-dimensional Schrödinger operator (Russian). Izv. Akad. Nauk SSSR. Ser. Mat. **28**, 1145–1172 (1964)
  307, 320 Mbakop–Mosco [16]
  cf. also Vol. III
- Burago, Yu.D., Maz'ya, V.G.: Certain questions of potential theory and function theory for irregular regions. Zap. Nauchn. Semin. LOMI **3** (1967); English transl.: Consultants Bureau, New York (1969)
  256, 257, 263, 271 Koskela–Miranda Jr.–Shanmugalingam [2]
  cf. also Vol. II
- Maz'ya, V.G.: Weak solutions of the Dirichlet and Neumann problems (Russian). Tr. Mosk. Mat. O-va. **20**, 137–172 (1969)
  207, 220 Gol'dshtein-Ukhlov [14]
  288, 298 Martín-M. Milman [19]
- Maz'ya, V.G., Khavin, V.P.: Nonlinear potential theory (Russian). Usp. Mat. Nauk **27**, 67–138 (1972); English transl.: Russ. Math. Surv. **27**, 71–148 (1973)
  221, 226, 238 Jacob–Schilling [19]
  cf. also Vol. II
- Maz'ya, V.G.: On certain integral inequalities for functions of many variables (Russian). Probl. Mat. Anal. **3**, 33–68 (1972); English transl.: J. Math. Sci., New York **1**, 205–234 (1973)
  16, 30, 79 Bobkov–Zegarlinski [28]
  81, 87, 103 Cianchi [33]
- Maz'ya, V.G.: The continuity and boundedness of functions from Soblev spaces (Russian). Probl. Mat. Anal. **4**, 46–77 (1973); English transl.: J. Math. Sci., New York **6**, 29–50 (1976)
  81, 87, 103 Cianchi [34]
- Maz'ya, V.G.: The continuity at a boundary point of solutions of quasilinear equations (Russian). Vestnik Leningrad. Univ. **25**, no. 13, 42–55 (1970). Correction, ibid. **27:1**, 160 (1972); English transl.: Vestnik Leningrad Univ. Math. **3**, 225–242 (1976)
  240, 254 Kinnunen–Korte [36]
  cf. also Vol. II
- Maz'ya, V.G., Plamenevskii, B.A.: Weighted spaces with nonhomogeneous norms and boundary value problems in domains with conical points. Transl., Ser. 2, Am. Math. Soc. **123**, 89–107 (1984)
  106, 136 Costabel–Dauge–Nicaise [6]
- Maz'ya, V., Shaposhnikova, T.: On the Bourgain, Brezis, and Mironescu theorem concerning limiting embeddings of fractional Sobolev spaces. J. Funct. Anal. **195**, no. 2, 230–238 (2002); Erratum: ibid **201**, no. 1, 298–300 (2003)
  162, 167 Frank–Seiringer [10]
- Maz'ya, V.G.: Lectures on isoperimetric and isocapacitary inequalities in the theory of Sobolev spaces. In: Heat Kernels and Analysis on Manifolds, Graphs, and Metric Spaces, pp. 307–340. Am. Math. Soc., Providence, RI (2003)
  242, 254 Kinnunen–Korte [38]
  375, 376, 384, 390 Xiao, [23]

- Filippas, S., Maz'ya, V., Tertikas, A.: Sharp Hardy–Sobolev inequalities. C. R., Math., Acad. Sci. Paris **339**, no. 7, 483–486 (2004)
  2, 11 Avkhadiev–Laptev [7]
  138, 159 Filippas–Tertikas–Tidblom [16]
- Maz'ya, V.G.: Conductor and capacitary inequalities for functions on topological spaces and their applications to Sobolev type imbeddings. J. Funct. Anal. **224**, 408–430 (2005)
  242, 254 Kinnunen–Korte [39]
  286, 298 Martín–M. Milman [21]
  301, 314, 320 Mbakop–Mosco [17]
  376, 383, 386, 387, 390 Xiao [24]
- Maz'ya, V.: Conductor inequalities and criteria for Sobolev type two-weight imbeddings. J. Comput. Appl. Math. **114**, no. 1, 94–114 (2006)
  30, 79 Bobkov–Zegarlinski [30]
  301, 314, 320 Mbakop–Mosco [18]
- Filippas, S., Maz'ya, V., Tertikas, A.: Critical Hardy–Sobolev Inequalities. J. Math. Pures Appl. (9) **87**, 37–56 (2007)
  138, 159 Filippas–Tertikas–Tidblom [17]
  cf. also Vol. II
- Maz'ya, V.: Integral and isocapacitary inequalities. arXiv:0809.2511v1 [math.FA] 15 Sep 2008
  375, 376, 381, 390 Xiao, [25]
- Maz'ya, V., Shaposhnikova, T.: A collection of sharp dilation invariant inequalities for differentiable functions. In: Maz'ya, V. (ed.), Sobolev Spaces in Mathematics. I: Sobolev Type Inequalities. Springer, New York; Tamara Rozhkovskaya Publisher, Novosibirsk. International Mathematical Series **8**, 223-247 (2009)
  138, 159 Filippas–Tertikas–Tidblom [23]